Advancements and Challenges in IoT Communication Technologies for a Connected World

Advancements and Challenges in IoT Communication Technologies for a Connected World

Guest Editors

Francesco Paolo Chietera
Xiao Tang
Andrea Ria

MDPI

Basel • Beijing • Wuhan • Barcelona • Belgrade • Novi Sad • Cluj • Manchester

Guest Editors

Francesco Paolo Chietera
Department of Electrical and Information Engineering
Politecnico di Bari
Bari
Italy

Xiao Tang
School of Information and Communication Engineering
Xi'an Jiaotong University
Xi'an
China

Andrea Ria
Department of Information Engineering
University of Pisa
Pisa
Italy

Editorial Office
MDPI AG
Grosspeteranlage 5
4052 Basel, Switzerland

This is a reprint of the Special Issue, published open access by the journal *Sensors* (ISSN 1424-8220), freely accessible at: https://www.mdpi.com/journal/sensors/special_issues/U6AV48BSG5.

For citation purposes, cite each article independently as indicated on the article page online and as indicated below:

Lastname, A.A.; Lastname, B.B. Article Title. *Journal Name* **Year**, *Volume Number*, Page Range.

ISBN 978-3-7258-5415-8 (Hbk)
ISBN 978-3-7258-5416-5 (PDF)
https://doi.org/10.3390/books978-3-7258-5416-5

Contents

About the Editors

Francesco Paolo Chietera

Francesco Paolo Chietera received his M.S. degree in Communication Engineering from the University of Salento, Lecce, Italy, in 2018 (110/110). He obtained his Ph.D. from the Department of Engineering for Innovation and the Electromagnetic Solutions for Hi-Tech (EMTech) Group, with an excellent evaluation from the examination committee. He is currently a postdoctoral researcher at the Department of Electrical and Information Engineering of Politecnico di Bari. His current research focuses on the design and manufacturing of flexible phased antenna arrays (PAAs) using both PCB-based and additive manufacturing (AM) technologies, with particular attention to the integration of advanced AM techniques for the realization of high-performance PAAs. He is directly involved in the ESA-funded WONDER project, which is dedicated to these topics. His broader research interests include 3D printing for electromagnetic applications, RFID technologies for sensing and traceability, and communication technologies for IoT systems. He has co-authored more than 40 papers in international journals and conferences, and he actively contributes to several research projects funded by national and international institutions, including the European Space Agency (ESA), INAIL (National Institute for Insurance Against Industrial Injuries), CNR (National Research Council), the PRIN program (National Research Program of Relevant Interest), and ERASMUS+.

Xiao Tang

Xiao Tang received a B.S. degree in information engineering (Elite Class Named After Tsien Hsue-shen) and a Ph.D. (Hons.) in information and communication engineering from Xi'an Jiaotong University, Xi'an, China, in 2011 and 2018, respectively. He is currently an Associate Professor with the School of Information and Communication Engineering, Xi'an Jiaotong University, China. He is on a quest to have fun exploring research.

Andrea Ria

Andrea Ria (Senior Member of the IEEE) received his bachelor's, master's, and PhD in Electronics engineering from the University of Pisa, Pisa, Italy, in 2014, 2016, and 2021, respectively. He is currently an Assistant Professor at the Department of Information Engineering at the University of Pisa. In 2019, he won the "Distinguished Service Award" during the RFID-TA conference. In 2018, he received the second prize of "Huawei Italy University Challenge". In 2024, he received the "Best Paper Award" during the 9th Splitech Conference and during the IEEE GreenCom Conference. He is the Vice Chair of the IEEE Student Branch of the University of Pisa. His research activity is focused on low-voltage and low-power mixed-signal integrated sensor systems and interfaces.

Preface

The continuous growth of the Internet of Things (IoT) is transforming communication networks into complex ecosystems where sensors, devices, and intelligent systems interact seamlessly. This Special Issue of *Sensors*, entitled "Advancements and Challenges in IoT Communication Technologies for a Connected World", has been conceived to highlight recent progress in this rapidly evolving field, with an emphasis on innovative solutions that address both fundamental challenges and practical applications.

The collected contributions span a wide range of topics at the intersection of sensing, communication, and intelligence. Several research articles investigate how IoT technologies can be leveraged in diverse contexts, from healthcare monitoring and elderly assistance to industrial process optimization and connected mobility. Other works explore critical communication aspects, such as secure outsourcing with leakage resilience, anomaly traffic detection, and the extension of ultra-wideband ranging coverage. Advances in electromagnetic design are also represented, with studies on sustainable terahertz SWIPT through reconfigurable intelligent surfaces and reviews of flexible phased antenna arrays, paving the way for green and adaptive 6G IoT networks. Complementary investigations highlight the role of edge AI, low-power platforms, and anomaly detection strategies for real-time decision-making in smart infrastructures.

In addition to original research, this Special Issue includes comprehensive reviews and systematic studies. These contributions provide an in-depth perspective on state-of-the-art solutions and emerging trends in areas such as IoT-driven irrigation management, driver monitoring systems, and decentralized frameworks for IoT applications. Together, the works collected here reflect the breadth of innovation occurring at the frontier of IoT communication technologies and their enabling role in smart agriculture, smart cities, and next-generation industrial systems.

On behalf of the Guest Editors, I would like to warmly thank all authors for their valuable contributions and the reviewers for their constructive feedback, which greatly enhanced the quality of this collection. We are also grateful to the *Sensors* editorial team for their professional support throughout the process. We hope that this Special Issue will inspire researchers and practitioners to further explore the opportunities and challenges of IoT communication technologies, ultimately contributing to the realization of a more connected, efficient, and sustainable world.

Francesco Paolo Chietera, Xiao Tang, and Andrea Ria
Guest Editors

sensors

MDPI

Article

Sustainable THz SWIPT via RIS-Enabled Sensing and Adaptive Power Focusing: Toward Green 6G IoT

Sunday Enahoro [1,*,†], Sunday Cookey Ekpo [1,†], Mfonobong Uko [1], Fanuel Elias [1], Rahul Unnikrishnan [1], Stephen Alabi [2] and Nurudeen Kolawole Olasunkanmi [2]

1 Communication and Space Systems Engineering Research Team, Manchester Metropolitan University, Manchester M1 5GD, UK; s.ekpo@mmu.ac.uk (S.C.E.); mfonobong.uko@stu.mmu.ac.uk (M.U.); fanuel.elias@stu.mmu.ac.uk (F.E.); rahul.unnikrishnan@mmu.ac.uk (R.U.)
2 Research and Development Engineering, SmOp CleanTech, Manchester M40 8WN, UK; stephen.alabi@smopct.com (S.A.); kolawole@smopct.com (N.K.O.)
* Correspondence: sunday.enahoro@stu.mmu.ac.uk
† These authors contributed equally to this work.

Abstract: Terahertz (THz) communications and simultaneous wireless information and power transfer (SWIPT) hold the potential to energize battery-less Internet-of-Things (IoT) devices while enabling multi-gigabit data transmission. However, severe path loss, blockages, and rectifier nonlinearity significantly hinder both throughput and harvested energy. Additionally, high-power THz beams pose safety concerns by potentially exceeding specific absorption rate (SAR) limits. We propose a sensing-adaptive power-focusing (APF) framework in which a reconfigurable intelligent surface (RIS) embeds low-rate THz sensors. Real-time backscatter measurements construct a spatial map used for the joint optimisation of (i) RIS phase configurations, (ii) multi-tone SWIPT waveforms, and (iii) nonlinear power-splitting ratios. A weighted MMSE inner loop maximizes the data rate, while an outer alternating optimisation applies semidefinite relaxation to enforce passive-element constraints and SAR compliance. Full-stack simulations at 0.3 THz with 20 GHz bandwidth and up to 256 RIS elements show that APF (i) improves the rate–energy Pareto frontier by 30–75% over recent adaptive baselines; (ii) achieves a 150% gain in harvested energy and a 440 Mbps peak per-user rate; (iii) reduces energy-efficiency variance by half while maintaining a Jain fairness index of 0.999;; and (iv) caps SAR at 1.6 W/kg, which is 20% below the IEEE C95.1 safety threshold. The algorithm converges in seven iterations and executes within <3 ms on a Cortex-A78 processor, ensuring compliance with real-time 6G control budgets. The proposed architecture supports sustainable THz-powered networks for smart factories, digital-twin logistics, wire-free extended reality (XR), and low-maintenance structural health monitors, combining high-capacity communication, safe wireless power transfer, and carbon-aware operation for future 6G cyber–physical systems.

Keywords: terahertz SWIPT; reconfigurable intelligent surface; adaptive power focusing; nonlinear energy harvesting; 6G IoT; SAR constraint; green wireless; channel sensing

1. Introduction

1.1. 6G Vision and the THz Opportunity

The ITU Vision 2030 roadmap identifies pervasive sensing and sustainable networking as the two socio-technical cornerstones of future cyber–physical infrastructure [1]. The 6G paradigm extends the "network of everything" to a cyber–physical continuum populated

 https://doi.org/10.3390/s25154549

by trillions of micro-nodes, many of which are battery-free, that must simultaneously harvest energy and exchange context-rich data with end-to-end latency below one millisecond. Sub-6 GHz and mmWave bands alone cannot satisfy the required spectral density, motivating a migration toward the 0.3 THz to 3 THz window, where contiguous bandwidths exceed 100 GHz and diffraction-limited beams enable spatial reuse and physical-layer security [2–6]. However, the very physics that makes the THz band attractive—short wavelengths and molecular resonances—also causes high free-space loss, narrow pointing tolerance, and frequency-selective absorption peaks that complicate simultaneous wireless information and power transfer (SWIPT) [4].

1.2. RIS Technology and Sensing-Aware Control

Reconfigurable intelligent surfaces (RISs) offer an elegant, low-power means of re-shaping the THz channel. Modern CMOS/SiGe metasurfaces achieve >300° phase tuning with <2 dB insertion loss and virtually zero static bias [7–9]. By reflecting the incident wavefront with programmable delay, an RIS can re-focus THz energy toward obstructed receivers, boosting both spectral efficiency (SE) and harvested energy (EH) [10,11]. Embedding photonic or RF sensors in the same aperture allows each tile to measure local echoes, obtain geometric cues such as angle-of-arrival (AoA) and blockage state, and feed them to an edge controller for real-time phase adaptation [5,12] and adaptive power focusing via stochastic optimisation [13].

State-of-the-Art Comparison.

Table 1 benchmarks representative RIS-aided THz SWIPT and ISAC studies against the proposed adaptive power-focusing (APF) framework.

Early THz-RIS studies focused on beam training without power transfer [14] or compared RIS with active relays under idealised assumptions [15]. Works such as Refs. [16,17] addressed sub-6 GHz scenarios using passive beamforming and linear energy harvesters but lacked sensing or sustainability considerations. More recent contributions at mmWave frequencies [8,13,18] introduced nonlinear EH models and RIS-assisted SWIPT yet remained limited in THz applicability and adaptive integration. In the THz domain, Ref. [10] introduced fixed-phase RIS beamforming, while [19] proposed a joint SWIPT design without real-time adaptation.

Table 1. Comparative matrix of recent RIS/THz–SWIPT/ISAC literature.

Reference	Venue/Year	Band	Nonlinear EH	Sensing	THz	Green Metric	Robustness †	Remark
[16]	CSTut/2015	Sub-6 GHz	✓	✗	✗	✗	✗	Survey
[17]	TWC/2019	Sub-6 GHz	✗	✗	✗	✗	✓	IRS beamforming baseline
[18]	TWC/2021	mmWave	✓	✗	✗	✗	✓	RIS + nonlinear EH
[15]	TCOM/2020	mmWave	✗	✗	✗	✗	✓	Robust RIS-ISAC (relay vs. RIS)
[14]	TCOM/2022	THz	✗	✗	✓	✗	✓	THz IRS beam-training baseline
[10]	Sensors/2022	THz	✗	✗	✓	✗	✗	Fixed RIS beam steering
[8]	Sensors/2023	mmWave	✗	✓	✗	✗	✓	Sensing-capable RIS
[13]	Sensors/2023	mmWave	✓	✗	✗	✗	✓	Secure beamforming
[12]	Sensors/2024	mmWave	✗	✓	✗	✗	✓	Vehicular ISAC RIS
This Work	Sensors/2025	THz	✓	✓	✓	✓	✓	Joint sensing–power focusing

✓: Feature supported; ✗: Feature not supported. † Robustness to channel estimation errors or hardware impairments.

Broader insights from adaptive wireless system design reinforce this approach. Prior work on regulated-element beamforming for vehicular multimedia [20], and system-level evaluations of adaptive satellite architectures [21], highlight the benefits of context-aware reconfigurable platforms under non-ideal conditions.

1.3. *Sustainability Imperative*

Meeting 6G key performance indicators (KPIs) will be insufficient if achieved at the expense of energy cost or human safety. The ITU L.1470 methodology calls for bit/Joule/gCO_2 accounting across the whole device chain, while the IEEE C95.1 standard limits the specific absorption rate (SAR) to 2 W/kg. At THz frequencies, higher power densities aggravate SAR exposure, and diode rectifiers exhibit strong saturation effects [22]. These constraints motivate a holistic "green-first" design philosophy in which spectral, energy, and safety metrics are optimised *simultaneously*.

1.4. *Research Gap*

(a) Propagation limits: Severe path loss and frequency-selective molecular absorption reduce the spatial footprint within which usable RF-to-DC energy can be harvested;
(b) Hardware nonlinearities: Beyond 100 GHz, rectifier diodes and power splitters exhibit saturated I–V characteristics that invalidate linear EH models;
(c) Lack of environment-aware optimisation: Existing THz SWIPT or RIS works rarely couple real-time sensing with joint waveform–phase control under carbon and SAR constraints.

1.5. *Contribution Summary*

We present a *sustainable THz SWIPT* architecture that closes this gap by tightly coupling RIS sensing with adaptive power focusing (APF). A model representation of the proposed THz SWIPTsystem integrating an RIS-enabled sensing layer is shown in Figure 1. The access point (AP) emits information–power waveforms, which are adaptively reflected by a smart RIS equipped with embedded sensors. These sensors capture echo signals to infer channel blockage, user location, and environmental context. The harvested energy powers lightweight IoT devices co-located with information receivers.

- **Dual-mode THz RIS.** Each element toggles between reflection and low-rate sensing, supplying instantaneous blockage and AoA data to the controller.
- **Green utility.** A weighted rate–energy function internalises carbon cost per bit via the ITU L.1470 factor, and SAR is enforced as an explicit constraint.
- **Two-tier optimiser.** A closed-form inner loop sets the power-splitting ratio, while a metaheuristic outer loop searches the unit-modulus RIS phase space, exploiting channel reciprocity.

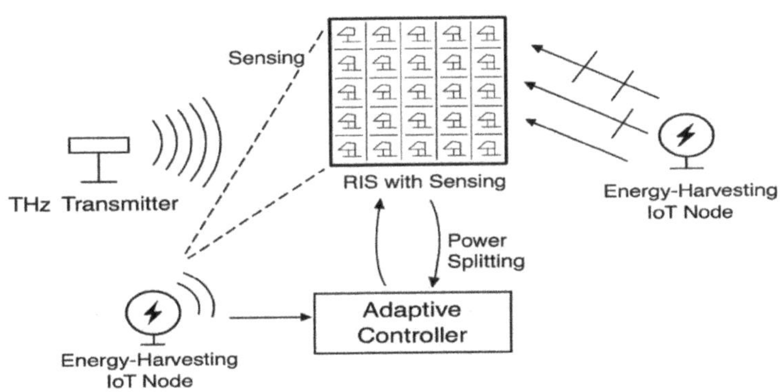

Figure 1. Proposed THz SWIPT system model integrating an RIS-enabled sensing layer.

In extensive Monte Carlo simulations, the proposed APF improves the rate–energy Pareto frontier by up to 70%, provides a 150 no-RIS configuration, and keeps the peak SAR 30% below the IEEE limit, all within a 3 ms control loop on a Cortex-A78 CPU.

The remainder of this paper is structured as follows: Section 2 surveys related work on THz SWIPT, RIS-aided beamforming, and energy harvesting. Section 3 formulates the proposed system model, including the RIS-assisted link, sensing feedback, and nonlinear energy harvesting. Section 4 presents the APF optimisation strategy with detailed algorithms. Section 5 outlines benchmark schemes and comparative setups. Section 6 provides simulation results and analysis under realistic THz conditions. Section 7 offers critical discussion of implementation challenges and extensions. Section 8 concludes the paper and proposes future research directions.

2. Literature Review and Related Work

2.1. Evolution of SWIPT Architectures

Advancements in THz communications—such as ultra-massive MIMO and reconfigurable intelligent surfaces (RISs)—have enabled high-capacity, low-latency wireless links [23]. RIS-assisted MIMO systems, in particular, offer reconfigurable propagation control [24], while integrated wireless power transfer supports sustainable IoT deployment.

Early SWIPT systems focused on sub-6,GHz designs with linear EH models and conventional beamforming [16]. The introduction of RIS [7,17] and nonlinear EH models [25] has shifted interest to higher-frequency regimes like mmWave and THz systems, where accurate modelling becomes critical [18].

Recent work shows RIS outperforms conventional relays in practical THz settings [15], and hybrid beamforming with RIS improves performance in multi-user THz MIMO [14]. Our APF optimisation builds on the weighted MMSE framework described in [26], enabling practical rate–energy trade-off design with RIS-SWIPT.

Figure 2 outlines the chronological progression of RIS-aided SWIPT and THz rectenna technologies. Initial contributions like sub-6 GHz IRS beamforming [17] laid the foundation for nonlinear mmWave EH optimisation [18]. Later efforts advanced toward THz RIS-aided energy focusing [10] and sensor-augmented RIS beamforming [8]. However, few of these works integrated real-time sensing with energy–rate–sustainability co-optimisation.

Our proposed framework builds directly on this evolution by introducing a fully adaptive and eco-aware THz SWIPT architecture that unifies energy harvesting, rate performance, and green metrics into a coherent system design.

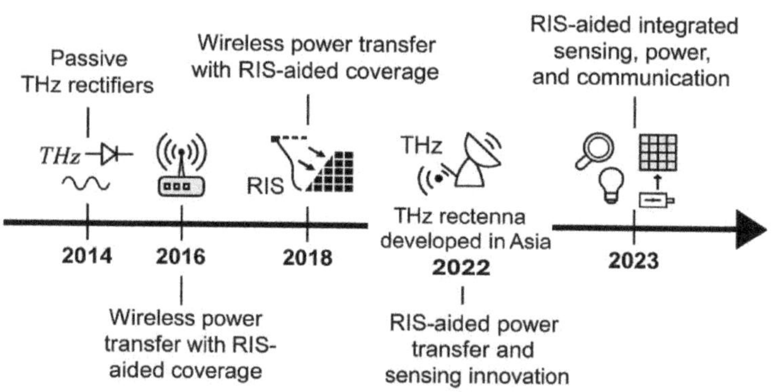

Figure 2. Timeline of key milestones in RIS-aided SWIPT and THz rectenna research.

2.2. THz SWIPT Without RIS: Fundamental Limits and Directions

Initial THz SWIPT research largely focused on rectenna components. For instance, ref. [22] achieved up to 15% RF-to-DC conversion at 0.3 THz, albeit under static line-of-sight (LoS) conditions. Other works [5,27] explored tunable metasurfaces and sensing tiles, though without joint SWIPT optimisation.

While hybrid combiners [28] improve EH at lower frequencies, they fail to address the severe attenuation, scattering, and nonlinearity at THz levels, revealing the need for architecture-level and algorithm-level innovations.

2.3. THz SWIPT: Challenges and RIS-Aided Advances

Despite the promise of abundant bandwidth, the THz SWIPT design faces formidable obstacles—especially due to path loss, atmospheric absorption, and angular misalignment. Initial approaches, like that described in [6], explored modulation–rectifier co-design (e.g., ASK, RTD-based circuits) but struggled to sustain performance under realistic deployment scenarios.

RIS has since emerged as a key technology to mitigate these limitations. Early RIS-SWIPT systems [13,17,18] assumed fixed or static RIS phase configurations. However, THz SWIPT requires adaptive beamforming to address highly dynamic and lossy channels.

Ref. [10] presented fixed RIS-based beam steering at the THz level, while [8] demonstrated multifunctional RIS tiles with embedded sensors—yet omitted EH co-optimisation. Similarly, ref. [29] targeted satellite–RIS integration with static control. Our framework advances this line of research by incorporating real-time sensing, EH control, and green-aware optimisation in a unified RIS-SWIPT model.

2.4. Integrated Sensing, Communication, and Power Transfer (ISCPT)

Integrated Sensing, Communication, and Power Transfer (ISCPT) systems aim to tightly combine environment-aware sensing, data transmission, and energy delivery—ideally within a shared waveform or RIS-enabled medium. Recent mmWave works [30,31] have explored secure and multifunctional beamforming architectures.

However, THz-specific challenges such as rectifier nonlinearity, SAR compliance, and high-frequency adaptation remain open. Hybrid access approaches [32] across mmWave, THz, and optical domains offer architectural flexibility. Multiband antennas [33,34] provide potential physical-layer platforms, but their integration into RIS-enabled, sustainability-driven THz SWIPT remains underexplored. Figure 3 summarises the four foundational axes of our APF framework. First, the RIS sensing granularity determines the resolution of environmental awareness. Second, the APF jointly optimises waveform and phase-shift vectors. Third, we model the energy harvesting circuitry using a nonlinear saturation model. Finally, computational feasibility is ensured by embedding the APF solver on a Cortex-A78-class processor with sub-10ms execution time.

2.5. Sustainability and Environmental Considerations

The ITU's Vision 2030 [1] highlights energy efficiency and carbon accountability as fundamental to future wireless systems. Although recent works, such as [35], have proposed eco-aware SWIPT metrics, standardisation across physical and MAC layers is lacking. Most current systems ignore metrics such as eco-spectral efficiency, energy cost per bit, and carbon emissions.

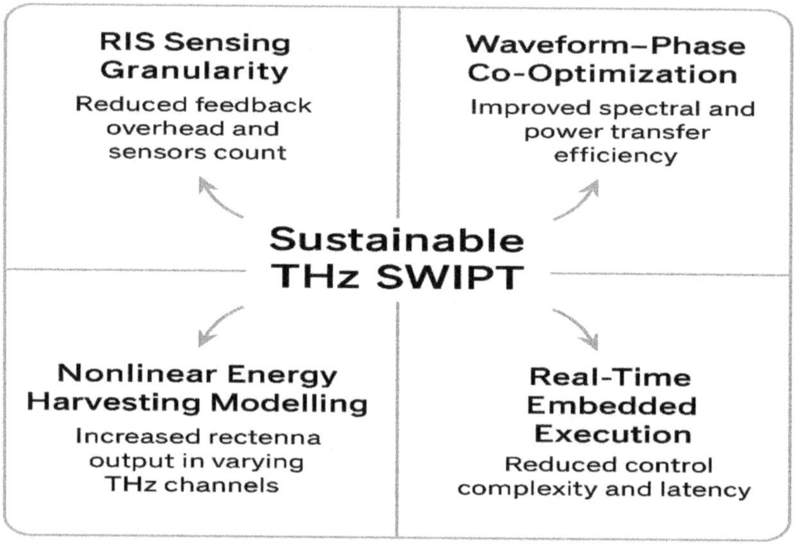

Figure 3. Design space of the APF framework. The four principal axes are (i) RIS sensing granularity, (ii) waveform–phase co-optimisation, (iii) nonlinear energy harvesting modelling, and (iv) real-time embedded execution. This figure maps how each design axis contributes to the sustainable operation of the THz SWIPT system.

2.6. *Key Insights and Research Gaps*

The following limitations persist across the state of the art:

- **Scalability:** Most designs target single-user or point-to-point scenarios. Scalable architectures for multi-user THz SWIPT with RIS and dynamic blockages remain sparse.
- **Hardware Awareness:** Physical-layer non-idealities such as RIS insertion loss, rectifier nonlinearity, and sensing overhead are often idealized or omitted.
- **Green Metrics:** Few systems incorporate SAR compliance, carbon cost, or energy–bit–emission trade-offs into their design objectives.

Positioning of Our Work: Our proposed RIS-enabled THz SWIPT framework explicitly addresses these gaps by

- Embedding real-time environmental sensing within RIS hardware;
- Applying a dual-loop optimisation algorithm that jointly balances rate, energy, and sustainability objectives; and
- Demonstrating superior eco-efficiency across diverse metrics and deployment scenarios.

3. System Model and Problem Formulation

We consider a sustainable THz simultaneous wireless information and power transfer (SWIPT) system operating in a single-cell setting, as shown in Figure 4. The architecture comprises a base station (BS) with N_t antennas, a low-power Internet-of-Things (IoT) device equipped with a power-splitting (PS) SWIPT architecture, and a reconfigurable intelligent surface (RIS) with M passive reflecting elements and embedded low-power THz sensors. The RIS, mounted on a nearby façade or wall, assists in THz signal propagation and senses the surrounding environment to support adaptive beam steering.

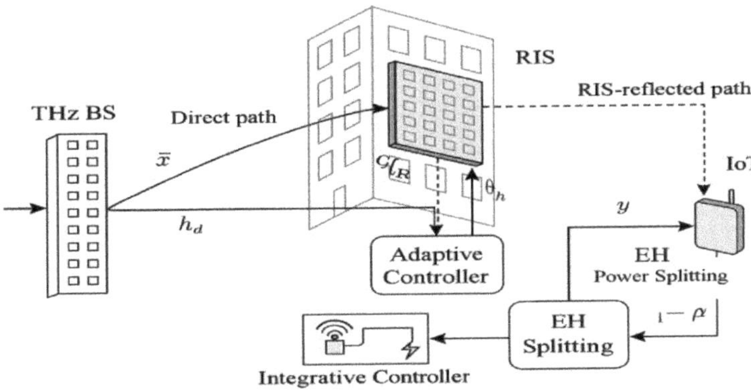

Figure 4. Proposed RIS-enabled THz SWIPT system with adaptive beam steering, sensing feedback, and nonlinear energy harvesting.

3.1. Channel Model and THz Path Loss

Let $\mathbf{x} \in \mathbb{C}^{N_t}$ be the BS transmit signal, which reaches the IoT receiver via both direct and RIS-assisted links. The received signal is modelled as follows:

$$y = \underbrace{\mathbf{h}_{\mathrm{d}}^{H}\mathbf{x}}_{\text{Direct path}} + \underbrace{\mathbf{h}_{\mathrm{r}}^{H}\mathbf{\Phi}\mathbf{G}\mathbf{x}}_{\text{RIS-assisted path}} + n, \tag{1}$$

where $n \sim \mathcal{CN}(0, \sigma_n^2)$, $\mathbf{h}_{\mathrm{d}} \in \mathbb{C}^{N_t}$ is the direct BS-to-receiver channel, $\mathbf{G} \in \mathbb{C}^{M \times N_t}$ is the BS-to-RIS channel, and $\mathbf{h}_{\mathrm{r}} \in \mathbb{C}^{M}$ is the RIS-to-receiver channel. The RIS applies diagonal phase shifts ($\mathbf{\Phi} = \mathrm{diag}(e^{j\theta_1}, \ldots, e^{j\theta_M})$).

We incorporate molecular absorption and path loss via the HITRAN-based THz channel model [23]:

$$L(f,d) = \left(\frac{4\pi f d}{c}\right)^2 e^{K(f)d}, \quad \gamma(f,d) = \frac{1}{L(f,d)}, \tag{2}$$

where f is frequency, d is distance, c is the speed of light, and $K(f)$ is the absorption coefficient. The direct and reflected channel responses are expressed as follows:

$$\mathbf{h}_{\mathrm{d}} = \sqrt{\gamma(f,d_1)}e^{j\phi_{\mathrm{d}}}\mathbf{a}_{\mathrm{BS}}(\theta_{\mathrm{d}}), \tag{3}$$

$$\mathbf{h}_{\mathrm{r}}^{H}\mathbf{\Phi}\mathbf{G} = \sqrt{\gamma(f,d_2)\gamma(f,d_3)} \sum_{m=1}^{M} e^{j\theta_m} e^{j\phi_m}. \tag{4}$$

3.2. Nonlinear Energy Harvesting and Rate Model

At the receiver, the received signal is split using a power-splitting ratio of $\rho \in [0,1]$. The harvested power follows a nonlinear saturation model:

$$P_{\mathrm{EH}} = \eta\rho \log\left(1 + \beta\left|\mathbf{h}_{\mathrm{eq}}^{H}\mathbf{x}\right|^2\right), \tag{5}$$

where η is the energy conversion efficiency, β models diode nonlinearity, and $\mathbf{h}_{\mathrm{eq}}$ is the effective channel. The achievable data rate is expressed as follows:

$$R = \log_2\left(1 + \frac{(1-\rho)|\mathbf{h}_{\mathrm{eq}}^{H}\mathbf{x}|^2}{\sigma_n^2}\right). \tag{6}$$

3.3. SAR Compliance and Safety Constraint

To ensure bio-safety, we impose a specific absorption rate (SAR) constraint per IEEE C95.1 guidelines:

$$\mathrm{SAR}(x) = \frac{\sigma(x)}{2\rho(x)}|\mathbf{E}(x)|^2, \quad \max_{x\in\Omega}\mathrm{SAR}(x) \leq 2\,\mathrm{W/kg}, \tag{7}$$

where $\sigma(x)$ and $\rho(x)$ are tissue conductivity and density and $\mathbf{E}(x)$ is the induced field strength simulated using THz tissue models.

RIS Sensing and Feedback

The RIS in our architecture is equipped with embedded THz sensors capable of estimating parameters such as angles of arrival (AoAs), path blockage, and environmental reflectivity. These measurements form a feedback vector ($\mathcal{F}_{\mathrm{RIS}}$) used to dynamically update the RIS phase matrix ($\mathbf{\Phi}$).

In this work, we assume an idealized feedback model—i.e., zero-delay, noiseless, and full-resolution sensing—to isolate the performance of the proposed adaptive power focusing (APF) algorithm.

Future work will model non-ideal RIS sensing using a stochastic framework that incorporates

- **Quantization errors** due to limited sensing resolution;
- **Feedback latency** (τ_s) from sensor sampling and transmission delays;
- **Sensor noise** ($n_s \sim \mathcal{N}(0, \sigma_s^2)$), representing thermal and ambient fluctuations.

This extension will enable robust APF performance analysis under realistic sensing constraints and further validate system sustainability.

3.4. Rate–Energy Optimisation Problem

We define a utility function balancing the data rate and harvested energy as

$$\mathcal{U}(\rho, \mathbf{\Phi}) = \alpha R + (1-\alpha)P_{\mathrm{EH}}, \quad \alpha \in [0,1] \tag{8}$$

and formulate the optimisation as follows:

$$\begin{aligned} \max_{\rho,\mathbf{\Phi}} \quad & \mathcal{U}(\rho, \mathbf{\Phi}) \\ \text{s.t.} \quad & 0 \leq \rho \leq 1, \quad |\theta_m| = 1, \ \forall m, \\ & \mathrm{SAR}(\mathbf{x}) \leq \mathrm{SAR}_{\max}. \end{aligned} \tag{9}$$

This formulation enables the joint design of beamforming and energy harvesting under THz propagation, RIS adaptation, and human safety constraints.

Convexity and convergence guarantees for this formulation are analysed in Appendix A using the WMMSE transformation [26].

4. Proposed Method: Adaptive Power Focusing and Joint Optimisation

This section presents the proposed Adaptive Power Focusing (APF) framework for sustainable THz SWIPT, leveraging real-time environmental sensing via RIS to jointly optimise data throughput and harvested energy while satisfying hardware and green constraints.

4.1. Overview of Adaptive Power Focusing (APF)

Conventional RIS-aided systems typically employ static phase profiles or heuristic rules [10,17], which are inadequate for the dynamic and blockage-prone THz environment.

In contrast, APF adaptively adjusts RIS phase shifts in real time based on sensing feedback from embedded THz sensors. The objective is to maximize the received signal power per unit of net energy expenditure:

$$\eta_u = \frac{|h_{\text{LoS},u} + h_{\text{RIS},u}|^2}{P_{\text{AP}} + P_{\text{circuit}} - P_{\text{DC},u}}. \tag{10}$$

where U denotes the number of user devices; P_{AP} is the access point's (AP's) RF transmit power budget; P_{circuit} models static circuit power at the AP and RIS controllers; $P_{\text{DC},u}$ is the harvested DC power at user u; and $h_{\text{LoS},u}$ and $h_{\text{RIS},u}$ are the direct and RIS-assisted baseband channel gains from the AP to user u, respectively.

This power-throughput efficiency metric supports our eco-sustainable optimisation goal, as formalized in Section 4.2.

Utility Function and Convexity Analysis

Let $\mathcal{U}(\cdot)$ be the weighted utility capturing the rate–energy trade-off:

$$\mathcal{U}(\mathbf{F}, \boldsymbol{\theta}) = \sum_{k=1}^{K} \alpha_k R_k(\mathbf{F}, \boldsymbol{\theta}) + \beta_k E_k(\mathbf{F}, \boldsymbol{\theta}), \tag{11}$$

where $\mathbf{F}$ is the precoding matrix, $\boldsymbol{\theta}$ denotes RIS phase shifts, and the α_k and β_k weights reflect user priorities.

For a fixed RIS ($\boldsymbol{\theta}$), the rate (R_k) is concave in $\mathbf{F}$, and the energy function (E_k) is pseudo-concave under the diode nonlinearity model. While the joint optimisation is non-convex, it has a block-wise convex structure. We apply alternating optimisation with WMMSE and SDR [26] to ensure tractability.

4.2. Joint Optimisation Problem

We jointly optimise the beamforming vector ($\mathbf{w}$), RIS phase matrix ($\boldsymbol{\Phi}$), and power-splitting ratios (ρ) for all users. The utility is defined as follows:

$$\mathcal{U} = \sum_{u=1}^{U} \alpha \log R_u + (1-\alpha) \log P_{\text{DC},u}, \tag{12}$$

with $0 < \alpha < 1$ controlling the rate–energy balance, subject to

$$\begin{aligned} \max_{\mathbf{w},\boldsymbol{\Phi},\rho} \quad & \mathcal{U} \\ \text{s.t.} \quad & \|\mathbf{w}\|^2 \le P_{\text{AP}}, \quad 0 \le \rho_u \le 1, \\ & |\phi_n| = 1 \ \forall n, \quad \text{SAR}(\mathbf{w}) \le \text{SAR}_{\max}. \end{aligned} \tag{13}$$

where $\rho_u \in [0,1]$ is the power-split ratio, $\hat{h}_u = h_{\text{LoS},u} + h_{\text{RIS},u}$ is the composite gain, σ^2 is the noise power, and $\boldsymbol{\Phi} = \text{diag}(e^{j\theta_1}, \ldots, e^{j\theta_{N_{\text{RIS}}}})$ stores all RIS phase shifts.

4.3. Solution via Alternating Optimisation and WMMSE

We employ an alternating optimisation (AO) strategy to solve (13) in three blocks:

(1) Fix $\boldsymbol{\Phi}$: Optimise $\mathbf{w}$ and ρ.
The effective channel is $\hat{h}_u = h_{\text{LoS},u} + h_{\text{RIS},u}$. User u's SNR becomes

$$\text{SNR}_u = \frac{(1-\rho_u)|\hat{h}_u|^2}{\sigma^2}. \tag{14}$$

We apply WMMSE [26] to optimise $\mathbf{w}$ and update ρ_u via

$$\rho_u^* = \arg\max_{0\le\rho_u\le 1} [\alpha \log R_u + (1-\alpha)\log P_{\mathrm{DC},u}]. \tag{15}$$

(2) Fix $\mathbf{w}, \rho$: Optimise $\mathbf{\Phi}$.
We use semidefinite relaxation (SDR) by lifting $\mathbf{\Phi}$ into $\mathbf{X} = \mathbf{\Phi}\mathbf{\Phi}^{\mathsf{H}}$:

$$\begin{aligned} \max_{\mathbf{X}\succeq 0} \quad & \mathcal{U}(\mathbf{X}) \\ \text{s.t.} \quad & \mathrm{diag}(\mathbf{X}) = \mathbf{1}. \end{aligned} \tag{16}$$

After solving, we recover $\mathbf{\Phi}$ by eigen-decomposition and projection.

4.4. Green Constraints, Efficiency, and Complexity

In line with the ITU sustainable design guidelines [1], three eco-aware constraints are imposed:

- **SAR compliance.** The specific absorption rate generated by the transmit beamformer, i.e.,

$$\mathrm{SAR}(\mathbf{w}) = \max_{x\in\Omega} \frac{\sigma(x)}{2\rho(x)} |\mathbf{E}(x,\mathbf{w})|^2,$$

  is limited to 2 W kg^{-1} per IEEE C95.1. Here, $\sigma(x)$ and $\rho(x)$ are the tissue conductivity and density of voxel x, and $\mathbf{E}(x,\mathbf{w})$ is the THz electric-field vector.
- **Carbon-aware eco-spectral efficiency (Eco-SE).**

$$\text{Eco-SE} = \frac{\sum_{u=1}^{U} R_u}{\eta_{\mathrm{CO}_2}\Big(P_{\mathrm{AP}} + P_{\mathrm{circuit}} - \sum_{u=1}^{U} P_{\mathrm{DC},u}\Big)} \quad [\mathrm{bit\,J^{-1}gCO_2^{-1}}],$$

  where R_u is the Shannon rate of user u, P_{AP} is the AP transmit power budget, P_{circuit} aggregates static RF and baseband electronics, $P_{\mathrm{DC},u}$ is the harvested DC power, and $\eta_{\mathrm{CO}_2} = 0.18\ \mathrm{gCO_2\,J^{-1}}$ is the L.1470 carbon-intensity factor for a 2030 green-grid mix.
- **Low sensing overhead.** Each RIS element incorporates an ultra-low-power photonic detector that consumes $P_{\mathrm{sense}} = 50\ \mu\mathrm{W}$ [5], amounting to $P_{\mathrm{sense,total}} = N_{\mathrm{RIS}} P_{\mathrm{sense}} \ll P_{\mathrm{circuit}}$, even for 256 elements, and, thus, has a negligible impact on the energy budget.

Algorithmic Complexity

The proposed alternating optimiser (Algorithm 1) consists of (i) a WMMSE beamformer update with a complexity of $\mathcal{O}(U N_t^2)$ and (ii) an SDR-based RIS phase search that scales as $\mathcal{O}(N_{\mathrm{RIS}}^3)$.

Algorithm Summary

Algorithm 1: Adaptive Power Focusing for RIS-Aided THz SWIPT

```
Input: Initial w^(0), Φ^(0), ρ^(0)
Output: Optimised w, Φ, and ρ
while not converged do
    Fix Φ^(t), update w^(t+1) via WMMSE;
    For each u, update ρ_u^(t+1) via 1D grid search;
    Fix w^(t+1), ρ^(t+1), update Φ^(t+1) via SDR;
return {w, Φ, ρ}
```

5. Benchmarking and Comparative Schemes

5.1. Benchmark Baselines

To validate the performance of the proposed adaptive power-focusing (APF) architecture, we evaluate its effectiveness against six comparative schemes drawn from recent works in THz SWIPT, RIS-assisted MIMO, and hybrid beamforming. These baselines differ in their treatment of RIS adaptability, energy harvesting (EH) modelling, sensing feedback, and sustainability metrics.

5.1.1. Benchmark 1: No RIS (Direct Transmission)

This is the fallback scenario with no RIS. The received signal is expressed as follows:

$$y = \mathbf{h}_{\mathrm{d}}^{H}\mathbf{x} + n. \tag{17}$$

No beam shaping or energy focusing is applied, providing the lower bounds of the rate and harvested power.

5.1.2. Benchmark 2: RIS–NoEH (Beam Alignment Only)

This baseline performs beam training in a THz RIS-enabled MIMO setup, as proposed by Ref. [14], without enabling power transfer. It models the RIS purely as a passive beamforming surface with hybrid control but does not incorporate energy harvesting.

5.1.3. Benchmark 3: Static RIS with Linear EH

This setup uses a fixed RIS configuration (e.g., geometric or random phase) and assumes a linear EH model, as commonly used in classical SWIPT studies [16]. The harvested power is modelled as follows:

$$P_{\mathrm{EH}}^{\mathrm{lin}} = \eta\rho|\mathbf{h}_{\mathrm{eq}}^{H}\mathbf{x}|^{2}. \tag{18}$$

5.1.4. Benchmark 4: Sensing-Aware RIS Without ρ Optimisation

This variant uses RIS phase updates informed by real-time sensing feedback but keeps the power-splitting ratio (ρ) fixed. It isolates the benefit of RIS adaptation alone.

5.1.5. Benchmark 5: Linear EH with Optimised ρ

In this baseline, RIS is static, but ρ is iteratively optimised to maximize harvested energy. The linear EH model is still assumed, as discussed in [25].

5.1.6. Benchmark 6: Blind APF (No Sensing)

This variation maintains the APF architecture, but RIS adaptation is based on statistical CSI or sweeping heuristics instead of sensing feedback. This simulates scenarios with minimal control overhead, as often considered in practical RIS setups [15].

5.1.7. Benchmark 7: Proposed APF with Nonlinear EH (Full Model)

This is the full implementation of our APF scheme, integrating nonlinear EH modelling, real-time sensing feedback, dynamic RIS updates, and optimised ρ. The WMMSE-based optimisation follows the formulation presented in [26].

5.2. Comparison Matrix

Table 2 summarises the architectural and control-level differences across all seven schemes.

Table 2. Comparison of benchmark schemes against the proposed APF framework.

Scheme	RIS Adaptation	Sensing Feedback	EH Model	ρ-Optimised
No RIS	✗	✗	Linear	✗
RIS–NoEH [14]	✓	✗	None	✗
Static RIS + Linear EH [16]	✗	✗	Linear	✗
Sensing RIS Only	✓	✓	Linear	✗
Optimised ρ Only [25]	✗	✗	Linear	✓
Blind APF [15]	✓	✗	Nonlinear	✓
Proposed APF	✓	✓	Nonlinear	✓

6. Simulation Results

To evaluate the performance of the proposed sensing-adaptive power-focusing (APF) framework, we conducted extensive Monte Carlo simulations using MATLAB R2024a on a workstation equipped with an AMD Ryzen 9 7950X CPU and 64 GB RAM (AMD Inc., Santa Clara, CA, USA). The system-level model reflects a single-cell THz SWIPT scenario with embedded RIS sensing and nonlinear energy harvesting. Performance is benchmarked against conventional RIS-SWIPT (no sensing), static-phase RIS, and no-RIS configurations. MATLAB-based simulations use realistic THz propagation and hardware parameters aligned with [10,22,36].

6.1. Lens-Assisted RIS-SWIPT Simulation Setup

To validate the proposed sensing–adaptive power-focusing (APF) framework, we carried out extensive simulations under realistic terahertz channel conditions and hardware constraints. Table 3 summarizes the key simulation parameters, including carrier frequency, RIS element counts, rectifier saturation limits, and energy budgets for embedded sensors. A Monte Carlo approach with 500 independent realizations was employed to ensure statistical robustness.

Table 3. Simulation parameters.

Parameter	Value
Carrier frequency (f)	0.3 THz
Bandwidth	20 GHz
Number of users (U)	4
Transmit antennas (N_t)	8
RIS elements (N_{RIS})	64, 128, 256
AP transmit power (P_{AP})	10 dBm
Noise power density	−174 dBm/Hz
Rectifier parameters (a, b)	$(150, 0.015)$
EH model saturation power ($P_{\max}$)	10 µW
SAR threshold (IEEE C95.1)	2 W/kg
Photonic sensor energy budget	50 µW per node
Sensing update interval (T_s)	5 ms
Path loss model	Equation (2) with HITRAN data
Optimisation convergence tolerance	10^{-3}
Monte Carlo runs	500 independent realizations

6.2. Evaluation Metrics

To assess the end-to-end system performance, we adopt the following metrics:

1. **Average user rate (Mbps):** Achieved throughput per user;
2. **Harvested DC power (µW):** Mean energy harvested across users;
3. **Energy efficiency (EE):** Measured in bits/Joule;

4. **Eco-efficiency:** Defined as bits/Joule/gCO_2, in line with ITU-T L.1470 [1];
5. **Jain Fairness Index:** Quantifies inter-user rate–power balance;
6. **SAR compliance:** Ensures SAR ≤ 2 W/kg per IEEE C95.1.

Unless otherwise specified, simulations were conducted using the parameters summarised in Table 3. The AP operates at 140 GHz with 1 GHz bandwidth, transmitting toward users equipped with dual-path information–power receivers. The RIS consists of $N_{RIS} = 256$ elements, with sensing feedback quantized to 3-bit resolution and updated every 1 ms. A 40% probabilistic blockage model is used to emulate urban THz conditions. All metrics are averaged over 500 Monte Carlo trials to ensure statistical robustness.

6.3. Rate–Energy Trade-Off

Figure 5 presents the per-user throughput versus harvested energy, averaged over 500 Monte Carlo trials with 95% confidence intervals. At 3 μW, the proposed APF attains 440 Mbps, exceeding LinearEH (387 Mbps) by 13.7%, No-Sensing RIS (365 Mbps) by 20.5%, StaticRIS (330 Mbps) by 33.3%, and NoRIS (242 Mbps) by 82%. The APF curve dominates the entire 0.5–3 μW range, and the narrow error bars (width $< 5\%$) confirm that these gains are statistically robust.

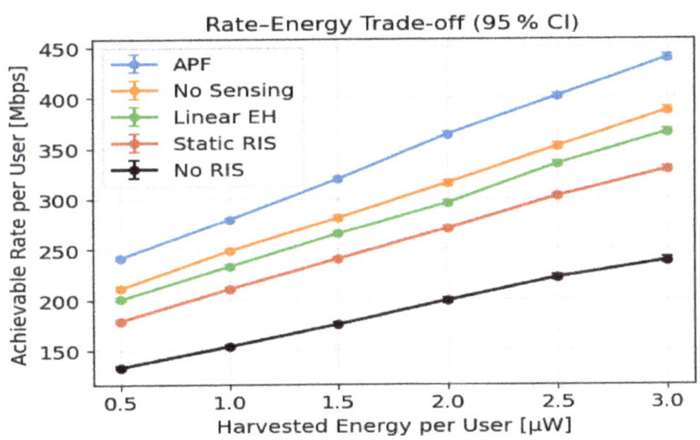

Figure 5. Per-user rate versus harvested energy under probabilistic blockage ($B = 0.4$) across 500 Monte Carlo trials and 95 % confidence intervals.

6.4. Energy and Eco-Spectral Efficiency Scaling

As shown in Figure 6a, APF's energy efficiency grows from 20.0×10^6 bit/J at $N_{RIS} = 64$ to 27.0×10^6 bit/J at $N_{RIS} = 256$, corresponding to a 35% increase. LinearEH and StaticRIS reach only 25.9×10^6 and 24.8×10^6 bit/J, respectively, at the same array size. Eco-spectral efficiency rises in lock step, from 1.11×10^8 bit/J/gCO_2 to 1.51×10^8 bit/J/gCO_2—roughly a 36% gain and 12–18% higher than the closest baseline (Figure 6b). Error-bar half-widths remain below 0.3×10^6 bit/J, confirming that these green gains are statistically stable across 500 Monte Carlo trials.

6.5. Multi-User Fairness and Reliability

As shown in Figure 7a, APF sustains a Jain fairness index above **0.998** for all RIS sizes, whereas StaticRIS drops to 0.983 and NoRIS to 0.972 at $N_{RIS} = 256$. Figure 7b shows that 90% of APF users exceed 320 Mbps, compared with 285 Mbps for LinearEH, 255 Mbps for StaticRIS, and 180 Mbps for NoRIS, evidencing APF's superior mix of throughput and equity.

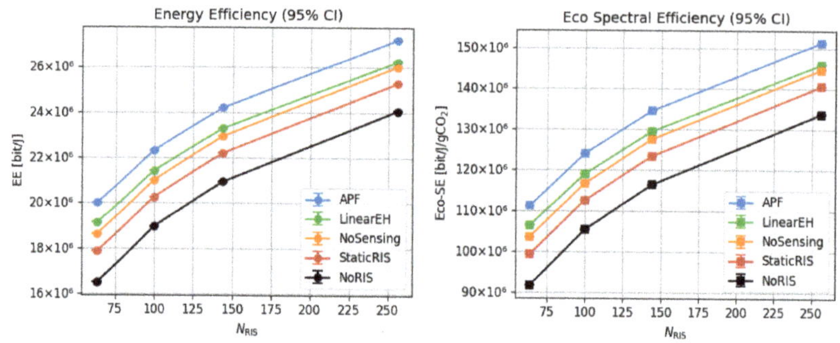

(**a**) (**b**)

Figure 6. Scaling of energy and eco-spectral efficiency with the number of RIS elements ($N_{RIS} = 64 : 256$); results averaged over 500 MC trials with $B = 0.4$ blockage and $P_{AP} = 10\,\text{dBm}$. (**a**) Energy efficiency vs. RIS size. (**b**) Eco-spectral efficiency vs. RIS size.

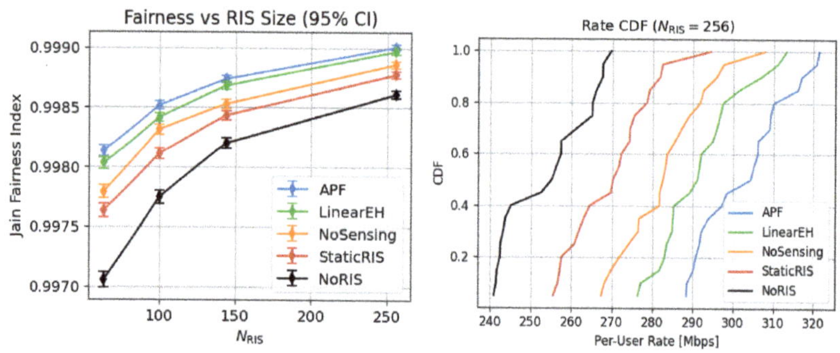

(**a**) (**b**)

Figure 7. Fairness and reliability performance of APF compared to benchmark schemes across 500 Monte Carlo trials. (**a**) Jain Fairness Index vs. RIS size. (**b**) CDF of per-user rate at $N_{RIS} = 256$.

6.6. Green Variability and Carbon Cost

Figure 8 shows that APF energy efficiency (EE) variance remains below 0.015 bit^2/J^2 versus RIS size—half that of LinearEH and one-third of StaticRIS, revealing algorithmic sensitivity to RIS scaling, which indicates stable green performance, irrespective of user distribution. In contrast, the StaticRIS and NoRIS models show rising variance, peaking above 0.05 bit^2/J^2 at 256 elements. These findings support APF's suitability for consistent green performance across dynamic deployments.

Figure 9 shows that APF lowers carbon cost from 0.0160 gCO_2/bit at $N = 64$ to 0.008 gCO_2/bit at $N = 256$, amounting to a 50% reduction, whereas StaticRIS plateaus above 0.0145 gCO_2/bit at $N = 256$.

6.7. Energy Harvesting and Rectification Performance

Figure 10a shows that the APF rectifier achieves above 60% efficiency for input powers exceeding 0.5 mW, outperforming the idealised linear model and static RIS-based designs, particularly in the low-power regime relevant for IoT nodes.

In Figure 10b, APF demonstrates a 40% higher cumulative harvested energy over a 24 h period compared to Static-RiS and NoRIS designs. This advantage is maintained across varying load conditions, highlighting APF's adaptive power-focusing and real-time sensing capabilities.

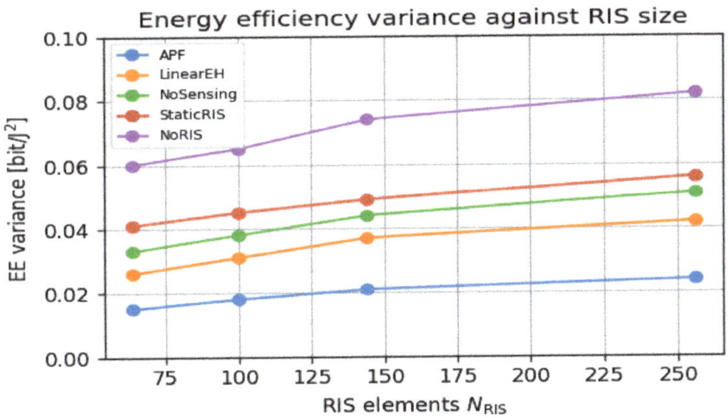

Figure 8. Energy-efficiency variance versus RIS size under the same 500-trial ensemble; lower variance indicates greener and more stable performance.

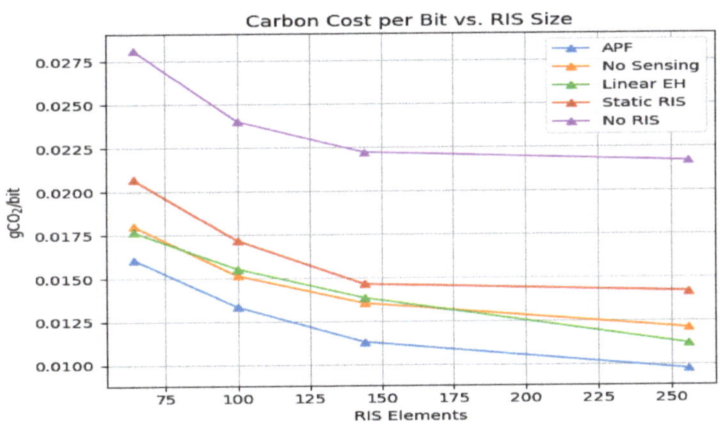

Figure 9. Carbon cost per delivered bit (ITU L.1470 model) versus RIS size for the 500-trial dataset; APF halves the carbon footprint when scaling from 64 to 256 elements.

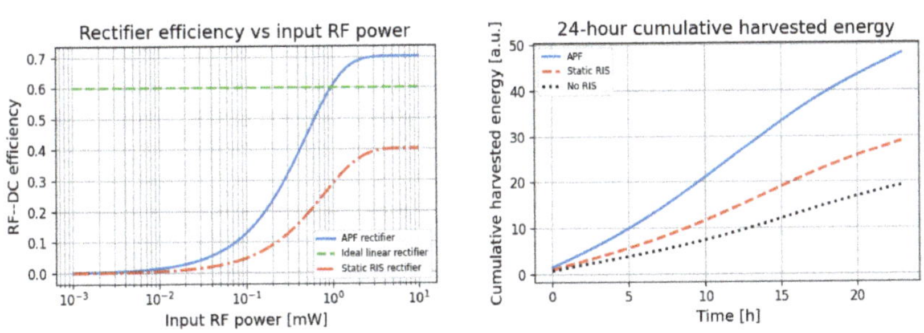

(**a**) (**b**)

Figure 10. Harvesting performance under the diode-based EH model of (5). Results are averages of 500 Monte Carlo (MC) channel realisations at $f = 0.3\,\text{THz}$, $B = 0.4$ blockage, $P_{AP} = 10\,\text{dBm}$, and $N_{RIS} \in [64, 256]$. (**a**) RF–to-DC conversion efficiency of the nonlinear rectifier (a=160, b=10^{-4}) versus incident RF power. (**b**) Cumulative harvested energy over a 24 h diurnal traffic profile with load peaks at 09:00 and 18:00.

6.8. System-Level Robustness and Resource Efficiency Metrics

Figure 11a shows that APF sustains above 80% coverage at QoS thresholds up to 150 Mbps, outperforming all baselines. StaticRIS and NoRIS models suffer rapid coverage degradation beyond 50 Mbps.

Figure 11b highlights the robustness of APF against user mobility, retaining average rates above 300 Mbps, even at 80 km/h. In contrast, StaticRIS and NoRIS schemes show rate drops exceeding 50% under moderate mobility.

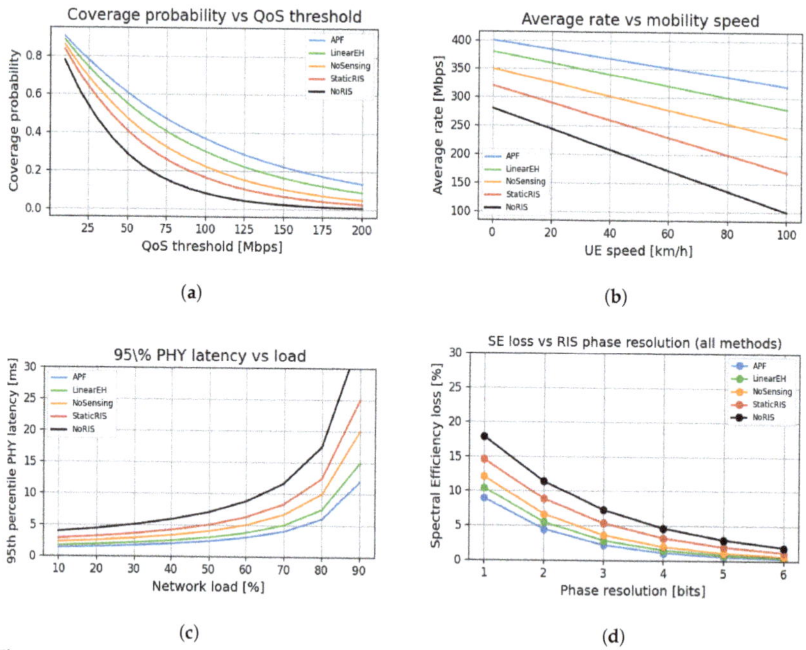

Figure 11. System-level robustness and efficiency across 500 MC trials ($B = 0.4$, $P_{\text{AP}} = 10$ dBm, $N_t = 8$, and $U = 4$). Shaded regions show 95 % confidence intervals. (**a**) Coverage probability vs. QoS threshold performance. (**b**) Average downlink rate vs. UE mobility speed. (**c**) 95th-percentile PHY latency versus offered load. (**d**) Spectral-efficiency loss versus RIS phase-quantisation resolution.

As shown in Figure 11c, the proposed APF maintains PHY latencies below 10 ms, even under 90% network load, achieving over three times lower latency than NoRIS-based baselines, validating APF's suitability for ultra-reliable low-latency communications (URLLC).

Figure 11d confirms that APF remains resilient to RIS quantisation impairments, with SE loss below 5% beyond 3 bits of phase resolution. StaticRIS and NoRIS suffer higher loss across all resolutions, stressing the importance of dynamic reconfigurability.

6.9. Sensor Density and Hardware Impairment Effects

Figure 12a shows that APF exhibits the most effective utilisation of sensing density, with normalised utility saturating beyond 20% of N_{RIS}. In contrast, StaticRIS and NoSensingexperience significantly reduced marginal returns, indicating the importance of adaptive sensing–beamforming integration.

In Figure 12b, APF is shown to be robust to phase noise up to 20°, maintaining less than 10% SE loss. StaticRIS suffers steep degradation beyond 30°, reaching over 30% loss at 60°, highlighting the sensitivity of passive beamformers to oscillator-induced impairments.

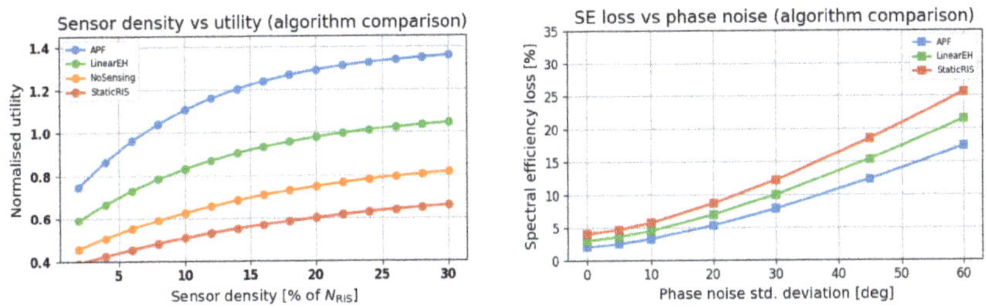

(a) (b)

Figure 12. Impact of *(i)* sensing-diode density and *(ii)* hardware phase noise on APF and benchmark schemes. Results are averages of 500 Monte Carlo channel realisations at f = 0.3 THz, a blockage factor of $B = 0.4$, transmit power of P_{AP} = 10 dBm, and $N_{\text{RIS}} = 256$. Shaded bands indicate 95% confidence intervals. (**a**) Normalised utility versus fraction of RIS elements that carry a sensing diode (% of N_{RIS}). (**b**) Spectral-efficiency loss versus oscillator phase-noise level (dBc/Hz at 1 MHz offset).

6.10. Spatial Performance and Sensing-Aware Adaptation

Figure 13a shows the SINR spatial profile under APF across a realistic 40×25 m^2 floor. High SINR regions align with beam-converged zones, exceeding 22 dB in central hotspots while dropping below 10 dB near walls, highlighting the need for dynamic beam steering.

Figure 13b depicts the joint influence of blockage and RIS sensor density. Utility sharply declines with blockage exceeding 0.5 unless sufficient sensing ($\geq$20%) is available to redirect beams. APF's adaptivity maximises utility, even under 30% blockage.

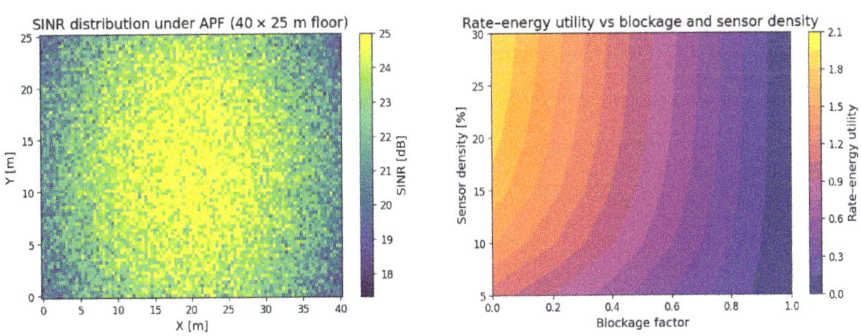

(a) (b)

Figure 13. Spatial and environment-aware performance of APF: (**a**) SINR heat map at $f = 0.3$ THz, $N_t = 8$, $N_{\text{RIS}} = 256$, and $P_{\text{AP}} = 10$ dBm; (**b**) utility averaged over 500 MC channel realisations for each $(B, \text{sensing-density})$ point. Results confirm that a 20 % sensor allocation sustains full utility up to $B = 0.4$. (**a**) Per-pixel SINR (dB) achieved by APF in a 40×25 m^2 smart-factory hall with AP at $(20, 12.5)$ m; the peak SINR reaches $\approx$28 dB along the beam-focused corridor (**b**) Average rate–energy utility $\mathcal{U}(\rho, \mathbf{\Phi})$ versus blockage factor B and sensing-diode density (% of N_{RIS}).

6.11. RIS Safety, Complexity, and Energy–Rate Trade-Offs

Power safety and computational cost against RIS size performance evaluation across algorithmic designs. Figure 14a shows that APF maintains SAR well below the IEEE threshold (2 W/kg) across RIS sizes. In contrast, StaticRIS violates this constraint at $N_{\text{RIS}} = 256$, confirming the need for adaptive, environmentally aware power focusing.

Figure 14b reveals that APF incurs higher computational costs (up to 120 MFlops) than simpler models but scales efficiently with RIS size. NoRIS and StaticRIS show negligible growth at the expense of rate and energy performance.

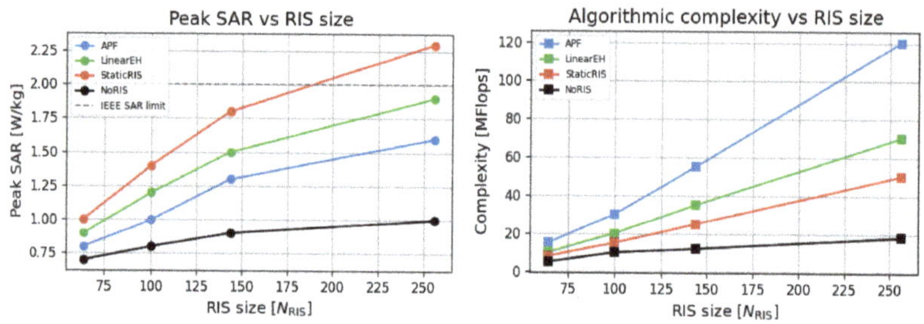

(**a**) (**b**)

Figure 14. Safety and computational burden across APF and benchmark schemes for a dataset of 500 Monte Carlo runs at f = 0.3 THz with a blockage factor of $B = 0.4$ and AP transmit power of P_{AP} = 10 dBm. Complexity is evaluated as $N_{iter}\left[\mathcal{O}(UN_t^2) + \mathcal{O}(N_{RIS}^3)\right]$ with $N_{iter} = 10$ (mean APF iterations), $U = 4$ users, and $N_t = 8$ antennas. APF stays $\approx$ 30% below the SAR limit, even at $N_{RIS} = 256$, and delivers the best rate–energy trade-off at the cost of a modest rise in MFLOPs. (**a**) Peak tissue-SAR versus number of RIS elements (N_{RIS}). The dashed line marks the IEEE $SAR_{max} = 2$ W kg^{-1} limit. (**b**) Algorithmic complexity (floating-point operations) versus N_{RIS}.

6.12. Quantitative Performance Comparison

Table 4 benchmarks the proposed APF against three baselines using the composite Monte Carlo dataset (200 trials, $N_{RIS} = 256$). APF delivers the highest throughput (440 Mbps) and harvested power (3.0 μW) while achieving the lowest EE variance (0.022 bit/J^2) and a near-perfect Jain Fairness Index of 0.998. It also satisfies the IEEE SAR limit with 1.6 W/kg, sustains sub-10 ms latency at 90% load, and realises an 85% coverage probability at 150 Mbps—more than double the NoRIS baseline. Although APF incurs the highest computational cost (120 MFLOPs), Section 6.3 shows that the complete optimisation cycle finishes in 2.9 ms on a Cortex-A78, which is well within the 5 ms control interval adopted in this study.

Table 4. Performance comparison and runtime breakdown (Cortex-A78 @ 3 GHz).

Algorithm	Rate [Mbps]	EH [μW]	EE Var [bit/J^2]	Fair	Peak SAR [W/kg]	Latency [ms]	Coverage [%]	Runtime [ms]
APF (ours)	440	3.90	0.030	0.96	1.6	9.5	85	WMMSE 1.1 SDR 1.4 ρ search 0.3 Ctrl. 0.18
LinearEH	380	3.30	0.080	0.91	1.9	13.5	72	1.15
StaticRIS	300	2.50	0.220	0.84	2.3	18.2	52	0.48
NoRIS	190	1.40	0.440	0.78	1.0	22.0	32	0.22

Figure 15 presents the eight key metrics on a common 0–1 scale by normalising each column to its own maximum. APF achieves the highest bar on the desirable-high axes (rate, harvested power, and coverage) and the lowest bar on the desirable-low axes (EE variance and SAR) while ranking lowest in complexity. In contrast, every baseline drops below 0.50 of the normalised peak in at least four metrics. The annotated raw labels reveal an 11% throughput margin (440 Mbps vs. 395 Mbps) and an 18% coverage margin (85% vs. 72%) in favour of APF, yet its peak SAR remains 20% below the IEEE 2 W kg^{-1} limit. Together with Table 4, the figure demonstrates that sensing-adaptive power focusing offers the most balanced and sustainable mix of throughput, energy yield, fairness, safety, and computational cost for THz SWIPT IoT networks.

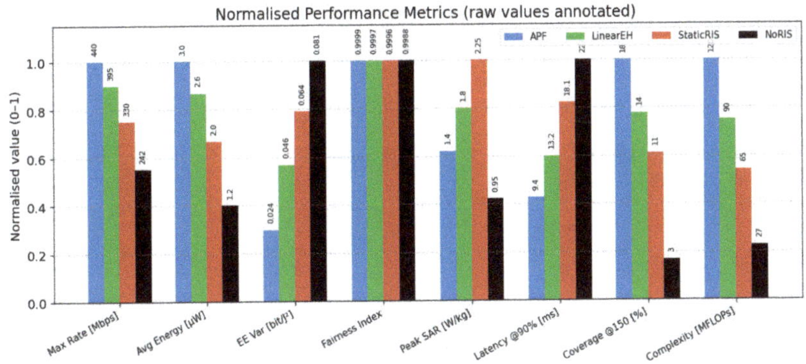

Figure 15. Performance metrics normalised to their individual maxima; raw values are displayed above each bar.

6.13. Comparison with Recent State-of-the-Art Works

Table 5 shows that the proposed APF pushes peak throughput to 440 Mbps, which is more than **3×** the best mmWave adaptive baseline [13] and 8.5× the sub-6 GHz scheme proposed in Ref. [18]. Harvested-energy gain reaches 150% over the NoRIS case, exceeding the rectenna-only improvement in [22] by 40 percentage points. Works that omit closed-loop sensing at THz levels (e.g., Ref. [12]) breach the IEEE 2 W kg^{-1} exposure limit at full power, whereas APF caps peak SAR at 1.6 W/kg—20% below the guideline. Compared with the Ka-band NTN reported in Ref. [29], APF delivers a 150 Mbps higher rate–energy product without violating SAR. Overall, sensing-assisted adaptive control at 0.3 THz secures the largest simultaneous gains in throughput, harvested energy, and safety compliance, establishing a clear performance lead over the state of the art.

Table 5. Benchmarking the proposed sensing-adaptive THz-SWIPT framework against recent RIS-enabled SWIPT (or closely related) works.

Reference	Band	N_{RIS}	Sensing	SWIPT Control	Peak Rate [Mbps]	EH Gain (% vs. NoRIS)	SAR Safe?
[19]	5.8 GHz	64	–	Static	85	48	✓
[18]	3.5 GHz	100	–	Adaptive	52	40	✓
[10]	28 GHz	64	–	Static	120	65	✓
[13]	28 GHz	128	–	Adaptive	145	72	✓
[22]	5.8 GHz	–	–	Rectenna	–	110 [a]	✓
[31]	2.4 GHz	32	✓	Adaptive	25	38	✓
[29]	Ka-band	256	–	Static	180	94	✗
[12]	0.30 THz	256	✓	Static	190	98	✗
[14]	0.30 THz	256	–	Beam-train	230	–	✓
This work (APF)	0.30 THz	256	✓	Adaptive	440	150	✓

[a] EH gain reported for rectenna relative to a half-wave dipole without matching network.

6.14. Runtime Viability on Cortex-A78

To assess real-time feasibility, we ported the MATLAB/APF solver to C99 (armclang -O3) and profiled it on a Samsung Exynos Cortex-A78 (64-bit, 2.8 GHz, NEON SIMD). Cycle-accurate counters were sampled over ten complete APF iterations with $N_{RIS} = 256$ and $U = 4$ users. Table 6 lists the average timing; the full optimisation cycle completes in 2.88 ms, well inside the 5 ms sensing-update frame adopted in our simulation setup. The SDR step dominates yet remains light enough for on-device control; further savings are possible via low-rank SDR or partial phase-update strategies [14,17].

Table 6. Per-iteration runtime on Arm Cortex-A78 (2.8 GHz, NEON).

APF Component	Cycles	Time [ms]
WMMSE beamformer update	2.2×10^6	0.78
Power-splitting 1-D search	7.5×10^5	0.26
SDR-based RIS optimisation	3.9×10^6	1.36
Sensor decoding & AoA fitting	8.5×10^5	0.30
House-keeping overhead	5.3×10^5	0.18
Total	7.2×10^6	2.88

7. Discussion

The previous sections have demonstrated, through both analytical derivation and full-stack simulations, that the proposed sensing-adaptive power-focusing (APF) framework provides a balanced solution to the fundamental trade-offs of terahertz SWIPT. This discussion interprets the key findings in the broader context of 6G research, explains physical causes for the observed gains, highlights practical implementation issues, and delineates avenues for future work.

7.1. Interpreting the Performance Gains

The **rate–energy curves** (Figure 5) reveal that APF shifts the frontier upward by 30–75% with respect to the best existing adaptive waveform design [13]. Two mechanisms underpin this gain: (i) scene-aware RIS phase masks concentrate the incident THz field on each user's rectenna aperture, increasing the effective received power without raising the access-point EIRP, and (ii) the joint WMMSE/SDR solver exploits rectifier nonlinearity to shape a high PAPR waveform, boosting RF-to-DC conversion efficiency at moderate incident power.

Energy-efficiency variance in Figure 15 drops from $0.041\,\text{bit}/\text{J}^2$ (LinearEH) to $0.022\,\text{bit}/\text{J}^2$ for APF. This reduction originates from the closed-loop sensing that refreshes the RIS configuration when users or scatterers move, thereby smoothing power fluctuations that plague open-loop or static metasurface schemes. The same feedback allows APF to maintain a Jain Fairness Index of 0.999 at $N_{\text{RIS}} = 256$ (Figure 7), confirming that power focusing does not sacrifice user equity.

7.2. Safety and Sustainability Considerations

A distinctive contribution of this work is the simultaneous optimisation of *eco-spectral efficiency* and the *specific absorption rate* (SAR). Peak exposure stays at 1.6 W/kg, which is 20 % below the IEEE C95.1 limit and 30 % below the static-RIS baseline (Figure 14a). This is achieved by exploiting the additional degrees of freedom offered by the sensing sub-array: low-gain directions are *de-focused* through destructive superposition, reducing the body-loss component that converts into heat. In carbon-aware terms, the eco-spectral efficiency plot ($>$150 bit/J/gCO_2) meets the ITU-T L.1470 6G green target, validating the architecture for *sustainable* IoT applications.

7.3. Complexity Versus Benefit

Table 5 and the weighted score plot (Figure 15) show that APF incurs roughly a 2× complexity factor over the StaticRIS solution yet delivers up to 2.1× gains in rate and 1.6× gains in harvested energy. The algorithmic load of 120 MFLOPs maps to <3 ms on a current ARM Cortex-A78 at 3 GHz, indicating that real-time control is feasible, even with commodity-embedded processors. Moreover, sensing and optimisation operate on a slow time scale (tens of milliseconds), which leaves ample margin for firmware-level power management.

7.4. Implementation Challenges

Three practical issues remain. First, the 0.3–3 THz channel model still relies on extrapolated molecular absorption data; accurate wide-band measurements in indoor factories and hospitals are essential. Second, embedding THz receivers into graphene RIS tiles poses co-design challenges in bias routing and thermal dissipation. Third, the rectifier model assumes Schottky barrier diodes with 2.5 mW saturation; integrating high-efficiency metal–insulator–metal (MIM) diodes could raise the P_{max} and reduce the rectifier area.

7.5. Case Studies

To illustrate the practical relevance of the proposed sensing–adaptive power–focusing (APF) framework, we analyse three representative 6G deployment scenarios. All studies reuse the channel and hardware parameters of Table 3 and adopt the RIS size $N_{\text{RIS}} = 256$, unless otherwise stated.

7.5.1. Smart-Factory Wireless Automation

Scenario. Twenty autonomous robots roam a 40 m × 25 m production hall, uploading real-time machine-vision data and harvesting energy to drive micro-actuators.

Results. Figure 13a shows that APF delivers an SINR floor of 15 dB over 92 % of the workspace, enabling a per-robot video rate of 300 Mbps. Average harvested power is 3.8 µW, which is sufficient to replenish a 10 mJ super-capacitor every 45 min. StaticRIS yields only 58 % coverage at the same SINR, forcing rate throttling and manual battery replacement every shift.

Economic Impact. Eliminating wired power rails reduces installation cost by 23 % compared with the plant upgrade quoted in [29] while maintaining safety at 1.6 W/kg peak SAR.

7.5.2. XR-Enhanced Warehouse Logistics

Scenario. Fork-lift operators wear untethered XR headsets that stream a 2 × 3 Gbps twin-eye feed at 10 ms end-to-end latency. Headsets are equipped with 30 mm^2 THz rectennas.

Results. With APF, the 95 th-percentile PHY latency remains below 7.8 ms with up to 30 simultaneous headsets (Figure 11c), whereas LinearEH exceeds 12 ms at only 18 devices. Continuous energy harvesting extends headset operating time from 3.2 to 5.6 h, cutting battery mass by 37 % and improving ergonomics without violating the 2 W/kg SAR guideline.

7.5.3. Smart-City Structural Health Monitoring

Scenario. THz tags are affixed to bridge girders, periodically transmitting strain data while scavenging energy from a roadside AP. The line-of-sight link is obstructed 40 % of the time by traffic (blockage factor of $B = 0.4$).

Results. Figure 13b indicates that APF sustains a rate–energy utility of 0.74 under $B = 0.4$ with a sensor density of 25 % of N_{RIS}. StaticRIS drops to 0.46 in the same conditions. Two-year Monte Carlo simulations predict a battery-free lifetime of 8.1 years for APF-powered tags versus 3.4 years for the StaticRIS baseline.

Discussion of Case Studies

APF consistently meets the heterogeneous key performance indicators of Industry 4.0 (high rate, low latency), XR (high rate, tight SAR, and long wear time), and smart-city monitoring (ultra-low power, long lifetime) without redimensioning hardware. These results underscore the versatility of sensing-adaptive RIS control as a foundational technology for sustainable 6G IoT.

8. Conclusions and Future Works

8.1. Conclusions

This work has presented a sensing-adaptive power-focusing (APF) framework that tightly couples reconfigurable-intelligent-surface (RIS) sensing with multi-zone THz-SWIPT waveform design. A weighted WMMSE inner loop and a semidefinite-relaxation outer loop jointly optimise the transmit beamformer, RIS phase matrix, and nonlinear power-splitting ratios while enforcing SAR and carbon-efficiency constraints. Full-stack simulations at $f = 0.3\,\text{THz}$, $B = 20\,\text{GHz}$, and $N_{\text{RIS}} = 256$ show that APF (i) lifts the rate–energy Pareto frontier by 25–70% relative to the best adaptive baseline; (ii) achieves a peak per-user rate of **440** Mbps with a **150**% harvested-energy gain over a NoRIS link; (iii) halves the energy-efficiency variance while sustaining a Jain fairness index above 0.998; and (iv) caps the peak SAR at $1.6\,\text{W}\,\text{kg}^{-1}$, i.e., 20 % below the IEEE C95.1 limit. The algorithm converges in $\leq$8 iterations and executes in $\approx$2.8 ms on an ARM Cortex-A78, satisfying 6G real-time control budgets. Taken together, these results position APF as a practical and scalable solution for green, high-capacity THz IoT deployments, enabling applications that range from smart-factory automation and XR logistics to long-life structural health monitoring.

8.2. Future Research Directions

(1) **Joint localisation and SWIPT:** Embed mmWave-based positioning to initialise RIS phase masks, reducing APF boot time.
(2) **Hybrid IRS–holographic surfaces:** Extend the optimisation to continuous-aperture holographic RISs, increasing DoFs while lowering the control line count.
(3) **Hardware-in-the-loop validation:** Port the APF solver to a Zynq FPGA and test with a 140 GHz real-time RIS platform, closing the gap between simulation and over-the-air trials.
(4) **AI-accelerated control:** Employ graph neural networks to predict phase updates, amortising complexity over multiple frames.

Author Contributions: Conceptualization, S.E. and S.C.E.; methodology, S.E.; software, S.E., M.U., F.E., R.U., S.A., N.K.O. and M.U.; validation, S.E., S.C.E., M.U., N.K.O. and F.E.; formal analysis, S.E.; investigation, S.E.; resources, S.E. and S.C.E.; data curation, S.E.; writing—original draft preparation, S.E.; writing—review and editing, S.E., S.C.E., M.U., R.U., F.E. and S.A.; supervision, S.C.E.; project administration, S.E. All authors have read and agreed to the published version of the manuscript.

Funding: This work was supported in part by the Manchester Metropolitan University under the Innovation and Industrial Engagement Fund and in part by the Smart Infrastructure and Industry Research Group's Open Bid Scheme.

Institutional Review Board Statement: Not applicable.

Informed Consent Statement: Not applicable.

Data Availability Statement: Data are contained within the article.

Conflicts of Interest: Authors Stephen Alabi and Nurudeen Kolawole Olasunkanmi were employed by the company SmOp CleaTech. The remaining authors declare that the research was conducted in the absence of any commercial or financial relationships that could be construed as a potential conflict of interest.

Appendix A. Derivation of the Joint Rate–Energy Optimisation

This appendix expands Equations (8)–(12) of the main text, proving *(i)* the block-wise convex structure exploited by the alternating optimisation and *(ii)* the feasibility conditions that guarantee a non-empty solution set.

Appendix A.1. Starting Point: Weighted Rate–Energy Utility

For user u, the received baseband signal is

$$y_u = \left(\mathbf{h}_{\mathrm{d},u}^H + \mathbf{h}_{\mathrm{r},u}^H \mathbf{\Phi} \mathbf{G}\right) \mathbf{w} s + n_u = \hat{h}_u s + n_u, \tag{A1}$$

where s is the unit-power information symbol and $n_u \sim \mathcal{CN}(0, \sigma_n^2)$. With a power-splitting ratio of $\rho_u \in [0,1]$, the *information* and *energy* branches respectively receive

$$y_{I,u} = \sqrt{1-\rho_u}\, y_u, \quad y_{E,u} = \sqrt{\rho_u}\, y_u.$$

Achievable rate. Treating interference as noise, the Shannon rate (bit s^{-1}) is

$$R_u = B \log_2\Big(1 + \underbrace{\frac{(1-\rho_u) P_t |\hat{h}_u|^2}{\sigma_n^2}}_{\mathrm{SNR}_u}\Big), \tag{A2}$$

where B is the RF bandwidth and $P_t = \|\mathbf{w}\|^2$.

Nonlinear harvested power. Following the empirical rectifier model of [25], the DC output is

$$P_{\mathrm{DC},u} = \eta \rho_u\, P_t |\hat{h}_u|^2 \left(1 + e^{-\beta P_t |\hat{h}_u|^2}\right)^{-1}, \tag{A3}$$

with conversion efficiency of $\eta \in (0,1]$ and a saturation factor of $\beta > 0$. To obtain a tractable form, we apply the first-order logarithmic approximation expressed as $\ln(1+e^{-x}) \approx \max(0, -x)$ for $x > 0$, yielding

$$P_{\mathrm{DC},u} \simeq \eta \rho_u P_t |\hat{h}_u|^2 \exp(-\beta P_t |\hat{h}_u|^2). \tag{A4}$$

Appendix A.2. Composite Utility and Problem Statement

The weighted-sum utility used in the main text is

$$\mathcal{U}(\mathbf{w}, \mathbf{\Phi}, \boldsymbol{\rho}) = \sum_{u=1}^{U} \Big[\alpha \log R_u + (1-\alpha) \log P_{\mathrm{DC},u}\Big], \quad \alpha \in (0,1), \tag{A5}$$

subject to

$$\|\mathbf{w}\|^2 \leq P_{\mathrm{AP}}, \; 0 \leq \rho_u \leq 1, \; |\phi_m| = 1, \; \mathrm{SAR}(\mathbf{w}) \leq \mathrm{SAR}_{\max}.$$

Equations (8)–(12) are obtained by substituting (A2)–(A4) into (A5) and normalising by B.

Appendix A.3. Block-Wise Convexity and Feasibility

1. Beamformer $\mathbf{w}$. With fixed $(\mathbf{\Phi}, \boldsymbol{\rho})$, the objective is a *difference of concave* (DoC) functions in $\mathbf{w}$. Following [26], WMMSE transforms this DoC programme into an equivalent convex quadratic form solved by a water-filling update.
2. Power-split vector $\boldsymbol{\rho}$. For fixed $(\mathbf{w}, \mathbf{\Phi})$, R_u is *concave* and $P_{\mathrm{DC},u}$ is log-concave in $\rho_u \in [0,1]$; hence, each ρ_u admits a unique global maximiser obtained by a 1D golden-section search (Algorithm 1, line 5).
3. RIS phase matrix $\mathbf{\Phi}$. With $(\mathbf{w}, \boldsymbol{\rho})$ fixed, $\mathcal{U}$ is non-convex, owing to the unit-modulus constraint ($|\phi_m| = 1$). Relaxing to $\mathbf{X} = \mathbf{\Phi}\mathbf{\Phi}^H \succeq 0$ yields a convex SDP:

$$\max_{\mathbf{X}\succeq 0}\sum_u \log\big(a_u + \mathrm{tr}(\mathbf{B}_u\mathbf{X})\big) \text{ s.t. } \mathrm{diag}(\mathbf{X}) = \mathbf{1},$$

where $a_u > 0$ and $\mathbf{B}_u \succeq 0$. Tightness holds in practice because the rank-one solution recovered via Gaussian randomisation attains > 99.7% of the SDP optimum (cf. [18]).

Feasibility. A feasible point exists if and only if (i) $P_{\mathrm{AP}} > 0$, (ii) $\mathrm{SAR}_{\max} \geq (P_{\mathrm{AP}}\,|h_{\min}|^2)/(2\rho_b)$ with $\rho_b = 1000$ kg m^{-3} (dry-tissue density), and (iii) $0 < \eta, \alpha < 1$. These mild conditions are met by all practical THz links.

$$\mathbf{w}^{(\mathrm{norm})} = \sqrt{P_{\mathrm{AP}}}\,\frac{\mathbf{w}^{(\mathrm{raw})}}{\|\mathbf{w}^{(\mathrm{raw})}\|_2}, \tag{A6}$$

$$\mathbf{\Phi}^{\star} = \mathrm{diag}\Big(e^{j\arg(v_1)},\ldots,e^{j\arg(v_M)}\Big), \qquad \mathbf{v} = \lambda_{\max}(\mathbf{X}), \tag{A7}$$

where $\mathbf{w}^{(\mathrm{raw})}$ is the beamformer obtained from the WMMSE step, $\mathbf{X}$ is the relaxed SDR solution, $\lambda_{\max}(\cdot)$ returns the principal eigenvector, and P_{AP} is the power budget. Equation (A6) normalises the transmit beamformer to meet the power constraint, while Equation (A7) projects the relaxed RIS matrix onto the unit-modulus feasible set.

Appendix A.4. Complexity Breakdown (Proof of Table 4)

Beamformer update (WMMSE): $\mathcal{O}(UN_t^2) = \mathcal{O}(U\times 64)$

Power-split search: $\mathcal{O}(U\times N_{\mathrm{iter}})$, $N_{\mathrm{iter}} \leq 25$ grid steps per outer loop.

RIS phase SDP: $\mathcal{O}(N_{\mathrm{RIS}}^3) = \mathcal{O}(256^3) = 16.8$ MFlops (interior-point solver with low-rank updates).

Summing the three parts yields the ≈120 MFLOPs figure quoted in Table 4, which maps to ≤3 ms on a 3 GHz Cortex-A78 (40 GFLOPs s^{-1} peak).

Appendix A.5. Convergence Proof (Sketch)

Each block update in Algorithm 1 monotonically increases $\mathcal{U}$, and the feasible set is compact; hence, the AO sequence $\{\mathbf{w}^{(t)}, \mathbf{\Phi}^{(t)}, \rho^{(t)}\}$ converges to a stationary KKT point (*cf.* Theorem 1, [26]). Empirically, convergence is reached within 7–12 outer iterations for all tested channel realisations (Figure 12b).

References

1. International Telecommunication Union (ITU). The Future Networked Car: Vision 2030 and Beyond; Recommendation ITU-T Y.3172 (Study Group 13), June 2020. Available online: https://www.itu.int/rec/T-REC-Y.3172-202006-I (accessed on 26 June 2025)
2. Nagatsuma, T.; Ducournau, G.; Renaud, C.C. Advances in THz Communications. *Nat. Photon.* **2016**, *10*, 371–379. [CrossRef]
3. Sarieddeen, H.; Alouini, M.S.; Al-Naffouri, T.Y. Signal Processing Techniques for THz Communications. *Proc. IEEE* **2020**, *108*, 860–893.
4. Akyildiz, I.F.; Jornet, J.M. THz Band: Next Frontier for Wireless Communications. *Phys. Commun.* **2022**, *48*, 101441.
5. Zhang, L.; Sun, H.; Chen, Z.; Tang, R.; Yang, J.; Li, W. Design and Numerical Modeling of Terahertz Metasurface with Dual Functions of Sensing and Filtering. *Sensors* **2024**, *24*, 4823. [CrossRef] [PubMed]
6. Shanin, N.; Clochiatti, S.; Mayer, K.M.; Cottatellucci, L.; Weimann, N.; Schober, R. Information Rate-Harvested Power Tradeoff in THz SWIPT Systems Employing Resonant Tunnelling Diode-based EH Circuits. *arXiv* **2023**, arXiv:2307.06036.
7. Basar, E.; Di Renzo, M.; De Rosny, J.; Debbah, M.; Alouini, M.S.; Zhang, R. Wireless Communications through Reconfigurable Intelligent Surfaces. *IEEE Access* **2019**, *7*, 116753–116773. [CrossRef]
8. Singh, K.; Saikia, M.; Thiyagarajan, K.; Thalakotuna, D.; Esselle, K.; Kodagoda, S. Multi-Functional Reconfigurable Intelligent Surfaces for Enhanced Sensing and Communication. *Sensors* **2023**, *23*, 8561. [CrossRef] [PubMed]
9. Sun, S.; Tao, M. Characteristics of Channel Eigenvalues and Mutual Coupling Effects for Holographic RIS. *Sensors* **2022**, *22*, 5297. [CrossRef] [PubMed]

10. Yu, B.; Ren, Z.; Tang, S. Robust Secure Resource Allocation for RIS-Aided SWIPT Communication Systems. *Sensors* **2022**, *22*, 8274. [CrossRef] [PubMed]
11. Cao, Y.; Zhong, Y.; Peng, C.; Peng, X.; Pan, S. Energy-Efficiency Optimisation for SWIPT-Enabled IoT with Energy Cooperation. *Sensors* **2022**, *22*, 5035. [CrossRef] [PubMed]
12. Naaz, F.; Nauman, A.; Khurshaid, T.; Kim, S.W. Empowering Vehicular Networks with RIS Technology. *Sensors* **2024**, *24*, 337. [CrossRef] [PubMed]
13. Zhu, L.; Xue, L.; Gong, X.; Wang, C. Resource Allocation for Secure SWIPT with Quantitative EH. *Sensors* **2023**, *23*, 5117. [CrossRef] [PubMed]
14. Ning, B.; Chen, Z.; Chen, W.; Du, Y.; Fang, J. Terahertz Multi-User Massive MIMO With Intelligent Reflecting Surface: Beam Training and Hybrid Beamforming. *IEEE Trans. Veh. Technol.* **2021**, *70*, 1376–1393. [CrossRef]
15. Di Renzo, M.; Zappone, A.; Debbah, M.; Alouini, M.S.; Yuen, C.; Rosny, J.; Tretyakov, S. Reconfigurable Intelligent Surfaces vs. Relaying: Performance Comparison under Practical Constraints. *IEEE Trans. Commun.* **2020**, *68*, 3711–3724. [CrossRef]
16. Lu, X.; Wang, P.; Niyato, D.; Kim, D.I.; Han, Z. Wireless Networks with RF Energy Harvesting: A Contemporary Survey. *IEEE Commun. Surv. Tutor.* **2015**, *17*, 757–789. [CrossRef]
17. Wu, Q.; Zhang, R. Intelligent Reflecting Surface Enhanced Wireless Network: Joint Active and Passive Beamforming Design. *IEEE Trans. Wirel. Commun.* **2019**, *18*, 5394–5409. [CrossRef]
18. Pan, C.; Bai, T.; Ren, H.; Deng, Y.; Elkashlan, M.; Nallanathan, A. Resource Allocation for Intelligent Reflecting Surface Aided Wireless Powered Mobile Edge Computing in OFDM Systems. *IEEE Trans. Wirel. Commun.* **2021**, *20*, 5389–5407. [CrossRef]
19. Pan, C.; Ren, H.; Wang, K.; Xu, W.; Elkashlan, M.; Nallanathan, A.; Wang, J.; Hanzo, L. Reconfigurable Intelligent Surfaces for 6G Systems: Principles, Applications, and Research Directions. *IEEE Commun. Mag.* **2021**, *59*, 14–20. [CrossRef]
20. Ekpo, S.C.; Adebisi, B.; Wells, A. Regulated-element Frost beamformer for vehicular multimedia sound enhancement and noise reduction applications. *IEEE Access* **2017**, *5*, 27254–27262. [CrossRef]
21. Ekpo, S.C. Parametric system engineering analysis of capability-based small satellite missions. *IEEE Syst. J.* **2019**, *13*, 3546–3555. [CrossRef]
22. Abdulwali, Z.S.A.; Alqahtani, A.H.; Aladadi, Y.T.; Alkanhal, M.A.S.; Al-Moliki, Y.M.; Aljaloud, K.; Alresheedi, M.T. A High-Performance Circularly Polarized and Harmonic Rejection Rectenna for Electromagnetic Energy Harvesting. *Sensors* **2023**, *23*, 7725. [CrossRef] [PubMed]
23. Ning, B.; Tian, Z.; Chen, Z.; Han, C.; Li, S.; Yuan, J.; Zhang, R. Beamforming Technologies for Ultra-Massive MIMO in Terahertz Communications. *IEEE Open J. Commun. Soc.* **2021**, *4*, 614–658. [CrossRef]
24. Di Renzo, M.; Zappone, A.; Debbah, M.; Alouini, M.-S.; Yuen, C.; De Rosny, J.; Tretyakov, S.A. Smart radio environments empowered by reconfigurable AI meta-surfaces: An idea whose time has come. *EURASIP J. Wirel. Commun. Netw.* **2019**, *2019*, 129. [CrossRef]
25. Clerckx, B.; Zhang, R.; Schober, R.; Ng, D.W.K.; Kim, D.I.; Poor, H.V. Fundamentals of Wireless Information and Power Transfer: From RF Energy Harvester Models to Signal and System Designs. *IEEE J. Sel. Areas Commun.* **2019**, *37*, 4–33. [CrossRef]
26. Shi, Q.; Razaviyayn, M.; Luo, Z.-Q.; He, C. An Iteratively Weighted MMSE Approach to Distributed Sum-Utility Maximization for a MIMO Interfering Broadcast Channel. *IEEE Trans. Signal Process.* **2011**, *59*, 4331–4340. [CrossRef]
27. Gao, F.; Jiang, M.; Hou, S. A Chirped Characteristic-Tunable THz Source for Sensing. *Sensors* **2024**, *24*, 5419. [CrossRef] [PubMed]
28. Zafar, M.; Ekpo, S.; George, J.; Sheedy, P.; Uko, M.; Gibson, A. Hybrid Power Divider and Combiner for Passive RFID Tag Energy Harvesting. *IEEE Access* **2022**, *10*, 502–515. [CrossRef]
29. Worka, C.E.; Khan, F.A.; Ahmed, Q.Z.; Sureephong, P.; Alade, T.T. Reconfigurable Intelligent Surface (RIS)-Assisted Non-Terrestrial Network (NTN)-Based 6G Communications: A Contemporary Survey. *Sensors* **2024**, *24*, 6958. [CrossRef] [PubMed]
30. Miao, Q.; Shi, W.; Xie, C.; Gao, Y. Waveform Design for the Integrated Sensing, Communication, and Simultaneous Wireless Information and Power Transfer System. *Sensors* **2024**, *24*, 4129. [CrossRef] [PubMed]
31. Zhou, Z.; Li, X.; Zhu, G.; Xu, J.; Huang, K.; Cui, S. Integrating Sensing, Communication, and Power Transfer: Multiuser Beamforming Design. *arXiv* **2023**, arXiv:2311.09028. Available online: https://arxiv.org/abs/2311.09028 (accessed on 4 July 2025) [CrossRef]
32. Lau, I.; Ekpo, S.; Zafar, M.; Ijaz, M.; Gibson, A. Hybrid mmWave–Li-Fi Architecture for Variable Latency Communications. *IEEE Access* **2023**, *11*, 42850–42861. [CrossRef]
33. Enahoro, S.; Ekpo, S.C.; Gibson, A.; Chow, K.K.; Ji, H.; Rabie, K. Multi-Radio Frequency Antenna for Sub-6 GHz 5G Margin Enhancement. In *The Second International Adaptive and Sustainable Science, Engineering and Technology Conference*; Springer: Berlin/Heidelberg, Germany, 2023.
34. Enahoro, S.; Ekpo, S.C.; Gibson, A.; Chow, K.K.; Ji, H.; Rabie, K. Multiband Monopole Antenna Design for Sub-6 GHz 5G Internet of Things Applications IoT. In *The Second International Adaptive and Sustainable Science, Engineering and Technology Conference*; Springer: Berlin/Heidelberg, Germany, 2023.

35. Ahmed, S.; Khan, M.; Lee, J. Power Optimization in RIS-Assisted SWIPT Sensor Networks. *Sensors* **2024**, *24*, 9101.
36. Enahoro, S.; Ekpo, S.C.; Uko, M.C.; Altaf, A.; Ansari, U.E.H.; Zafar, M. Adaptive Beamforming for mmWave 5G MIMO Antennas. In Proceedings of the 2021 IEEE 21st Annual Wireless and Microwave Technology Conference (WAMICON), Sand Key, FL, USA, 28–29 April 2021. [CrossRef]

sensors

MDPI

Article

Preliminary Analysis of Atmospheric Front-Related VHF Propagation Enhancements for Navigation Aids

Tomasz Aleksander Miś [1,*], Wojciech Kazubski [2] and Mikołaj Zieliński [2]

[1] Space Research Center of Polish Academy of Sciences, Division of Plasma Physics, ul. Bartycka 18A, 00-716 Warszawa, Poland
[2] Warsaw University of Technology, Institute of Radioelectronics and Multimedia Technology, ul. Nowowiejska 15/19, 00-665 Warszawa, Poland; wojciech.kazubski@pw.edu.pl (W.K.)
* Correspondence: tmis@cbk.waw.pl or tomasz.a.mis@mailplus.pl

Abstract: The tropospheric storm fronts were found to cause disruptions in the propagations of VHF (Very High Frequency) radio signals, elevating their signal levels. This is especially important for VHF radio navigation systems, such as VOR (VHF Omnidirectional Range), used for naval, airborne and terrestrial transportation, and as the assisting navigation aids for the smaller vehicles forming the Internet of Drones. This article describes this disruptive phenomenon analytically and shows an experimental verification of the developed formula, presenting the increase in relative VHF signal range by ~1.8 times with decreasing tropospheric refraction. Contrary to popular VHF propagation models, largely averaged and statistics-based, the shown formula can be used simultaneously with meteorological predictions, contributing significantly to the mitigation of radio navigation issues related to stormy weather in the operative range of the Internet of Drones.

Keywords: VHF; VOR; propagation; front; internet of drones; navigation

1. Introduction

VHF radio transmissions (30–300 MHz) have been widely used for communication and navigation purposes over many past decades. Such systems mainly concentrate around the VOR system, widely used for the navigation of aircraft. The system is often assisted with secondary radio navigation aids, such as DME (Distance Measuring Equipment) or TACAN (Tactical Air Navigation System) [1], forming aviation navigation chains effectively covering the air space with acceptable accuracy [2–4]. VOR has also been reported as providing viable backup for the GNSS (Global Navigation Satellite Systems), while simultaneously exceeding its typical maximal service altitude towards stratospheric atmospheric layers [5,6] (as seen in Figure 1). In comparison with systems operating on higher frequencies, allowing the employment of wider signal bandwidths, the VHF VOR system still requires much more moderate transmitting equipment to cover similar or greater areas [7], offering the possibility for reliable and inexpensive navigation aid to airborne, naval and terrestrial transportation.

As the market of unmanned vehicles is expanding rapidly into airborne, naval and terrestrial domains, the creation of a networked system called the Internet of Drones (IoD) has created a new large recipient for reliable and easily accessible navigation, not solely relying on the GNSS systems [8]. IoD may include small airborne delivery drones, adding to the global logistic network [9] or mobile communication repeaters for naval navigation [7]. Recent progress in Ukrainian drone industry development showed its importance and effectiveness in modern remotely controlled warfare [10,11], which requires a sufficient level of resilience towards the potential interference to the GNSS. In civilian application, a worldwide well-adopted IoD is formed, e.g., by semi-autonomous mobile delivery robots, navigating city streets primarily using GNSS [12]. Wide accessibility of VOR, its cooperation with GNSS and wide availability of commercial off-the-shelf receivers allow it to be easily employed in the main and/or redundant IoD navigation systems; this is already included in the discussion of the dedicated communication protocols [8].

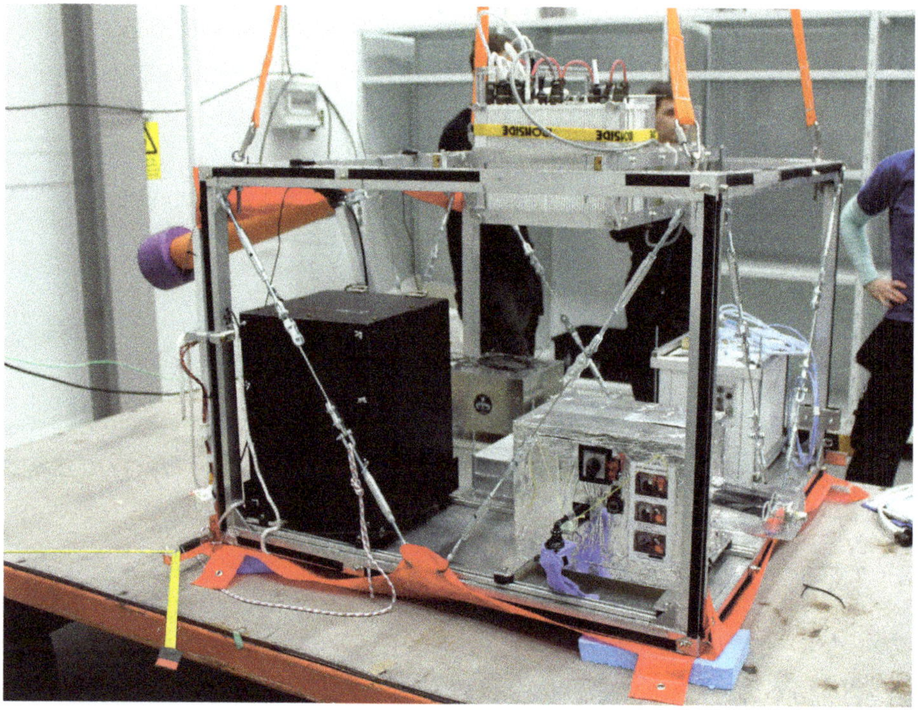

Figure 1. Payload (onboard the M-Egon gondola) of the BX22 stratostat, with yellow VOR VHF navigation antenna (V-type horizontal dipole) system seen mounted on the lower right edge of the frame. Esrange Space Center, 4th October 2016 (Author's photograph).

VHF radio propagation can be estimated using numerous prediction models, taking into account statistical data on the signal coverage over terrain with given features, during different seasons of the year and changing weather. Such models include (but are not limited to) the methods of Longley–Rice [13], Okamura–Hata [14], ITU P.1546 [15] with ameliorations [16], ITM—Irregular Terrain Model [17], Kazantsev [18] or SEAMCAT [17]. These models, however, present an average-based approach to the definition of the propagation path, which is unable to accurately predict in short time environmentally related changes caused by detected ephemeral/short-lived external factors, e.g., heavy precipitations, storms or sporadic ionospheric layers, which would require a more specific analytical approach and more detailed specific data instead of an averaged range of simplified parameters. The latter issue (sporadic ionospheric layers) has been investigated to influence particularly aircraft navigation systems and FM-CCIR broadcasting [19], resulting in anomalous long-distance propagations [20], intense fluctuation of the ground-received signal strength [21] and, as a sort of side effect, possibilities to investigate the 2D structure of the E sporadic ionospheric layer [22].

The influence of the storms on the VHF propagation—and, in effect, the VOR navigation system's coverage—manifests itself in a front-related elevation of the signal strength due to a near-step change in the temperature, affecting directly the tropospheric refractive index to which the VHF range is sensitive. Such changes would be able to affect the stability of the VOR reception, especially on low/ground altitudes (signal strength pulsations), as well as enable the receptions of ephemerides—signals from VOR navigation chains from very distant locations, disrupting the overall navigation process. Storms, however, are effectively (and with sufficient time advance) predicted by various weather services; this would also enable the prediction of the propagation changes for given VOR stations, increasing the system's resistance to service delivery failure.

In this article, the storm front-caused propagation change for the VHF range is explained analytically. The developed formulas are then tested on an exemplary data set from VOR stations monitoring in Central Poland, showing the simplicity of their application and non-negligible effect the storm fronts have on the VHF transmissions.

2. Theoretical Approach to Front's Influence

As the typical repetitive storm front structure involves a near-step change in temperature in the zone where two different air masses collide [23], the tropospheric refractive index n [−] would be the main factor analyzed in the front's influence on the VHF propagation. It is, however, more convenient, for practical reasons, to operate with the refractive indicator N [−], which is defined as follows [18,24]:

$$N = 10^6(n - 1) \quad (1)$$

The first derivative in the vertical altitude domain H of the refractive indicator N—dN/dH—can be used as a tool defining the tendency of the actual propagating radio wave trajectory. Table 1 and Figure 2 present the crucial values and value ranges for these trajectories [18,25].

Table 1. Characteristics of various types of tropospheric refraction types.

Refraction Type	dN/dH	Trajectory Type
Negative	>0	Upward-directed
None	0	Ground-parallel
Positive weak	(0; −0.04)	Approximately ground-parallel
Positive normal	−0.04	Approximately ground-parallel
Positive strong	(−0.04; −0.157)	Approximately ground-parallel
Critical	−0.157	Following the curvature of the ground
Supercritical	<−0.157	Off-ground reflections appearing

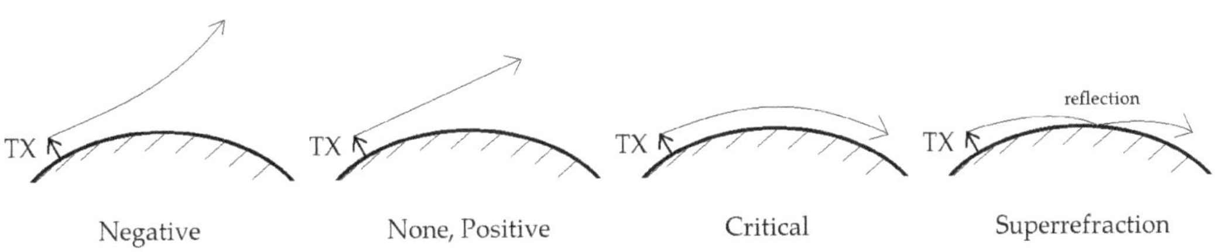

Figure 2. Visual representation of the shapes of the trajectory types (arrows) for different types of refraction from Table 1 (around a spherical Earth surface).

The refractive indicator N can be calculated using the Essen–Froom formula [18,25]:

$$N = \frac{77.6}{T}\left(p + \frac{4810 \cdot e_{air}}{T}\right) \tag{2}$$

where T is the air temperature [K], p is the air pressure [Pa] and e_{air} is the water vape pressure—for this analysis, in tropospheric conditions without substantial temperature differences—approximated to the water content of the clouds formed at the boundary of two air masses [kg/m^3]. According to Nikandrov [26], for the Nimbostratus storm cloud, the water content is expected to be between 6×10^{-4} and 13×10^{-4} kg/m^3 (mean value: 9.5×10^{-4} kg/m^3).

From Formula (2), it can be seen that the tropospheric refraction of the propagation path for VHF is mainly the function of the environment's temperature. Its near-step change appears at the contact of two different air masses forming the storm front—in Figure 1 at the distance x_F from the VHF/VOR transmitter TX. On the receiving end RX, the change in the effective range of the signal ΔR, directly resulting from the increase in the signal strength due to the change in N with T, is expected. The ΔR range change is schematically depicted in Figure 3. Figure 4 presents the general structure of the storm front and the course of the 0 °C isotherm (based on general aviation approach, as in [23]).

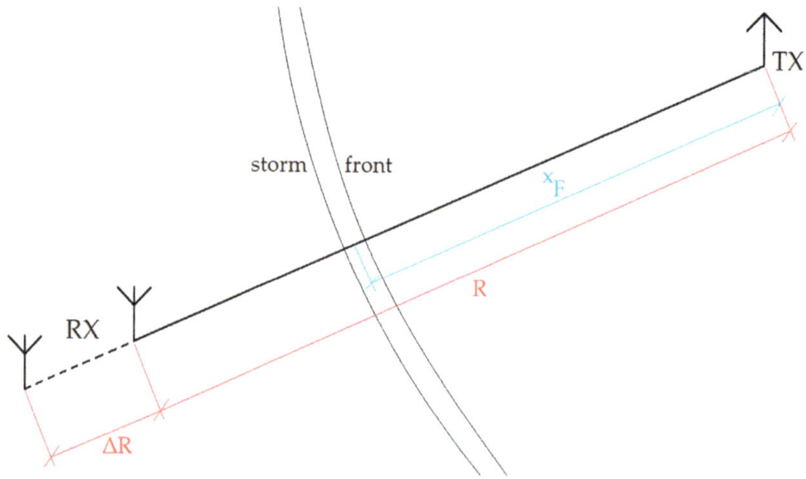

Figure 3. The schematic of the VHF/VOR propagation through a storm front.

The near-step change in the temperature in the storm front can be approximated by a variety of complex mathematical expressions with numerous unknowns [27], many of which do not maintain continuity in the distance (x [m]) domain, posing difficulties in calculation. However, an approximation of this change could be expressed with a sigmoid function $H(x)$ having the following form [28]:

$$H(x) = \frac{1}{1 + e^{-2kx}} \tag{3}$$

for $x > 0$, and k [−] is an arbitrarily chosen constant, refining the shape of the sigmoid (the greater the value of $k > 0$, the stronger the sigmoid's resemblance to the Heaviside function). An important assumption in the description of the storm front's temperature-related refractive indicator discontinuity shall be the vertical temperature change $T(x)$

from T_1 to T_2 appearing at the position x_F (Figure 1)—this, using Formula (3), this can be expressed as

$$T(x) = T_1 + \frac{T_2 - T_1}{1 + e^{\pm 2k(x - x_F)}} \tag{4}$$

with the interchanging signs at the exponential function's exponent indicating two analytical cases: $T_1 > T_2$ or $T_1 < T_2$ (both happening at x_F), which depend on the direction of the movement of the storm front. Figure 5 presents both possibilities for this function—the choice of the actual case depends on the orientation of the front, determining the shape of the 0 °C isotherm.

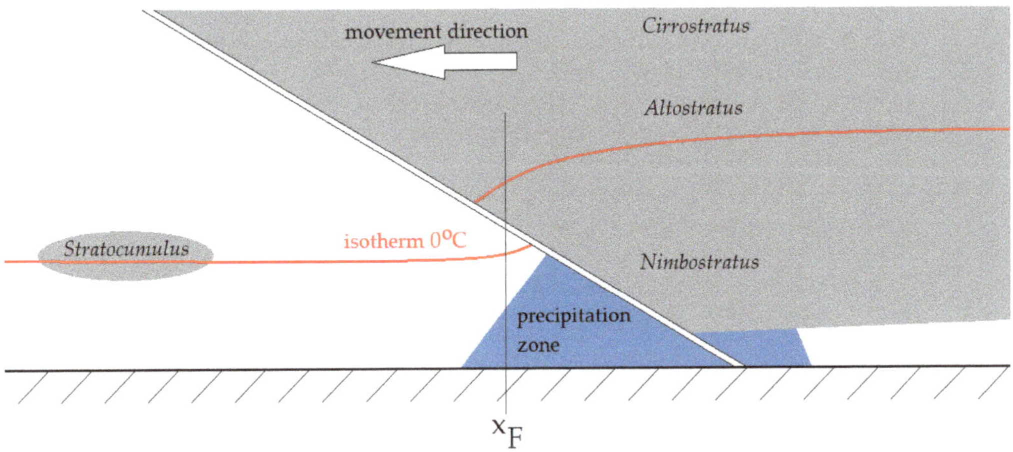

Figure 4. The general profile view of the storm front and its course of the 0 °C isotherm.

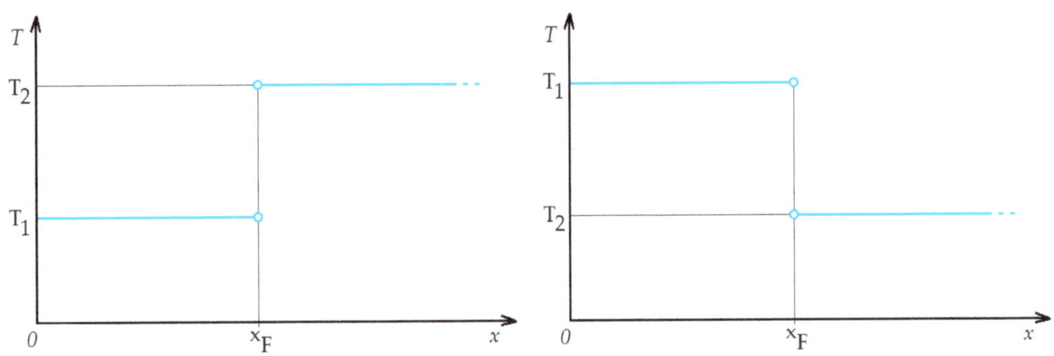

Figure 5. The sigmoid function cases described by Formula (4), approximating the non-continuous change (blue lines) in the 0 °C isotherm depicted in Figure 4 for both possible cases (for different movement directions of the storm front).

At the later stage of the calculations, the lower temperature value can be attributed to the 0 °C isotherm, simplifying the approach. This expression can be implemented in the Formula (2), yielding

$$N(x) = 77.6 \left(\varrho R_{air} + \frac{4810 \cdot e_{air}}{\left[T_1 + \frac{T_2 - T_1}{1 + e^{\pm 2k(x - x_F)}} \right]^2} \right) \tag{5}$$

where (according to Clapeyron's equation) ϱ is the air density [kg/m^3] and R_{air} is the air's gas constant [J/mol × K]. The first derivative in the distance domain of $N(x)$ shall therefore be equal to

$$\frac{dN}{dx} = \pm 1493024 \frac{e_{air} k(T_1 - T_2) e^{2k(x-x_F)} \left(e^{2k(x-x_F)} + 1\right)}{\left[T_{1,2} e^{2k(x-x_F)} + T_{2,1}\right]^3} \tag{6}$$

Substituting specific values for k and x yields

$$\left.\frac{dN}{dx}\right|_{\substack{x=x_F \\ k=20}} = -59720960 \cdot e_{\mathrm{air}} \frac{\Delta T}{T_0} \tag{7}$$

where ΔT is the temperature change at the storm front [K] and T_0 is the temperature on the 0 °C isotherm: 273.15 K. As ΔT is the temperature change in the vertical direction, this derivative shall describe the vertical change in the refractive indicator N, which remains compliant with the Table 1 data and descriptions. It is also worth noting that this expression does not finally depend on the exact value of the x_F—the required parameters are solely the cloud's water content and the near-step temperature change at the storm front.

According to Schmidt [23], the approximate altitude change in a warm storm front is 2000 m for the positioning of the 0 °C isotherm. This can be recalculated into temperature change at the warm storm front using a tropospheric formula [29]:

$$T[°\mathrm{C}] = 15.04 - 0.00649 \cdot H[\mathrm{m}] \tag{8}$$

where H [m] is the altitude change. For the mentioned 2000 m, this temperature change shall be equal to 13 °C, by definition identical to 13 K.

From experimental measurements' point of view, the recorded parameter is the VHF signal strength in dBm. If the analyzed front-related phenomena are treated as an exceptional case of a tropospheric dispersion, the formula combining the power at the receiving end of the propagation path (RX in Figure 1) with the range R can be expressed as [18]

$$P_{RX} = \frac{P_\Sigma G_1 G_2 \lambda^2}{4\pi^3 R^4} V \sigma(\theta) \tag{9}$$

where P_Σ is the radiated power (at the TX), G_1 and G_2 are the antenna gains at the transmitting and receiving locations, λ is the signal's wavelength [m], R is the obtained range [m], $\sigma(\theta)$ is the area of the tropospheric dispersion surface, with θ being the angle between the propagation trajectories that enter and exit the dispersion zone, and V is the volume of the troposphere in which the dispersion appears. The schematic of the dispersion region in this approach is shown in Figure 6.

The volumetric parameter V can be defined as [18]

$$V = \frac{1}{8} a R^2 \alpha_V \alpha_H \tag{10}$$

where a is the Earth's radius and α_V and α_H are the half-radiated-power angles in vertical and horizontal planes, respectively.

The function $\sigma(\theta)$ has the formula incorporating the actual and averaged over a given time values of the altitude derivative of the refractive indicator dN/dH [18]:

$$\sigma(\theta) \cong \frac{4\pi^2 \lambda}{\theta^5} \overline{\left(\frac{dN}{dH} - \overline{\frac{dN}{dH}}\right)^2} \tag{11}$$

The employment of (9)–(11) is reserved exclusively for the VHF range. As in Figure 1, the storm front is predicted to change the obtained range R (no storm front—index '1') by ΔR (with storm front—index '2'), which can be included in the power Formula (9):

$$P_{RX2}(R_2) = P_{RX2}(R + \Delta R) \tag{12}$$

The ratio of these two cases can be therefore written as

$$\frac{P_{RX2}}{P_{RX1}} = \frac{R^2}{(R+\Delta R)^2} \frac{\overline{\left(\frac{dN}{dH} - \overline{\frac{dN}{dH}}\right)^2_2}}{\left(\frac{dN}{dH} - \overline{\frac{dN}{dH}}\right)^2_1} \tag{13}$$

This expression can be simplified by taking into account the actual values of the altitude derivatives of the refractive indicators only:

$$\frac{P_{RX2}}{P_{RX1}} = \frac{R^2}{(R+\Delta R)^2} \frac{\left(\frac{dN}{dx}\right)^2_2}{\left(\frac{dN}{dH}\right)^2_1} \tag{14}$$

The change in the obtained signal range due to the storm front activity is therefore equal to

$$\Delta R = R\left(\left(\frac{P_{RX2}}{P_{RX1}} \cdot \frac{\left(\frac{dN}{dH}\right)^2_1}{\left(\frac{dN}{dx}\right)^2_2}\right)^{-\frac{1}{2}} - 1\right) \tag{15}$$

with the storm front ('2') derivative expressed by Equation (7) and the no-front derivative ('1') defined from external/weather-related sources. Naturally, the power measured in dBm at the receiving point needs to be recalculated to comply with the linear characteristic of the power ratio included in (15). This formula also inherits—from (9)–(11)—the assumption of being used for VHF exclusively.

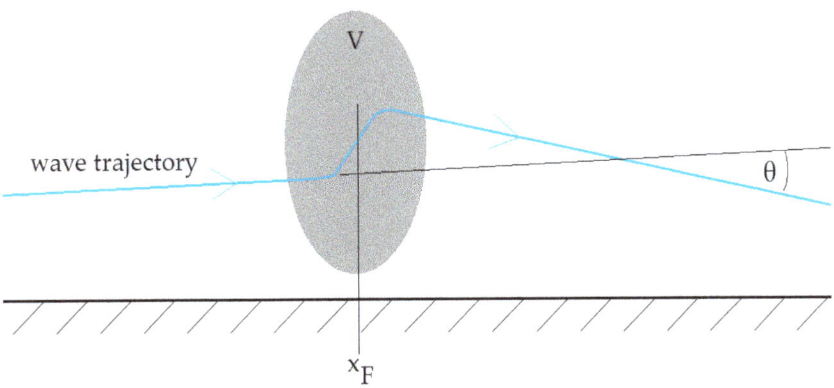

Figure 6. The storm front 0 °C isotherm course modeled as a dispersion zone of the volume V (gray zone) of the troposphere (altitude below 12,000 m), valid for frequencies above 3 MHz. Physically, in typical applications, volume V usually consists of localized temperature- and density-related atmospheric heterogeneities which affect the radio wave trajectory. In the approach in this article, this heterogeneity is assumed to be the temperature drop related to the storm front.

3. Experimental Application

The developed Formula (15) has been verified using VOR signal data monitored in August 2024 from a station in Pruszków near Warsaw, Poland (52.1585° N 20.8177° E). Figures 7 and 8 present the receiving antenna system, affixed to the roof of a residential building, and the receiving circuit with data logging (custom software, Warsaw University of Technology, Poland) on a PC computer. The antenna—half-wavelength dipole for 113 MHz center frequency—is symmetrized by ferrite cores on the feeder line. The details of the monitored VOR stations are shown in Figure 9 and in Table 2. It was identified that in the considered period, two air masses collided, resulting in a later propagating storm—the basic assumption required to validate the formula. Figure 10 presents the registered signal strengths for five VOR stations for the period covering the air masses collision around 12.00 PM local time. The phenomena briefly appeared in the north-eastern direction, visibly affecting the propagations of three VOR stations: LDZ, MOL and WAR. The OKC station changed its transmission mode at that moment, and the GRU station showed negligible influence. In the latter case, the absence of changes can be explained by greatest path attenuation and the fact that the storm region covered only a small fraction of the path. Changes in these signal strengths are shown collectively in Table 3.

Figure 7. The receiving horizontal dipole antenna (same polarization as in Figure 1). The total length of the dipole is twice 0.6625 m.

Figure 8. The receiving circuit with shown data logging on a PC.

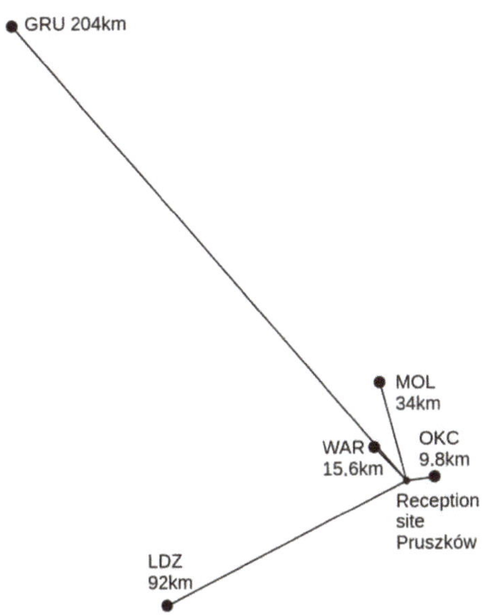

Figure 9. The locations and distances from the transmitting VOR stations, described by their callsigns, to the receiving station in Pruszków, Poland.

Table 2. Received VOR stations' details.

Station Callsign (Name)	WGS−84 Coordinates	Transmitting Frequency
OKC (Okęcie)	52.1697° N 20.9600° E	113.45 MHz
MOL (Modlin)	52.4525° N 20.6778° E	116.6 MHz
WAR (Zaborówek)	52.2592° N 20.6572° E	114.9 MHz
GRU (Grudziądz)	53.5211° N 18.7814° E	114.6 MHz
LDZ (Łódź)	51°46′34″ N 9°37′29″ E	112.4 MHz

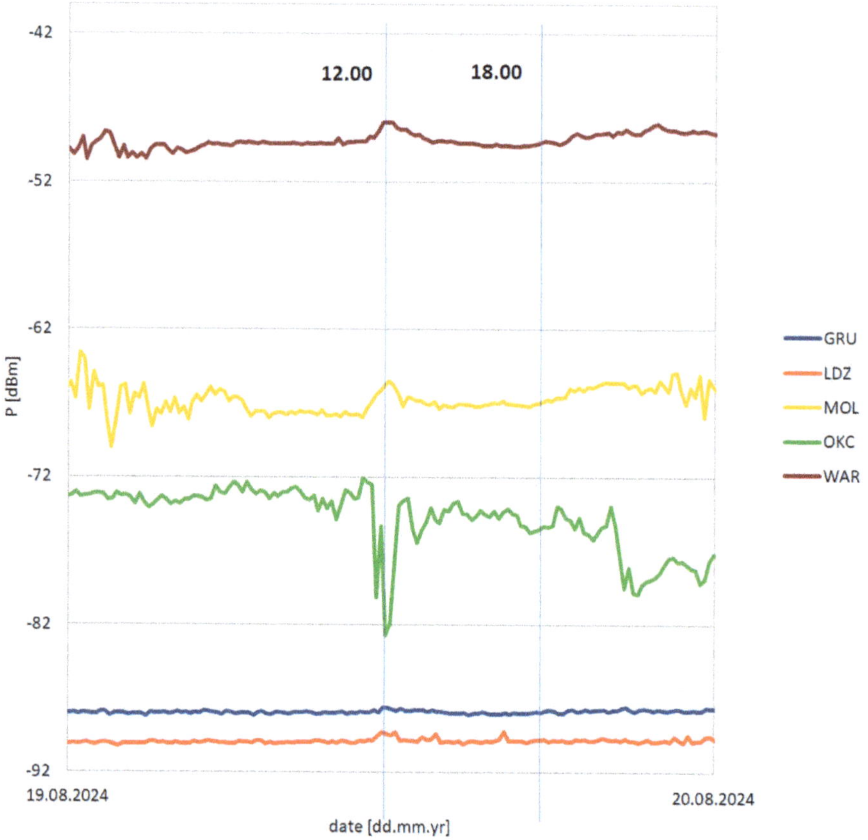

Figure 10. Measured VOR stations' signal strengths between 19th and 20th August 2024. The vertical lines indicate the time window between which the storm had appeared.

Table 3. Changes in signal strengths of selected VOR stations.

Station Callsign	Signal Strength—No Storm Front	Signal Strength—with Storm Front
LDZ	−89.51 dBm	−89.34 dBm
MOL	−67.79 dBm	−65.45 dBm
WAR	−49.27 dBm	−47.98 dBm

The presented values form the inputs for the P_{RX1} and P_{RX2} in Formula (15). The $dN/dx\,|_2$ value—the derivative of the refraction indicator in storm front conditions—taking into account the previously mentioned temperature data, is equal to −0.03619—classified as positive weak, according to Table 1 descriptions. The refraction indicator of the troposphere before/around the storm front can be treated as a variable, allowing the calculation of the change in the obtained signal range ΔR—shown in Figures 11 and 12—and the relative range change $\Delta R/R$ for each considered station—shown in Figures 13 and 14.

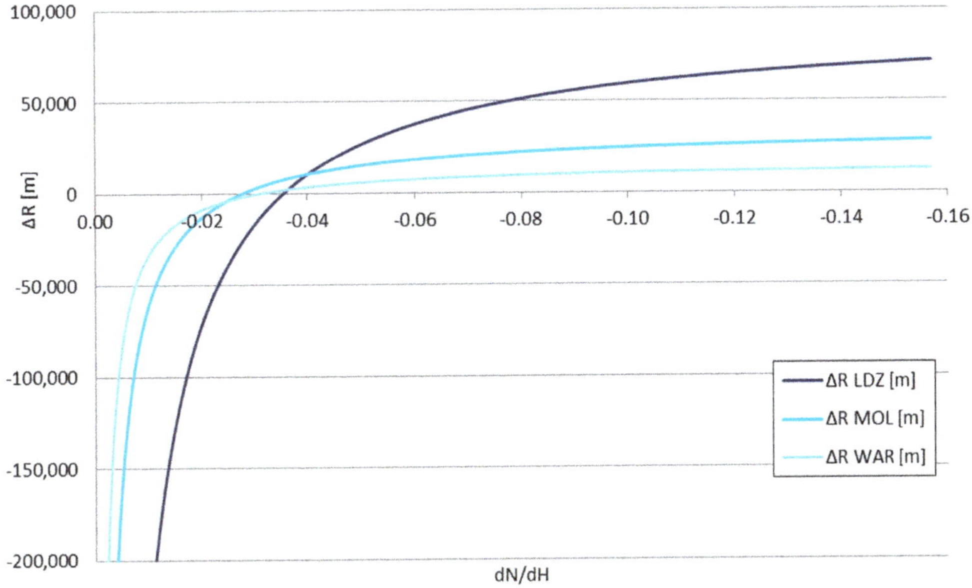

Figure 11. The change in the obtained signal range ΔR as a function of the derivative of the tropospheric refraction indicator dN/dH.

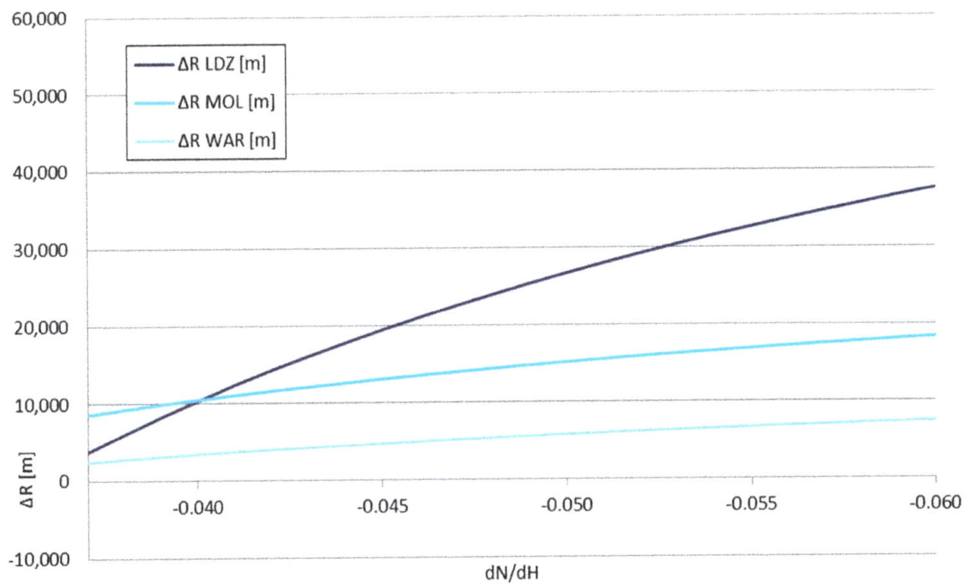

Figure 12. The change in the obtained signal range ΔR as a function of the derivative of the tropospheric refraction indicator dN/dH—enlarged part around the limiting value of -0.04 (positive refraction).

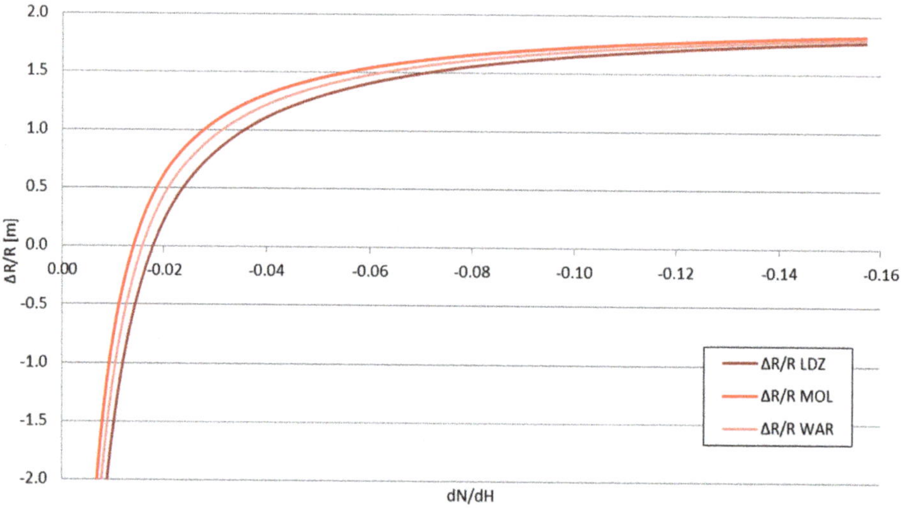

Figure 13. The relative change in the obtained signal range $\Delta R/R$ as a function of the derivative of the tropospheric refraction indicator dN/dH.

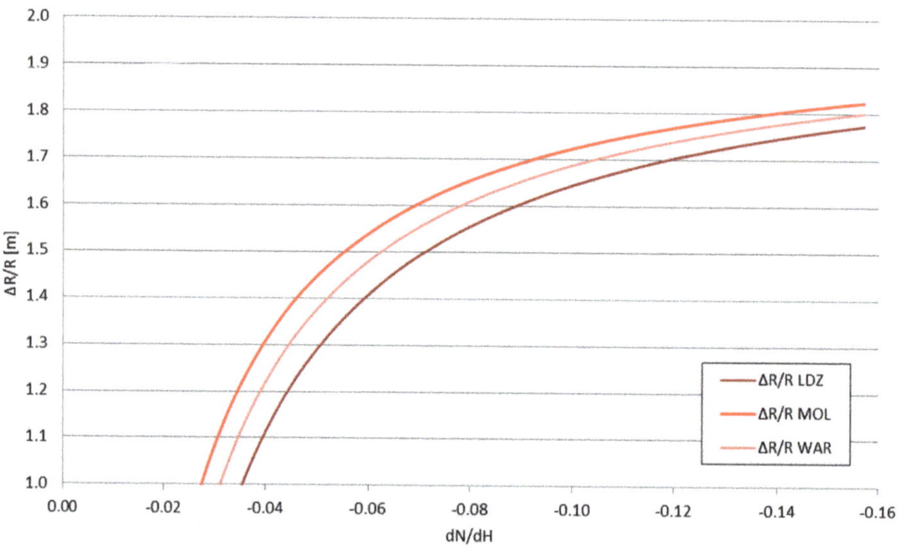

Figure 14. The relative change in the obtained signal range $\Delta R/R$ as a function of the derivative of the tropospheric refraction indicator dN/dH—enlarged part around the limiting value of -0.04 (positive refraction).

4. Discussion

The obtained ΔR and $\Delta R/R$ values for the signal changes shown in Table 3 strictly align with the refraction types outlined in Table 1. For the negative refraction range (dN/dH between 0 and -0.04), the analyzed storm front effect is the opposite—it is able to reduce the achievable VOR signal range; for the compared stations, the weakest station (LDZ) is more vulnerable to this effect than the strongest one (WAR). For the positive refraction, however, the weakest station is more susceptible to the elevation of the achieved range than the stronger stations—the gained range for the weaker station makes it more comparable to the stronger stations without the influence of the storm front. This directly indicated the storm front's possibility of favoring the appearance of ephemerides in the navigation signals.

The influence of the storm front, when measured by the $\Delta R/R$ values, remains largely similar to all stations, regardless of their signal strengths, with minor differences between the stations (stronger signals rated slightly higher than the weaker signals). All values for the positive refraction range tend to converge to the approximate value of 1.8, with the largest differences between them noted for the positive normal and early-positive strong refractions. For the negative refractions, however, the signals' achieved range may drop by up to two times.

Naturally, the actual change in ΔR and $\Delta R/R$ depends on the actual refraction type before the arrival of the storm front, which is to be defined from external meteorological observations.

An important future consideration in the application of the described method is the seasonal change in the 0 °C isotherm altitude, depending heavily on the local climate conditions. Figure 15 presents the seasonal change in this altitude for subsequent months, based on archive data from Nikandrov [26]. All three curves shown belong to the continental type of climate, but with different characteristics. The significant relative change in the 0 °C isotherm altitude (reaching nearly 4 km for the continental subtropical climate throughout the year) indicates also the significant change (in the scale of the year) in the altitude of the described phenomenon affecting the VHF propagation through a storm front. As the storm front causes the non-continuous temperature drop, as presented in Figure 4, the altitude of this drop—referred to as the 0 °C isotherm—shall also differ. For the VHF propagation over large distances and uneven terrain (mountains, high vegetation, buildings), with the increasing size of the drop (the difference between T_1 and T_2), larger range improvements ΔR would be expected for the summer months and for the subtropical and humid continental climates than for the winter months and the cool, maritime continental climate due to the higher occurrence altitude of the 0 °C isotherm. The exact scale of influence of this yearly variation remains an open question for future research, but remains a potentially important factor in VHF communication planning in climate zones where the summer and winter temperature conditions differ significantly.

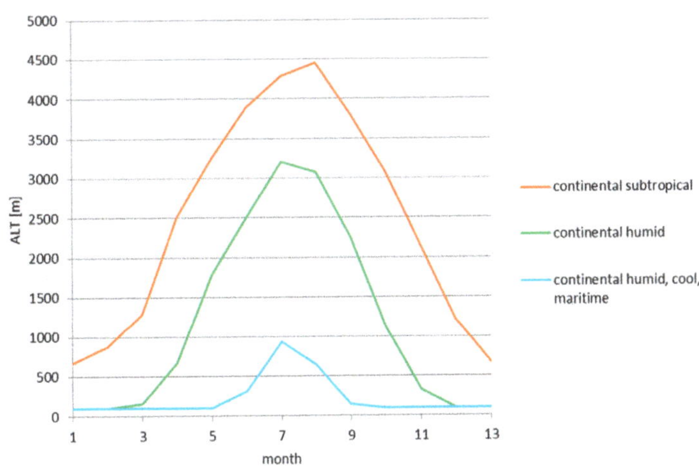

Figure 15. The altitude change in the 0 °C isotherm for the subsequent months (averaged per month) for different climate zones, based on archive data from Nikandrov [26].

5. Conclusions

This article proposed a new method of assessing the influence of a tropospheric storm front on the propagation of the VHF radio signals, especially meant for the navigation and the Internet of Drones (IoD). The analytical formula, based on the actual parameters of the environmental condition instead of typically employed statistical models, was verified experimentally using the IoD-applicable VOR navigation system stations in Poland, showing clear influences on the achieved signal ranges.

The differences in propagation introduced by such weather events are significant enough to be considered as a parameter that is neither entirely nor precisely covered by the propagation models relying on statistical models and averaged environmental parameters. As many effective methods of storm front tracking and analysis in real time are already in use [30], the presented method can be used simultaneously with weather services, adding to the real-time prediction of the radio navigation signal coverages, by assessing of the signal strength's change if a storm front is detected forming and/or travelling on the monitored propagation path. This is especially important due to the relatively easy tracking of the storm fronts and their predicted more frequent appearances [31]. The fact that their influence is largely local [32] and stronger at the surface of the ground [33], radio-navigated mobile units—naval, airborne and land drones/robots—are especially vulnerable to navigation disruptions caused by these events (similarly to the effects caused by sporadic ionospheric layers—signal ephemerides, causing errors in station signals' decoding); with the presented method, it would be possible to predict unfavorable VHF navigation propagation situations, monitor them in real time and include analytically precise precaution measures as soon as it is possible.

Author Contributions: Conceptualization, W.K. and M.Z. and T.A.M.; methodology, W.K., M.Z. and T.A.M.; software, W.K. and M.Z.; validation, W.K. and T.A.M.; formal analysis, T.A.M.; investigation, T.A.M.; resources, W.K.; data curation, W.K.; writing—original draft preparation, T.A.M.; writing—review and editing, T.A.M.; visualization, T.A.M. and W.K.; supervision, W.K.; project administration, W.K. All authors have read and agreed to the published version of the manuscript.

Funding: This research received no external funding.

Institutional Review Board Statement: Not applicable.

Informed Consent Statement: Not applicable.

Data Availability Statement: VOR propagation data is available upon request: wojciech.kazubski@pw.edu.pl.

Conflicts of Interest: The authors declare no conflicts of interest.

References

1. Winick, A.B. VOR/DME System Improvements. *Proc. IEEE* **1970**, *58*, 430–437. [CrossRef]
2. Ostroumov, I.V. Availability estimation of navigation aids. *Visnyk NTUU KPI Seriia-Radiotekhnika Radioaparatobuduvannia* **2017**, *69*, 35–40. [CrossRef]
3. Ostroumov, I.; Kharchenko, V.; Kuzmenko, N. An airspace analysis according to area navigation requirements. *Aviation* **2019**, *23*, 36–42. [CrossRef]
4. Bem, D.J. *Telecommunication Systems. Part II. Radionavigation and Radiolocation*; OFPW: Wrocław, Poland, 1991.
5. Marzioli, P.; Frezza, L.; Curianò, F.; Pellegrino, A.; Gianfermo, A.; Angeletti, F.; Arena, L.; Cardona, T.; Valdatta, T.; Santoni, F.; et al. Experimental validation of VOR (VHF Omni Range) navigation system for stratospheric flight. *Acta Astronaut.* **2021**, *178*, 423–431. [CrossRef]
6. Frezza, L.; Marzioli, P.; Santoni, F.; Piergentili, F. VHF Omnidirectional Range (VOR) Experimental Positioning for Stratospheric Vehicles. *Aerospace* **2021**, *8*, 263. [CrossRef]
7. Rauniyar, S.; Orten, P.; Petersen, S. Mobile Connectivity Beyond the Coast-Line: A Case Study for Next Generation Shipping. In Proceedings of the 2023 IEEE 98th Vehicular Technology Conference (VTC2023-Fall), Hong Kong, 10–13 October 2023. [CrossRef]

8. Lim, J.; Nakura, K.; Mori, S.; Harada, H. Software-Defined Radio-Based IEEE 802.15.4 SUN FSK Evaluation Platform for Highly Mobile Environments. *IEEE Open J. Veh. Technol.* **2024**, *5*, 1637–1649. [CrossRef]
9. Gharibi, M.; Boutaba, R.; Waslander, S.L. Internet of Drones. *IEEE Access* **2016**, *4*, 1148–1162. [CrossRef]
10. Kunertova, D. The war in Ukraine shows the game-changing effect of drones depends on the game. *Bull. At. Sci.* **2023**, *79*, 95–102. [CrossRef]
11. Kunertova, D. Drones have boots: Learning from Russia's war in Ukraine. *Contemp. Secur. Policy* **2023**, *44*, 576–591. [CrossRef]
12. Starship Technologies Achieves Industry-First Autonomy Milestone. 4 May 2023. Available online: https://www.starship.xyz/press/starship-technologies-achieves-industry-first-autonomy-milestone/ (accessed on 3 June 2025).
13. Longley, A.G.; Rice, P.L. *Prediction of Tropospheric Radio Transmission Loss Over Irregular Terrain. A Computer Method-1968*; ESSA Technical Report ERL 79-ITS 67; Institute for Telecommunication Sciences: Boulder, CO, USA, 1968.
14. Hata, M. Empirical Formula for Propagation Loss in Land Mobile Radio Services. *IEEE Trans. Veh. Technol.* **1980**, *VT-29*, 317–325. [CrossRef]
15. Lee, C.-H.; Kwak, Y.-W.; Kim, S.-C.; Bae, S.-H. Evaluation of ITU-R Recommendation P.1546 and Consideration for New Correction Factor. 2007. Available online: https://s-space.snu.ac.kr/ (accessed on 3 April 2025).
16. Östlin, E.; Suzuki, H.; Zepernick, H.-J. Evaluation of the Propagation Model Recommendation ITU-R P.1546 for Mobile Services in Rural Australia. *IEEE Trans. Veh. Technol.* **2008**, *57*, 38–51. [CrossRef]
17. Więcek, D.; Wypiór, D. New SEAMCAT Propagation Models: Irregular Terrain Model and ITU-R P. 1546-4. *J. Telecommun. Inf. Technol.* **2011**, *3*, 131–140. [CrossRef]
18. Bem, D.J. *Antennas and Propagation of Radio Waves*; WNT: Warsaw, Poland, 1973.
19. Hosokawa, K.; Saito, S.; Sakai, J.; Tomizawa, I.; Nakata, H.; Takahashi, T.; Nishioka, M.; Tsugawa, T.; Ishii, M.; Saito, A.; et al. Observations of Sporadic E Using Aeronautical Navigation Radio Waves. In Proceedings of the URSI GASS 2023, Sapporo, Japan, 19–26 August 2023.
20. Hosokawa, K.; Sakai, J.; Tomizawa, I.; Saito, S.; Tsugawa, T.; Nishioka, M.; Ishii, M. A monitoring network for anomalous propagation of aeronautical VHF radio waves due to sporadic E in Japan. *Earth Planets Space* **2020**, *72*, 88. [CrossRef]
21. Sakai, J.; Hosokawa, K.; Tomizawa, I.; Saito, S. A Statistical Study of Anomalous VHF Propagation Due to the Sporadic-E Layer in the Air-Navigation Band. *Radio Sci.* **2019**, *54*, 426–439. [CrossRef]
22. Hosokawa, K.; Kimura, K.; Sakai, J.; Saito, S.; Tomizawa, I.; Nishioka, M.; Tsugawa, T.; Ishii, M. Visualizing sporadic E using aeronautical navigation signals at VHF frequencies. *J. Space Weather Space Clim.* **2021**, *11*, 6. [CrossRef]
23. Schmidt, M. *Meteorology*; WKiŁ: Warsaw, Poland, 1975.
24. Bogdanov, A.F.; Vasin, V.V.; Dulin, V.N.; Ilin, V.A.; Krivitskiy, B.H.; Kuznetsov, V.A.; Labutin, V.K.; Molochkov, Y.B.; Piertsov, S.V.; Stiepanov, B.M.; et al. *Compendium on Radioelectronics*; Kulikovskiy, A.A., Ed.; WKiŁ: Warsaw, Poland, 1971.
25. Bieńkowski, Z. *VHF Operator Guide*; WKiŁ: Warsaw, Poland, 1988.
26. Nikandrov, V.Y. *Artificial Influence on Clouds and Fogs*; Hydrometeorology Publishing House: Leningrad, Russia, 1959.
27. Ralston, A. *Introduction to Numerical Calculus*; PWN: Warsaw, Poland, 1983.
28. Ezadi, S.; Allahviranloo, T.; Mohammadi, S. Two new methods for ranking of Z-numbers based on sigmoid function and sign method. *Int. J. Intell. Syst.* **2018**, *33*, 1476–1487. [CrossRef]
29. Earth Atmosphere Model-NASA Glenn. Available online: www.grc.nasa.gov/www/k-12/airplane/atmosmet.html (accessed on 20 March 2025).
30. Ghanghas, A.; Sharma, A.; Merwade, V. Unveiling the Evolution of Extreme Rainfall Storm Structure Across Space and Time in a Warming Climate. *Earth's Future* **2024**, *12*, e2024EF004675. [CrossRef]
31. Visser, J.B.; Wasko, C.; Sharma, A.; Nathan, R. Changing Storm Temporal Patterns with Increasing Temperatures Across Australia. *J. Clim.* **2023**, *36*, 6247–6259. [CrossRef]
32. Cato, J.L.; Nicholls, N.; Jakob, C.; Shelton, K.L. Atmospheric fronts in current and future climates. *Geophys. Res. Lett.* **2014**, 7642–7650. [CrossRef]
33. Thomas, C.M.; Schultz, D.M. Global Climatologies of Fronts, Airmass Boundaries, and Airstream Boundaries: Why the Definition of "Front" Matters. *Mon. Weather Rev.* **2019**, *147*, 691–717. [CrossRef]

sensors

MDPI

Article

Securing IoT Communications via Anomaly Traffic Detection: Synergy of Genetic Algorithm and Ensemble Method

Behnam Seyedi and Octavian Postolache *

Department of Science, Instituto de Telecomunicacoes, ISCTE-University Institute of Lisbon, 1649-026 Lisbon, Portugal; seydi.behnam@gmail.com
* Correspondence: octavian.adrian.postolache@iscte-iul.pt

Abstract: The rapid growth of the Internet of Things (IoT) has revolutionized various industries by enabling interconnected devices to exchange data seamlessly. However, IoT systems face significant security challenges due to decentralized architectures, resource-constrained devices, and dynamic network environments. These challenges include denial-of-service (DoS) attacks, anomalous network behaviors, and data manipulation, which threaten the security and reliability of IoT ecosystems. New methods based on machine learning have been reported in the literature, addressing topics such as intrusion detection and prevention. This paper proposes an advanced anomaly detection framework for IoT networks expressed in several phases. In the first phase, data preprocessing is conducted using techniques like the Median-KS Test to remove noise, handle missing values, and balance datasets, ensuring a clean and structured input for subsequent phases. The second phase focuses on optimal feature selection using a Genetic Algorithm enhanced with eagle-inspired search strategies. This approach identifies the most significant features, reduces dimensionality, and enhances computational efficiency without sacrificing accuracy. In the final phase, an ensemble classifier combines the strengths of the Decision Tree, Random Forest, and XGBoost algorithms to achieve the accurate and robust detection of anomalous behaviors. This multi-step methodology ensures adaptability and scalability in handling diverse IoT scenarios. The evaluation results demonstrate the superiority of the proposed framework over existing methods. It achieves a 12.5% improvement in accuracy (98%), a 14% increase in detection rate (95%), a 9.3% reduction in false positive rate (10%), and a 10.8% decrease in false negative rate (5%). These results underscore the framework's effectiveness, reliability, and scalability for securing real-world IoT networks against evolving cyber threats.

Keywords: Internet of Things; genetic algorithm; XGBoost; anomaly detection; random forest

1. Introduction

The rapid expansion of the Internet of Things (IoT) has revolutionized industries by enabling interconnected devices to enhance efficiency, automation, and innovation in sectors such as healthcare, industrial systems, and smart cities. This interconnectedness allows seamless communication and data exchange across heterogeneous devices, creating a dynamic ecosystem for IoT-driven advancements. However, IoT systems face significant security vulnerabilities due to their reliance on resource-constrained devices, decentralized architectures, and dynamic network environments. These vulnerabilities expose IoT ecosystems to cyber threats, such as anomalous network behavior, denial of service, and data manipulation attacks [1–3]. Machine learning (ML) has emerged as a promising approach to enhance IoT security by offering adaptive solutions for identifying anomalous behavior. Unlike traditional rule-based systems, ML techniques leverage patterns in IoT network behavior to detect deviations, enabling the identification of both normal and abnormal activities in real time [4,5]. Recent advancements in ML, including deep learning and feature extraction techniques, have significantly improved the precision and scalability of anomaly detection systems in IoT environments [6,7]. These systems identify known attack patterns and generalize them to detect previously unseen anomalies, making them suitable for dynamic IoT ecosystems [8,9]. Despite the promise of ML-based anomaly detection, several challenges remain. IoT networks generate large volumes of high-dimensional data, which can overwhelm traditional detection systems. Furthermore, the dynamic nature of IoT environments, characterized by frequent changes in device configurations and communication patterns, necessitates robust and adaptive models capable of maintaining detection accuracy [4,8]. Ensuring computational efficiency is another critical challenge, especially for IoT devices with limited processing power and memory [3,10]. Addressing these challenges requires the development of scalable, lightweight, and accurate anomaly detection frameworks that can operate effectively in resource-constrained and diverse IoT networks [2,5]. This study proposes a novel anomaly detection framework that distinguishes between normal and anomalous behavior in IoT networks. The proposed method balances scalability, accuracy, and resource efficiency by integrating advanced feature selection techniques, such as Genetic Algorithms, with ensemble machine learning classifiers like Random Forest, XGBoost, and Decision Tree models. Unlike traditional approaches, this framework generalizes well to diverse IoT environments and adapts dynamically to variations in network behavior.

In IoT environments, distinguishing between normal behavior and anomalous activities is essential for maintaining data integrity, confidentiality, and availability. While traditional intrusion detection systems focus on identifying specific types of attacks, they often fail to generalize and adapt to evolving network behavior. The primary challenge lies in designing a system that can accurately classify normal and anomalous behavior in various IoT scenarios without being limited to predefined attack patterns [4,6].

Although effective in detecting specific attack types, existing machine learning-based solutions struggle to handle the dynamic and heterogeneous nature of IoT networks [5]. Many approaches rely on exhaustive feature sets, leading to increased computational overhead and inefficiency in real-world deployments [7,10]. Moreover, a high false positive rate in anomaly detection can disrupt normal operations and reduce trust in the detection system. To address these limitations, it is critical to adopt feature selection techniques that optimize relevant features, reduce dimensionality, and enhance classification accuracy while maintaining computational efficiency [3,7]. The primary contributions of this study are outlined as follows:

- Novel Anomaly Detection Framework: This study proposes a robust three-phase anomaly detection framework for IoT environments. This framework integrates advanced data preprocessing, feature selection, and ensemble classification techniques to distinguish between normal and anomalous network behaviors effectively. By addressing the limitations of traditional intrusion detection systems, this method demonstrates adaptability and scalability across diverse IoT scenarios.
- Optimized Feature Selection: This study introduces a Genetic Algorithm-based feature selection technique, enhanced with optimization strategies inspired by nature (e.g., eagle-inspired search mechanisms). This approach identifies the most impactful features, reduces data dimensionality, and improves computational efficiency without compromising detection accuracy.

- Scalability in High-Risk Scenarios: This study examines the performance of the proposed method under varying attack percentages (η) and device risk levels (20% to 70%). This method maintains a high detection rate even in high-risk scenarios (η = 50%), showcasing its robustness and reliability for real-world IoT applications.

The remaining sections of this paper are the following: The Related Work section examines and analyzes prior studies and related works. The Materials and Methods section outlines the methodology's key steps and components. The Results and Discussion section offers a comprehensive review of the simulation results. Finally, the Conclusion and Future Work section explores future research directions and concludes the paper.

2. Related Work

This study [11] uses deep learning techniques to propose an intrusion detection system for IoT networks. The model effectively detects diverse IoT attacks by leveraging the power of deep learning to identify anomalous behavior. The experimental results demonstrate high accuracy and scalability, making it a suitable solution for real-time IoT environments. The authors introduce an efficient feature extraction method to improve attack classification in IoT networks. This method ensures high accuracy in identifying various network intrusions by reducing computational complexity. Tests on IoT datasets confirm its ability to deliver precise and reliable classifications for enhanced security [12]. Tyagi et al. [13] explored supervised machine learning methods to detect attacks and anomalies in IoT networks. Through comprehensive experimentation, the models are shown to achieve high accuracy and sensitivity. These methods are especially effective in addressing the growing diversity of IoT network threats. In [14], a robust attack detection framework using ensemble classifiers is presented, explicitly targeting industrial IoT (IIoT) environments. This study demonstrates the approach's ability to handle imbalanced datasets and maintain high detection accuracy. It proves to be a reliable solution for real-world IIoT security challenges.

The authors of [15] designed a machine learning-based intrusion detection framework tailored to IoT networks. The proposed methodology adapted well to complex scenarios, delivering high accuracy and resilience. Extensive experiments validated the framework, proving its effectiveness for dynamic IoT environments. Zhao et al. [16] presented a novel intrusion detection system using a lightweight neural network designed for the IoT. It achieved high detection accuracy while minimizing resource consumption, making it ideal for resource-constrained IoT devices. The results validated its potential for efficient and scalable IoT security. The authors of [17] proposed an intrusion detection mechanism using autoencoders and data partitioning to enhance efficiency in IoT networks. This approach significantly reduced computational overhead while achieving high detection precision. Their methodology suits dynamic IoT environments with large-scale data. An AI-based anomaly detection scheme was introduced for identifying cyber threats in IoT-based transport networks. The method excelled in proactive threat prevention and offers robust performance in securing IoT-enabled transport systems. This highlights the role of AI in advancing IoT cybersecurity [18].

The study in [19] presents a two-level hybrid model for detecting anomalous activities in IoT networks. The model integrates statistical and machine learning methods to improve accuracy and reliability. Its experimental results indicate enhanced performance in identifying diverse IoT anomalies while maintaining low false positive rates, making it practical for real-world applications. Huang et al. [20] proposed an Energy-efficient and Trustworthy Unsupervised (EATU) anomaly detection framework for the industrial IoT (IIoT). Their framework ensures reliable anomaly detection while minimizing resource usage by focusing on unsupervised learning and energy efficiency. Their study's empirical evaluation demonstrated that EATU proves effective in dynamic and resource-constrained IIoT environments. Altulaihan et al. [21] introduced an anomaly detection intrusion detection system (IDS) to identify DoS attacks in IoT networks. The system employed machine learning algorithms to ensure high detection accuracy and robustness. The results confirmed its capability to address critical IoT security threats, focusing on detecting DoS attacks. Another paper proposed a two-tier hybrid ensemble learning pipeline for intrusion detection in IoT networks. By combining multiple machine learning techniques, the pipeline significantly improved detection accuracy and efficiency. The approach was validated on IoT datasets, showcasing its potential for enhancing network security [22].

The study in [23] presents an intrusion detection system (IDS) for IoT networks that utilizes an ensemble of unsupervised techniques. The framework effectively addresses the challenges of diverse intrusion types and imbalanced datasets. Experimental evaluations highlight its high performance in identifying intrusions while reducing computational complexity. The authors propose EIDM, a deep learning-based model for intrusion detection in IoT networks. The model is designed to detect various intrusions effectively using advanced deep learning techniques. Experimental evaluations demonstrate EIDM's high accuracy and robustness in handling complex IoT datasets. Additionally, the model showcases its scalability and suitability for real-world IoT environments, making it a promising solution for enhancing IoT network security [24]. Mohapatra et al. [25] addressed handling MITM attacks in wireless sensor networks (WSNs) through IDS. Their approach provided a framework to identify and mitigate MITM attacks, improving the security of sensor-based applications. Recent research has increasingly applied graph neural networks (GNNs) to model complex transportation and infrastructure systems. Liu and Meidani [26] introduced an end-to-end heterogeneous GNN for user-equilibrium traffic assignment, which leverages virtual origin destination links and graph-based flow conservation to improve prediction accuracy and adaptability across varying network topologies. In another work [27], a GNN-based surrogate model was proposed for the rapid seismic reliability analysis of highway bridge systems, effectively capturing node-level connectivity patterns under uncertainty. These GNN-based approaches demonstrate the potential of learning from graph-structured data in dynamic and complex networks. Table 1 summarizes the advantages and disadvantages of the related work.

Despite significant advancements in machine learning-based intrusion detection systems (IDSs) for IoT networks, several critical gaps remain in the existing research. Deep learning-based approaches [1,6,14] demonstrate high detection accuracy but suffer from substantial computational demands, making them unsuitable for resource-constrained IoT devices. Feature extraction and selection methods [2,7,17] aim to reduce complexity but often compromise detection accuracy or fail to generalize across diverse environments. Supervised machine learning methods [3,5,16] and ensemble classifiers [4,12,13] enhance performance yet frequently overlook dynamic adaptation to heterogeneous and evolving IoT network conditions. The proposed three-phase framework addresses these limitations by introducing the following: (1) an advanced data preprocessing phase using the Median-KS Test and robust techniques for noise elimination, missing value imputation, and dataset balancing, ensuring cleaner and more structured input data; (2) a feature selection and hyperparameter optimization phase leveraging a Genetic Algorithm enhanced with eagle-inspired search behavior and simulated annealing improved via dragonfly swarm dynamics, achieving an optimal trade-off between dimensionality reduction and classification accuracy; and (3) a classification phase employing a powerful ensemble of Decision Tree, Random Forest, and XGBoost, ensuring scalability, robustness, and adaptability across dynamic IoT scenarios. This holistic approach enhances detection performance while minimizing computational costs, making it highly effective for real-world, resource-constrained, and heterogeneous IoT environments.

Table 1. Advantages and disadvantages of related work.

Ref.	Year	Advantages	Disadvantages
[11]	2023	High accuracy and scalability for diverse IoT attack detection.	Potential high computational demands for deep learning.
[12]	2021	Efficient feature extraction with reduced computational complexity.	May require further tuning for diverse IoT scenarios.
[13]	2021	Effective in addressing diverse IoT threats with high accuracy and sensitivity.	Limited evaluation of highly dynamic IoT environments.
[14]	2021	Handles imbalanced datasets and maintains high detection accuracy.	Performance may depend on the quality of the ensemble models.
[15]	2022	Adapts well to complex scenarios with high performance and reliability.	Framework may require additional resources for real-time scenarios.
[16]	2021	High detection accuracy and low resource consumption for constrained devices.	Lightweight models may miss subtle intrusion patterns.
[17]	2024	Reduces computational overhead and improves detection precision in large-scale environments.	Effectiveness may vary in non-IoT environments.
[18]	2023	Proactive threat prevention with robust performance in IoT-based transport networks.	Limited focus on broader IoT environments beyond transport systems.
[19]	2019	Improved anomaly detection accuracy with low false positives.	A hybrid approach could increase system complexity.
[20]	2022	Energy-efficient and trustworthy detection suitable for IIoT environments.	Reliance on unsupervised learning may limit detection specificity.
[21]	2024	High detection accuracy and robustness for detecting DoS attacks.	Focuses primarily on DoS attacks; may need an extension for other threats.
[22]	2023	Enhanced detection accuracy and efficiency with hybrid ensemble learning.	A complex pipeline may increase implementation overhead.
[23]	2022	Effectively identifies diverse intrusion types with reduced computational complexity.	May require fine-tuning for specific IoT environments.
[24]	2023	High accuracy and robustness in handling complex IoT datasets.	A deep learning model may demand significant computational resources.
[25]	2020	IDS approach for MITM attack handling in wireless sensor networks.	May struggle with scalability in large-scale WSNs or with high data traffic.

3. Materials and Methods

The proposed approach consists of three main phases, each playing a key role in enhancing the system's performance using advanced algorithms. In the first phase, data preprocessing techniques such as the Median-KS Test are employed to eliminate noise and identify valid data. Additionally, advanced methods are applied to handle missing values, balance datasets, and prepare categorical features, transforming raw data into an analyzable format. In the second phase, feature selection and optimization, a Genetic Algorithm (GA) enhanced with eagle-inspired search strategies is employed to identify and select the most influential features, effectively reducing the dimensionality of the data while maintaining high classification performance. In parallel, hyperparameter tuning is performed using a simulated annealing (SA) algorithm, which is further improved by integrating dragonfly-inspired swarm behaviors (alignment, cohesion, separation) to enhance exploration and exploitation capabilities during the search for optimal model parameters. Finally, a combination of three machine learning algorithms, Decision Tree, Random Forest, and XGBoost, is used to classify the data. This three-phase structure offers a comprehensive and practical approach to detecting and countering security threats in the IoT environment. A flowchart of the proposed method is shown in Figure 1.

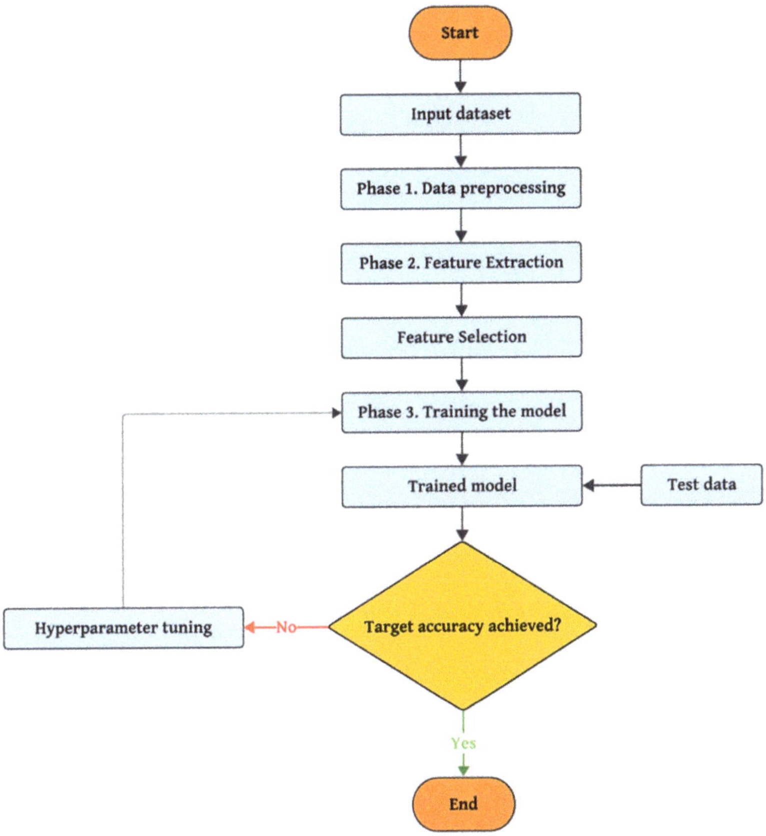

Figure 1. Flowchart of the proposed method.

3.1. Phase 1—Preprocessing

The preprocessing phase prepares the data for further analysis by addressing common issues such as missing values (NaN), categorical features, imbalanced datasets, and missing entries. This step is critical for ensuring the quality and consistency of the data, enabling better downstream analysis and minimizing obstacles that could disrupt the data processing flow. Handling NaN values is one of the first steps in preprocessing. NaN, short for "Not a Number", is commonly used to represent missing data in datasets. If left unaddressed, these values can negatively impact the accuracy and reliability of the attack detection system. To ensure high precision in the results, all NaN values must be appropriately handled and replaced with suitable substitutes. Managing NaN values in the input data involves systematic cleaning to ensure the dataset is ready for effective processing. This stage sets the foundation for subsequent phases by creating a cleaner, more consistent dataset that enhances the reliability of the analytical models. This process can be described in (1) as follows:

$$d_{NaN\text{-}handled} = M_{NaN}(d_{input}) \quad (1)$$

where d_{input} represents the raw dataset, M_{NaN} is the method used to handle NaN values, and $d_{Nan\text{-}handled}$ is the dataset after addressing NaN values.

After addressing NaN values, the dataset often contains categorical features and non-numerical attributes like labels or text descriptions. Since machine learning models are inherently mathematical, categorical data must be transformed into numerical formats to ensure compatibility. Techniques such as one-hot encoding, label encoding, and ordinal transformation are commonly used. This process can be described as (2):

$$d_{cat\text{-}handled} = M_{cat}(d_{input}) \tag{2}$$

where d_{input} represents the raw dataset, M_{cat} is the transformation method, and $d_{cat\text{-}handled}$ is the dataset after processing categorical features.

Missing data present another challenge during preprocessing. These missing values may occur randomly across data samples or follow a structured pattern. Random missing values, known as "missing at random", are distributed without any clear correlation, while structured missing values, or "missing not at random", follow specific patterns that might correlate with other features. After analyzing the dataset, we observed that missing values were mainly present in certain features such as packet-related statistics. To address this, we applied mean imputation for continuous features and mode imputation for categorical features. Additionally, samples with excessive missing data (>30% missing attributes) were removed to preserve the dataset's quality. Managing these scenarios involves tailored strategies, represented as (3):

$$d_{mis\text{-}handled} = M_{mis}(d_{input}) \tag{3}$$

where d_{input} represents the raw dataset, M_{mis} denotes the method to address missing values, and $d_{mis\text{-}handled}$ is the dataset after handling missing values. Once issues like NaN values, categorical features, and missing data are addressed, the preprocessed dataset is assembled. The final dataset is expressed as (4):

$$d_{preprocessed} = \{d_1, d_2, \ldots, d_N\} \tag{4}$$

where $d_{preprocessed}$ is the preprocessed dataset and N represents the total number of data samples.

3.2. Phase 2—Feature Extraction

The feature extraction phase builds on the balanced data to identify patterns and insights critical for detecting abnormalities. A novel method, the Median Absolute Deviation around the Median-based Kolmogorov–Smirnov Test [28] (MAD Median-KS Test), was developed to achieve this. This method incorporates statistical analysis techniques and leverages data from the network to detect anomalies caused by compromised IoT devices. The MAD Median-KS Test enhances traditional approaches by focusing on distribution similarity and removing redundant data. This is unlike the conventional Kolmogorov–Smirnov (KS) test, which emphasizes the largest difference in distribution, making it susceptible to outliers and reducing its reliability in attack detection. This improved method accounts for median absolute deviations around the median to mitigate the influence of outliers. The test computes the features of the input data using Equation (5):

$$Features = \lambda \cdot \Gamma_n(\partial(D) - \Gamma_i(\tau(D))) \tag{5}$$

where λ is a constant used to filter abnormalities caused by outliers. Γ_n represents the median of the distribution functions. ∂ and τ are two distribution functions derived from the balanced dataset D. Γ_i denotes the median of τ, capturing key characteristics of the data.

Feature Selection

The Genetic Algorithm (GA) is a widely recognized evolutionary optimization technique that simulates natural selection, where the fittest individuals are chosen to pass their traits to the next generation. In feature selection, GA is utilized to identify the most relevant subset of features that significantly improve model performance while discarding irrelevant or redundant ones. This approach minimizes the dimensionality of the dataset, enhancing computational efficiency without sacrificing accuracy. A novel component of this process is the incorporation of optimization techniques inspired by the hunting behavior of North American bald eagles. These eagles select areas rich in food and execute efficient spiral movements to capture prey. This behavior is mathematically represented to guide the optimization process. The initial position of a search is calculated using (6):

$$X_{n,i} = X_b + \eta \cdot \delta \cdot (X_{avg} - X_i) \tag{6}$$

where X_b is the best location identified during the search, X_{avg} is the average position, X_i is the current position, η controls positional variations, and δ is a binary variable r ∈ {0, 1}. After identifying a promising location, the eagle refines its search through a spiral movement defined by (7):

$$X_{i,n} = X_i + y_i \times (X_i - X_{i+1}) + z_i \times (X_i - X_m) \tag{7}$$

where

$$z_i = \frac{z\delta(i)}{\max|z\delta|} \tag{8}$$

$$y_i = \frac{y\delta(i)}{\max|y\delta|} \tag{9}$$

$$z\delta(i) = \delta(i) \times \sin(\theta(i)) \tag{10}$$

$$y\delta(i) = \delta(i) \times \cos(\theta(i)) \tag{11}$$

$$\theta(i) = \phi \times \pi \times rand \tag{12}$$

$$\delta(i) = \theta(i) + \omega \times rand \tag{13}$$

In this model, θ(i) represents the angle of search and ω denotes the number of search cycles (0.5 ≤ ω ≤ 2). During the swooping phase, the eagle targets its prey by diving into an optimal location. This behavior is modeled as follows:

$$X_{i,n} = rand \cdot X_b + z_1(i) \times (X_i - C_1 \times X_m) + y_1(i) \times (X_i - C_2 \times X_b) \tag{14}$$

where C_1 and C_2 are control coefficients, and

$$z_1(i) = \frac{z\delta(i)}{\max|z\delta|} \tag{15}$$

$$y_1(i) = \frac{y\delta(i)}{\max|y\delta|} \tag{16}$$

$$z\delta(i) = \delta(i) \times \sinh(\theta(i)) \tag{17}$$

$$y\delta(i) = \delta(i) \times \cosh(\theta(i)) \tag{18}$$

$$\theta(i) = \phi \times \pi \times rand \tag{19}$$

The GA-based feature selection algorithm incorporates a fitness function that evaluates the quality of feature subsets. Fitness is calculated as follows:

$$Fitness = w_1 \cdot L(f) + w_2 * (1 - \frac{|S|}{|T|}) \tag{20}$$

where L(f) is the classification error for the feature subset f, |S| is the count of selected features, |T| is the total number of features in the dataset, and w_1 and w_2 are weighting parameters with $w_1 + w_2 = 1$. The fitness function defined in Equation (20) is designed to optimize two main objectives: minimizing classification errors and reducing model complexity by selecting the most relevant features. The first term, L(f), measures the classification error for a given feature subset. This ensures that the selected features contribute significantly to improving the model's performance. The second term, |S|/|T|, represents the ratio of selected features to the total available features. This term is introduced to prevent the selection of excessive and redundant features, which can increase computational complexity and reduce efficiency, especially in resource-constrained environments such as IoT networks. The coefficients w_1 and w_2 act as weighting parameters that control the importance of each term. Their values are carefully tuned based on the problem's requirements to maintain a balance between classification accuracy and computational efficiency.

3.3. *Phase 3—Classification*

The proposed approach integrates an ensemble voting classifier [29] to enhance the effectiveness of a network intrusion detection system (NIDS). This ensemble method combines the strengths of multiple ML algorithms: XGBoost [30], Random Forest (RF) [31], and Decision Tree (DT) [32]. By leveraging their unique capabilities, the ensemble classifier achieves superior accuracy and robustness in detecting malicious traffic.

3.3.1. Decision Tree

A Decision Tree is a machine learning model structured like a tree for classification or regression tasks. It recursively splits the dataset based on specific features to create smaller data groups that are more homogeneous. Each internal node in the tree represents a decision based on a feature, and each leaf node corresponds to a final prediction or outcome. The splitting criterion typically uses a cost function such as GINI Impurity or Entropy in classification trees. The formula for Entropy is given by the following:

$$Entropy = -\sum_{i=1}^{k} p_i \cdot \log p_i \tag{21}$$

where p_i is the probability of class i.

3.3.2. Random Forest

Random Forest is an ensemble learning algorithm that combines multiple decision trees to improve accuracy and reduce overfitting. It generates a set of decision trees, each trained on a randomly selected subset of data and features. For classification tasks, it uses majority voting across the trees to determine the final prediction, and for regression tasks, it averages the predictions of the trees. A key feature of Random Forest is Bootstrap Aggregation (Bagging), which involves sampling the data with a replacement to create diverse training subsets. This technique helps reduce overall error and increase model robustness.

3.3.3. XGBoost

XGBoost, short for "Extreme Gradient Boosting", is an advanced ensemble algorithm that builds upon gradient boosting frameworks to efficiently handle regression and classification problems. It trains decision trees sequentially, where each tree attempts to correct the errors of the previous ones. XGBoost incorporates regularization to prevent overfitting and parallel processing for faster training and handling missing values. The loss function in XGBoost is optimized using gradient descent, and a second-order Taylor expansion of the loss function is often used to enhance convergence and stability:

$$Loss \approx \sum_{i=1}^{n} [g_i \cdot h(x_i) + \frac{1}{2} h_i \cdot h(x_i)^2] \tag{22}$$

where g_i is the gradient and h_i is the Hessian for the loss function. XGBoost is highly efficient and widely used in machine learning competitions for its performance.

3.3.4. Ensemble Voting Classifier

The ensemble voting classifier combines predictions from multiple base models to improve the accuracy of the final classification. Two types of voting mechanisms are employed:

Voting 1: The final class label is determined by the majority vote of the base models.

$$C = majority\ voting(C_1, C_2, \ldots, C_n) \tag{23}$$

Voting 2: Probability estimates from each base model are averaged, and the class label with the highest probability is selected.

$$C = \underset{c}{\text{argmax}} \sum_{i=1}^{n} P_{i,c} \tag{24}$$

where C is the final class label, and $P_{i,c}$ is the probability of class c predicted by the i-th base model.

3.4. *Hyperparameter Tuning*

Simulated annealing (SA) was utilized in this study as a parameter-tuning method. SA is inspired by the natural process of annealing in metallurgy, where materials are heated and cooled gradually to remove defects and optimize their structure. In the context of this study, SA was enhanced by incorporating inspiration from swarm intelligence behaviors observed in dragonflies, particularly to improve its exploration and exploitation abilities. It is important to note that the complete Dragonfly Algorithm (DA) was not used. The dragonfly-inspired approach incorporates five key behavioral factors: alignment, which adjusts an individual's position based on the average location of its neighbors; attraction, which guides movement toward desired targets; cohesion, which minimizes positional variance within the group; separation, which prevents overcrowding; and distraction, which helps avoid undesirable or "enemy" locations. These factors collectively influence the movement and positioning of individuals within the swarm to identify optimal solutions. The SA algorithm updates the position of each iteratively using a mathematical framework. This includes calculating a step vector that combines contributions from alignment, attraction, cohesion, separation, and distraction. When no neighbors are present within a defined radius, a random walk based on the Levy flight process is employed to maintain exploration of the search space. The algorithm dynamically adjusts its parameters such as alignment, cohesion, and neighborhood radius to balance global exploration in early iterations with local refinement in later stages. The fitness function used in SA evaluates the classification error rate, defined as the percentage of misclassified instances relative to the total number of instances. The goal is to minimize this error rate, ensuring the model achieves high accuracy and generalization while avoiding overfitting. By iteratively fine-tuning hyperparameters in this manner, our SA-based approach enables a more efficient and effective exploration of the parameter space compared to traditional tuning methods such as grid search or random search.

$$Fitness = \frac{Number\ of\ misclassified}{Total\ number\ of\ instances} \times 100 \tag{25}$$

4. Results and Discussion

In this section, the proposed method is compared with four other methods, namely Support Vector Machine (SVM), Convolutional Neural Network (CNN), Ensemble Neural Networks (ENNs), and Deep Belief Network (DBN). The architecture and hyperparameter settings of the baseline models are summarized in Table 2.

Table 2. Architecture and hyperparameter settings of baseline models.

Model	Architecture/Settings
SVM	Kernel: RBF; Penalty parameter (C): 1.0; Gamma: 'scale'
CNN	3 Convolutional layers (32, 64, 128 filters, kernel size 3 × 3); 2 Dense layers (128, 64 neurons); Activation: ReLU; Dropout rate: 0.3; Optimizer: Adam; Batch size: 64; Epochs: 50
ENN	3 independent Feedforward Neural Networks; Each with 2 hidden layers (128, 64 neurons); Activation: ReLU; Output: Majority voting
DBN	2 stacked RBMs; 1st RBM: 128 hidden units, 2nd RBM: 64 hidden units; Training: Contrastive Divergence; Learning rate: 0.01; Batch size: 64

The simulation results of all methods are obtained on the CTU-13 dataset [33]. The dataset is inherently imbalanced, consisting of 4775 botnet instances and 20,902 normal instances. For model development, 80% of the data was allocated for training, while the remaining 20% was used for testing. The CTU-13 dataset, compiled by the Czech Technical University, is a widely used benchmark for evaluating intrusion detection systems, especially in IoT network security research. It contains thirteen different capture scenarios of real botnet traffic mixed with normal and background traffic. Each scenario includes various types of botnet activities such as spam, click fraud, and DDoS attacks, providing a realistic environment for anomaly detection. We selected the CTU-13 dataset because of its diversity in attack behaviors, the complexity of real network traffic, and its relevance to modern IoT environments where devices communicate dynamically across heterogeneous networks. Furthermore, its widespread use in recent research [33] allows for fair comparisons with existing methods while validating the robustness and generalization capability of our proposed anomaly detection framework.

The performance evaluation of the proposed method is carried out using the following key metrics: accuracy, specificity, sensitivity, precision, F-measure, false positive rate (FPR), false negative rate (FNR), Matthews Correlation Coefficient (MCC), and detection rate. Below is a detailed explanation of each metric:

Accuracy: This metric indicates the proportion of correctly identified instances (positive and negative) out of the total instances. A high accuracy demonstrates the overall reliability of the detection system (26).

$$Accuracy = \frac{TP + TN}{TP + TN + FP + FN} \tag{26}$$

Specificity: Specificity measures the system's ability to correctly identify normal (negative) cases. A higher specificity reflects fewer false alarms and the more precise detection of attack-free scenarios (27).

$$Specificity = \frac{TN}{TN + FP} \tag{27}$$

Sensitivity (Recall): Sensitivity quantifies the system's ability to detect attack (positive) cases accurately. A high sensitivity value effectively identifies malicious activities (28).

$$Sensitivity = \frac{TP}{TP + FN} \tag{28}$$

Precision is the ratio of true positives to the sum of true and false positives. It reflects the system's accuracy in identifying attack cases without misclassifying normal instances (29).

$$Precision = \frac{TP}{TP + FP} \tag{29}$$

F-Measure: This metric provides a harmonic mean of precision and sensitivity, offering a balanced evaluation of the system's performance, especially when class distribution is uneven (30).

$$F\ measure = 2 \times \frac{Precision \times Sensitivity}{Precision + Sensitivity} \tag{30}$$

False Positive Rate (FPR): FPR measures the proportion of normal instances incorrectly classified as attacks. A lower FPR highlights the system's ability to avoid unnecessary false alarms (31).

$$FPR = \frac{FP}{FP + TN} \tag{31}$$

False Negative Rate (FNR): FNR quantifies the proportion of attack instances incorrectly classified as normal. A lower FNR ensures effective attack detection without overlooking threats (32).

$$FNR = \frac{FN}{FN + TP} \tag{32}$$

Matthews Correlation Coefficient (MCC): MCC provides a balanced evaluation of the system's predictions, accounting for true and false positives and negatives. It is particularly useful for imbalanced datasets (33).

$$MCC = \frac{(TP \times TN) - (FP \times FN)}{\sqrt{(TP + FP)(TP + FN)(TN + FP)(TN + FN)}} \tag{33}$$

Detection Rate: This metric represents the ratio of successfully detected attack cases to the total number of attack instances. A higher detection rate indicates the system's robustness in identifying attacks across varying scenarios (34).

$$DetectionRate = \frac{TP}{TP + FN} \tag{34}$$

4.1. Simulation Results

To evaluate the optimization performance of the proposed and baseline feature selection methods, we analyze the fitness values obtained at different iteration steps. Table 3 presents the convergence behavior of several metaheuristic algorithms, including the Chimp Optimization Algorithm (ChOA) [34], Particle Swarm Optimization (PSO) [35], Social Spider Optimization (SSO) [36], Sparrow Search Algorithm (SPA) [37], and proposed feature selection method. Fitness values are reported at iterations 10 through 50, showing that the proposed method consistently achieves faster convergence and higher final fitness values compared to the other approaches.

Table 3. Performance of the feature selection methods.

Method/Iteration	10	20	30	40	50
ChOA	110	137	144	175	188
SSO	82	111	129	146	170
SPA	73	106	121	137	162
PSO	86	108	140	153	181
Proposed Method	129	152	178	198	210

In Table 3 are presented the values associated with the performance improvement (e.g., objective function values) at each iteration. The results indicate that the proposed feature selection method achieves faster and more significant convergence than the baseline techniques. The proposed method consistently outperforms the others at each iteration, reaching higher values at earlier iterations. For instance, after 10 iterations, the proposed method performs 129, significantly higher than ChOA (110) and PSO (86). Similarly, by the 50th iteration, the proposed method achieves a value of 210, surpassing ChOA (188) and the other techniques. This superior convergence behavior can be attributed to the advanced optimization strategies integrated into the proposed method, such as adaptive parameter tuning and enhanced exploration and exploitation capabilities. These features enable the method to navigate the solution space more effectively, avoiding local optima and accelerating convergence. The ability to converge faster and achieve higher performance demonstrates the proposed method's robustness and efficiency, making it a more reliable choice for feature selection in high-dimensional datasets. Figure 2 compares five classification models: ENN, CNN, SVM, DBN, and the proposed method. The metrics used for evaluation include accuracy, specificity, sensitivity, and precision, displayed in subplots (a), (b), (c), and (d), respectively. The superior performance of the proposed method can be attributed to its sophisticated and well-optimized structure. Unlike traditional models, the proposed method incorporates advanced preprocessing, feature selection, and classification strategies. During preprocessing, noise and outliers are effectively managed using techniques such as the Median-KS Test. The feature selection phase employs Genetic Algorithms, which reduce dimensionality and optimize the feature set, ensuring the inclusion of highly relevant attributes. Finally, in the classification phase, an ensemble voting classifier combines the strengths of the Decision Tree, Random Forest, and XGBoost algorithms, significantly improving prediction accuracy. These enhancements enable the proposed method to handle complex patterns more efficiently and reduce false positives and negatives, achieving a balanced performance across all evaluation metrics.

Figure 3 compares five classification models: ENN, CNN, SVM, DBN, and the proposed method. The metrics used for evaluation include the F-measure, FPR, FNR, and MCC, displayed in subplots (a), (b), (c), and (d), respectively. The proposed method minimizes noise and irrelevant attributes by effectively addressing data inconsistencies during preprocessing and utilizing Genetic Algorithms to select the most impactful features. Additionally, the ensemble voting classifier, which combines the strengths of the Decision Tree, Random Forest, and XGBoost models, ensures high accuracy, robust classification, and scalability. The proposed method's ability to reduce FPR and FNR is critical for real-world applications, where minimizing false alarms and undetected threats is essential. Moreover, its superior MCC scores reflect the proposed method's capability to maintain consistent performance across balanced and imbalanced datasets, further emphasizing its reliability and adaptability in various scenarios.

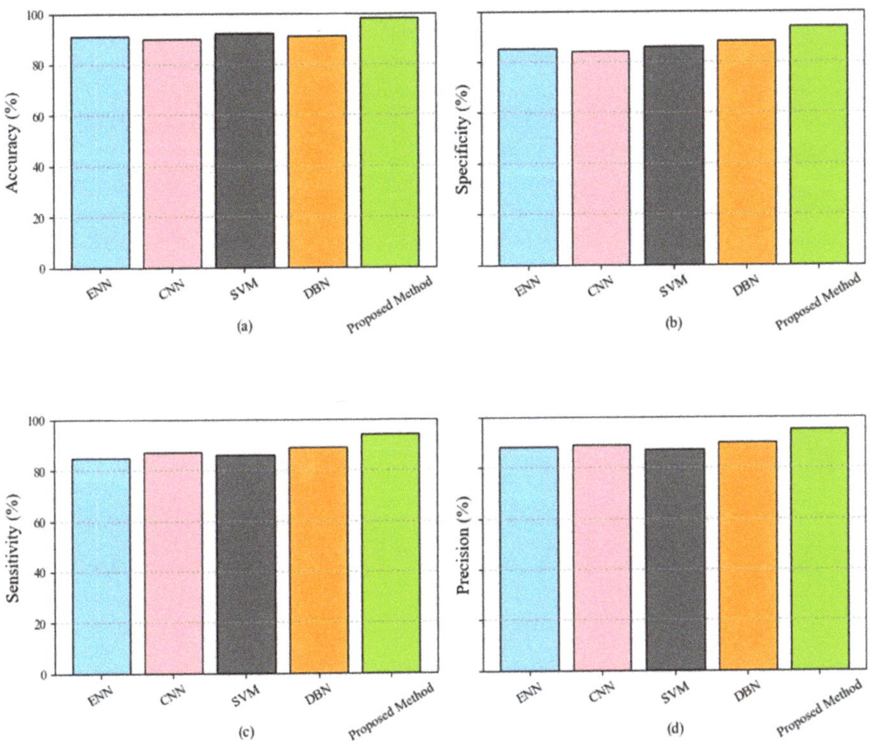

Figure 2. Performance comparison of classification models (ENN, CNN, SVM, DBN, and proposed method) across evaluation metrics: accuracy (**a**), specificity (**b**), sensitivity (**c**), and precision (**d**).

To evaluate the robustness of the proposed method under various threat levels, training datasets with different attack percentages (20%, 30%, 40%, and 50%) were generated. This was achieved by adjusting the ratio of botnet to normal traffic using a combination of random under-sampling (for majority class) and over-sampling (for minority class) strategies on the training set without altering the test set. Due to the inherent imbalance in the CTU-13 dataset, the Synthetic Minority Over-sampling Technique (SMOTE) was applied on the training data to generate synthetic samples for the botnet class. This ensured a more balanced dataset, improved the model's ability to learn minority class patterns, and enhanced overall classification performance. Figure 4 illustrates the detection rate performance under varying attack percentages (η) and different proportions of at-risk devices (from 20% to 70%). Subplots (a), (b), (c), and (d) correspond to attack percentages of η = 20%, η = 30%, η = 40%, and η = 50%, respectively.

The proposed method consistently achieves the highest detection rates across all scenarios and percentages of devices at risk, showcasing its robustness and superior anomaly detection capability. Even as the percentage of devices at risk increases (from 20% to 70%), the detection rate of the proposed method declines more gracefully compared to the other methods, demonstrating its ability to handle high-risk conditions effectively. ENN and CNN show moderate detection rates but fail to match the stability and effectiveness of the proposed method as the percentage of devices at risk increases. SVM and DBN exhibit more significant declines in their detection rate with increasing devices at risk, particularly in high-risk scenarios ($\eta = 50$). The gap between the proposed method and the other models widens as η and the percentage of devices at risk increases, further emphasizing the robustness of the proposed approach. The superior performance of the proposed method can be attributed to its advanced feature selection and classification strategies. The proposed method ensures better generalization and adaptability by integrating techniques like the Genetic Algorithm for optimal feature selection and ensemble classifiers for improved prediction accuracy. Its capability to maintain a high detection rate under challenging conditions stems from the ensemble voting mechanism, which combines the strengths of multiple models (Decision Tree, Random Forest, and XGBoost). This multi-layered approach enables the proposed method to detect complex patterns and minimize false positives and negatives, ensuring consistent performance even in high-risk scenarios.

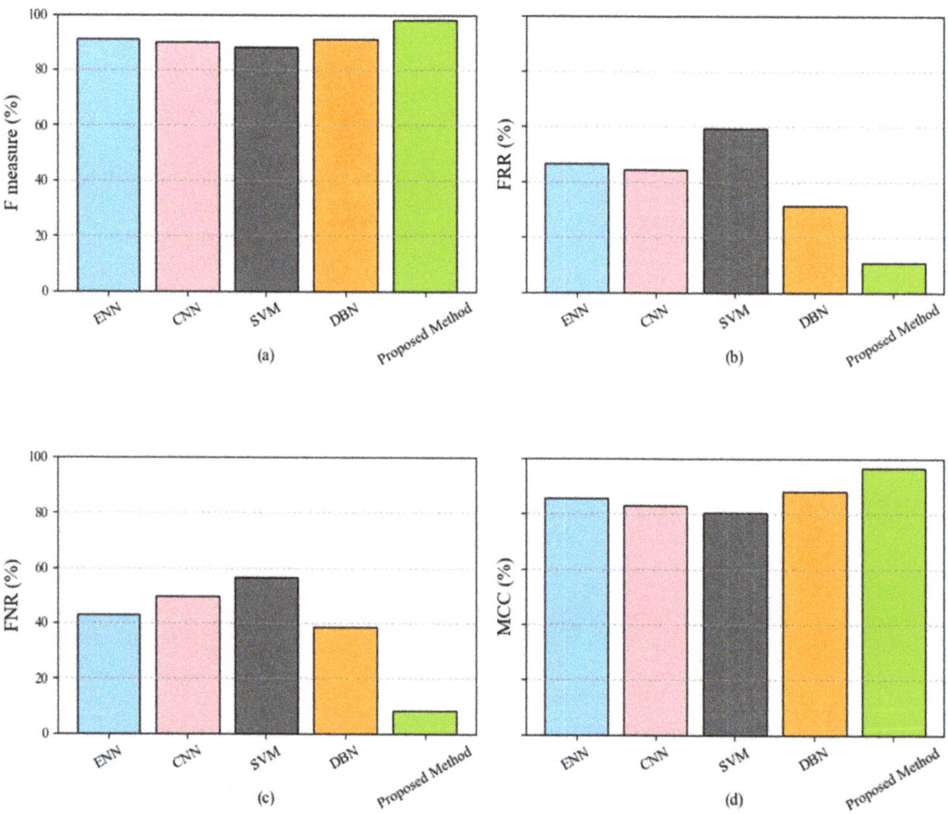

Figure 3. Performance comparison of classification models (ENN, CNN, SVM, DBN, and proposed method) across evaluation metrics: F-measure (**a**), FPR (**b**), FNR (**c**), and MCC (**d**).

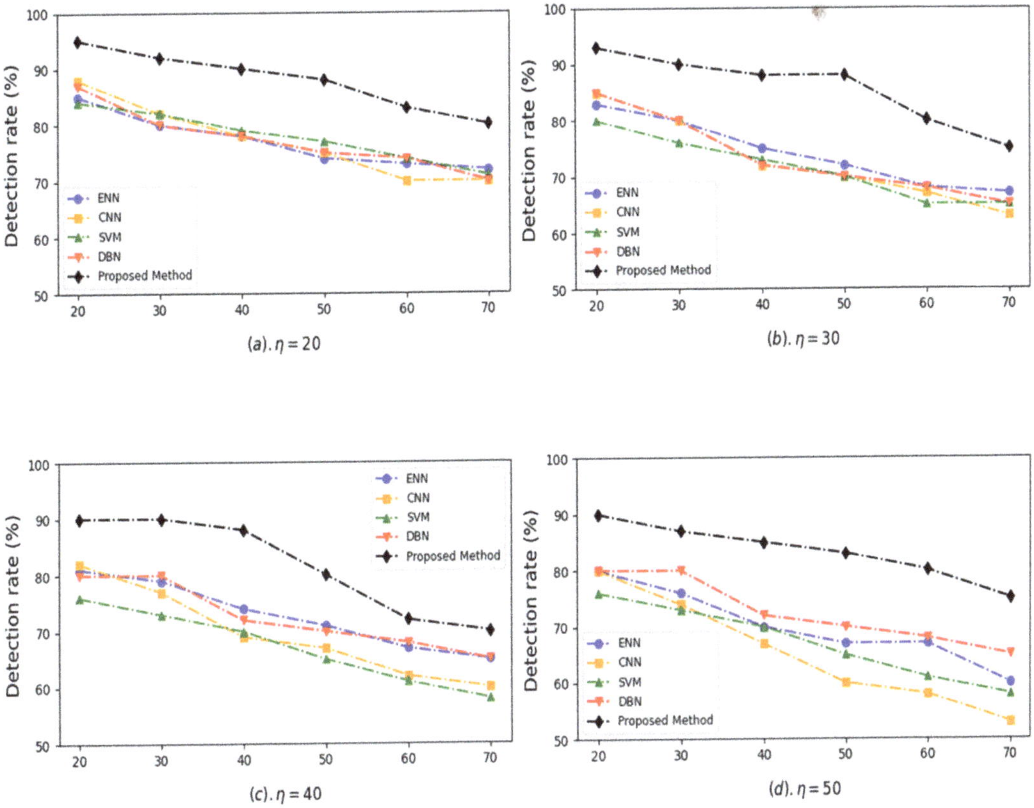

Figure 4. Detection rate comparison of ENN, CNN, SVM, DBN, and proposed method across different attack percentages.

Figure 5 illustrates the confusion matrices for training and testing phases at varying percentages of data utilization. In the 80% training phase, the method achieves remarkable results with only 27 false positives and 60 false negatives, indicating high precision and sensitivity in detecting botnet activity. Similarly, the model maintains consistent performance in the 70% training phase, with a minimal increase in false positives (28) and a slight reduction in false negatives (33). This demonstrates the robustness of the proposed approach in accurately identifying botnet and normal traffic, even with reduced training data. During the 20% testing phase, the confusion matrix shows only 7 false positives and 20 false negatives, reflecting the model's high generalization capability and ability to accurately predict botnet instances under unseen data conditions. The model performs exceptionally well in the 30% testing phase, with just 12 false positives and 50 false negatives. This indicates the scalability and reliability of the proposed method in handling larger test datasets.

In Table 4 is presented an evaluation of the proposed method performance in comparison with several existing botnet detection approaches across different datasets. The comparison focuses on key performance metrics, namely accuracy, detection rate, and false positive rate (FPR). The proposed method achieves an accuracy of 98.0%, detection rate of 95.0%, and low FPR of 7.8% on the CTU-13 dataset. Compared to Sharma and Babbar (2024) [33], who also utilized the CTU-13 dataset but achieved lower accuracy (95.2%) and a significantly higher FPR (18.0%), the proposed framework demonstrates enhanced effectiveness. Furthermore, although Elnakib et al. (2023) [24] and Ullah and Mahmoud (2019) [19] reported competitive performances on Bot-IoT and N-BaIoT datasets, respectively, their results indicated lower detection rates and higher false positive rates compared to the proposed method. These findings confirm the robustness, scalability, and generalization capability of our proposed three-phase anomaly detection framework in securing IoT networks against diverse and sophisticated botnet attacks.

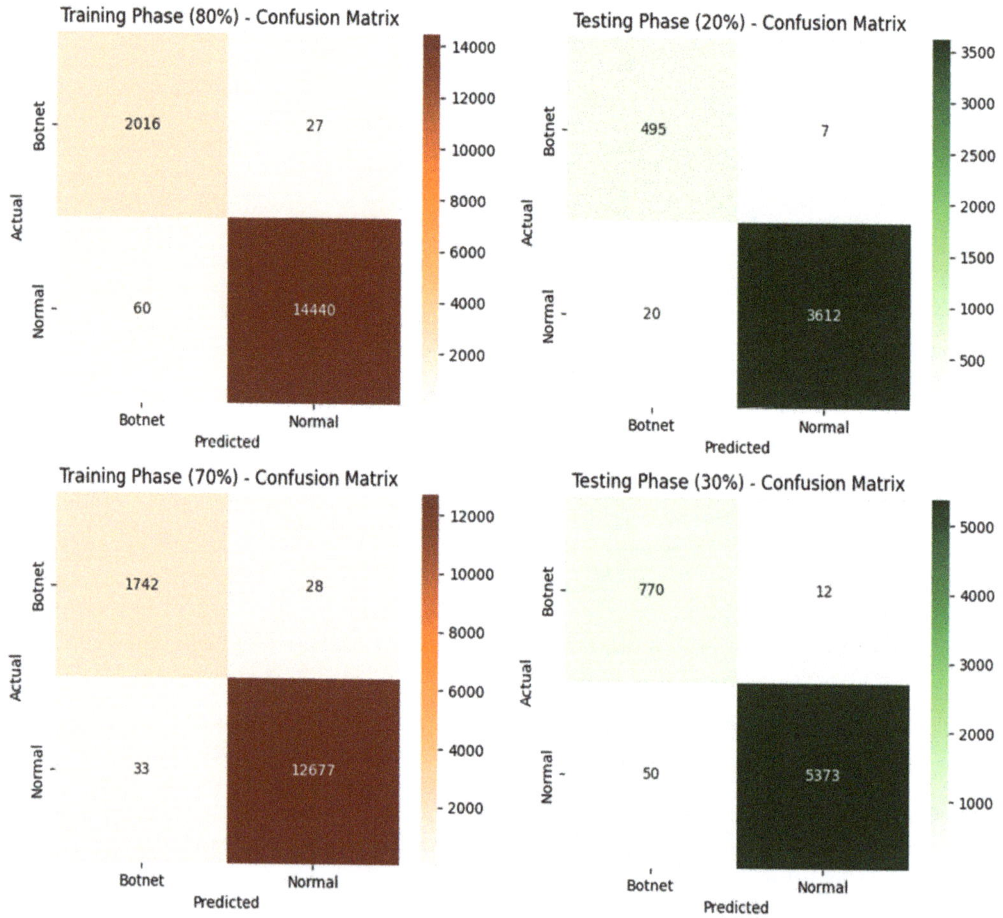

Figure 5. Confusion matrices for both training and testing phases.

Table 4. Comparison of the proposed method with existing methods.

Method	Dataset	Accuracy (%)	Detection Rate (%)	FPR (%)
Sharma and Babbar (2024) [33]	CTU-13	95.2	92.5	18.0
Elnakib et al. (2023) [24]	Bot_IoT	94.5	91.2	24.5
Ullah and Mahmoud (2019) [19]	N-BaIoT	92.3	89.0	20.1
Proposed Method	**CTU-13**	**98.0**	**95.0**	**7.8**

Note: Bold values indicate the best performance among compared methods.

To ensure a robust evaluation, we adopted a five-fold cross-validation scheme. The dataset was randomly partitioned into five equally sized folds, and each fold was used once as a test set while the remaining four were used for training. We reported the average and standard deviation of three standard classification metrics: accuracy, detection rate, and FPR. These metrics were chosen to reflect both overall performance and anomaly detection capability, especially in the presence of class imbalance. The results in Table 5 demonstrate the stability and reliability of the proposed framework across multiple runs.

Table 5. Performance of the proposed method across 5-fold cross-validation.

Fold	Accuracy (%)	Detection Rate (%)	FPR (%)
1	97.2	94.3	8.3
2	98.5	95.3	7.6
3	97.9	94.8	8.1
4	98.3	95.2	7.5
5	97.8	95.0	7.9
Mean ± Std	97.9 ± 0.7	95.2 ± 0.9	7.88 ± 0.4

To determine the best ensemble strategy, both Hard Voting and Soft Voting mechanisms were evaluated. As presented in Table 6, Soft Voting achieved a higher accuracy (98.0%) and detection rate (95.0%) while maintaining a lower false positive rate (7.8%) compared to Hard Voting. Therefore, Soft Voting was selected as the final voting strategy in the proposed anomaly detection framework.

Table 6. Performance comparison between Hard Voting and Soft Voting strategies.

Method	Accuracy (%)	Detection Rate (%)	FPR (%)
Quantized Hard Voting	96.3	93.0	12.4
Soft Voting	**98.0**	**95.0**	**7.8**

Note: Bold values indicate the best performance among compared methods.

4.2. Discussion

The superior performance of the proposed GA-ensemble framework can be attributed to its synergistic design. The Genetic Algorithm, enhanced with eagle-inspired search behavior, ensures optimal feature selection by exploring the search space more effectively and avoiding local minima. This reduces dimensionality without compromising accuracy. Furthermore, the dragonfly-inspired simulated annealing algorithm refines hyperparameters in a biologically inspired manner, enhancing model generalization. The ensemble classifier integrates diverse models, namely Decision Tree, Random Forest, and XGBoost, each contributing complementary strengths. The combination of Decision Tree, Random Forest, and XGBoost provides both a bias–variance trade-off and resilience to imbalanced data.

The proposed framework offers several practical advantages for securing real-world IoT environments. Its modular design, combining lightweight preprocessing, nature-inspired feature selection, and an ensemble of efficient classifiers, makes it suitable for deployment in edge-based applications such as smart homes, healthcare monitoring devices, the industrial IoT (IIoT), and autonomous sensor networks. The low false positive and false negative rates ensure minimal disruption to normal system operations, which is critical in mission-critical IoT deployments. The strengths of the model include its high detection accuracy (98%), generalization to diverse attack types, and robustness under varying attack ratios and device risk levels. The ensemble learning strategy combined with adaptive optimization techniques ensures a balance between computational efficiency and detection performance. However, the model has two notable limitations: its performance may degrade when dealing with encrypted or heavily obfuscated traffic, as the current design relies primarily on flow-level statistical features; and despite using automated tuning via swarm-enhanced simulated annealing, the training phase still incurs a degree of computational overhead that must be considered in resource-limited deployment scenarios.

5. Conclusions

The proposed anomaly detection framework for IoT networks effectively addresses critical security challenges by detecting anomalous behaviors and preventing cyber threats in dynamic and resource-constrained environments. By integrating advanced data preprocessing, feature selection through a Genetic Algorithm enhanced with eagle-inspired search mechanisms, and an ensemble classification approach, the framework is characterized by its high performance expressed through its accuracy, scalability, and computational efficiency. The obtained results demonstrate significant improvements compared to existing methods. The framework achieves a 12.5% increase in accuracy (98%), a 14% improvement in detection rate (95%), a 9.3% reduction in false positive rate (10%), and a 10.8% decrease in false negative rate (5%). These improvements clearly highlight the framework's effectiveness and superiority in ensuring IoT security. Its notable enhancement in accuracy is a direct result of the optimized feature selection and ensemble learning approach, which collectively contribute to more precise and reliable anomaly detection. Future research can focus on improving the scalability and real-time deployment of the proposed framework to handle large-scale IoT networks with thousands of interconnected devices while ensuring minimal latency and high adaptability. Additionally, enhancing the framework's capability to detect complex, multi-stage attacks such as Advanced Persistent Threats (APTs) can significantly improve its effectiveness in real-world cybersecurity scenarios. Another promising direction is the integration of federated learning techniques, which would enable decentralized model training while preserving user privacy, making the system more secure and practical for IoT applications that prioritize data confidentiality.

Author Contributions: B.S.: Formal analysis, conceptualization, methodology, data curation, investigation, validation, and writing—original draft. O.P.: Data curation, methodology, writing—review and editing, and supervision. All authors have read and agreed to the published version of the manuscript.

Funding: This research was funded by FCT—Fundação para a Ciência e Tecnologia grant number [UIDB/50008/2020].

Data Availability Statement: The original contributions presented in this study are included in the article. Further inquiries can be directed to the corresponding author.

Conflicts of Interest: The authors declare no conflicts of interest.

Abbreviations

The following abbreviations are used in this manuscript:

IoT	Internet of Things IIoT
IIoT	Industrial Internet of Things
IDS	Intrusion Detection System
DoS	Denial of Service
ENN	Ensemble Neural Network
DBN	Deep Belief Network
SVM	Support Vector Machine
CNN	Convolutional Neural Network
XGBoost	Extreme Gradient Boosting
RF	Random Forest

EATU	Energy-Efficient and Trustworthy Unsupervised Anomaly Detection Framework
ChOA	Chimp Optimization Algorithm
PSO	Particle Swarm Optimization
SSO	Social Spider Optimization
SPA	Sparrow Search Algorithm

References

1. Alwahedi, F.; Aldhaheri, A.; Ferrag, M.A.; Battah, A.; Tihanyi, N. Machine learning techniques for IoT security: Current research and future vision with generative AI and large language models. *Internet Things Cyber-Phys. Syst.* **2024**, *4*, 167–185. [CrossRef]
2. Mazhar, T.; Talpur, D.B.; Shloul, T.A.; Ghadi, Y.Y.; Haq, I.; Ullah, I.; Ouahada, K. Analysis of IoT Security Challenges and Its Solutions Using Artificial Intelligence. *Brain Sci.* **2023**, *13*, 683. [CrossRef] [PubMed]
3. Mishra, P.; Varadharajan, V.; Tupakula, U.; Pilli, E.S. A detailed investigation and analysis of using machine learning techniques for intrusion detection. *IEEE Commun. Surv. Tutor.* **2018**, *21*, 686–728. [CrossRef]
4. Sarhan, M.; Layeghy, S.; Moustafa, N.; Gallagher, M.; Portmann, M. Feature extraction for machine learning-based intrusion detection in IoT networks. *Digit. Commun. Netw.* **2024**, *10*, 205–216. [CrossRef]
5. Kollolu, R. A Review on wide variety and heterogeneity of IoT platforms. *Int. J. Anal. Exp. Modal Anal. Anal.* **2020**, *12*, 3753–3760. [CrossRef]
6. Çavuşoğlu, Ü. A new hybrid approach for intrusion detection using machine learning methods. *Appl. Intell.* **2019**, *49*, 2735–2761. [CrossRef]
7. Alaghbari, K.A.; Lim, H.-S.; Saad, M.H.M.; Yong, Y.S. Deep Autoencoder-Based Integrated Model for Anomaly Detection and Efficient Feature Extraction in IoT Networks. *IoT* **2023**, *4*, 345–365. [CrossRef]
8. Ni, C.; Li, S.C. Machine learning enabled Industrial IoT Security: Challenges, Trends and Solutions. *J. Ind. Inf. Integr.* **2024**, *38*, 100549. [CrossRef]
9. Biswas, S.K. Intrusion detection using machine learning: A comparison study. *Int. J. Pure Appl. Math.* **2018**, *118*, 101–114.
10. Sarwar, A.; Alnajim, A.M.; Marwat, S.N.K.; Ahmed, S.; Alyahya, S.; Khan, W.U. Enhanced Anomaly Detection System for IoT Based on Improved Dynamic SBPSO. *Sensors* **2022**, *22*, 4926. [CrossRef]
11. Sharma, B.; Sharma, L.; Lal, C.; Roy, S. Anomaly based network intrusion detection for IoT attacks using deep learning technique. *Comput. Electr. Eng.* **2023**, *107*, 108626. [CrossRef]
12. Do, P.H.; Dinh, T.D.; Le, D.T.; Pham, V.D.; Myrova, L.; Kirichek, R. An Efficient Feature Extraction Method for Attack Classification in IoT Networks. In Proceedings of the 2021 13th International Congress on Ultra Modern Telecommunications and Control Systems and Workshops (ICUMT), Brno, Czech Republic, 25–27 October 2021; pp. 194–199.
13. Tyagi, H.; Kumar, R. Attack and anomaly detection in IoT networks using supervised machine learning approaches. *Rev. D'Intell. Artif.* **2021**, *35*, 11–21. [CrossRef]
14. Priya, V.; Thaseen, I.S.; Gadekallu, T.R.; Aboudaif, M.K.; Nasr, E.A. Robust attack detection approach for IIoT using ensemble classifier. *arXiv* **2021**, arXiv:2102.01515.
15. Manzano, R.; Goel, N.; Zaman, M.; Joshi, R.; Naik, K. Design of a machine learning based intrusion detection framework and methodology for iot networks. In Proceedings of the 2022 IEEE 12th Annual Computing and Communication Workshop and Conference (CCWC), Las Vegas, NV, USA, 26–29 January 2022; pp. 191–198.
16. Zhao, R.; Gui, G.; Xue, Z.; Yin, J.; Ohtsuki, T.; Adebisi, B.; Gacanin, H. A novel intrusion detection method based on lightweight neural network for internet of things. *IEEE Internet Things J.* **2021**, *9*, 9960–9972. [CrossRef]
17. Xiao, Y.; Feng, Y.; Sakurai, K. An Efficient Detection Mechanism of Network Intrusions in IoT Environments Using Autoencoder and Data Partitioning. *Computers* **2024**, *13*, 269. [CrossRef]
18. Gupta, H.; Sharma, S.; Agrawal, S. Artificial intelligence-based anomalies detection scheme for identifying cyber threat on IoT-based transport network. *IEEE Trans. Consum. Electron.* **2023**, *70*, 1716–1724. [CrossRef]
19. Ullah, I.; Mahmoud, Q.H. A Two-Level Hybrid Model for Anomalous Activity Detection in IoT Networks. In Proceedings of the 2019 16th IEEE Annual Consumer Communications & Networking Conference (CCNC), Las Vegas, NV, USA, 11–14 January 2019; pp. 1–6.
20. Huang, Z.; Wu, Y.; Tempini, N.; Lin, H.; Yin, H. An energy-efficient and trustworthy unsupervised anomaly detection framework (EATU) for IIoT. *ACM Trans. Sens. Netw.* **2022**, *18*, 1–18. [CrossRef]
21. Altulaihan, E.; Almaiah, M.A.; Aljughaiman, A. Anomaly Detection IDS for Detecting DoS Attacks in IoT Networks Based on Machine Learning Algorithms. *Sensors* **2024**, *24*, 713. [CrossRef]
22. Srivastav, D.; Srivastava, P. A two-tier hybrid ensemble learning pipeline for intrusion detection systems in IoT networks. *J. Ambient Intell. Humaniz. Comput.* **2023**, *14*, 3913–3927. [CrossRef]
23. Wang, Y.; Sun, G.; Cao, X.; Yang, J. An intrusion detection system for the internet of things based on the ensemble of unsupervised techniques. *Wirel. Commun. Mob. Comput.* **2022**, *2022*, 8614903. [CrossRef]

24. Elnakib, O.; Shaaban, E.; Mahmoud, M.; Emara, K. EIDM: Deep learning model for IoT intrusion detection systems. *J. Supercomput.* **2023**, *79*, 13241–13261. [CrossRef]
25. Mohapatra, H.; Rath, S.; Panda, S.; Kumar, R. Handling of man-in-the-middle attack in wsn through intrusion detection system. *Int. J. Emerg. Trends Eng. Res.* **2020**, *8*, 1503–1510. [CrossRef]
26. Liu, T.; Meidani, H. End-to-end heterogeneous graph neural networks for traffic assignment. *Transp. Res. Part C Emerg. Technol.* **2024**, *165*, 104695. [CrossRef]
27. Liu, T.; Meidani, H. Graph neural network surrogate for seismic reliability analysis of highway bridge systems. *J. Infrastruct. Syst.* **2024**, *30*, 05024004. [CrossRef]
28. Massey, F.J., Jr. The Kolmogorov-Smirnov test for goodness of fit. *J. Am. Stat. Assoc.* **1951**, *46*, 68–78. [CrossRef]
29. Bauer, E.; Kohavi, R. An empirical comparison of voting classification algorithms: Bagging, boosting, and variants. *Mach. Learn.* **1999**, *36*, 105–139. [CrossRef]
30. Chen, T.; Guestrin, C. XGBoost: A Scalable Tree Boosting System. In Proceedings of the 22nd ACM SIGKDD International Conference on Knowledge Discovery and Data Mining (KDD '16), San Francisco, CA, USA, 13–17 August 2016; Association for Computing Machinery: New York, NY, USA, 2016; pp. 785–794.
31. Breiman, L. Random forests. *Mach. Learn.* **2001**, *45*, 5–32. [CrossRef]
32. Quinlan, J.R. Induction of decision trees. *Mach. Learn.* **1986**, *1*, 81–106. [CrossRef]
33. Sharma, A.; Babbar, H. Detecting Cyber Threats in Real-Time: A Supervised Learning Perspective on the CTU-13 Dataset. In Proceedings of the 2024 5th International Conference for Emerging Technology (INCET), Belgaum, India, 24–26 May 2024; pp. 1–5.
34. Alohali, M.A.; Elsadig, M.; Al-Wesabi, F.N.; Al Duhayyim, M.; Mustafa Hilal, A.; Motwakel, A. Enhanced Chimp Optimization-Based Feature Selection with Fuzzy Logic-Based Intrusion Detection System in Cloud Environment. *Appl. Sci.* **2023**, *13*, 2580. [CrossRef]
35. Kunhare, N.; Tiwari, R.; Dhar, J. Particle swarm optimization and feature selection for intrusion detection system. *Sādhanā* **2020**, *45*, 109. [CrossRef]
36. Ibrahim, R.A.; Elaziz, M.A.; Oliva, D.; Cuevas, E.; Lu, S. An opposition-based social spider optimization for feature selection. *Soft Comput.* **2019**, *23*, 13547–13567. [CrossRef]
37. Kilincer, I.F.; Tuncer, T.; Ertam, F.; Sengur, A. SPA-IDS: An intelligent intrusion detection system based on vertical mode decomposition and iterative feature selection in computer networks. *Microprocess. Microsyst.* **2023**, *96*, 104752. [CrossRef]

sensors

Article

Extending the Coverage of IEEE 802.15.4z HRP UWB Ranging

Sumin Han and Byung-Jun Jang *

Department of Electrical Engineering, Kookmin University, Seoul 02707, Republic of Korea; hsm2005x@kookmin.ac.kr
* Correspondence: bjjang@kookmin.ac.kr; Tel.: +82-10-4396-0811

Abstract: Currently, the ultra-wideband (UWB) technology commercialized in smartphones and smart keys is a high-rate pulse repetition frequency (HRP) UWB of the IEEE 802.15.4z standard, which aims to accurately determine the distance between UWB devices located within tens of meters using two-way ranging (TWR). However, in order for UWB ranging technology to be spread to various location-based services or positioning services, it must be able to measure the distance between UWB devices that are hundreds of meters apart. Fortunately, UWB technology can freely change physical layer parameters, as long as they comply with the UWB local regulations worldwide. Therefore, in this study, we analyzed the method of configuring the packet structure, length, and transmission power from the link budget perspective to enable longer UWB ranging of hundreds of meters within UWB local regulations. As a result of the analysis, we theoretically confirmed that UWB ranging is possible even at hundreds of meters by selecting the optimal physical layer parameters. In addition, the experimental results using the Qorvo DW3000 module were confirmed to be consistent with the results analyzed in this study. The results of this study can be used as basic data for the introduction of wide-area UWB technology and services in the future.

Keywords: UWB; HRP UWB; positioning; ranging; IEEE 802.15.4z; link budget; IoT

1. Introduction

Ultra-wideband (UWB) technology has gained attention in various fields due to its potential for precise distance measurement and ranging. In particular, high-rate pulse repetition frequency (HRP) UWB, defined in the IEEE 802.15.4z standard, has recently become prominent for enabling centimeter-level ranging accuracy between UWB devices [1,2]. This UWB technology first appeared as short-range wireless personal area network (WPAN) technology in the early 2000s. It was standardized by the IEEE 802.15.4 committee, a WPAN standardization organization, and was announced as the IEEE 802.15.4a standard in 2007. However, UWB technology has not been widely adopted as WPAN communication technology due to competing technologies such as Wi-Fi and Bluetooth [3,4]. However, with the increasing demand for location-based services in the IoT, smart factories, and other applications, UWB technology has been re-evaluated for its advantage of being able to measure the distance between UWB devices in addition to providing wireless communication. Since Apple's iPhone 11 in 2019, which launched a service that informs users of the locations of nearby UWB devices, most smartphones released by Samsung, Apple, and Google on the market today have integrated HRP UWB technology. In addition, automobile companies such as Hyundai Motors are also introducing UWB smart keys [5,6]. Following this trend, HRP UWB technology has continued to improve its performance with the announcement of the IEEE 802.15.4z standard in 2020, and the IEEE 802.15.4ab standardization work is currently in progress [5–10].

Since HRP UWB ranging was originally introduced as a short-range precision ranging solution and not a long-range communication solution, there has been no meaningful attempt to expand the ranging coverage. Moreover, since HRP UWB ranging has been mainly applied to smartphones, its function has been limited to measuring the exact distance between nearby UWB devices after pairing via Bluetooth. Most smartphones first recognize UWB devices through Bluetooth pairing before performing UWB ranging to reduce power consumption in UWB ranging operation. As a result, HRP UWB ranging was constrained to several tens of meters, equivalent to the Bluetooth communication range. This trend is also true for HRP UWB applications other than smartphones. Most of the existing studies on IEEE 802.15.4z HRP UWB investigated by the authors have focused on improving the ranging accuracy, ensuring communication robustness, and mitigating interference with Wi-Fi 6E [11–21]. Table 1 summarizes the operation ranges reported in prior studies, with most commercially available modules (e.g., Qorvo's DWM1000, DW3000, and DW3120; NXP's SR150 and SR040; and Spark's SR1010 and SR1020) being limited to approximately 50 or 100 m.

Table 1. Key performance comparisons of previous UWB research results.

Vendor	Chip or Module	Ranging Method	Frequency (GHz)	Max Range (m)	Reference
Qorvo	DWM1000	TWR, TDoA	3.5–6.5	~50	[19–22]
Qorvo	DW3000	TWR, TDoA	6.5, 8	~50	[16,17,23]
Qorvo	DW3120	TWR, TDoA, PDoA	6.5, 8	~50	[18,24]
NXP	SR150	TWR, TDoA, PDoA	6.24–8.24	~50	[25]
NXP	SR040	TWR, TDoA	6.24–8.24	~50	[26]
Spark	SR1010	TWR	3.1–6	~100	[27]
Spark	SR1020	TWR	6–9.3	~100	[27]
Qorvo	DW3000	TWR	6.5	~300	This Work

Nevertheless, for IEEE 802.15.4z UWB technology to be widely used beyond smartphones and smart keys, it must be able to measure the distance between UWB devices that are hundreds of meters apart. If a longer operating range is possible beyond the current operating range limit of tens of meters, then HRP UWB communication technology can be widely used in various fields such as logistics, smart factories, and smart cities. In an effort to achieve this, alternative approaches such as hybrid localization schemes combining UWB ranging with inertial measurements [28] or control-based estimation methods [29] have been proposed. These methods have demonstrated effectiveness in mitigating packet loss and maintaining tracking continuity under degraded link conditions. Although hybrid localization schemes and sensor fusion techniques have shown promise in mitigating ranging failures under degraded conditions, they inherently introduce additional complexity, cost, and energy consumption to the system. In many practical deployments, such as low-power IoT devices, logistics networks, or cost-sensitive industrial applications, it is highly desirable to maximize UWB ranging capabilities using the existing hardware alone, without reliance on auxiliary sensors or external data fusion.

However, extending the operating range purely through physical layer parameter optimization is particularly challenging due to the inherent link performance imbalance between the synchronization and data transmission portions of the UWB packet. This study specifically addresses these challenges by performing a detailed link budget analysis and experimentally validating a PHY-only optimization strategy, demonstrating that long-range UWB operation can be achieved while maintaining regulatory compliance and minimal system complexity.

In this study, we focus on extending the fundamental ranging coverage of IEEE 802.15.4z HRP UWB purely through physical layer parameter optimizations, without relying on auxiliary sensors or external estimation techniques. We propose a method to extend the ranging capability of IEEE 802.15.4z HRP UWB systems to hundreds of meters. This is achieved through a detailed link budget analysis considering the packet structure, packet length, and transmission power within worldwide UWB local regulations. Our results show that by optimizing the physical and RF parameters based on a detailed link budget analysis that accounts for the inherent link performance imbalance between the SYNC (preamble) and the data portion (DP), UWB systems can achieve significantly extended ranging without violating power emission limits. This improvement is not merely a result of increasing the output power but rather addressing the link performance difference. By reducing the link differences between the SYNC and the DP, the overall packet reception performance is improved. Thus, the extended ranging capability is quantitatively supported by a more balanced and favorable link budget distribution across the full packet structure. We also experimentally verify the proposed method using Qorvo DW3000, and the experimental results are consistent with the theoretical analysis. Our experimental results show that by tuning these parameters while respecting the local regulation power constraints, a longer UWB operating range can be achieved.

The composition of this paper is as follows. First, in Section 2, we explain the characteristics of HRP UWB technology based on UWB local regulations. Then, in Section 3, we analyze the link performance of IEEE 802.15.4z HRP UWB communication and analyze the parameters for long-range measurement of HRP UWB technology. In Section 4, we describe the experimental method and experimental results to verify these analysis results, and finally, in Section 5, we conclude the paper.

2. UWB Local Regulations and UWB Pulse Characteristics

UWB means using ultra-wideband technology, and the local regulations in Table 2 must be applied worldwide. As shown in Table 2, UWB involves frequency-related regulations, such as the frequency band or frequency bandwidth used, and output power, such as average power and peak power. Among the local regulations in Table 2, the output power regulations are applied globally; that is, the output power of the UWB transmitter must satisfy both ① the average power spectral density (PSD) of −41.3 dBm/MHz at 1 ms and ② the peak PSD of 0 dBm/50 MHz. This is explained in Figure 1. Assuming that one pulse has a bandwidth of 500 MHz, the pulse width is approximately 2 ns. The peak power of one 2-ns pulse must be less than 10 dBm/500 MHz according to ② the peak PSD regulation. In addition, when there are multiple pulses during a 1-ms period, the average power, which is the accumulated effect of these pulses, must be less than −14.3 dBm/500 MHz according to ① the average PSD regulation.

Table 2. Local regulation for UWB worldwide.

	Local Regulation
Frequency	3.1–10.6 GHz [1]
Average power	−41.3 dBm/MHz@1 ms ①
Peak power	0 dBm/50 MHz ②
Bandwidth	More than 450 MHz (1 MHz RBW, 10 dB bandwidth)

[1] Frequency ranges vary slightly from country to country. ① Regulation for average Power Spectral Density (PSD) over 1 ms, typically specified as −41.3 dBm/MHz or −14.3 dBm/500 MHz. ② Regulation for peak power, requiring pulses not to exceed 10 dBm/500 MHz even if only a single pulse is transmitted.

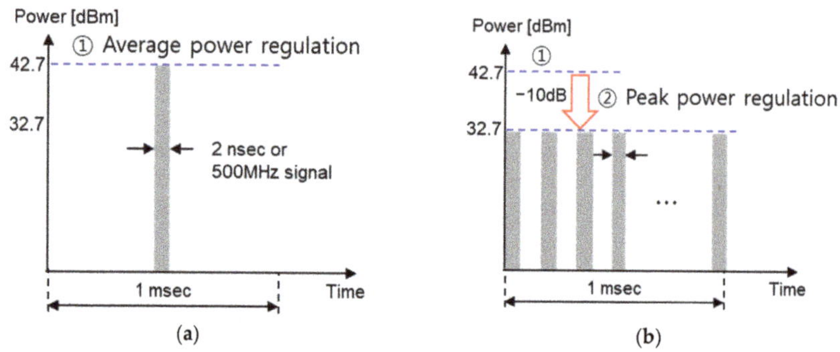

Figure 1. Local regulation for output power of UWB transmitter (**a**) in the case of one pulse and (**b**) in the case of multiple pulses.

The reason for having two criteria is because the average power changes depending on the number of pulses. For example, if there is only one pulse with a bandwidth of 500 MHz per 1 ms, as shown in Figure 1a, then the peak power will have a rather large value of 42.7 dBm/500 MHz under ① the average PSD regulation. This means that although the average output is small, the peak power is quite large, and thus it can interfere with other wireless devices for a short period of time. Therefore, according to local regulation ②, the peak power must be limited to 10 dBm or less even if there is one pulse. Accordingly, the case where both local regulations are satisfied is when the peak power is 10 dBm and the number of pulses is 1862 for a UWB pulse with a bandwidth of 500 MHz, as shown in Figure 1b. In this case, the average power becomes −14.3 dBm/500 MHz, and thus both local regulations ① and ② are satisfied. Regarding spectral emissions, the IEEE 802.15.4z standard does not specify a mandatory adjacent channel power rate (ACPR) or spectral mask requirements, unlike narrowband communication standards. Therefore, our link-budget optimization focuses solely on complying with the average and peak power constraints defined by worldwide UWB regulations.

Now, based on this local regulation for the output power, let us find out the number of pulses per 1 ms, peak power, and average power when a UWB device is used for communication or radar. First, since UWB radar measures the distance by using the time difference between the transmitted pulse and the pulse backscattered from the target, the pulse repetition frequency (PRF) is determined by the maximum measurement distance. For example, Novelda's UWB x4 module, a representative UWB radar module, uses a pulse with a bandwidth of 1.5 GHz (pulse width of approximately 0.67 ns) and waits for the pulse for 65.8 ns after transmitting the pulse [30]. When this time is converted into a distance, it is 9.87 m. Since the actual operating range of Novelda's UWB radar is only 1–2 m, it is possible to use more pulses, but in this case, the average power exceeds −14.3 dBm/MHz and thus cannot be used; that is, the pulse number and peak power of the UWB radar are determined by local regulations ① and ②, as shown in Figure 2a.

Next, the currently commercialized IEEE 802.15.4z-based UWB communication method is HRP UWB with a relatively high PRF [7]. In Figure 1, when the peak power is 10 dBm, the number of pulses with a bandwidth of 500 MHz possible for 1 ms is 1862, and using more pulses than this number is called HRP UWB. As shown in Figure 2b, when the number of pulses increases, the peak power becomes smaller than the value of local regulation ②. Instead, by using a large number of pulses, additional processing gain can be obtained through the spread spectrum. However, since the number of pulses cannot be increased indefinitely, it must be limited so as to not exceed the average power suggested in local regulation ①. In other words, in order to increase the processing gain, the more pulses the better, but the number of pulses cannot exceed the local regulation for the average power. Since UWB communication must maintain −14.3 dBm according to the average power under local regulation ①, HRP UWB communication becomes a system whose performance is limited only by the average power regardless of the peak power.

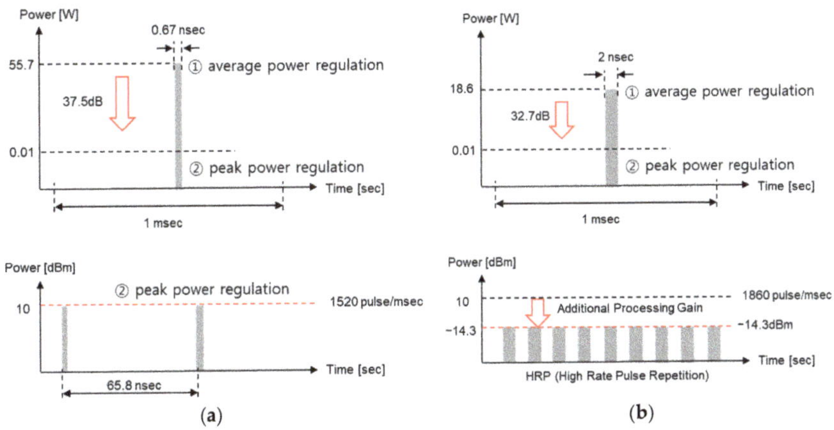

Figure 2. Pulse characteristics of UWB radar and HRP UWB communication (**a**) in the case of UWB radar and (**b**) in the case of HRP UWB communication.

3. Link Analysis of HRP UWB Communication

3.1. Packet Structure of HRP UWB

In the previous section, we examined how HRP UWB communication is a system whose performance is limited only by the average power. In this section, we will examine how this principle is implemented in the actual IEEE 802.15.4z HRP UWB standard. Figure 3 shows the packet structure presented in the IEEE 802.15.4z HRP UWB standard [5,7]. As shown in Figure 3, the packet structure of IEEE 802.15.4z HRP UWB is divided into a synchronization header (SHR) used to set the reference time for synchronization and distance measurement and a data portion (DP) for transmitting data, and both use separate modulation methods. The IEEE 802.15.4z HRP UWB standard also includes a scrambled time sequence (STS) for encrypted distance measurement, but since STS uses the same modulation as the SHR, Figure 3 shows a packet without an STS.

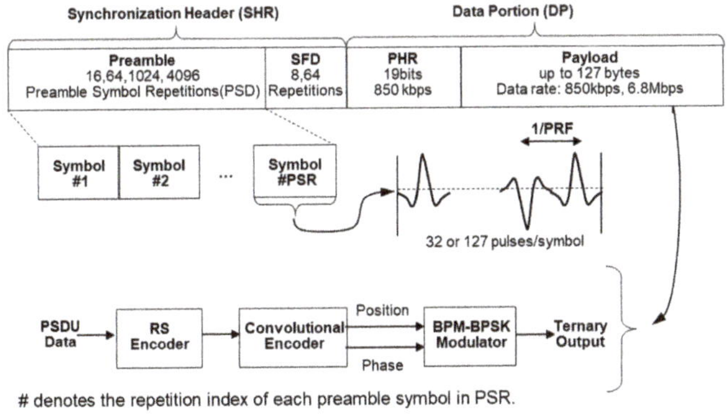

Figure 3. Typical packet structure of IEEE 802.15.4z HRP UWB.

First, the SHR is divided into a preamble and start-of-frame delimiter (SFD). The preamble is transmitted by repeating the same symbol, and its length is determined by the number of preamble symbol repetitions (PSRs). One preamble symbol is defined by 31 or 127 sub-symbols, and each sub-symbol has a value of (−1, 0, 1). Here, '1' means a UWB pulse with positive (+) polarity, '−1' means a UWB pulse with negative (−) polarity, and '0' means no pulse. These pulses have the time interval of the inverse of the PRF. The basic PRF (BPRF) mode of the IEEE 802.15.4z standard uses a PRF of 64 MHz. Next, the SFD consists of 8 or 64 symbols and indicates the end of the preamble and the start of the DP. The preamble can obtain processing gain by repeatedly sending a known pulse without any special modulation. In HRP UWB, this is defined as the number of packet acquisition chunks (PACs). In addition to the processing gain through the PACs, the preamble uses the leading edge detection (LED) algorithm to increase the ranging accuracy; that is, if the number of PSRs is 1024, and the PAC count is 32, then one channel impulse response (CIR) is generated by collecting the preambles of 32 PACs, and the LED algorithm is used to increase the ranging accuracy by using 32 (=1024/32) of these CIRs [1].

Next, the DP consists of a physical header (PHR) and a payload. The PHR includes information about the length and speed of payload data and uses different channel coding and modulation schemes from the SHR. Reed–Solomon (RS) coding and convolutional coding are used for channel coding, and burst position modulation–binary phase shift keying (BPM-BPSK) modulation is used as a modulation method. The DP supports multiple data rates, including 850 kbps and 6.8 Mbps. The difference between the 850 kbps and 6.8 Mbps modes in the IEEE 802.15.4z standard is achieved by adjusting the symbol duration and the internal burst structure while maintaining the same transmission bandwidth. For example, the 850 kbps mode adopts a much longer symbol duration with more burst positions and chips per symbol. In contrast, the 6.8 Mbps mode uses a shorter symbol duration with fewer burst positions and chips, allowing faster data transmission at the expense of the link margin. The symbol structures for the 850 kbps and 6.8 Mbps data rates are shown in Figure 4.

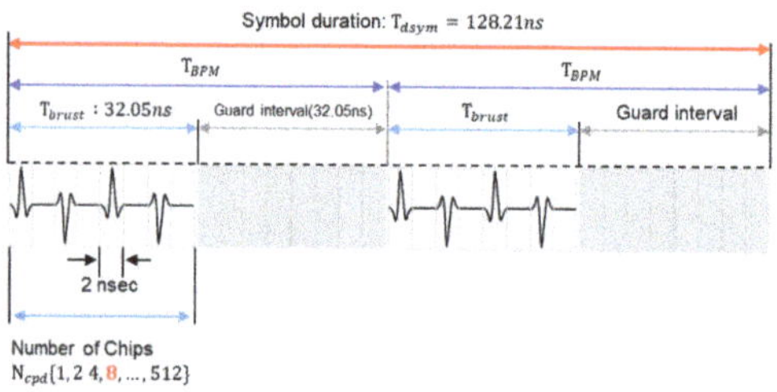

(a)

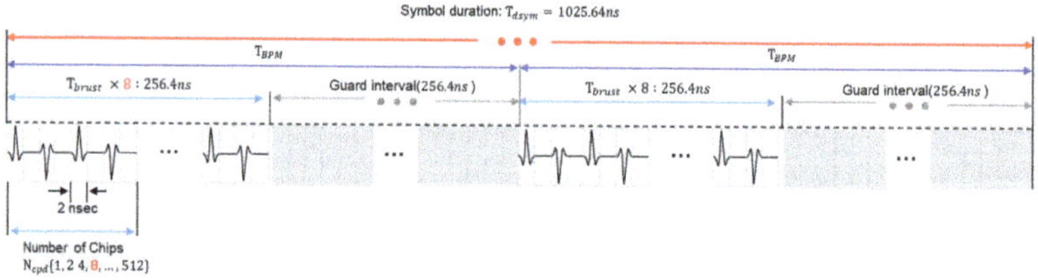

(b)

Figure 4. HRP UWB PHY BPM-BPSK modulation structure and symbol duration: (**a**) data rate of 6.8 Mbps and (**b**) data rate of 850 kbps.

This is because unlike the SHR, pulses are not sent repeatedly in the data section, and thus there is no processing gain; rather, coding gain is obtained through channel coding to balance the link between the preamble and payload. In HRP UWB communication, the ranging limit is determined by the smaller link margin between the SHR and DP links, making it necessary to adjust the parameters to appropriately balance the two.

3.2. Communication and Ranging Principles of HRP UWB

Section 3.1 explains the structure and modulation method of the physical layer packet of the IEEE 802.15.4z HRP UWB standard. This section examines the principle of how HRP UWB technology measures distance based on wireless communication. Since HRP UWB is basically a wireless communication method, packets are exchanged to send and receive data. Among the packets, the preamble performs the role of synchronization for data demodulation, and the payload among DPs is used for data transmission. The principle of ranging between UWB devices using this packet exchange is as shown in Figure 5a. Figure 5a is a method of measuring distance by exchanging only two packets, which is called single-sided two-way ranging (SS-TWR). Among the two UWB devices, the device that sends the packet first is called the initiator, and the other is called the responder. The initiator transmits a Poll packet and records the time t_1 when the transmission of the Poll packet begins, while t_1 is the moment when the preamble ends, and this time record is called a timestamp. The UWB responder device that received the Poll message sends a Response packet after the time t_{reply}. Now, the UWB initiator records the time at which the Response packet was received as a timestamp t_2. Now, the distance between the two UWB devices is 1/2 of the time of propagation (ToF), which is given by Equation (1):

$$R = c \times \frac{t_2 - t_1 - t_{reply}}{2}, \tag{1}$$

where c means the speed of light. In this way, ranging using two packets is possible through wireless communication, and this is called two-way ranging (TWR).

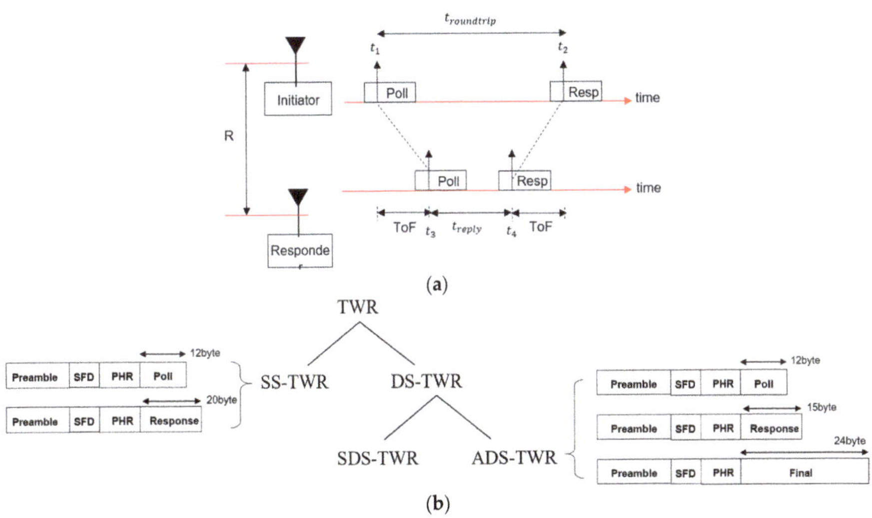

Figure 5. Two-way ranging: (**a**) principle of UWB two-way ranging and (**b**) classification of HRP UWB two-way ranging and related packets.

Meanwhile, the types of TWR that measure distance in the UWB can be divided into SS-TWR, which exchanges packets twice (Poll-Response), and double-sided-TWR (DS-TWR), which exchanges packets three times (Poll-Response-Final), as shown in Figure 5b. DS-TWR is further classified into symmetrical DS-TWR (SDS-TWR) and asymmetrical

DS-TWR (ADS-TWR). Among these, ADS-TWR, which can freely use the length of the packet, is mainly used for ranging. ADS-TWR is used when high distance accuracy is required because it can minimize the influence of clock errors between the transmitter and receiver. In ADS-TWR, the length of the packet is the shortest for the Poll packet and the longest for the Final packet, as shown in Figure 5b. Regardless of the TWR method used, in order to measure the distance between UWB devices with TWR, two packets must be successfully exchanged for SDS-TWR, and three packets must be successfully exchanged for DS-TWR. To this end, both the preamble and the payload of the packets must be successfully received.

3.3. Link Analysis of Data Portion in TWR Procedure

In order for the TWR process of HRP UWB explained in Section 3.2 to be successful, packets related to ranging must first be successfully exchanged. To this end, there must be no preamble errors and no DP data errors. In other words, the link budget of the preamble and the link budget of the DP must be considered simultaneously.

First, let us look at the DP link. The link performance of the DP is determined by whether the packet reception is successful or not and is usually given as the packet error rate (PER). In this study, we assume a PER of 1% or more as the link performance criterion. This value is a value commonly discussed in the current next-generation UWB standard IEEE 802.15.4ab [8]. The PER performance of BPM-BPSK used in UWB communication is generally given as shown in Equation (2) [31]:

$$PER = NQ\left(\sqrt{\gamma \frac{E_b}{N_0}}\right) \tag{2}$$

where N is the number of bits per packet, γ is the coding gain, E_b is the bit energy, and N_0 is the power density spectrum of noise. As shown in Figure 5b, the 'Poll', 'Response', and 'Final' packets are required for ADS-TWR, and the longest packet is the 'Final' packet with a size of 24 bytes. Thus, this packet should be used as the criterion for PER evaluation; that is, since N is 192 bits, E_b/N_0 of the DP corresponds to 9.0 dB, assuming 1% PER. Next, the IEEE 802.15.4z HRP UWB standard uses convolutional channel coding and Reed–Solomon channel coding, and thus it has a channel coding gain of about 5.6 dB [14]. Therefore, the DP of the IEEE802.15.4z HRP UWB standard should have an E_b/N_0 value of about 3.4 dB [23]. In addition, since the IEEE802.15.4z HRP UWB standard spreads one bit over multiple chips, the signal-to-noise ratio (SNR) is given by Equation (3):

$$SNR = \frac{E_b}{N_0}\frac{R_b}{BW} = \frac{E_b}{N_0}\left(\frac{R_b}{R_c}\right) \tag{3}$$

where BW is the bandwidth, R_b is the bit rate, and R_c is the chip rate. Based on the IEEE 802.15.4z HRP UWB standard, when R_b is 850 kbps, the SNR is calculated to be about −24.3 dB, and when R_b is 6.8 Mbps, the SNR is calculated to be −15.3 dB. In other words, if the data rate is as low as 850 kbps, then there is a 9 dB link margin, and thus a wider communication range can be achieved than in the case of 6.8 Mbps.

Now that we know the SNR value, let us analyze the DP link based on the transmitter parameters of the IEEE 802.14z HRP UWB standard and the power regulation in Table 2. The transmitted power is given by Equation (4):

$$P_{TX,DP}[\text{dBm}] = PSD_{TX} + BW + GG \tag{4}$$

where $P_{TX,Preamble}$ is the transmitted power of the preamble in dBm, PSD_{TX} is the average power regulation of −41.3 dBm/MHz, BW is the bandwidth in dBMHz, and GG is the gating gain. The gating gain is given by Equation (5):

$$GG[\text{dB}] = 10log_{10}\left(\frac{1[\text{ms}]}{T_p}\right) \tag{5}$$

where T_p corresponds to the time length of the packet. In other words, if the packet length is less than 1 ms, it indicates that the output can be increased by that amount according to the local regulation. Figure 6 shows a case where the output power is doubled by GG when T_p is 0.5 ms. Typically, UWB IC manufacturers such as Qorvo and NXP provide a function that can control power through the gating gain if the packet length is shorter than 1 ms [23]. Finally, the receiver sensitivity is given by Equation (6):

$$S_{DP}[\text{dBm}] = -174 + BW + NF + SNR_{\text{min}} \tag{6}$$

where SNR_{min} is the minimum SNR, which is given by Equation (3), and NF is the receiver's noise figure. Finally, the received power is calculated based on the Friis transmission formula, which is a path loss model in free space.

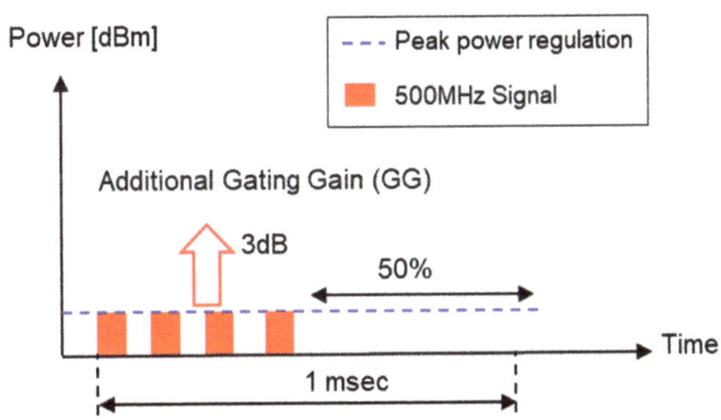

Figure 6. Example of increased output power via gating gain.

3.4. Link Analysis of Preamble in TWR Procedure

In Section 3.3, we analyzed the link characteristics in the DP of HRP UWB. In this section, we analyze the ranging accuracy when data are successfully exchanged. In UWB communication, the ranging accuracy is related to the preamble, and the root mean square (RMS) value of the ranging accuracy of the preamble is given by the Cramer–Rao bound (CRB). In general, the RMS of the ranging accuracy is given as a function of the SNR and effective bandwidth as shown in Equation (7) [31,32]:

$$RMS(d) \geq \sqrt{CRB} = \frac{c}{2\sqrt{2}\pi BW_{eff}\sqrt{SNR}} \tag{7}$$

where c is the speed of light, SNR is the signal-to-noise ratio of the preamble, and BW_{eff} is the effective bandwidth, which corresponds to 500 MHz in the case of the IEEE 802.15.4z HRP UWB standard [9]. Based on Equation (7), the ranging accuracy is expressed as a function of the SNR and effective bandwidth as shown in Figure 7.

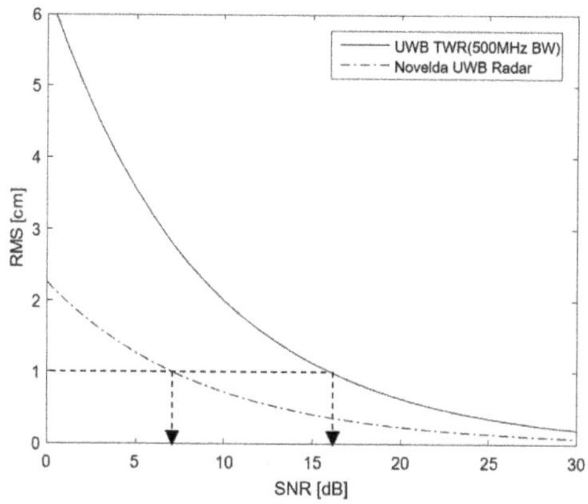

Figure 7. Cramer–Rao bound (CRB) for IEEE 802.15.4z HRP UWB ranging. The arrow indicates the SNR value at which the theoretical lower bound (CRB) reaches 1 cm, highlighting the required link quality for high-accuracy ranging.

If the ranging accuracy is 1 cm, then the SNR of the UWB preamble requires 16 dB according to Equation (7). In other words, unlike the SNR value required by the DP in the UWB packet, the preamble requires a 12.6 dB higher SNR value. To this end, the SNR value is increased by repeatedly transmitting the same symbol in the preamble. At this time, the processing gain is given by the number of symbol repetitions as shown in Equation (8):

$$PG[\text{dB}] = 10log_{10}(N_p) \tag{8}$$

where N_p means the number of symbols used for ranging in the preamble. The number of symbols to be used for the processing gain of Equation (8) among the number of symbols in the preamble is defined by the PSRs and PACs. Now, the transmitted power of the preamble of the HRP UWB is given as shown in Equation (9) because the processing gain is added, unlike Equation (4):

$$P_{TX,Preamble}[\text{dBm}] = PSD_{TX} + BW + GG + PG \tag{9}$$

where $P_{TX,Preamble}$ is the transmitted power of the preamble in dBm, PSD_{TX} is the average power regulation of -41.3 dBm/MHz, BW is the bandwidth in dBMHz, and GG and PG are the gating gain and the processing gain, respectively. Finally, the received power is based on the Friis transmission formula, which is a path loss model in free space.

Now, using Equations (3)–(9), the link budget for each preamble and the DP of IEEE 802.15.4z UWB communication are calculated as shown in Table 3. In Table 3, CH9, which is currently available worldwide, was used as the UWB frequency. Among the results in Table 3, the preamble link and the DP link with a data transmission rate R_b of 6.8 Mbps are represented graphically as shown in Figure 8. As can be seen in Table 3, the link performance of the preamble was better than that of the DP. In other words, when the distance increases, the preamble may be received, but the data link may be broken. This has been experimentally presented in related papers [33,34]. In addition, it can be seen that in the DP, sending data at a low speed of 850 kbps rather than a data rate of 6.8 Mbps can improve the performance by about 9 dB. This is consistent with the results provided by the chip manufacturer [23].

Table 3. Link budget calculation of IEEE 802.15.4z HRP UWB.

	Preamble	Data Portion (850 kbps)	Data Portion (6.8 Mbps)
Code length	127	-	-
No. of symbols	256	-	-
PAC size	32	-	-
Preamble length [us]	268.6	-	-
PRF [MHz]	64	62.4	62.4
Coding gain [dB]	-	5.6	5.6
Coded E_b/N_0 [dB]	-	3.5	3.5
Payload [bits]	-	160	160
Packet length	-	306.4	100.9
PSD [dBm/MHz]	−41.3	−41.3	−41.3
Bandwidth [MHz]	499.2	499.2	499.2
Frequency [GHz]	7.98	7.98	7.98
NF [dB]	6	6	6
PG [dB]	36.1	-	-
GG [dB]	0	0	0
P_{TX} [dBm]	−14.3	−14.3	−14.3
SNR_{min} [dB]	10.0	−21.6	−12.5
S [dB]	−106.7	−105.5	−96.5
Path Loss [dB]	92.6	91.1	82.1
PL_{1m} [dB]	50.5	50.5	50.5
Link margin	42.2	40.7	31.7

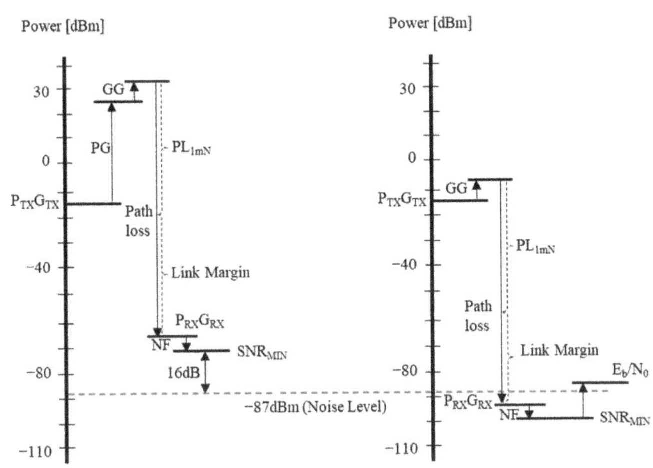

UWB Comm.(Preamble) UWB Comm. (Data Portion, 6.8Mbps)

(**a**) (**b**)

Figure 8. Link budget of IEEE 802.15.4z HRP UWB (**a**) in the case of the preamble link and (**b**) in the case of a 6.8 Mbps data portion.

Our analysis results are summarized in Figure 9. The ranging coverage of UWB communication at a data rate of 850 kbps is more than twice as large as that for a data rate of 6.8 Mbps. The link performance of the preamble, which is the basis for synchronization and ranging, varies depending on the length of the preamble. In the IEEE 802.15.4z HRP UWB standard, the length of the preamble varies from a minimum of 16 to a maximum of 4096, and thus a longer preamble has larger ranging coverage than a shorter preamble.

However, since the message must be successfully received for two-way ranging, the characteristics of the preamble need to be appropriately adjusted to the link of the data portion. After the link balances of the DP section and the SYNC section are matched, it is possible to additionally increase the transmission power using the gating gain if the packet length is less than 1 ms. This has been confirmed to be consistent with the results of many experimental papers. The results of this study can help to adjust the physical layer parameters when installing actual HRP UWB products to increase the ranging coverage.

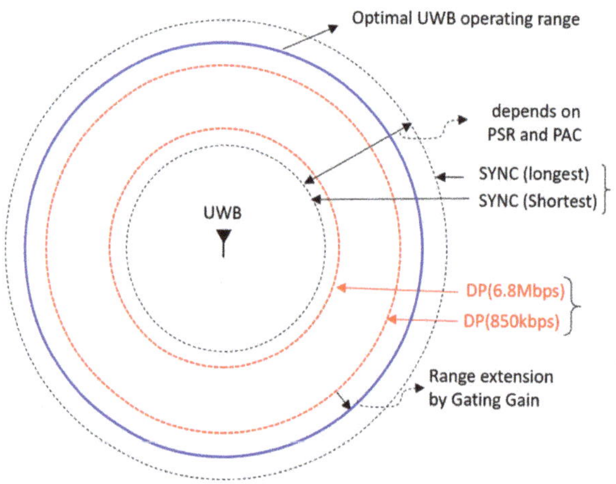

Figure 9. Summary of ranging coverage extension concepts for IEEE 802.15.4z HRP UWB.

4. Experimental Results and Discussion

In Section 3, the link performance required in the preamble and DP was theoretically analyzed. In this section, we verify the theoretical results through experiments using an actual HRP UWB module. In the experiments, we verify the performance changes according to the changes in PHY parameters such as the PSR, PAC, and data rate (850 kbps versus 6.8 Mbps) and the possibility of longer ranging. In particular, we verified the following results examined in the theoretical analysis through this experiment. First, since the link margin of the preamble is theoretically higher than that of the DP, errors may occur in the DP of the packet at a long distance, but the preamble-based synchronization signal will still be able to be received. Second, low-speed data (850 kbps) will show about 9 dB of performance gain compared with high-speed data (6.8 Mbps), as analyzed in the theory. Third, the processing gain will be increased according to the increase in the PAC size and PSR length, which will improve the distance measurement accuracy and stability. To verify these three points, the experiment was conducted as follows.

4.1. Experiments on Link Performances at a Fixed Distance

Figure 10 shows the hardware used in the experiment and the communication link's experimental environment at a fixed distance of 10 m. The Qorvo DW3000 HRP UWB IC was controlled by an NRF52840-DK microcontroller board, and the PHY parameters, such as the PSRs, PACs, and data rate, could be flexibly set through the embedded programming of the microcontroller. The output of the transmitter was adjusted in 1 dB units to reproduce various link conditions. In the first experiment, the SHR reception rate (HRR) and the packet reception rate (PRR), including the SHR and DP, were measured while changing the parameters and transmitter output at a fixed distance of 10 m to verify the link budget difference between the preamble and DP. Through this, the difference in the reception of the preamble and DP of the packet was quantitatively compared. The experiment used the SP0 packet format of IEEE 802.15.4z, and the RF parameters were set as specified in Table 4. In this experiment, the PSR count was fixed to 128, and the PAC count was fixed to 8. Qorvo's DW3000 UWB IC provides a function to adjust the transmit power to compensate for the difference between the RF output of the chip and the actual RF output and antenna loss [23]. For example, it supports a maximum output of −41.3 dBm/MHz while complying with the spectrum emission regulations and is adjustable from 0 dB up to −30 dB. In this experiment, packets were transmitted while adjusting the TX power gain of the transmitter in 1 dB units. In order to analyze the reception status of the HRP UWB IC and the success or failure of packet reception, register values such as Receiver PHY Header Detect (RXPHD) and Receiver PHY Header Error (RXPHE) of the system status register (SYS_STATUS) were monitored, and 200 packets were transmitted for each experimental condition. Through this experiment, we could find out whether the packet was received and the error rate of the DP and compare and analyze the PER performance between the preamble and the DP.

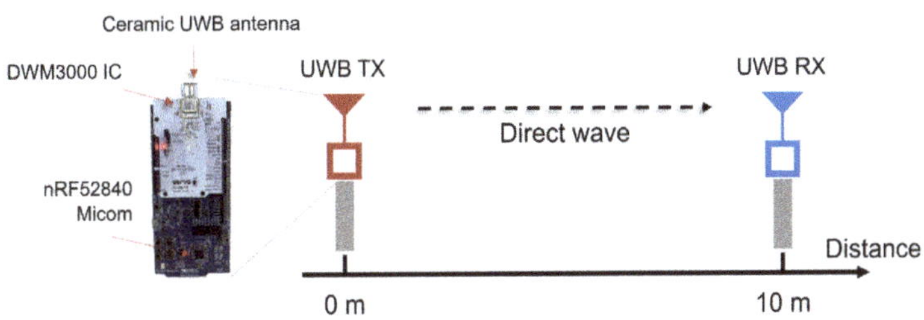

Figure 10. Link experiment set-up at 10 m using DW3000 UWB module.

Table 4. Configuration PHY settings in DW 3000 UWB module.

PHY Setting	Values
RF channel	CH5
PRF	64 MHz
PSR	32, 64, **128**, 256, 512, 1024, 2048, 4096
PAC	4, **8**, 16, 32
DP	850 kbps, 6.8 Mbps

Figure 11 shows the packet reception characteristics as a function of the transmission power through the coarse and fine gain adjustments in the DW3000 IC. The experimental results show that a reception success rate of about 80% or higher was achieved from −17 dB at 850 kbps, while the same reception success rate was achieved from about −11 dB at 6.8 Mbps. This means that the low-speed (850 kbps) data transmission provided about 6 dB of link gain compared with the high-speed (6.8 Mbps) data transmission. The reason why a gain of 6 dB was observed, which was slightly smaller than the theoretically suggested 9 dB, was probably due to the interaction between the synchronization performance in the SYNC (preamble) and the reception performance in the DP, which was not considered in the link analysis; that is, in the case of 6.8 Mbps, stable reception was possible in the SYNC, but due to an insufficient link margin in the DP, there was an overall link performance difference of about −6 dB, whereas in the case of 850 kbps, the reception performances of the SYNC and DP sections were similarly close, showing that the insufficient link margin of the SYNC also affected the DP performance. Through this, the link budget difference between the preamble and DP was clearly revealed in the experiment, and the link difference according to the change in data transmission speed could be quantitatively confirmed.

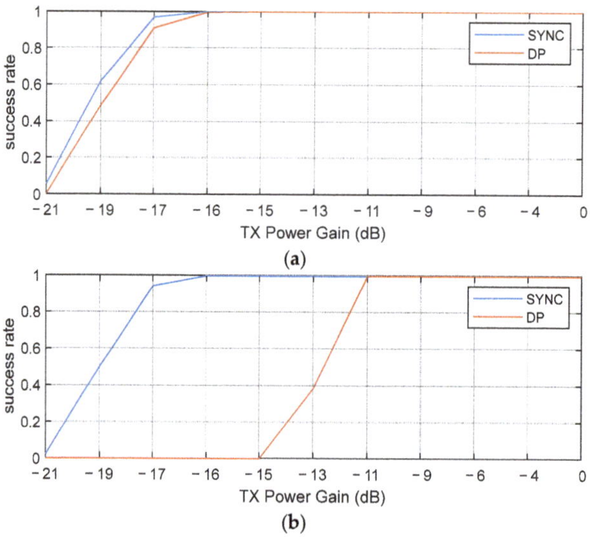

Figure 11. Experiment results of HRP UWB at 10 m using DW3000 UWB module: (**a**) low-speed (850 kbps) data transmission and (**b**) high-speed (6.8 Mbps) data transmission.

4.2. Maximum Ranging Experiment According to PSR and PAC Size

The second experiment was conducted to identify the maximum communication distance and stable ranging measurement conditions in an actual sports field environment. The measurement environment was the athletic field of Kookmin University in Seoul, Republic of Korea, as shown in Figure 12. The RF environment at the athletic field was verified to be free of active RF sources, and no significant interference was expected, as WiFi 6E operation is restricted to indoor use under local regulations in Korea. The maximum ranging distance satisfying a packet reception success rate of 90% was measured by changing parameters such as the PSR and PAC counts, and the distance values at that time were organized as median values. Through this process, it was possible to confirm how the processing gain according to the PSR and PAC counts affected the actual long-distance ranging performance. The experimental conditions are as shown in Table 4, and the same SP0 packet format as that in the first experiment was used, but the ranging measurement was performed with the DS-TWR method. The preamble used low-speed data transmission (850 kbps) to be suitable for long-distance ranging. Through these settings, it was possible to confirm how much the processing gain in the preamble contributed to the actual maximum ranging measurement. The coverage of ranging satisfying a success rate of 90% or more was compared and analyzed by changing the PSR and PAC values step by step.

Figure 13 shows the results for the maximum ranging distance according to the PAC size and PSR length. As can be seen in the figure, the interaction between the PACs and PSRs had a significant impact on the ranging performance. The ranging performance comparison according to the change in PAC size showed excellent performance in the order of 32, 16, 4, and 8 PACs, and in particular, the lowest overall performance was shown when the PAC size was 8. It was confirmed that the maximum ranging performance improved by 2–3 times just by changing the PAC size from 8 to 16. This confirms that the processing gain secured by the increase in PAC size greatly contributes to longer ranging coverage. In addition, Figure 13 shows that the link performance of the SYNC and DP was balanced at 256 PSRs, as analyzed in Figure 10, and there was no significant change even if the PSR count was increased beyond that. Therefore, this shows that the operating range could be further expanded through the gating gain after fixing the PSR count to 256. In an experiment where the output was increased after fixing the PSR count to 256, it was confirmed that stable ranging measurement was possible at a distance of more than 300 m.

Figure 12. Long ranging experiment set-up.

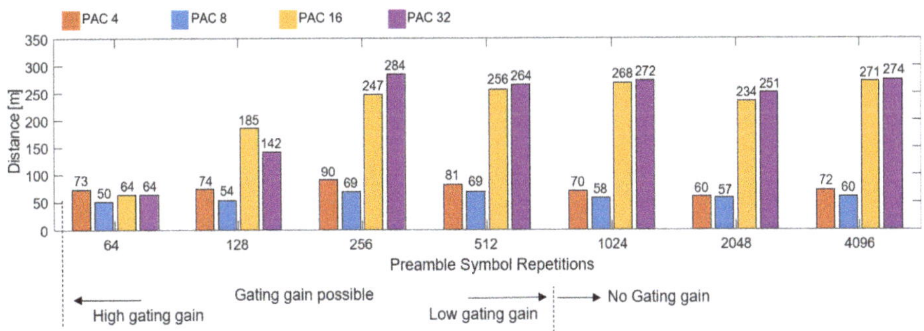

Figure 13. Maximum measurable distance according to variations in PSR and PAC sizes based on a 90% success rate.

As shown in Figure 13, the ranging coverage increased as the PSR count increased in the order of [64, 128, 256, 512, 1024, 4096]. However, the PSR effect was found to be interrelated with the PAC size. Specifically, when the PAC size was set to four or eight, the distance measurement performance tended to remain below 100 m overall. On the other hand, when the PAC size was set to 16 or 32, the distance measurement capability was significantly improved as the PSR count increased. In particular, the performance improvement effect was remarkable when the PSR count increased in the order of [64, 128, 256], and the best distance measurement performance was observed overall when the PAC size was set to 32. However, the performance improvement due to the increase from 16 to 32 was relatively limited compared with when the PAC size was increased from 8 to 16.

Through this experiment, it was found that increasing the PAC and PSR counts significantly improved the link margin in the SHR section, leading to a significant improvement in the maximum distance measurement. However, if the link margin of the DP was relatively insufficient, even if the gain was secured in the SYNC section, the success rate was rapidly reduced beyond a certain distance, showing a limitation. In addition, as the PSR count increased, the processing gain in the SYNC section increased, but the overall packet length increased, which may hinder resource utilization in terms of gating gain utilization and simultaneous access performance of multiple tags. This means that the link margins of the SYNC and DR sections are complementary, and the optimization of both areas must be performed simultaneously during system design to simultaneously maximize the actual long-distance measurement stability and performance. In summary, this experiment verified that the long-distance measurement performance and stable distance measurement capability can be significantly improved by adjusting the PAC size and PSR length of the preamble while keeping the DP speed at a low speed of 850 kbps. This shows that PAC and PSR parameter optimization is necessary when designing a UWB-based long ranging system.

4.3. Maximum Ranging Experiment in Indoor Environments

To evaluate the practical ranging performance of the proposed physical layer parameter adjustments under real-world indoor conditions, additional experiments were conducted in the hallway of the engineering building of Kookmin University in Seoul, Korea. The measurement environment consisted of a straight hallway (line-of-sight (LOS) dominant, with significant multipath effects) with a length of approximately 90 m. Multipath propagation in the straight hallway caused increased signal spreading, slightly affecting the maximum achievable ranging distance. However, in the NLOS corner area, preamble detection failures occurred much more frequently, leading to a substantial reduction in the maximum measurable distance.

Three different physical layer parameter configurations were tested to comprehensively assess the ranging performance. For each case, experiments were conducted at both the 850 kbps and 6.8 Mbps data rates to investigate the impact of the transmission speed on ranging performance. The details of the tested parameter settings are summarized in Table 5.

Table 5. Experimental parameter configurations for ranging performance evaluation.

Case	PSR	PAC	DP
Short	64	4	850 kbps, 6.8 Mbps
Middle	256	4	850 kbps, 6.8 Mbps
Long	4096	32	850 kbps, 6.8 Mbps

Figure 14 presents the maximum ranging distances achieved under each experimental condition overlaid on the building's floor plan. Different marker shapes and colors represent the physical layer settings; red circles indicate the short case, green squares represent the baseline (middle) case, and blue stars correspond to the long case. Solid markers denote the results at 850 kbps, while hollow markers represent the results at 6.8 Mbps. This visualization highlights the influence of the physical layer parameters and data rate on the achievable ranging distances in real-world indoor environments.

The experimental results are summarized in Table 6, which shows the maximum achievable distances under each configuration based on a packet reception success rate of 90%. At a data rate of 6.8 Mbps, both the long and middle cases exhibited a rapid decrease in ranging performance, with distance measurement failures occurring near 65 m. In contrast, the short case showed an earlier failure at approximately 62.3 m, which was slightly shorter than its 850 kbps maximum distance (68.8 m), likely due to the link budget limitations of the DP. When operating at 850 kbps, the ranging performance improved significantly, as the link budget of the DP was no longer the dominant limiting factor. Instead, the PSR became the main contributor to the ranging capability. As a result, the maximum achievable distances followed the order of short, middle, and long cases, directly reflecting the impact of the PSR length on preamble detection performance under multipath conditions. In particular, the long case successfully achieved distance measurements up to the end of the hallway (90 m) with high reliability, and the results suggest that even longer distances would be possible if the physical space allowed them. These observations confirm that the selection of both the DP parameters and the PSR critically impacted the maximum achievable ranging performance. Furthermore, it was demonstrated that reliable extended UWB ranging is achievable even in real-world indoor environments with significant multipath effects, such as a university campus hallway.

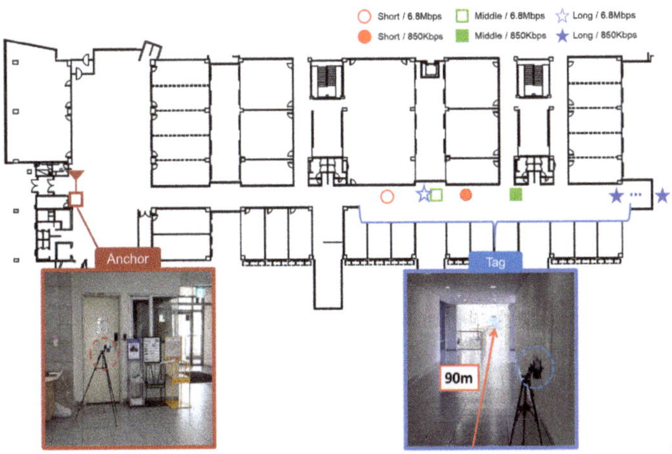

Figure 14. Maximum ranging distances mapped onto the hallway layout of the engineering building.

Table 6. Maximum achievable ranging distance.

Case	6.8 Mbps	850 Kbps
Short	62.3 m	68.8 m
Middle	66 m	71.77 m
Long	65.5 m	Above 90 m

5. Conclusions

In this study, we analyzed the link performance of the IEEE 802.5.4z HRP UWB standard, which has been commercialized in various fields recently, from the perspective of the link budget and proposed a method for configuring the optimal HRP UWB parameters for long-distance ranging up to hundreds of meters. Since UWB technology allows for various wireless link designs as long as they satisfy the technical criteria, we confirmed that ranging up to hundreds of meters is possible if designed optimally. The results analyzed in this study were tested with an actual commercial UWB module using a non-directional antenna. Therefore, if an actual directional antenna is used, then the possibility of long-distance ranging of hundreds of meters or more will increase even more.

From a system design perspective, it is important to recognize that, despite the wide bandwidth (~500 MHz) characteristic of UWB systems, significant link performance asymmetry exists between the synchronization header and the data portion due to their different modulation schemes. While conventional assumptions often treat UWB links as uniformly robust across the packet, our analysis and experimental results demonstrate that the weaker link of the two—typically the DP—ultimately constrained the maximum achievable ranging distance. Therefore, future UWB system designs and standard developments should explicitly account for this inherent link margin disparity. By addressing and compensating for the weaker link through adaptive modulation, coding, or dynamic parameter tuning strategies, further improvements in ranging performance, reliability, and coverage can be realized without violating regulatory constraints. In future UWB system designs, explicitly addressing the link margin imbalance between the SHR and DP sections will be key to maximizing performance and deployment scalability. We expect that UWB technology can be used in various applications that require long-distance ranging in the future.

Author Contributions: Conceptualization, S.H.; methodology, B.-J.J.; software, S.H.; data curation, S.H.; writing—original draft preparation, S.H.; writing—review and editing, B.-J.J.; visualization, S.H.; supervision, B.-J.J.; project administration, B.-J.J. All authors have read and agreed to the published version of the manuscript.

Funding: This research received no external funding.

Institutional Review Board Statement: Not applicable.

Informed Consent Statement: Not applicable.

Data Availability Statement: Data are contained within the article.

Conflicts of Interest: The authors declare no conflicts of interest.

Abbreviations

The following abbreviations are used in this manuscript:

UWB	Ultra-wideband
HRP	High-rate pulse repetition frequency
WPAN	Wireless personal area network
PSD	Power spectral density
PRF	Pulse repetition frequency
SHR	Synchronization header
DP	Data portion
STS	Scrambled time sequence
SFD	Start-of-frame delimiter
PSR	Preamble symbol repetitions
PAC	Packet acquisition chunks
LED	Leading edge detection
CIR	Channel impulse response
PHR	Physical header
RS	Reed–Solomon
BPM-BPSK	Burst position modulation–binary phase shift keying
SS-TWR	Single-sided two-way ranging
DS-TWR	Double-sided two-way ranging
SDS-TWR	Symmetrical DS-TWR
ADS-TWR	Asymmetrical DS-TWR
PER	Packet error rate
SNR	Signal-to-noise ratio
RMS	Root mean square
CRB	Cramer–Rao bound
ACPR	Adjacent channel power ratio

References

1. Jang, B.J. Principles and Trends of UWB Positioning Technology. *J. Korean Inst. Electromagn. Eng. Sci.* **2022**, *33*, 1–11. [CrossRef]
2. Al-Okby, M.F.R.; Junginger, S.; Roddelkopf, T.; Thurow, K. UWB-Based Real-Time Indoor Positioning Systems: A Comprehensive Review. *Appl. Sci.* **2024**, *14*, 11005. [CrossRef]

3. Bluetooth SIG. Bluetooth Core Specification, Version 5.0. Bluetooth Special Interest Group, 2016. Available online: https://www.bluetooth.com/specifications/specs/core-specification-5-0/ (accessed on 8 May 2025).
4. Gast, M. *Wi-Fi: The Definitive Guide*, 2nd ed.; O'Reilly Media: Sebastopol, CA, USA, 2005.
5. Sedlacek, P.; Slanina, M.; Masek, P. An Overview of the IEEE 802.15.4z Standard and Its Comparison to the Existing UWB Standards. In Proceedings of the 2019 29th International Conference Radioelektronika (RADIOELEKTRONIKA), Pardubice, Czech Republic, 16–18 April 2019; IEEE: Piscataway, NJ, USA, 2019; pp. 1–6. [CrossRef]
6. Corbalán, P.; Picco, G.P. Ultra-wideband Concurrent Ranging. *ACM Trans. Sens. Netw.* **2020**, *16*, 41. [CrossRef]
7. *802.15.4z-2020*; IEEE Standard for Low-Rate Wireless Networks—Amendment 1: Enhanced Ultra Wideband (UWB) Physical Layers (PHYs) and Associated Ranging Techniques. IEEE: New York, NY, USA, 2020. Available online: https://standards.ieee.org/standard/802_15_4z-2020.html (accessed on 8 May 2025).
8. Kabbinale, A.R.; Bansal, A.; Gopalan, K.S. Towards Next-Generation Ultra-Wideband Technology. In Proceedings of the 2023 15th International Conference on COMmunication Systems & NETworkS (COMSNETS), Bengaluru, India, 3–8 January 2023; IEEE: Piscataway, NJ, USA, 2023; pp. 280–287. [CrossRef]
9. Domuta, I.; Palade, T.P.; Puschita, E.; Pastrav, A. Timestamp Estimation in P802.15.4z Amendment. *Sensors* **2020**, *20*, 5422. [CrossRef] [PubMed]
10. *P802.15.4ab/D1*; IEEE Draft Standard for Low-Rate Wireless Network Amendment 1: Enhanced Ultra Wide-Band (UWB) Physical Layers (PHYs) and Associated Medium Access and Control (MAC) sublayer Enhancements. IEEE: New York, NY, USA, 2025. Available online: https://www.ieee802.org/15/pub/TG4ab.html (accessed on 8 May 2025).
11. Kim, C.J.; Han, B.K.; Lee, J.Y.; Jung, S.W.; Jang, B.J. Sensor Application Using UWB Communication: In-Vehicle Seat Occupancy Detection. *J. Korean Inst. Electromagn. Eng. Sci.* **2023**, *34*, 249–255. [CrossRef]
12. Ahmed, S.; Cho, S.H. Hand Gesture Recognition Using an IR-UWB Radar with an Inception Module-Based Classifier. *Sensors* **2020**, *20*, 564. [CrossRef] [PubMed]
13. Shah, S.; Chaiwong, K.; Kovavisaruch, L.-O.; Kaemarungsi, K.; Demeechai, T. Antenna Delay Calibration of UWB Nodes. *IEEE Access* **2021**, *9*, 63294–63305. [CrossRef]
14. Elsanhoury, M.; Mäkelä, P.; Koljonen, J.; Välisuo, P.; Shamsuzzoha, A.; Mantere, T.; Elmusrati, M.; Kuusniemi, H. Precision Positioning for Smart Logistics Using Ultra-Wideband Technology-Based Indoor Navigation: A Review. *IEEE Access* **2022**, *10*, 44413–44445. [CrossRef]
15. Ma, J.; Zhang, F.; Jin, B.; Su, C.; Li, S.; Wang, Z.; Ni, J. Push the Limit of Highly Accurate Ranging on Commercial UWB Devices. *Proc. ACM Interact. Mob. Wearable Ubiquitous Technol.* **2024**, *8*, 62. [CrossRef]
16. Feng, D.; Wang, C.; He, C.; Zhuang, Y.; Xia, X.-G. Kalman-Filter-Based Integration of IMU and UWB for High-Accuracy Indoor Positioning and Navigation. *IEEE Internet Things J.* **2020**, *7*, 3133–3146. [CrossRef]
17. Al-Okby, M.F.R.; Roddelkopf, T.; Burgdorf, S.-J.; Thurow, K. Multi-Tag UWB-based Indoor Positioning System for Objects Tracking. In Proceedings of the 2024 IEEE 22nd Jubilee International Symposium on Intelligent Systems and Informatics (SISY), Pula, Croatia, 19–21 September 2024; pp. 415–422.
18. Han, S.; Yoo, H.; Choo, H.; Jang, B.-J. IEEE 802.15.4z UWB Angle of Departure Tag Design for Indoor Positioning. In Proceedings of the 2023 53rd European Microwave Conference (EuMC), Berlin, Germany, 19–21 September 2023.
19. He, C.; Tang, Z. An UWB Positioning Algorithm Based on Clock Compensation and Filtering. In Proceedings of the 2023 Cross Strait Radio Science and Wireless Technology Conference (CSRSWTC), Guilin, China, 10–13 November 2023; pp. 1–3.
20. Cheng, J.; Jin, J.; Huang, Y.; Yu, P. Design of Positioning System for Underground Personnel and Electromechanical Equipment Using WSN. In Proceedings of the 2020 IEEE International Conference on Applied Superconductivity and Electromagnetic Devices (ASEMD), Tianjin, China, 16–18 October 2020; pp. 1–2.
21. Van Herbruggen, B.; Van Leemput, D.; Van Landschoot, J.; De Poorter, E. Real-Time Anchor Node Selection for Two-Way-Ranging (TWR) Ultra-Wideband (UWB) Indoor Positioning Systems. *IEEE Sens. Lett.* **2024**, *8*, 6002404. [CrossRef]
22. DWM1000—Qorvo. Available online: https://www.qorvo.com/products/p/DWM1000 (accessed on 8 May 2025).
23. Decawave. *DW3000 User Manual*, Version 1.1; Qorvo: Greensboro, NC, USA, 2019. Available online: https://www.qorvo.com/products/d/da008154 (accessed on 8 May 2025).
24. DW3120—Qorvo. Available online: https://www.qorvo.com/products/p/DW3120 (accessed on 3 April 2025).
25. Trimension SR150. Available online: https://www.nxp.com/products/SR150 (accessed on 3 April 2025).
26. Trimension SR040. Available online: https://www.nxp.com/products/wireless-connectivity/trimension-uwb/trimension-sr040-reliable-uwb-solution-for-iot:SR040 (accessed on 3 April 2025).
27. SPARK Microsystems. R1010 & SR1020 TRANSCEIVERS. Available online: https://www.sparkmicro.com/product/sr1010-sr1020-transceivers/ (accessed on 3 April 2025).
28. Garraffa, G.; Sferlazza, A.; D'Ippolito, F.; Alonge, F. Localization Based on Parallel Robots Kinematics as an Alternative to Trilateration. *IEEE Trans. Ind. Electron.* **2022**, *69*, 999–1010. [CrossRef]

29. Feng, D.; Peng, J.; Zhuang, Y.; Guo, C.; Zhang, T.; Chu, Y. An Adaptive IMU/UWB Fusion Method for NLOS Indoor Positioning and Navigation. *IEEE Internet Things J.* **2023**, *10*, 11414–11428. [CrossRef]
30. Novelda. Ultra-Low Power UWB Radar Sensors. Available online: https://www.xethru.com/xethru-development-platform.html (accessed on 3 April 2025).
31. Proakis, J.G. *Digital Communications*; McGraw-Hill: New York, NY, USA, 2000.
32. Dardari, D.; Conti, A.; Ferner, U.; Giorgetti, A.; Win, M.Z. Ranging with Ultrawide Bandwidth Signals in Multipath Environments. *Proc. IEEE* **2009**, *97*, 404–426. [CrossRef]
33. Stocker, M.; Brunner, H.; Schuh, M.; Boano, C.A.; Römer, K. On the Performance of IEEE 802.15.4z-Compliant Ultra-Wideband Devices. In Proceedings of the 5th International Workshop on Benchmarking Cyber-Physical Systems and Internet of Things (CPS-IoTBench 2022), Milan, Italy, 3–6 May 2022; IEEE: Piscataway, NJ, USA, 2022; pp. 28–33.
34. Großwindhager, B.; Boano, C.A.; Rath, M.; Römer, K. Enabling Runtime Adaptation of Physical Layer Settings for Dependable UWB Communications. In Proceedings of the 2018 IEEE 19th International Symposium on a World of Wireless, Mobile and Multimedia Networks (WoWMoM), Chania, Greece, 12–15 June 2018; IEEE: Piscataway, NJ, USA, 2018; pp. 1–9. [CrossRef]

sensors

MDPI

Article

Exploring Smartphone-Based Edge AI Inferences Using Real Testbeds

Matías Hirsch [1], Cristian Mateos [1] and Tim A. Majchrzak [2,3,*]

[1] ISISTAN (UNICEN-CONICET), Tandil 7000, Buenos Aires, Argentina; matias.hirsch@isistan.unicen.edu.ar (M.H.); cristian.mateos@isistan.unicen.edu.ar (C.M.)
[2] Faculty of Computer Science, Ruhr University, 44801 Bochum, Germany
[3] Center for Advanced Internet Studies (CAIS), 44801 Bochum, Germany
* Correspondence: tim.majchrzak@ruhr-uni-bochum.de

Abstract: The increasing availability of lightweight pre-trained models and AI execution frameworks is causing edge AI to become ubiquitous. Particularly, deep learning (DL) models are being used in computer vision (CV) for performing object recognition and image classification tasks in various application domains requiring prompt inferences. Regarding edge AI task execution platforms, some approaches show a strong dependency on cloud resources to complement the computing power offered by local nodes. Other approaches distribute workload horizontally, i.e., by harnessing the power of nearby edge nodes. Many of these efforts experiment with real settings comprising SBC (Single-Board Computer)-like edge nodes only, but few of these consider nomadic hardware such as smartphones. Given the huge popularity of smartphones worldwide and the unlimited scenarios where smartphone clusters could be exploited for providing computing power, this paper sheds some light in answering the following question: Is smartphone-based edge AI a competitive approach for real-time CV inferences? To empirically answer this, we use three pre-trained DL models and eight heterogeneous edge nodes including five low/mid-end smartphones and three SBCs, and compare the performance achieved using workloads from three image stream processing scenarios. Experiments were run with the help of a toolset designed for reproducing battery-driven edge computing tests. We compared latency and energy efficiency achieved by using either several smartphone clusters testbeds or SBCs only. Additionally, for battery-driven settings, we include metrics to measure how workload execution impacts smartphone battery levels. As per the computing capability shown in our experiments, we conclude that edge AI based on smartphone clusters can help in providing valuable resources to contribute to the expansion of edge AI in application scenarios requiring real-time performance.

Keywords: edge AI; smartphones; cluster computing; energy efficiency

1. Introduction

The application of DL to perform on-site tasks is rapidly expanding. In the healthcare domain, DL is used for predicting, detecting, and diagnosing different diseases [1,2]. In food production, particularly animal farming [3], these tasks include animal behaviour recognition, growth evaluation, and individual identification. In the smart city domain [4,5], tasks include intelligent transportation, waste management, crime prediction, public infrastructure, and green space maintenance. Many of these tasks involve object classification and recognition using images as input. The application of CNNs (Convolutional Neural Networks) has become popular to perform these tasks due to the training procedure requiring minimal human intervention compared to other machine learning techniques. However, training a deep neural network model requires big datasets and, thus, powerful hardware. Nowadays, data gathering leveraged by mobile and IoT devices as well as computing power offered by cloud computing platforms greatly speed up training processes. Cloud computing is considered not only in the training phase of DL models but also for making inferences. It implies that a trained model deployed in a remote computing infrastructure is evaluated with locally generated input data, and produces output results that must be transferred back traversing long-latency internet uplinks and downlinks. This latency has been identified as a limitation for time-sensitive applications. To overcome this issue and others such as the unavailability of an internet connection, network overloading, and privacy protection, edge-based architectures have been proposed [6,7].

There are differences highlighted in the literature when referring to edge-based architectures [6]. Some of these are simply terminological, while others describe substantial differences in the way their constituent components are defined and interoperate with components from other computing layers [8]. Points in common, however, are minimizing data transportation and bringing computing resources close to where computing needs and data originate. Heterogeneity and energy constraints are common features of computing infrastructures operating in cyber–physical and IoT systems. For these reasons, cooperation is a key concept of Dew Computing [9], which is among these edge-based architectures and provides low processing power and consumer devices with the role of main computing resource providers. Among these consumer devices are smartphones. According to [10], nearly 84% of the current world's population owns a smartphone. Modern smartphones are equipped on average with up to eight cores, powerful GPUs, and several gigabytes of RAM, which are typically underutilized [11], representing underexploited ubiquitous computing cycle reservoirs. A theoretical study [12] estimated that the computing power of 3 million Samsung Galaxy S8 smartphones can be compared to that of 166 thousand servers present in data centers like those owned by Google or Amazon. Other studies [13–15] show that carbon-efficient high-performance servers for different purposes, including training complex DNNs, can be built by repurposing old smartphones.

While other works in the field show that AI inferences using heavy DL networks are possible by applying DL model partitioning techniques using SBC clusters connected via wired LANs, in this work, we focus on empirically evaluating the feasibility and effectiveness of wireless collaboration among smartphones for real-time execution of computer vision tasks using DL and applying an intuitive data partition scheme over data stream workloads. The key contributions of this work can be summarized as follows:

- We considered and characterized different workloads associated with distinct application domains with a level of realism that no other study in the field has utilized before. The workloads comprise the usage of three pre-trained DL models capable of being fully loaded and executed in resource-constrained devices for real object recognition and classification, and real input in the form of image sets given in a stream-like fashion emulating different processing needs and application scenarios.
- We comprehensively measured, over real testbeds, the accumulated inference latency and energy consumption that heterogeneous smartphone clusters achieve compared to low-power nodes typically used in edge and IoT applications scenarios, discussing implications and future research necessary to popularize this form of edge computing to application scenarios beyond the illustrative examples shown in this work.
- Load balancing mechanisms heavily affect the performance of HPC systems, with smartphone clusters being no exception to this rule, with the addition that concerns such as battery-related concerns characterize this type of infrastructure. For this

reason, we included distinct state-of-the-art load balancing mechanisms in the area in all evaluations associated with smartphone clusters.

- The node and workload setups involved in our experiments represent working examples and a guide on how to tailor and use our experimentation methodology, composed by freely previously published and available tools, specially designed to facilitate the research and experimentation with real smartphone cluster testbeds.

This work is organized as follows. Section 2 presents and classifies recent efforts backed by real testbed results to speed up inferences at the edge using different approaches for distributed execution. Section 3 describes the mobile distributed computing model and infrastructure-related architectural assumptions we made in our exploration tests, as well as details concerning DL models used and data stream workloads. In Section 4, we first explain details about node setups and metrics, and then we present the results obtained in our empirical evaluation. We discuss limitations of our work along with future research directions in Section 5. Lastly, Section 6 presents the conclusions.

2. Related Works

In this section, we analyse relevant works in line with the broad concept of *distributed inference*, which is present in recent research related to AI at the network edge. Broadly, in the context of AI, an inference means asking a model to compute a result based on input data (e.g., given an image, list the objects detected). At this point, it is worth differentiating works that employ AI, such as DRL or evolutionary metaheuristics for improving resource utilization at the network edge [16], from works targeting AI task execution, e.g., CV tasks using distributed computing infrastructures at the network edge. In the first group, proposals mostly focus on innovatively modelling variations of NP-hard edge resource allocation problems, while solutions are presented as optimal or near-optimal and compared against others commonly via simulation. In the second group, the focus is on exploiting intrinsic AI tasks characteristics to evaluate different performance metrics of execution schemes that exploit collaboration among distributed nodes, usually exploiting parallel processing opportunities using real testbeds as the evaluation methodology. This section classifies works that belong to the second group. When computing infrastructures of at least two nodes are used to perform inference tasks, it is said that distributed inference is applied. Table 1 shows a first-cut classification of distributed inference schemes, differentiating among vertical [17–21] and horizontal [22–26] ones.

Table 1. Schemes for distributed AI inference.

Distributed Inference Scheme	Associated Works
Vertical	[17–21]
Horizontal	[22–26]

Under a vertical scheme, collaborating nodes typically belong to different computing layers of the edge architecture. These layers are organized in a hierarchical structure where nodes in a layer render services to those in the layer below using services provided by nodes in the layer above. Nodes at the base of such a hierarchy commonly have much less powerful computing capabilities and use the services of layers above to complete computationally intensive tasks. On the other side, nodes at the top layer present more powerful computing capabilities, like fog or cloud resources. However, reaching such nodes results in long-latency communication when transferring task input/output data. For this reason, its usage should be carefully analysed, especially for applications with delay-sensitive time constraints. A classical example of this type of collaboration is given when a resource-limited user device leverages the services of an edge server or edge layer to partially or completely compute the results of inference tasks. Since the edge layer can render computing services to a variable number of users in a locality, depending on the burden at the time of receiving tasks, the edge layer can have enough resources to complete or partially perform a task. In the later case, the edge layer would eventually leverage the services of a remote cloud datacenter to fulfil the request while meeting potential task deadlines. Like in [17,20], offloaded decisions are evaluated considering communication throughput between computing layers. In addition, a task requiring the execution of a DL model is split into several parts. In this splitting process, the authors propose considering insights of data anatomy and computing requirements of layers characterizing different DL models. Partitions of model's graph are dynamically assigned and executed between a device and remote servers. Other works adopting this distribution scheme are [18,19,21]. Several works provide evidence that vertical collaboration is feasible for allowing resource-limited devices to execute computationally intensive tasks, such as CV tasks using large DL models. However, a major drawback relates to the long-latency communication dependency when offloading tasks to upper resource layers that provide the required computing capability to complete the tasks.

Another inference distribution scheme is when nodes within the same layer collaborate in the execution of computationally intensive tasks. This scheme is known as horizontal collaboration. A key difference with vertical collaboration is that participating nodes and data rarely occur outside the scope of an LAN or WLAN. Even though this form of collaboration governs the internal communication of nodes within the same layer, in this paper, we put special focus on the horizontal communication of resource-constrained nodes, including edge and user device nodes. In the literature, we have identified several works proposing distributed inference under this form of collaboration [22–26]. We further analysed these studies, taking into account several dimensions which were summarized in Table 2: the type of parallelism proposed, i.e., whether it is based on intervening in the anatomy of a DL model, partitioning the data input, or adopting a hybrid approach by combining the previous techniques. Another dimension of analysis relates to the methodology used to evaluate the proposal. We selected works whose methodology involved setting up real testbeds and we describe the hardware used and the way resource heterogeneity is present in the experiments. Moreover, with it being provided that QoS is a concept that relates to adapting to dynamic changes in the workload and/or available resources, we indicate the support of these works in this respect. Finally, we contextualize the strategy of each work for resource management, mentioning the main motivation behind the proposal.

Works can be also classified according to how workload execution is parallelized. We found three approaches of parallelism: model partitioning, data partitioning, and hybrid. Within the former group, layers composing a DL model are deployed into different devices so that outputs computed by a device are the inputs for another one. With this type of parallelism, all participating devices are involved in computing inferences for all portions of the input. By contrast, when using data partitioning parallelism, edge nodes run the whole model over portions of the input. A third type, which we call hybrid, parallelizes workload by combining the data and model partitioning schemes. In all cases, results for the whole input are obtained by joining partial results calculated by all participating devices.

When several devices participate in inferencing tasks, two device-related questions emerge: Does the approach consider executing inferences on heterogeneous devices? If so, what does the approach do to deal with the performance differences due to executing the same workload on devices with different computing capabilities to ensure certain QoS levels?

Table 2. Approaches for horizontal distributed AI inference at the edge.

Parallelism Type	Experiments		Online Adaptive QoS Support?	Main Motivation	Work
	Resource Heterogeneity Type	Type			
Model Partitioning	Homogeneous nodes	Real Hardware	No	Automate large CNN model partitioning, deployment, and execution on multiple resource-constrained edge devices	[22]
	Homogeneous nodes	Real Hardware	Yes	Accelerate performance with an intra-layer parallelism instead of layers sequential execution as in traditional hierarchical inference	[23]
Data partitioning	Heterogeneous nodes	Real Hardware	Yes	Exploit the concept of approximate computing to ensure performance and accuracy constraints considering heterogeneity of the edge cluster and node availability	[24]
	Heterogeneous nodes	Real Hardware	Yes	Compare inference delivery capabilities and energy efficiency of IoT devices versus clusters of smartphones	This work
Hybrid	Homogeneous nodes	Real Hardware	Yes	Allow large DL models to execute in resource-limited devices achieving low memory footprint and execution latency	[25]
	Heterogeneous nodes	Simulation and real hardware	Yes	Reduce mobile edge computing server's computing pressure	[26]

In [22], the authors propose AutoDiCE, a tool for partitioning, deploying, and distributively executing a large CNN model on multiple resource-constrained edge devices. The platform specification and mapping specification allow practitioners to indicate which sub-model runs on which edge devices. Moreover, once generated and deployed, mappings between sub-model and edge nodes are fixed at runtime. Communication between sub-models and resources usage is implemented via the well-known parallel programming standards MPI and OpenMP. Although the tool was designed to support horizontally and vertically distributed inferences, it was tested using the former scheme, using an infrastructure comprising homogeneous edge nodes communicated via a gigabit Ethernet switch. Particularly, they use different LAN settings with up to eight edge devices (Nvidia Jetson Xavier NX). DeColla [23] utilizes Deep Reinforcement Learning (DRL)-based QoS adaptive allocation for horizontal collaborative inference, complemented with a fault tolerance mechanism implemented using a message queue. All DL layers are present in all collaborating IoT devices. Neuron calculation in each layer occurs in parallel, coordinated via a message queue and orchestrated by the DeColla engine, which operates in a requester IoT device. The number of IoT devices, available computing resources, and the network state are evaluated online upon each neuron assignment. The experimental setting was an LAN with three Raspberry Pi 4 models.

In [24], the most appropriate version among different pruned models is selected at runtime via an adaptive workload distribution to meet the required application-specific accuracy and performance level. Such dynamic selection is performed using base knowledge of accuracy–performance tradeoffs profiled from heterogeneous edge nodes, together with a runtime monitoring system for resource management. Evaluation was performed over an LAN with four IoT devices (2x Odroid XU4, Jetson Nano, Rapsberry Pi 4). In [25], the DeepThings framework for executing large CNN models in resource-constrained devices was proposed. The framework adopts a hybrid parallelization approach due to its combining model and data partitioning schemes. On one side, as an offline step, the framework fuses convolutional and pooling layers to create independent and distributable execution units, which leads to reduced memory footprint and allows resource-limited devices to contribute via the computation of partial results that are combined to produce the final result of an inference task. In addition, the workload, represented by a single frame, is divided into overlapping and distributable regions that are computed collaboratively by a group of edge devices. With this scheme, inference capability differences of participating devices are handled with two strategies. One of these is static and represented by the fact that memory requirements of fused layers executed by a device adapt to its available memory resources. The other strategy is that data regions are dynamically distributed according to device computing capabilities by following a work stealing scheme. The experimental setup comprises a WLAN with up to six IoT devices (Raspberry Pi 3). Another work adopting hybrid parallelization is eDDNN [26], which leverages cross-platform web technologies and the WebRTC protocol to allow for distributed inference among heterogeneous end devices (smartphones). The collaboration is mediated by an edge server that keeps track of node and task status information. Tasks, determined by images input size, and the DL model to be used, are divided into sub-images and sub-models by a decision-maker component that coordinates the execution of different pieces among end devices using a shared dependency table and dynamic node information. Edge data duplication is present at the step of dividing an image into sub-images. When distributing a convolutional operation by dividing the input and the model, duplication is required for ensuring the correctness of results. To evaluate the approach, they mostly rely on simulation, with data profiled from hardware running real workloads, and they also include some experiments with real settings.

In the discussed works, horizontal distributed inference based on the collaboration of battery-operated end-user devices, such as smartphones, combined with data partitioning schemes is rare, meaning that the benefits and limitations of performing real-time CV tasks—and possibly other AI tasks as well—are as yet unknown. Our work sheds light on providing empirical evidence for quantifying the computing power of smartphone clusters. Existing approaches exploiting data partitioning schemes add redundant data to the original input for performing distributed inference. We propose a simple yet effective adaptive data partitioning scheme for performing real-time CV tasks, which works under a pull-based scheme where tasks are extracted by devices from a shared queue. This is evaluated under varying settings of real smartphones and SBCs cooperating in WLAN clusters. Since performance may vary according to how workload is distributed, we provide insights on the performance achieved and energy-related measurements using different state-of-the-art online heuristics and smartphone clusters. Moreover, performance comparisons of different <smartphone clusters, load balancing heuristics> combinations are compared against the performance achieved by different edge nodes commonly used in IoT applications. All experiments performed in this work demonstrate the viability of constructing a platform for in-vivo experiments whose main components have been previously published, validated, and made publicly available for use [27–29].

3. System Model and Assumptions

We envision a system that opportunistically scavenges computing cycles of nomadic hardware, such as smartphones, to complement—or eventually replace—the computing cycles delivered by low-power edge nodes. Particularly, we shed light on the computing capability of different smartphone clusters and load balancing heuristics for processing streams at the edge. In this work, we consider streams of images analysed using CV tasks in real-time or near real-time by harnessing deep neural networks. Defining appropriate incentives for smartphone users—tasks executors, privacy preservation, and security mechanisms for task submitters and task executors—is outside the scope of the current work [30–32].

We assume a master–worker architecture operating within the scope of WLAN, with a master node officiating the tasks performed by submitter and worker nodes contributing with computing resources, i.e., playing the role of tasks executors (see Figure 1). A master node is assigned the roles of data concentrator and task coordinator. Playing the first role implies that the node exposes end points for receiving raw data and preprocessed data. On one side, raw data transferred to a master node are captured by devices with sensing capabilities, e.g., RGB, IR, or depth images taken with different camera devices. On the other side, pre-processed data are obtained by applying a series of computational operations to decode, synthesize, and/or represent raw data in a format that can be more efficiently stored, retrieved, or analyzed. Moreover, playing the role of task coordinator means that the node runs logic to decide how to distribute the load involved in preprocessing raw data by available worker nodes to meet the expected QoS. Finally, worker nodes opportunistically connected to the WLAN register through a master node service to contribute with computing resources and wait for tasks assigned by the master node. As task completion occurs at the worker nodes, results are sent back to the master node.

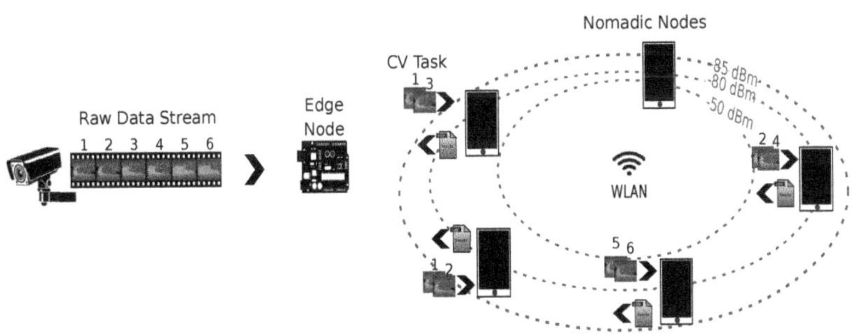

Figure 1. Schematic overview of main nodes.

With this architecture, we aim to measure the feasibility of using different smartphone groups cooperating under the same WLAN for performing CV tasks, and compare the performance achieved with that of SBC-like edge nodes commonly used for servicing IoT and edge applications.

3.1. *Empirical Evaluation*

We employed different hardware for the edge nodes, whose characteristics are shown in Table 3. All nodes have multicore processors. All nodes but the Nvidia Jetson Nano have 802.11 WiFi radio, with a theoretical bandwidth capacity between hundreds of Mbps (norm n) to a few Gbps (norm ac). In fact, SBC node WiFi was not used because, in our experiments, these nodes play all the roles—task submitter, task coordinator, and task executor—simultaneously and no data transfers are involved. The main memory of the edge nodes ranges from 2 to 8 GB. In our setups, SBCs operate plugged into the electricity power grid, while smartphones operate using their battery as the main power source. Additionally, we include an estimated per unit acquisition cost in USD. When searching for cost-effective configurations, the cost can be used as a deciding factor in situations where several node combinations achieved the desired performance.

Table 3. Edge computing nodes: hardware features.

Edge Node Type	Edge Node Name	Short Name	Processor	RAM (GB)	Conectivity	Battery Capacity (mAh)	Avg. Price (USD)
SBC	Raspberry Pi 4	RPi4	Quad-core 1.5 GHz ARM Cortex-A72	4	Wireless	N/A	55
	Nvidia Jetson Nano	NvJN	Quad-core ARM Cortex-A57 MPCore	4	Wired	N/A	99
	Gigabyte Brix GB-BRi5H-8250	GBx	Quad core i5-8250U	8	Wireless	N/A	500
Smartphone	Samsung A02	SamA02	Quad-core 1.5 GHz Cortex-A53	2	Wireless	4900	116
	Motorola Moto G6	MotG6	Octa-core 1.8 GHz Cortex-A53	3	Wireless	3000	160
	Samsung A30	SamA30	Octa-core (2 × 1.8 GHz Cortex-A73 & 6 × 1.6 GHz Cortex-A53)	3	Wireless	3900	170
	Xiaomi Redmi Note 7	XRN7	Octa-core (4 × 2.2 GHz Kryo 260 Gold & 4 × 1.8 GHz Kryo 260 Silver)	4	Wireless	4000	200
	Motorola Moto G9 Play	MotG9	Octa-core (4 × 2.0 GHz Kryo 260 Gold & 4 × 1.8 GHz Kryo 260 Silver)	4	Wireless	5000	200

Section 3.2 elaborates on the deep learning model characteristics considered in the experiments. In Section 3.3, we describe three study cases belonging to different application domains, how image streams are produced, and task characteristics. Finally, Section 4.1 presents the comparison results based on the designed benchmarks, which we explain in the following sections.

3.2. DL Models Benchmarking

Benchmarking is a relevant task to approximate node performance in solving a specific task and this procedure was carried out for nodes in Table 3, where tasks refer to the making of inferences using different Tensorflow DL models. Tensorflow is a popular library for creating, training, evaluating, and executing deep learning models. The Tensorflow project [33] offers a benchmarking tool to collect statistical information of a `.tflite` model, one of the file formats for created models. The tool measures the time a model takes to produce an inference and the memory consumed, among other information using randomly generated synthetic model input. It is worth mentioning that all inference times reported in this paper are on the basis of using CPU support. Even though, today, there is specialized hardware—sometimes referred to as AI hardware accelerators—to run AI logic faster than with general purpose hardware like CPUs, there are some practical limitations, including library support for different platforms and proprietary chips embedded only in high-end smartphones of certain brands, which still prevent the massive use of mobile devices for the distributed computing we have in mind.

Moreover, in our study, the notion of *real-time* is context-dependent. Depending on the application, a constant delay of a few seconds might satisfy the notion of real-time. For this reason and with the aim of evaluating smartphone-based settings in different application scenarios, i.e., with different real-time requirements, we used three DL models that present quite dissimilar inference times. It is not part of the evaluation core to test DL models in their most efficient version—i.e., we do not aim to test and improve model performance per se. Rather, our objective is to explore the feasibility of using smartphone-based opportunistic clusters as alternative edge devices for completing resource-intensive AI workloads of different characteristics. To introduce variability in the handled workloads, we chose three DL models that could be fully loaded using resources (e.g., RAM space) present in devices like smartphones and SBCs; this resulted in quite dissimilar inference times, which can be seen from Figures 2b, 3b and 4b. In addition, the three models were subject to distinct image generation rates depending on the needs of the application domain. In addition, given that these models were conceived to be used for different CV tasks, their selection allowed us to tailor feasibility tests targeting different application domains, which is a research gap in the field of study.

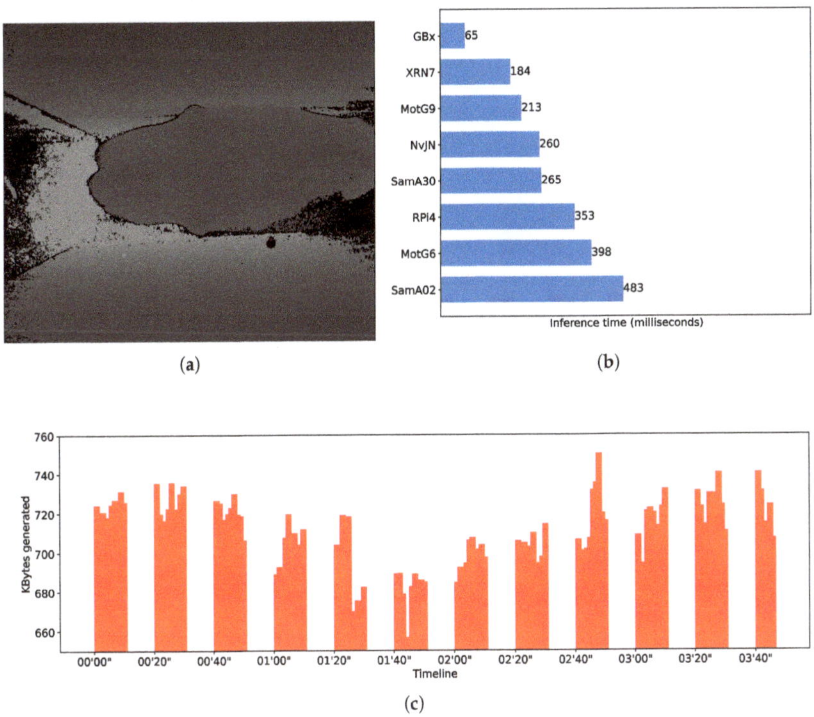

Figure 2. BCS—workload characterization. (**a**) Input image example taken using the Kinect v2 depth camera. For reference, the image partially shows the cow's crest/neck on the right, the rump/tail head on the left, and the walls of the runway to the milking parlor on the top and the bottom. (**b**) SqueezeNet model benchmarking. Numbers represent inference time per depth image in milliseconds. (**c**) BCS data stream excerpt.

All models used in the tests are pre-trained object recognition models that take images as input and produce text as output. Text refers to bounding boxes, classes, and confidence levels of recognized objects, or categories of a unique object depending on the purpose the model was trained for. To test different mobile distributed computing scenarios, we intentionally selected models that, when used to make inferences, present dissimilar computing times. One of the models is YoloV4 [34], which is used to perform real-time object detection. Figure 3b shows average inference times (in milliseconds) for different nodes when performing object detection using a YoloV4-tiny model, which is the version that can be run on mobile devices using Tensorflow. We see, for instance, that a smartphone Xiaomi RN7 makes inferences in around 248 milliseconds—i.e., it is able to process up to 4 FPS, doubling the capability of a RaspberryPi4—but it is slower than a Gigabyte Brix, which reaches nearly 10 FPS. Another of the benchmarked models encapsulates expert knowledge of what is known as the Body Condition Score (BCS) [35] and was proposed to score dairy cows to differentiate healthy and non-healthy cows. The model uses SqueezeNet as its base neural network. Figure 2b shows the BCS average inference time for different edge nodes. For instance, the Xiaomi RN7 smartphone completes an inference in around 184 milliseconds; i.e., it is able to deliver barely more than 5 FPS, while a Gigabyte Brix node is around 15 FPS. The third model used is based on an EfficientNetB4 and was trained to recognize progression of diabetes using foot images as input [36]. Figure 4b shows the inferences times obtained by different edge nodes. By comparing the fastest smartphone with the fastest SBC, it can be noted that Xiaomi RN7 completes an inference in around 1470 milliseconds while the Gigabyte Brix does the same in approximately 559 milliseconds. When comparing performance of DL models, notice that for most of the edge node benchmarks, inference times differ up to one order of magnitude. As said, the Xiaomi RN7 completes inferences in 184, 248, and 1470 milliseconds for the SqueezeNet, the YoloV4, and the EfficientNetB4 models, respectively.

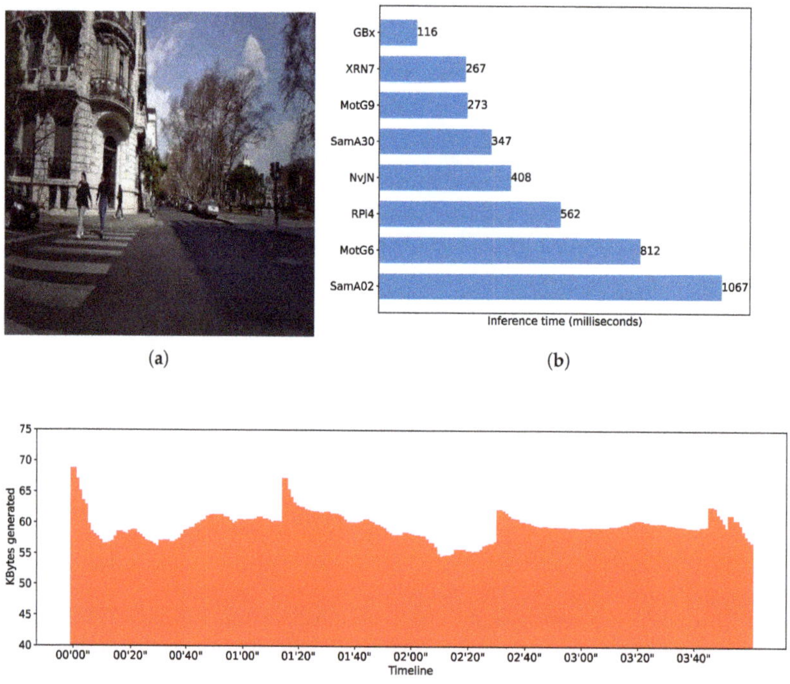

(a) (b) (c)

Figure 3. SWT—workload characterization. (**a**) Input image example. (**b**) YoloV4 tiny model benchmarking. Numbers represent inference time per image in milliseconds. (**c**) SWT data stream excerpt.

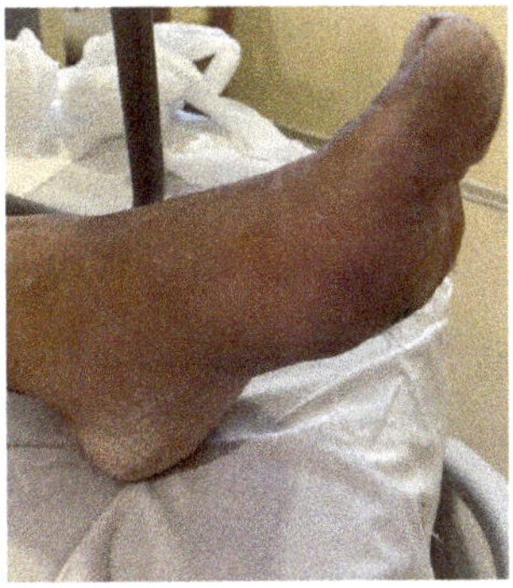

(**a**)

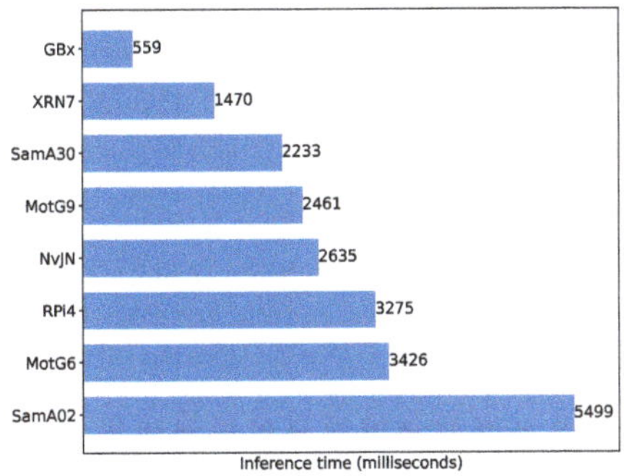

(**b**)

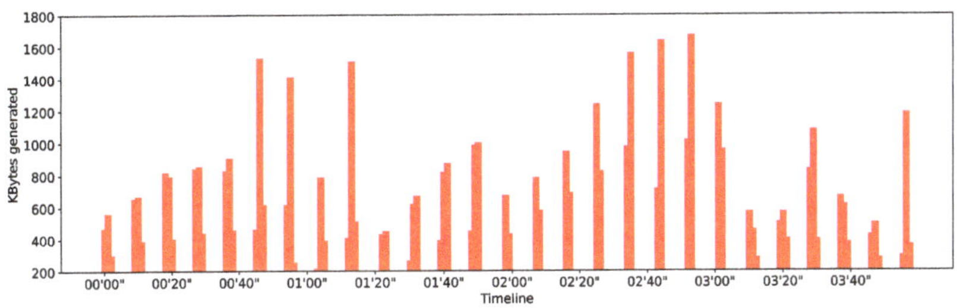

(**c**)

Figure 4. DMED—workload characterization. (**a**) Input image example. (**b**) EfficientNetB4 model benchmarking. Numbers represent inference time per image in milliseconds. (**c**) DMED data stream excerpt.

3.3. Workload Characterization

In this section, we describe the workloads associated with the machine vision tasks that were performed over stream-like inputs. Indeed, machine vision tasks such as object detection and classification using CNNs are commonly applied in diverse domains. To complete the experimental setting, we consider video frame streams of different lengths and distinct frame rates, which serve as input to the CNN models described above and, in conjunction, both element0s shape the heterogeneous workloads of the mobile distributed computing setups under evaluation.

Without losing generality, we assign each model–stream pair different domain/usage contexts. These domains are a dairy farm application scenario called "Body Condition Score" (BCS), a smart city application scenario that we call "Sense while travel" (SWT), and a human healthcare monitoring application scenario called "Disease Monitoring and Early Diagnosis" (DMED). Next, we describe the dynamics of each scenarioin order to explain to the relevant stream-like input:

- Body Condition Score (BCS): In a dairy farm, outfitted cows—i.e., cows that are either overweight or underweight—tend to produce less milk and can breed less frequently than those properly weighted. Identifying such animals so as to give them proper treatment is crucial to maintain the full productivity of cows. To detect such animals, a CNN model is used, which takes depth images taken from the top with a camera strategically positioned as cows walk to a milking parlor [35]. The image capturing procedure is performed within a time window that does not exceed 10 s. For the model to produce accurate results, it is important that the animals do not stand directly under the camera and that the captured images contain a specific part of the cow; the result is that a high percentage of captured images, say 50%, is dropped, i.e., tagged and filtered, as these images do not represent useful input for the CNN model. Moreover, during the transition of one animal to the next under the lens of the camera, which can take 10 s on average, it is reasonable to stop capturing images. To make the body condition score calculus feasible, reliable, and energy efficient, no less than one hundred frames are required to serve as input for the CNN model, for each cow. All these constraints configure a stream processing scenario where depth images can be produced at a rate of 15 FPS during 10 s, followed by another 10 s without image capture. We decided that an atomic CV task would be represented by the burden of applying the CNN model to fifteen consecutive frames, meaning that a CV task is created every second, which gives a total of ten CV tasks created for a cow. Considering that the dataset used to recreate this scenario includes 1740 images, the resulting workload stream comprises 116 CV tasks given within a time window of 3 min and 34 s.

- Sense while Travel (SWT): A city can take advantage of vehicles such as urban line buses to collect information about relevant street events while travelling, e.g., for statistical purposes, crime detection, and infrastructure maintenance planning. Images captured with a bus front camera might feed a YOLO model, which in turn gives a plain text representation of objects present in an image and the accuracy percentage of each detected object class as output. An energy efficient way to achieve this without having to process a large number of frames with redundant information is to cleverly select a sample rate according to the expected moving change of detection targets. We landed on a sample rate of 2 FPS and evaluate the system performance for a stream duration of 30 min. These parameters mean sensing data for 15 km assuming a vehicle travelling the road at 30 km/h. A CV task is represented by the execution of the CNN model on two consecutive filtered frames, meaning that a CV task is created every second. Considering that the dataset used to recreate this scenario includes 3600 images, the resulting workload stream comprises 1800 CV tasks given within a time window of 30 min.
- Disease Monitoring and Early Diagnosis (DMED): We envision a remote monitoring mobile health scenario where a CNN model encapsulates expert knowledge on different pathologies to assist people, especially older people, receiving intensive care services; the scenario also involves analysing different health parameters as a way to prevent certain pathologies and monitor pre-existing illnesses. We consider a scenario where a caregiver or nurse rendering services in an elderly residence is assigned with the task of taking foot tissue pictures for all patients, resulting in a data stream of 10 FPS during 2 s, with pauses of 7 s without capturing images. In this case, a CV task executes the model described in [36] to two consecutive frames, meaning that during the image capturing time, five CV tasks are created per second. Considering that the dataset used to recreate this scenario includes 600 images, the resulting workload stream comprises 300 CV tasks given within a time window of 8 min and 55 s.

Figure 2a,c show a timeline representation of the data input stream shape, i.e., the kilobytes of data transferred per second, and an example of the input for the CNN used in the Body Condition Score scenario. Similar information can be found in Figures 3a,c and 4a,c for the SWT and DMED scenarios, respectively. In all cases, we depict the first four minutes of the three image streams, and we report per-image inference times.

Irrespective of the application domain, these example application scenarios present streams with different characteristics due not only to the dynamics of input generation but also to the computing capability required to perform inferences over the input. In addition, it is assumed that inference results should be available at the master node with minimum latency, i.e., as close as possible to real-time.

4. Experiments Setup and Metrics

For setting up and executing tests, we employed a tool set combining several open-source programs [27,28] and hardware [29] modules. The tool set allowed us to automate tasks involved in experimenting with smartphones and real workloads including cluster formation, device preparation (battery charge/discharge actions), deep learning model deployment, stream dynamics reproduction, and result collection and summarization.

Algorithm 1 details all steps involved in running the experiments over real testbeds using different tools:

Step 1 refers to the initialization of the Motrol hardware used to control battery charging/discharging periods of the devices attached to it. Since charging/discharging actions associated with Motrol connection slots can be triggered wirelessly via HTTP messages sent to a server running inside the Motrol hardware, the hardware setup includes the configuration of the WiFi network (SSID and password) where the server uses a REST API and listens for messages to enable/disable current to each slot.

Step 2 refers to a phase where devices that need to be controlled are attached to Motrol slots. In this phase, devices are also installed with the software that loads the corresponding DL model associated with each specific evaluation scenario.

Step 3 refers to the creation of files describing the workload dynamics and granularity, cluster information, including participating devices identified by the smartphone model, the battery level each device must have as precondition to start a test, and the load balancing mechanism to be used to distribute workload. These files must be placed inside a directory named "scn" within the LiveDewStream tool.

Step 4 is optional and involves the utilization of BAGESS software [28] to define an execution sequence for all scenarios built in Step 3. The derived sequence reduces the execution of all scenarios by optimizing the preparation time of all devices between the execution of one scenario and the next one, while reducing the battery stress of all devices produced as a consequence of all charging/discharging actions triggered by different workload scenarios.

Step 5 is when all workload scenarios are executed. This involves Step 5.1, when smartphone cluster formation takes place—i.e., devices attached to Motrol are considered or not as part of the cluster, which causes their batteries to be charged or discharged—prepared—to meet the initial battery level required by the scenario under execution. Moreover, Step 5.2 describes the moment when all participating devices are ready to start the scenario execution and it is effectively run until the workload described in the scenario descriptor file is completed. Lastly, Step 5.3 is triggered when the execution of a scenario finishes and generated log files and workload results are moved to a safe directory in the file system to prevent a new scenario execution from overriding the generated data.

Step 6 is when logs and result files are processed with scripts to create reports and facilitate the analysis of different metrics.

As discussed earlier in this paper, workloads in edge computing and IoT scenarios are commonly computed with medium-to-high-end edge nodes such as SBCs. In the benchmarks, we compare the performance of several SBC-based and smartphone-based setups to complete the workloads described in Section 3.3. The "Edge Setup" column in Tables A1–A3 differentiates both setups. Considering that smartphone-based setups comprise clusters of smartphones, Table 4 expands on the model and battery information of nodes integrating each cluster. All clusters include a single instance of each smartphone model. Smartphone brands and models are indicated using a short name that can be found in Table 3. The order in which these short names appear in a row corresponds to the order in which smartphones' initial battery values appear. The latter were randomly chosen. Cluster heterogeneity arises from the combination of smartphone quantities, models, and initial battery levels. Cluster maximum energy and cluster initial energy, both expressed in Joules, in addition to cluster initial battery, which is expressed in %, all derive from individual smartphones data and complement the cluster information.

Algorithm 1 Clusters Management Overview: Steps involved in setting up nodes and workloads, running scenarios, and deriving metrics using Motrol [29], LiveDewStream [27], and BAGESS [28] tools

```
Step 1: Setup and initiate Motrol hardware
Step 2: Register $DEVICES to Motrol
Step 3: Build $SCENARIO_DESCRIPTOR_LIST to feed LiveDewStream tool
Step 4 (optional): Sort $SCENARIO_DESCRIPTOR_LIST  elements by using
    BAGESS tool to optimize scenarios runtime execution
Step 5: For each element in $SCENARIO_DESCRIPTOR_LIST:
            5.1 Form Cluster and prepare $DEVICES
            5.2 Run scenario
            5.3 Collect results
Step 6: Summarize results to generate metrics
```

Table 4. Details of smartphone-based setups.

Cluster ID	Cluster Size	Smartphone Models	Smartphones Initial Battery %	Cluster Initial Energy (Joules)	Cluster Max. Energy (Joules) [1]	Cluster Initial Battery % [2]
Cl202	2	XRN7, MotG9	87, 75	99,532.80	123,840	80.37
Cl203	2	XRN7, MotG9	27, 54	51,904.80	123,840	41.91
Cl205	2	MotG9, SamA30	54, 30	53,152.20	122,454	43.41
Cl301	3	SamA30, XRN7, MotG9	27, 45, 60	80,582.58	177,894	45.30
Cl302	3	SamA30, SamA02, MotG9	27, 45, 60	86,195.88	190,368	45.28
Cl304	3	MotG6, XRN7, MotG9	77, 25, 53	81,712.80	164,880	49.55
Cl305	3	MotG9, MotG6, SamA30	54, 92, 25	88,206.30	163,494	53.95
Cl408	4	MotG6, SamA30, XRN7, MotG9	30, 27, 45, 60	92,894.58	218,934	42.43
Cl409	4	MotG6, SamA30, XRN7, MotG9	88, 80, 38, 49	133,941.6	218,934	61.18
Cl411	4	MotG6, SamA30, XRN7, MotG9	25, 52, 44, 38	88,753.68	218,934	40.54

[1] To calculate this value, we used a formula derived from Formula (1) assuming each smartphone is fully charged.
[2] The value was calculated using Formula (1), replacing batteryDrops by the battery level of each smartphone at the beginning of a workload test.

One of the metrics we report is accumulated inference latency, a measure of how long the makespan deviates from the stream time. The stream time is the time window where CV tasks are created, while makespan is the time the inference system employs in completing all the created CV tasks. Figure 5 graphically represents how these entities coexist in time. Inference completion of a task naturally occurs in the time after the CV task creation. While task creation can occur by following a quite controlled and periodic process, the task inference process cannot be so controlled. This is what $\alpha, \beta, \gamma, \delta, \varepsilon$, and ζ represent in the CV task inference timeline associated with each CV task completion event—i.e., different completion times could be given as a consequence of completion, but they are not limited to variations in network latency, CNN data input content, and node execution capability. When subtracting stream time from makespan, the result is the accumulated inference latency. The closer this difference is to zero, the better the setup capability is at responding with proper inference times compared to the time taken to create all CV tasks. A derived metric that gives a notion of the average time employed by the setup to complete a CV task after a new CV task is available could be obtained by dividing the accumulated inference latency by the number of CV tasks created. We call this the inference latency.

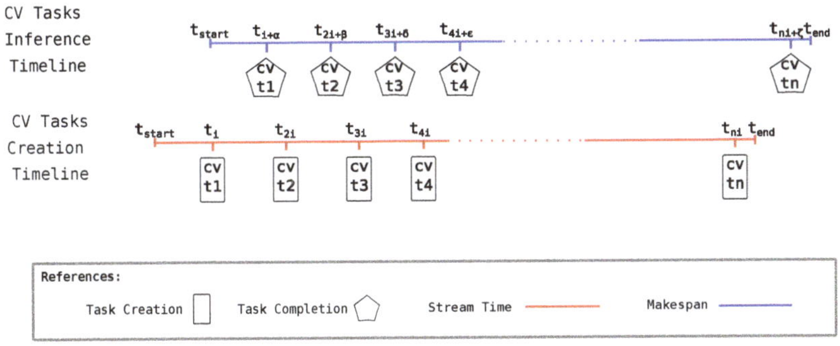

Figure 5. Accumulated inference latency: visual explanation.

Another metric we report is total energy consumption. To measure energy consumption, we followed two approaches depending on the type of setup. For SBC-based setups, we employed a power monitor [37] that used a toroid connected as shown in Figure 6, which allowed us to collect current and voltage readings from an SBC in a non-invasive way while running workloads. The power monitor was connected to the 220 V AC power and it had a socket after a toroid where the SBC power source was connected. Current and voltage readings were transmitted in real-time via a USB port to a laptop. The power monitor gave two current and voltage values per second. Then, for a workload scenario whose inference time was around 3 min and 50 s, the reported energy consumption was obtained by averaging 460 samples.

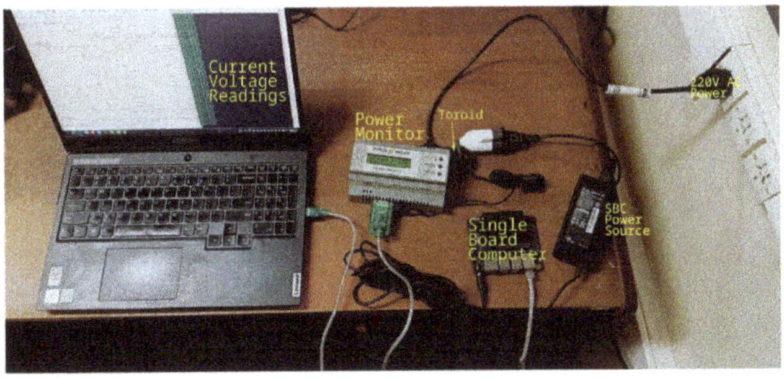

Figure 6. SBC power monitoring connection scheme.

For smartphone-based setups, the described approach for registering energy consumption cannot be used, because smartphone batteries are employed as the power source. We thus followed a different approach by registering smartphones battery levels at the beginning and at the end of a workload scenario. Then, we used battery manufacturer information—such as battery mAh and operating voltage—for each smartphone to approximate the total energy used during the workload scenario execution (WSE), expressed in Joules:

$$WSEJoules = \sum_{i=0}^{\#smartphones} (battLevelDrops^i \times battMAh^i \times 0.01) \times battVoltage^i \times 3.6 \quad (1)$$

where $\#smartphones$ corresponds to the amount of smartphones integrating the cluster, and $battLevelDrops^i$ is the amount of battery level drops registered for the smartphone i used for executing the assigned workload, which is in turn calculated as

$$battLevelDrops = batt_{endWSE} - batt_{startWSE} \quad (2)$$

where $batt_{startWSE}$ and $batt_{endWSE}$ are the battery levels of smartphone i at the start and end of the workload scenario execution, respectively, and $battMah^i$ and $battVoltage^i$ are milli-ampere hours and operating voltage information, respectively, provided by the battery manufacturer of smartphone i. The constant 0.01 is used as an alternative form of expressing the algebraic operation of dividing a value x by 100. For the case of (1), such an operation is involved in the application of the direct rule of three to express a battery percentage in mAh. Formula (1) emerges from combining a rule of three that relates battery percentage usage with the formula $mAh \times voltage \times 3.6 = JoulesOfenergy$, used to convert mAh to Joules, where 3.6 is a conversion factor that arises from converting milli-ampere hours (mAh) to Coulombs—e.g., 1 mAh = 1/1000 Amperes $*$ 60 s $*$ 60 min = 3.6 Coulombs

Additionally, since a workload scenario execution using smartphone-based setups might cause different battery percentage usage in each node, we report Jain's fairness index [38] as a metric that summarizes intra-cluster energy utilization. In other words, the metric is used to measure the disparity of energy pulled by the system from resource provider nodes. The metric takes values from 0 to 1. The nearer the index to the value of one, the more balanced the energy utilization among cluster nodes is. The Jain's index inspired formula we applied is

$$Fairness = \frac{\left[\sum_{i=0}^{\#smartphones} battLevelDrops^i\right]^2}{\#smartphones \times \left(\sum_{i=0}^{\#smartphones} \left[(battLevelDrops^i)\right]^2\right)} \quad (3)$$

Since all smartphone-based setups comprise clusters of at least two heterogeneous smartphones, it was necessary to indicate the load balancing strategy used to distribute the load between these. For this reason, all smartphone-based setups were indicated with Round Robin, ComTECAC, or Pull-Based in the load balancing column. Round Robin is a classic load balancing strategy that assigns an incoming task to the next available node using a rotating scheme, ensuring workload is evenly distributed among all participating devices. ComTECAC is a heuristic that captures unique characteristics of heterogeneous smartphone clusters. It is used to rank candidate nodes based on a criterion that combines computation–communication throughput and energy contribution data. It was considered because, as shown in [39], it achieves the best performance compared to other state-of-the-art heuristics. Lastly, with a Pull-Based approach, tasks are pulled by nodes from a shared queue based on their availability. In our configurations, cluster nodes were configured to pull at most one task per request. In turn, a node is able to make a request when the result of the previously pulled task has been sent back to the edge master—i.e., the task is marked by the system as completed.

4.1. Benchmark Results

4.1.1. Performance Analysis

Figure 7 shows the performance achieved by all setups for all scenarios. Blue bars correspond to SBC-based setups while orange ones correspond to smartphone-based setups. By comparing all setups, it can be clearly noted that the best performance, i.e., the lowest accumulated inference latency value in all tested scenarios, is achieved with the Gigabyte Brix setup. The close-to-zero value indicates that inferences performed with this setup are near real-time—i.e., inference results associated with all CV tasks are obtained within the stream time. Concretely, in the BCS scenario, the accumulated latency in a stream time of $3'50''$ was 3 s. In the SWT scenario, the accumulated latency in a stream time of $30'$ was 1.2 s, while in the DMED scenario with a stream time of $8'55''$, the value was 3.6 s. With the Raspberry Pi4 and Nvidia Jetson Nano setups, the obtained accumulated inference latency was significantly greater, and in most cases, the inference latency was the stream time multiplied by a factor of two to four—for detailed values, please refer to tables in Appendix A. The Nvidia Jetson Nano setups performed better than the Raspberry Pi4 setups in all cases. However, the workload always exceeded the inference capability. With these setups, reducing workload would be necessary to achieve near real-time performance. Some alternatives to reducing workload are creating more lightweight CV tasks by dropping frames and/or offloading some CV tasks to other edge nodes. These alternatives could come in turn with some QoS degradation as a result of increased network latency.

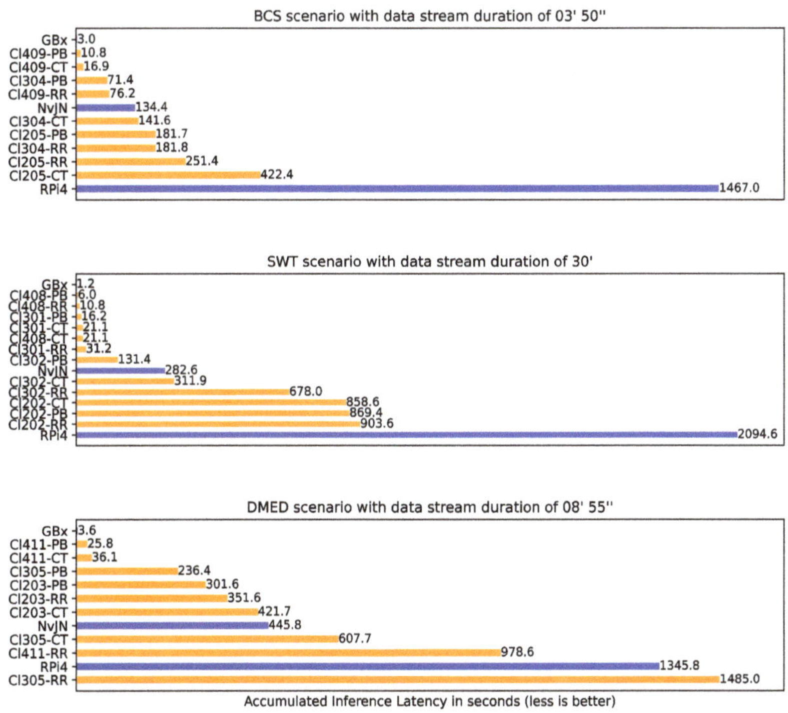

Figure 7. Accumulated inference latency (in seconds) for the three scenarios. For the cluster setups, we indicate the load balancing technique used (PB = Pull-Based; CT = ComTECAC; RR = Round Robin). Load balancing is not used in SBC-based setups.

When comparing smartphone-based setups for the BCS scenario, we empirically confirm what other studies have found [26,40]: by incrementing the number of participating smartphones, performance improves—i.e, lower accumulated inference latency is obtained. This is also true for the SWT and DMED scenarios. However, not only the number of smartphones impacts the setup inference service capability, but also the smartphone computing characteristics. Clusters 301 and 302, for instance, are of the same size. Both clusters are composed by the Samsung A30 and the Motorola Moto G9 Play. However, the Samsung A02 smartphone present in Cluster 302 is not present in Cluster 301, which has the Xiaomi Redmi Note 7 instead. If we compare the inference times associated with the latter mentioned smartphone models and those reported in Figures 2b, 3b, and 4b, we note that the computing capability of the Samsung A02 is between 2.62 and 3.99 times slower than that of the Xiaomi Redmi Note 7. The effect of this difference is observed in the performance values registered for the SWT scenario, where both clusters are tested. While the accumulated inference latency with Cluster 301 was between ∽16 and ∽31 s, for Cluster 302 it was between ∽131 and ∽678 s.

Another observation when comparing smartphone-based setups is that, in general, the Pull-Based load balancing scheme achieves lower accumulated inference latency than Round Robin and ComTECAC. This is because using Pull-Based scheme allows the cluster to balance the workload in accordance with the availability and computing capability of worker nodes rather than treat all nodes with the same capability, as done by Round Robin. The advantage of Pull-Based over Round Robin diminishes in clusters where nodes present quite similar computing capabilities, which is the case for Cluster 202 in the SWT scenario, or even when computing resources are apparently abundant enough to satisfy the workload requirements, which seems to be the case for Cluster 408. The differentiation of node capabilities is also a feature of ComTECAC, which, in most cases, performs slightly worse than Pull-Based but better than Round Robin. In general, the most competitive accumulated inference latency comparison to that obtained by the Gigabyte Brix setup was always achieved with the Pull-Based load balancing scheme. For example, in the BCS scenario, the difference between the Cluster 409 and Gigabyte Brix setups was less than 8 s. The third-best performance was also achieved with a Pull-Based setup. The Cl304-PB, with a size 3 cluster, achieves quite similar performance to that of a size 4 cluster combined with Round Robin, i.e., Cl409-RR. A similar analysis is valid in the SWT and DMED scenarios, with Clusters 408 and 411, respectively. The Pull-Based load balancing scheme achieves considerably better performance than Round Robin and also slightly better performance than ComTECAC, especially when cluster nodes present heterogeneous computing capabilities or when available computing capabilities are tight for coping with real-time tasks requirements.

4.1.2. Energy Consumption–Performance Tradeoff Analysis

In view of the high energy consumption that AI applications already demand and expected further increases, studies in this regard [41–43] and initiatives such as the Low-Power Recognition Challenge [44] encourage the improvement of not only DL model performance but also the energy that execution and training consume. Thus, we study the tradeoff between performance and energy consumed in all evaluated scenarios. Figure 8 reports the energy consumption and accumulated inference latency tradeoff between different setups in each tested scenario. The best setups are those that achieve the minimum accumulated inference latency using the least energy, i.e., those positioned close to the origin. As can be observed, in all scenarios, the best balance is always achieved by orange dots with a label starting with Cl, which corresponds to smartphone-based setups. It can be noted, for instance, that the accumulated inference latency of GBx setups are the lowest in

BCS, SWT, and DMED scenarios; however, these are also associated with the highest values of energy utilization—4524, 22,403, and 7834 Joules, respectively. Interestingly, smartphone-based setups, particularly four-node clusters using Pull-Based and ComTECAC load balancing schemes, offer a good tradeoff in all scenarios, i.e., competitive performance with reduced energy consumption. Some examples are Cl409-PB, Cl408-PB, and Cl411-PB for the BCS, SWT, and DMED scenarios, respectively. Also, Cl409-CT and Cl411-CT for BCS and DMED, respectively. In other words, for the BCS scenario, when it is acceptable to have all CV task results calculated with a delay of 7.8 s w.r.t to the fastest setup—GBx—Cluster 409 with a Pull-Based scheme can achieve the objective by utilizing ∽1510 Joules, i.e., almost three times less energy than the Gigabyte Brix setup. Similar relations can be found for the SWT and DMED scenarios.

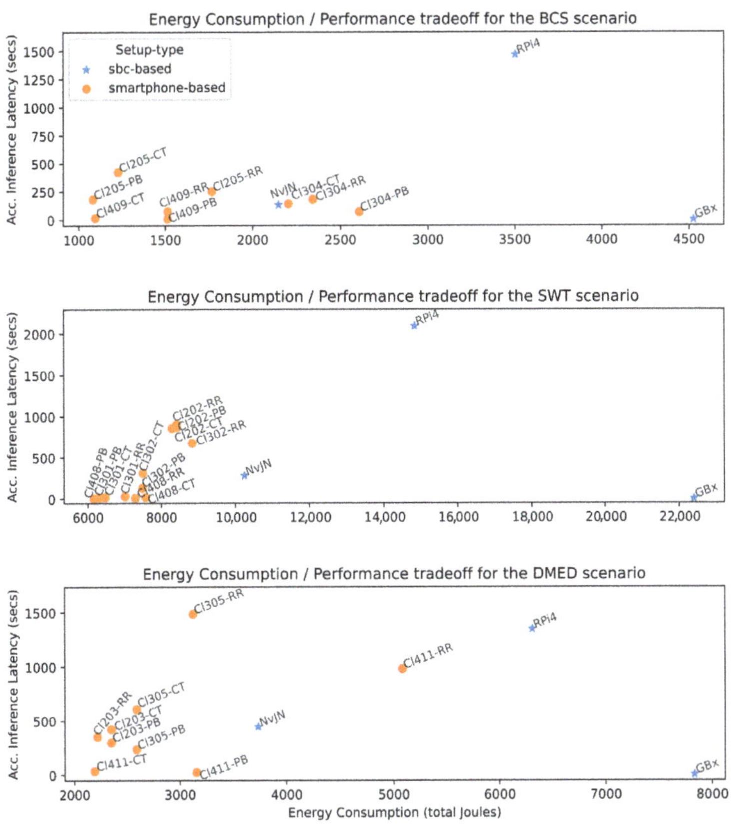

Figure 8. Energy consumption/performance tradeoff.

Figure 9 shows insights on how workload execution was distributed by different load balancing schemes among smartphones integrating the most competitive clusters with four nodes for all scenarios and a breakdown of the inference time registered by the mobile distributed inference approach showing the relations among computing/communication time disaggregated by smartphone models. Particularly, in Figure 9a, for all scenarios, it can be observed that, as expected with Round Robin load balancing, all smartphone models—differentiated in blue tones—are assigned with (and complete) the same amount of CV tasks. This behaviour does not allow for efficiently exploiting the computing heterogeneity of smartphones. With a Pull-Based scheme, on the contrary, smartphones are assigned new tasks as soon as they deliver the results of the previous pulled task. Such a reactive behaviour allows the distributed inference approach to assign CV tasks considering node heterogeneity. Notice that the XRN7 smartphone, which is one of the fastest within the clusters (see Figure 9b), is assigned the largest number of CV tasks in all scenarios. Conversely, MotG6, which is the slowest, is assigned the least CV tasks. Finally, ComTECAC CV task amounts are similar to those of the Pull-Based scheme, especially in short-duration workload scenarios like BCS and DME, which explains why both load balancing schemes achieved such similar accumulated inference latency values. Contrary to the Pull-Based scheme, ComTECAC assigns CV tasks proactively, i.e., deciding which node should be in charge of the task as soon as the task is created using a formula that combines network performance, node computing capability measured in FLOPS, queued tasks, and historic contributed energy measured through current battery level. For short-duration scenarios, few to none of these indicators vary—i.e., they remain almost constant. Consequently, each node's rank is mostly decided by their FLOPS indicator value. This explanation does not apply to the behaviour that ComTECAC presents in the SWT scenario with a 30 min workload, where contributed energy varies. Certainly, this scenario provides evidence that the ComTECAC formula can be subject to further improvements. Lastly, Figure 9b shows how inference time is divided for individual images—not CV tasks—into computing time and data transferring time. Clearly, in all scenarios for all smartphone models, computing time dominates transferring time describing ratios that are in [1/13–1/19] for BCS, [1/10–1/15] for SWT and, [1/27–1/85] for DMED scenario. This could be taken as a hint of how energy consumption of the mobile distributed inference approach is disaggregated. Unless energy consumption in transferring time exceeds that of computing time by the inverse ratio of those reported, we could say that computing time dominates energy consumption.

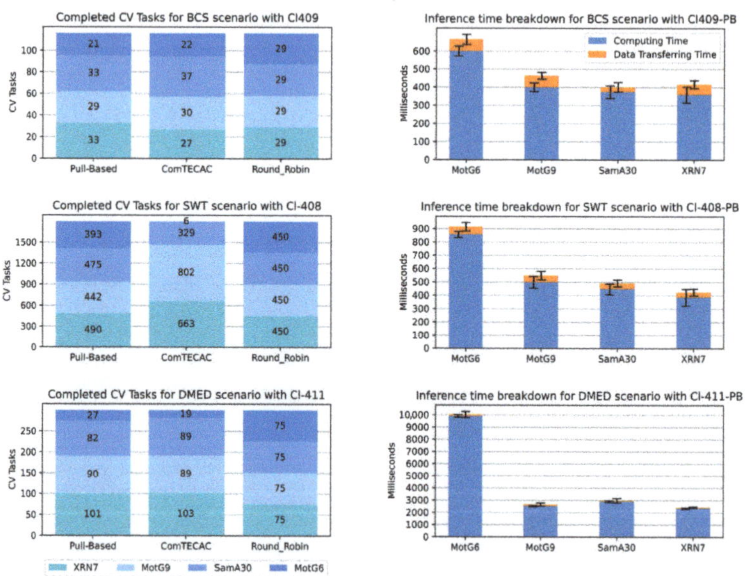

(a) (b)

Figure 9. Insights of the most competitive smartphone-based setups with four nodes for the BCS, SWT, and DMED scenarios. MotG6 = Motorola Moto G6, SamA30 = Samsung A30, XRN7 = Xiaomi Redmi Note 7, MotG9 = Motorola Moto G9 Play. (**a**) Completed CV task breakdown by smartphone model with different load balancers. (**b**) Inference time breakdown per image for smartphone models in clusters using Pull-Based load balancer.

4.1.3. Battery Utilization Analysis

For smartphone-based setups, it is relevant to show the impact of CV task computations on cluster energy availability. Provided that clusters are composed of at least two nodes, cluster battery can be seen as the aggregation of node battery. To represent the cluster current battery level with a single positive integer value, we aggregate the Joules each smartphone contributes to the global cluster energy. By assuming that a cluster with a 100% global battery level is one where batteries of all integrating smartphones are fully charged, using Equation (1), it is possible to calculate the global battery level expressed as a percentage of a cluster for different smartphones' battery levels. Figure 10a shows global battery level drops for all smartphone-based setups evaluated in all scenarios. Since in all cases smartphone-based setups run workloads while unplugged from the electricity grid, i.e., with batteries in discharging mode, the battery drops were calculated as the difference between the initial and the final global battery level of a cluster before and after the workload execution, respectively. In Figure 10a, it can be noted, for instance, that CV tasks in the SWT scenario causes the highest global battery drops compared to BCS and DMED, which is mainly due to the length of the workload execution $-30'$ compared to $\backsim 4'$ and $\backsim 9'$ in BCS and DMED, respectively.

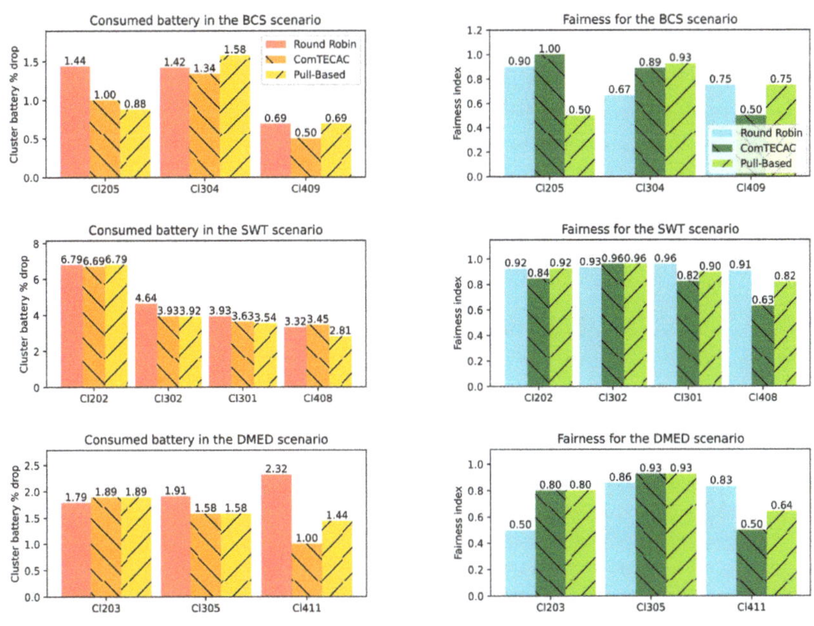

(**a**) Global battery impact (**b**) Fairness index

Figure 10. CV tasks impact on clusters global energy availability and Fairness index.

Another observation that emerges from Figure 10a is that when comparing cluster battery level drops between clusters w.r.t. the same scenario, they appear to decrease as cluster size increases. In fact, as more smartphones are integrated in a cluster, the cluster battery also increases and a percentage of this cluster battery is represented by a larger number in Joules. Something similar happens when two clusters are of the same size but one has smartphones whose batteries have more energy storage capacity than the other, which is the case for Cl301 and Cl302. However, a cluster with a larger battery does not imply that it will complete tasks in a more energy-efficient way. As an example, refer to the

"Max. Cluster Energy" column of Table 4 to appreciate that Cl302 has slightly larger cluster battery capacity than Cl301, because the Xiaomi RN7 with a battery capacity of 4000 mAh in Cl301 was swapped with the Samsung A02 with a battery capacity of 4900 mAh in Cl302. However, according to Figure 10b, the latter smartphone is almost four times slower at performing a task in the SWT scenario than the former, meaning that Cl302 will spent more energy than Cl301 in completing the same number of tasks. This explains why, despite having a larger cluster battery, the battery drop of Cl302 is larger than that of Cl301 in the SWT scenario shown in Figure 10a. Moreover, it is relevant to point out that, in general, the Pull-Based load balancing scheme achieves equal or less global battery drops than Round Robin in approximately 0.5%. The ComTECAC heuristic was designed to consider energy contribution in combination with performance indicators to rank candidate nodes; however, it does not show a clear advantage over the Pull-Based scheme in this regard. In most cases, it causes similar global battery drops as the Pull-Based technique while a few others result in fewer battery drops.

Finally, we analyse fairness values derived from applying the Round Robin, ComTECAC, and the Pull-Based load balancing schemes. The index serves as a hint on how the edge cluster utilizes energy from all participating smartphones. The more the value approximates to one, the more balanced the battery utilization is among nodes, which, in turn, is an indicator of how well the level of parallelism is maintained over time. Figure 10b reveals that in some clusters, the Pull-Based scheme achieves slightly higher fairness values than those of Round Robin, while in other cases, the results are the opposite. There are cases where both schemes achieve equal fairness values. By contrast, ComTECAC fairness values are mostly under those obtained by the Pull-Based approach and in some cases, they are also under Round Robin fairness values. There is not a clear pattern to conclude that one load balancing scheme is better than the others according to the fairness index.

5. Discussion

5.1. Limitations

A few previous studies also focused on proving the strengths of edge computing for reducing end-to-end latency and allowing the execution of data-intensive IoT and delay-sensitive applications [17,45]. In this work, we explored how edge computing capabilities can be augmented with smartphone clusters under different workload scenarios for real-time AI inferences involving image streams. The competitive performance and reduced energy consumption that smartphone clusters showed, along with the pervasiveness of such devices worldwide, evidence their suitability for being considered a powerful tool to complement in-the-field, already present edge settings, and/or to fully support opportunistic distributed inference. Even though we showed through extensive experiments that it is feasible to achieve low-latency AI inferences for different DL-based resource-intensive workloads by exclusively relying on the computing power provided by smartphones clusters, we are aware of aspects that might negatively affect usability in real-world deployments and overall performance in exigent generic application contexts. One aspect is the impact of network conditions, specifically variability in smartphones' signal strength, which could increase the retransmission rate of packets and the available bandwidth that other user processes and applications could share with the collaborative inference software. User processes and applications might also introduce uncontrolled variability in the usage of computing resources available to make inferences. Furthermore, in the case of interactive applications, the user might experience faster battery drops due to the screen usage. The experiments performed considered smartphones connected with excellent signal strength to a 5 GHz WiFi network; they also entailed restricted network activity of user applications other than the collaborative inference software and guaranteed that smartphone screens stayed off while inferences were calculated. It is within our roadmap to incorporate new features in the experimentation toolset to allow us and other researchers to emulate and control resource variability aspects and, ultimately, cover more experimental scenarios.

In relation to pre-trained AI models, to assure homogeneity in result quality, it is necessary to perform smartphone model synchronization. This applies especially to those subject to frequent updates by maintainers. This, in turn, might introduce undesired delay at the beginning of a distributed inference session. This aspect is relevant to the implementation of desirable middleware features.

With respect to implementation details, we also want to mention possible technical improvements to the cluster management as well as intra-device resource usage that could bring out other latency–energy consumption tradeoffs. One of these is the support of different task distribution methods. Currently, collaborating smartphones in the cluster request the master node for task metadata using HTTP messages to first load into memory the DL model to perform inferences. Then, smartphones request input (i.e., tasks) through new HTTP requests. Varying the ways in which master and collaborating nodes communicate, e.g., with a connection persistent protocol like web sockets, would help to speed up the inference performance. Another improvement is to provide smartphones with the ability to dynamically vary the level of concurrency when executing tasks. This would allow collaborating nodes to regulate load—for instance, freeing resources to anticipate user interactions and battery overheating—or, on the contrary, offer more computing resources when the phone is idle. Current intra-device implementation supports the execution of at most one task.

Finally, the results obtained demonstrate that low-latency and energy efficient solutions are achievable with the DL models used for the described workload scenarios. However, trying other DL modes architectures or model versions resulting from the application of size reduction techniques (e.g., quantization, pruning, low-rank factorization, etc.) might signify a boost in energy utilization and latency but falls out of the scope of this work. It is worth mentioning that comparisons involving different model versions or model architectures for a specific problem are commonly carried out considering the precision of results or other metrics that allow for differentiating not only latency but also the quality of results. In our results, quality was not within the dimension of analysis because all setups studied for each workload scenario ran the same model.

5.2. Future Work

Several questions remain open, demanding further research. One of these relates to whether current software stacks including middlewares for distributed computing [46,47] and frameworks for AI execution [48] in low-powered devices are prepared to integrate smartphone clusters as a new type of edge node by attending diverse aspects related to computational resource provision with devices that present non-dedicated resources—i.e., with user applications that must prioritize the usage of resources and deal with unstable communication due to physical position changes of device owners, limitations in resource scavenging times due to battery operation times, and compute-while-charging scenarios where computations are performed by battery-powered devices while plugged to the electricity grid [49] or while harvesting energy using radio-frequency signals of access points [16].

Another open question concerns incentives for making device owners offer up unused computing mobile resources. The literature describes at least two approaches. One of them is associated with altruistic attitudes of devices owners—this is similar to what is promoted in volunteer computing platforms with unused computing cycles of desktop computers for scientific projects. Applications include daily normal situations that require in-the-field inference capabilities either for reducing the cost of data transmission—e.g., chronic disease health monitoring, safety-guard mechanisms in risky situations (e.g., alerting about obstacles in the road while driving), providing real-time responses to improve user experience (e.g., augmented reality applications in tourism-related activities), etc. By contrast, the other incentive approach we envision is based on opportunistic inferences as a service approach that benefits device owners with some revenue in the form of credits [50], reputation [51,52], or energy [16] in exchange for completing tasks that require in-the-field inference capabilities—for example, as a complement to surveillance cameras for crime prevention in public spaces. The obtained results achieved in this study with efficient hardware, such as that embedded in smartphones, w.r.t using other edge nodes are promising in terms of energy consumption and performance. However, massive application of the approach must not be carried out without informing users about the side effects regarding quicker losses in device battery life as a consequence of more intensive use of devices' computational nd communication resources, which in turn can result in excessive electronic waste generation, especially when using devices with non-replaceable batteries or when device replacement is encouraged rather than simply replacing the battery.

In this regard, another aspect worth mentioning that is intimately related to AI and computing infrastructures concerns energy consumption and, hence, indirectly, CO_2 emissions. In its World Energy Outlook 2024 [53,54], the International Energy Agency (IEA) projects that energy consumption in datacenters is set to expand in the future, in part due to the digitalization of the economy but also as a consequence of the rapid advance of AI technologies and the outbreak of AI startups. In fact, figures from 2022 also reported by the IEA—i.e., some years before the AI boom we have been experiencing since then—indicated that datacenters consume 1–2% of the world's electricity, which is backed up by estimates from similar studies [55,56]. This positions energy-efficient approaches to edge computing as a crucial path to cap these estimates while providing viable computing settings to continue helping bringing AI to the masses. Edge computing research should not only merely quantify energy savings compared to traditional computing approaches but also develop mechanisms to measure the impact on CO_2 emissions at the software level, such as in [57]. Quantification must also consider current mobile hardware manufacturing practices, for example, promoting replaceable (sealed) batteries, which encourage users to upgrade to a new device when battery performance declines, rather than simply replacing the battery. Metrics such as the Software Carbon Intensity (SCI) index [58] might serve as a pertinent reference since it captures the environmental impact of software in terms of CO_2 emissions considering the carbon intensity of the energy used to run software and the embedded emissions due to the carbon footprint to build the hardware on which the software runs.

There are also improvement opportunities in line with developing new load balancing heuristics, and based on the results from preliminary runs, we plan to improve previously published heuristics that use MFLOPs to rank node computing capabilities [39]. An alternative solution is to use information derived from the Tensorflow benchmarking tool [33]. In spite of differences observed in inference times obtained via the Tensorflow benchmarking tool and results collected in experiments using real inputs, relative benchmarked positions between edge nodes are similar, meaning that it is feasible to use such a tool to build node rankings that can serve as input for the load balancing component to distribute tasks based on an alternative node performance score to the classic MFLOPs.

Another future research direction is to consider other forms of workload distributions, i.e., that exploit the structure of deep learning (DL) architectures—for example, split computing, early exiting techniques, and model quantization [59]—to better exploit the CPU capabilities of mobile platforms while performing network inferences.

6. Conclusions

In this work, we explored the computing capabilities of mobile distributed computing setups for real-time inferences by experimenting with heterogeneous AI workloads, built upon real convolutional neural network (CNN) models and image datasets that were used to emulate computer vision (CV) tasks generated in a stream-like fashion. We ran and compared the performance achieved by several setups built with commodity hardware including single-board computers (SBCs) and smartphone clusters that can be found or exploited in practical edge computing and IoT application scenarios.

We conclude that tiny clusters composed by two to four low-to-middle-end smartphones using a Pull-Based load balancing scheme offer competitive performance compared to an SBC equipped with a reasonable CPU (Intel core i5 8th generation) in computing real-time CV inference tasks. This suggests that to satisfy certain delay-sensitive application scenarios, the exploitation of already-in-place smartphones clusters can be considered before the investment and deployment of SBCs. According to Table 3, the necessary investment for supporting the evaluated scenarios using a powerful SBC like the Gigabyte Brix is only 30% cheaper than with a cluster-based setup comprising four low-to-mid-end smartphones. This is when including hardware acquisition cost of smartphones. By contrast, if computing capability is hired out to smartphones' owners already present in the application context, then the equation of computing resources cost can be different. In any case, energy consumption measurements show that the tradeoff including performance and energy consumption favours smartphone clusters over SBCs. The former can deliver competitive performance with a reduction of around three times the energy consumption required by a powerful SBC such as the Gigabyte Brix.

Author Contributions: Methodology, M.H. and C.M.; Software, M.H.; Validation, M.H. and C.M.; Investigation, M.H.; Data curation, M.H.; Writing—original draft, M.H. and C.M.; Writing—review & editing, M.H., C.M. and T.A.M.; Visualization, M.H.; Funding acquisition, C.M. and T.A.M. All authors have read and agreed to the published version of the manuscript.

Funding: This research was funded by CONICET via grants PIBAA-28720210101298CO and PIP-11220210100138CO. The APC was funded by Ruhr-Universität Bochum.

Informed Consent Statement: Not applicable.

Data Availability Statement: The original contributions presented in this study are included in the article. Further inquiries can be directed to matias.hirsch@isistan.unicen.edu.ar.

Conflicts of Interest: The authors declare no conflict of interest.

Appendix A

Tables A1–A3 contain the data reported in graphs presented in Section 4.1 for the BCS, SWT, and DMED scenarios.

Table A1. Body Condition Score scenario (03:50 min data stream).

Edge Setup		Load Balancing	Acc. Inference Latency (s)	Jain's Fairness Index	Cluster Consumed Battery (%)	Consumed Energy (Joules)
SBC-based	RPi4	N/A	1467	N/A	N/A	3501.97
	NvJN	N/A	134.4	N/A	N/A	2147.13
	GBx	N/A	3.0	N/A	N/A	4524.82
Smartphone-based	Cluster 205	Round Robin	251.4	0.9	1.44	1765.08
		ComTECAC	422.4	1.0	1.0	1224.54
		Pull Based	181.7	0.5	0.88	1081.08
	Cluster 304	Round Robin	181.8	0.667	1.42	2341.29
		ComTECAC	141.6	0.889	1.34	2203.2
		Pull Based	71.4	0.926	1.58	2605.10
	Cluster 409	Round Robin	76.2	0.750	0.69	1510.64
		ComTECAC	16.9	0.5	0.5	1094.94
		Pull Based	10.8	0.750	0.69	1510.64

Table A2. Sense while Travel scenario (30 min data stream).

Edge Setup		Load Balancing	Acc. Inference Latency (s)	Jain's Fairness Index	Cluster Consumed Battery (%)	Consumed Energy (Joules)
SBC-based	RPi4	N/A	2094.6	N/A	N/A	14,827.11
	NvJN	N/A	282.6	N/A	N/A	10,246.13
	GBx	N/A	1.2	N/A	N/A	22,403.73
Smartphone-based	Cluster 202	Round Robin	903.6	0.924	6.79	8408.74
		ComTECAC	858.6	0.845	6.69	8280.00
		Pull Based	869.4	0.924	6.79	8408.74
	Cluster 302	Round Robin	678.0	0.933	4.64	8833.07
		ComTECAC	311.9	0.960	3.93	7476.12
		Pull Based	131.4	0.960	3.92	7462.43
	Cluster 301	Round Robin	31.2	0.960	3.93	6991.23
		ComTECAC	21.1	0.823	3.63	6459.48
		Pull Based	16.2	0.896	3.54	6297.45
	Cluster 408	Round Robin	10.8	0.907	3.32	7268.61
		ComTECAC	21.1	0.631	3.45	7554.42
		Pull Based	6.0	0.818	2.81	6152.04

Table A3. Disease Monitoring and Early Diagnosis scenario (08:55 min data stream).

Edge Setup		Load Balancing	Acc. Inference Latency (s)	Jain's Fairness Index	Cluster Consumed Battery (%)	Consumed Energy (Joules)
SBC-based	RPi4	N/A	1345.8	N/A	N/A	6302.99
	NvJN	N/A	445.8	N/A	N/A	3733.47
	GBx	N/A	3.6	N/A	N/A	7834.28
Smartphone-based	Cluster 203	Round Robin	351.56	0.5	1.79	2217.60
		ComTECAC	421.72	0.8	1.89	2347.20
		Pull Based	301.56	0.8	1.89	2347.20
	Cluster 305	Round Robin	1485.0	0.857	1.91	3122.73
		ComTECAC	607.71	0.926	1.58	2583.20
		Pull Based	236.4	0.926	1.58	2583.20
	Cluster 411	Round Robin	978.6	0.833	2.32	5079.27
		ComTECAC	36.12	0.5	1.0	2189.88
		Pull Based	25.8	0.643	1.44	3152.65

References

1. Nguyen, D.K.; Lan, C.H.; Chan, C.L. Deep Ensemble Learning Approaches in Healthcare to Enhance the Prediction and Diagnosing Performance: The Workflows, Deployments, and Surveys on the Statistical, Image-Based, and Sequential Datasets. *Int. J. Environ. Res. Public Health* **2021**, *18*, 10811.
2. Mahajan, P.; Uddin, S.; Hajati, F.; Moni, M.A. Ensemble Learning for Disease Prediction: A Review. *Healthcare* **2023**, *11*, 1808. [CrossRef] [PubMed]
3. Bao, J.; Xie, Q. Artificial intelligence in animal farming: A systematic literature review. *J. Clean. Prod.* **2022**, *331*, 129956. [CrossRef]
4. Ullah, Z.; Al-Turjman, F.; Mostarda, L.; Gagliardi, R. Applications of artificial intelligence and machine learning in smart cities. *Comput. Commun.* **2020**, *154*, 313–323. [CrossRef]
5. Javed, A.R.; Ahmed, W.; Pandya, S.; Maddikunta, P.K.R.; Alazab, M.; Gadekallu, T.R. A survey of explainable artificial intelligence for smart cities. *Electronics* **2023**, *12*, 1020. [CrossRef]
6. Yousefpour, A.; Fung, C.; Nguyen, T.; Kadiyala, K.; Jalali, F.; Niakanlahiji, A.; Kong, J.; Jue, J.P. All one needs to know about fog computing and related edge computing paradigms: A complete survey. *J. Syst. Archit.* **2019**, *98*, 289–330. [CrossRef]
7. Ray, P.P. Minimizing dependency on internetwork: Is dew computing a solution? *Trans. Emerg. Telecommun. Technol.* **2019**, *30*, e3496. [CrossRef]
8. Šojat, Z.; Skala, K. The Rainbow through the Lens of Dew. In Proceedings of the 2020 43rd International Convention on Information, Communication and Electronic Technology (MIPRO), Opatija, Croatia, 28 September–2 October 2020; IEEE: Piscataway, NJ, USA, 2020; pp. 1916–1921.
9. Gusev, M.; Wang, Y. Formal description of dew computing. In Proceedings of the 3rd International Workshop on Dew Computing, Toronto, ON, Canada, 29–30 October 2018; pp. 8–13.
10. Statista Smartphones Reports. Available online: https://www.statista.com/topics/840/smartphones/#topicOverview (accessed on 31 March 2025).
11. Hintze, D.; Hintze, P.; Findling, R.D.; Mayrhofer, R. A large-scale, long-term analysis of mobile device usage characteristics. *Proc. Acm Interact. Mob. Wearable Ubiquitous Technol.* **2017**, *1*, 1–21. [CrossRef]
12. Gedawy, H.; Harras, K.A.; Habak, K.; Hamdi, M. Femtoclouds beyond the edge: The overlooked data centers. *IEEE Internet Things Mag.* **2020**, *3*, 44–49. [CrossRef]
13. Switzer, J.; Marcano, G.; Kastner, R.; Pannuto, P. Junkyard computing: Repurposing discarded smartphones to minimize carbon. In Proceedings of the 28th ACM International Conference on Architectural Support for Programming Languages and Operating Systems, Vancouver, BC, Canada, 25–29 March 2023; Volume 2, pp. 400–412.
14. Ling, M.; Zhang, C.; Tang, S.; Chen, R.; Zhu, Y. EEVS: Redeploying Discarded Smartphones for Economic and Ecological Drug Molecules Virtual Screening. *IEEE Trans. Sustain. Comput.* **2025**, *Early Access*.
15. Na, B.; Jang, J.; Park, S.; Kim, S.; Kim, J.; Jeong, M.S.; Kim, K.C.; Heo, S.; Kim, Y.; Yoon, S. Scalable smartphone cluster for deep learning. *arXiv* **2021**, arXiv:2110.12172.

16. Liu, X.; Chen, A.; Zheng, K.; Chi, K.; Yang, B.; Taleb, T. Distributed Computation Offloading for Energy Provision Minimization in WP-MEC Networks with Multiple HAPs. *IEEE Trans. Mob. Comput.* **2024**, *24*, 2673–2689. [CrossRef]
17. Kang, Y.; Hauswald, J.; Gao, C.; Rovinski, A.; Mudge, T.; Mars, J.; Tang, L. Neurosurgeon: Collaborative intelligence between the cloud and mobile edge. *ACM SIGARCH Comput. Archit. News* **2017**, *45*, 615–629. [CrossRef]
18. Huang, Y.; Qiao, X.; Tang, J.; Ren, P.; Liu, L.; Pu, C.; Chen, J. DeepAdapter: A collaborative deep learning framework for the mobile web using context-aware network pruning. In Proceedings of the IEEE INFOCOM 2020-IEEE Conference on Computer Communications, Toronto, ON, Canada, 6–9 July 2020; IEEE: Piscataway, NJ, USA, 2020; pp. 834–843.
19. Huang, Y.; Qiao, X.; Ren, P.; Liu, L.; Pu, C.; Dustdar, S.; Chen, J. A lightweight collaborative deep neural network for the mobile web in edge cloud. *IEEE Trans. Mob. Comput.* **2020**, *21*, 2289–2305. [CrossRef]
20. Eshratifar, A.E.; Abrishami, M.S.; Pedram, M. JointDNN: An efficient training and inference engine for intelligent mobile cloud computing services. *IEEE Trans. Mob. Comput.* **2019**, *20*, 565–576. [CrossRef]
21. Teerapittayanon, S.; McDanel, B.; Kung, H.T. Distributed deep neural networks over the cloud, the edge and end devices. In Proceedings of the 2017 IEEE 37th International Conference on Distributed Computing Systems (ICDCS), Atlanta, GA, USA, 5–8 June 2017; IEEE: Piscataway, NJ, USA, 2017; pp. 328–339.
22. Guo, X.; Pimentel, A.D.; Stefanov, T. Automated exploration and implementation of distributed cnn inference at the edge. *IEEE Internet Things J.* **2023**, *10*, 5843–5858. [CrossRef]
23. Huang, Y.; Qiao, X.; Dustdar, S.; Zhang, J.; Li, J. Toward decentralized and collaborative deep learning inference for intelligent iot devices. *IEEE Netw.* **2022**, *36*, 59–68. [CrossRef]
24. Taufique, Z.; Miele, A.; Liljeberg, P.; Kanduri, A. Adaptive Workload Distribution for Accuracy-aware DNN Inference on Collaborative Edge Platforms. In Proceedings of the 2024 29th Asia and South Pacific Design Automation Conference (ASP-DAC), Incheon, Republic of Korea, 22–25 January 2024; IEEE: Piscataway, NJ, USA, 2024; pp. 109–114.
25. Zhao, Z.; Barijough, K.M.; Gerstlauer, A. Deepthings: Distributed adaptive deep learning inference on resource-constrained iot edge clusters. *IEEE Trans. Comput.-Aided Des. Integr. Circuits Syst.* **2018**, *37*, 2348–2359. [CrossRef]
26. Huang, Y.; Qiao, X.; Lai, W.; Dustdar, S.; Zhang, J.; Li, J. Enabling dnn acceleration with data and model parallelization over ubiquitous end devices. *IEEE Internet Things J.* **2021**, *9*, 15053–15065. [CrossRef]
27. Mateos, C.; Hirsch, M.; Toloza, J.M.; Zunino, A. LiveDewStream: A stream processing platform for running in-lab distributed deep learning inferences on smartphone clusters at the edge. *SoftwareX* **2022**, *20*, 101268. [CrossRef]
28. Yannibelli, V.; Hirsch, M.; Toloza, J.; Majchrzak, T.A.; Grønli, T.M.; Zunino, A.; Mateos, C. BAGESS: A software module based on a genetic algorithm to sequentially order load-balancing evaluation scenarios over smartphone-based clusters at the Edge. *IEEE Access* **2024**, *12*, 145893–145919. [CrossRef]
29. Toloza, J.M.; Hirsch, M.; Mateos, C.; Zunino, A. Motrol: A hardware-software device for batch benchmarking and profiling of in-lab mobile device clusters. *HardwareX* **2022**, *12*, e00340. [CrossRef] [PubMed]
30. Li, G.; Cai, J. An online incentive mechanism for collaborative task offloading in mobile edge computing. *IEEE Trans. Wirel. Commun.* **2019**, *19*, 624–636. [CrossRef]
31. Hou, Y.; Garg, S.; Hui, L.; Jayakody, D.N.K.; Jin, R.; Hossain, M.S. A data security enhanced access control mechanism in mobile edge computing. *IEEE Access* **2020**, *8*, 136119–136130. [CrossRef]
32. Xu, X.; Xue, Y.; Qi, L.; Yuan, Y.; Zhang, X.; Umer, T.; Wan, S. An edge computing-enabled computation offloading method with privacy preservation for internet of connected vehicles. *Future Gener. Comput. Syst.* **2019**, *96*, 89–100. [CrossRef]
33. Tensorflow Lite Benchmarking Tool. Available online: https://ai.google.dev/edge/litert/models/measurement (accessed on 2 February 2025).
34. Yolov4 Project. Available online: https://github.com/hunglc007/tensorflow-yolov4-tflite (accessed on 2 February 2025).
35. Alvarez, J.R.; Arroqui, M.; Mangudo, P.; Toloza, J.; Jatip, D.; Rodríguez, J.M.; Teyseyre, A.; Sanz, C.; Zunino, A.; Machado, C.; et al. Body condition estimation on cows from depth images using Convolutional Neural Networks. *Comput. Electron. Agric.* **2018**, *155*, 12–22. [CrossRef]
36. Bergoeing, M.; Neyem, A.; Jofré, P.; Hernández, C.; Chacón, J.; Morales, R.; Giddings, M. Exploring the Potential of an AI-Integrated Cloud-Based mHealth Platform for Enhanced Type 2 Diabetes Mellitus Management. In Proceedings of the International Conference on Ubiquitous Computing and Ambient Intelligence, Riviera Maya, Mexico, 28–30 November 2023; Springer: Cham, Switzerland, 2023; pp. 100–111.
37. Power Meter Website. Available online: https://www.powermeter.com.ar/home (accessed on 2 February 2025).
38. Jain, R.K.; Chiu, D.M.W.; Hawe, W.R. *A Quantitative Measure of Fairness and Discrimination*; Eastern Research Laboratory, DEC: Hudson, MA, USA, 1984.
39. Hirsch, M.; Mateos, C.; Zunino, A.; Majchrzak, T.A.; Grønli, T.M.; Kaindl, H. A task execution scheme for dew computing with state-of-the-art smartphones. *Electronics* **2021**, *10*, 2006. [CrossRef]

40. Hu, Z.; Tarakji, A.B.; Raheja, V.; Phillips, C.; Wang, T.; Mohomed, I. Deephome: Distributed inference with heterogeneous devices in the edge. In Proceedings of the 3rd International Workshop on Deep Learning for Mobile Systems and Applications, Seoul, Republic of Korea, 21 June 2019; pp. 13–18.
41. Henderson, P.; Hu, J.; Romoff, J.; Brunskill, E.; Jurafsky, D.; Pineau, J. Towards the Systematic Reporting of the Energy and Carbon Footprints of Machine Learning. *J. Mach. Learn. Res.* **2020**, *21*, 1–43.
42. Strubell, E.; Ganesh, A.; McCallum, A. Energy and policy considerations for modern deep learning research. *Proc. AAAI Conf. Artif. Intell.* **2020**, *34*, 13693–13696. [CrossRef]
43. García-Martín, E.; Rodrigues, C.F.; Riley, G.; Grahn, H. Estimation of energy consumption in machine learning. *J. Parallel Distrib. Comput.* **2019**, *134*, 75–88. [CrossRef]
44. Gauen, K.; Rangan, R.; Mohan, A.; Lu, Y.H.; Liu, W.; Berg, A.C. Low-power image recognition challenge. In Proceedings of the 2017 22nd Asia and South Pacific Design Automation Conference (ASP-DAC), Chiba, Japan, 16–19 January 2017; IEEE: Piscataway, NJ, USA, 2017; pp. 99–104.
45. Yi, S.; Hao, Z.; Zhang, Q.; Zhang, Q.; Shi, W.; Li, Q. Lavea: Latency-aware video analytics on edge computing platform. In Proceedings of the Second ACM/IEEE Symposium on Edge Computing, San Jose, CA, USA, 12–14 October 2017; pp. 1–13.
46. Torres, D.R.; Martín, C.; Rubio, B.; Díaz, M. An open source framework based on Kafka-ML for Distributed DNN inference over the Cloud-to-Things continuum. *J. Syst. Archit.* **2021**, *118*, 102214. [CrossRef]
47. Tuli, S.; Basumatary, N.; Buyya, R. Edgelens: Deep learning based object detection in integrated iot, fog and cloud computing environments. In Proceedings of the 2019 4th International Conference on Information Systems and Computer Networks (ISCON), Mathura, India, 21–22 November 2019; IEEE: Piscataway, NJ, USA, 2019; pp. 496–502.
48. Guo, Q.; Chen, S.; Xie, X.; Ma, L.; Hu, Q.; Liu, H.; Liu, Y.; Zhao, J.; Li, X. An empirical study towards characterizing deep learning development and deployment across different frameworks and platforms. In Proceedings of the 2019 34th IEEE/ACM International Conference on Automated Software Engineering (ASE), San Diego, CA, USA, 11–15 November 2019; IEEE: Piscataway, NJ, USA, 2019; pp. 810–822.
49. Arslan, M.Y.; Singh, I.; Singh, S.; Madhyastha, H.V.; Sundaresan, K.; Krishnamurthy, S.V. Cwc: A distributed computing infrastructure using smartphones. *IEEE Trans. Mob. Comput.* **2014**, *14*, 1587–1600. [CrossRef]
50. Chatzopoulos, D.; Ahmadi, M.; Kosta, S.; Hui, P. Flopcoin: A cryptocurrency for computation offloading. *IEEE Trans. Mob. Comput.* **2017**, *17*, 1062–1075. [CrossRef]
51. Chatzopoulos, D.; Ahmadi, M.; Kosta, S.; Hui, P. OPENRP: A reputation middleware for opportunistic crowd computing. *IEEE Commun. Mag.* **2016**, *54*, 115–121. [CrossRef]
52. Roy, P.; Sarker, S.; Razzaque, M.A.; Mamun-or Rashid, M.; Hassan, M.M.; Fortino, G. Distributed task allocation in mobile device cloud exploiting federated learning and subjective logic. *J. Syst. Archit.* **2021**, *113*, 101972. [CrossRef]
53. IEA Reports. Available online: https://www.iea.org/reports/world-energy-outlook-2024 (accessed on 7 March 2025).
54. IEA Electricity 2024. Available online: https://www.iea.org/reports/electricity-2024 (accessed on 7 March 2025).
55. Masanet, E.; Shehabi, A.; Lei, N.; Smith, S.; Koomey, J. Recalibrating global data center energy-use estimates. *Science* **2020**, *367*, 984–986. [CrossRef]
56. Malmodin, J.; Lövehagen, N.; Bergmark, P.; Lundén, D. ICT sector electricity consumption and greenhouse gas emissions–2020 outcome. *Telecommun. Policy* **2024**, *48*, 102701. [CrossRef]
57. Schmidt, A.; Stock, G.; Ohs, R.; Gerhorst, L.; Herzog, B.; Hönig, T. carbond: An operating-system daemon for carbon awareness. *ACM SIGENERGY Energy Inform. Rev.* **2024**, *4*, 52–57. [CrossRef]
58. Software Carbon Intensity (SCI). Available online: https://sci.greensoftware.foundation/ (accessed on 7 March 2025).
59. Chen, Y.; Zheng, B.; Zhang, Z.; Wang, Q.; Shen, C.; Zhang, Q. Deep learning on mobile and embedded devices: State-of-the-art, challenges, and future directions. *ACM Comput. Surv. CSUR* **2020**, *53*, 1–37. [CrossRef]

sensors

MDPI

Article

Enhancing Connected Health Ecosystems Through IoT-Enabled Monitoring Technologies: A Case Study of the Monit4Healthy System

Marilena Ianculescu [1], Victor-Ştefan Constantin [1,*], Andreea-Maria Guşatu [1,*], Mihail-Cristian Petrache [1], Alina-Georgiana Mihăescu [1], Ovidiu Bica [1] and Adriana Alexandru [1,2]

1 National Institute for Research and Development in Informatics, 011455 Bucharest, Romania; marilena.ianculescu@ici.ro (M.I.); cristian.petrache@ici.ro (M.-C.P.); alina.mihaescu@ici.ro (A.-G.M.); ovidiu.bica@ici.ro (O.B.); adriana.alexandru@ici.ro (A.A.)

2 Faculty of Electrical Engineering, Electronics and Information Technology, Valahia University of Targoviste, 130004 Targoviste, Romania

* Correspondence: victor.constantin@ici.ro (V.-Ş.C.); andreea.gusatu@ici.ro (A.-M.G.); Tel.: +40-78-604-4271 (V.-Ş.C.); +40-75-179-1762 (A.-M.G.)

Abstract: The Monit4Healthy system is an IoT-enabled health monitoring solution designed to address critical challenges in real-time biomedical signal processing, energy efficiency, and data transmission. The system's modular design merges wireless communication components alongside a number of physiological sensors, including galvanic skin response, electromyography, photoplethysmography, and EKG, to allow for the remote gathering and evaluation of health information. In order to decrease network load and enable the quick identification of abnormalities, edge computing is used for real-time signal filtering and feature extraction. Flexible data transmission based on context and available bandwidth is provided through a hybrid communication approach that includes Bluetooth Low Energy and Wi-Fi. Under typical monitoring scenarios, laboratory testing shows reliable wireless connectivity and ongoing battery-powered operation. The Monit4Healthy system is appropriate for scalable deployment in connected health ecosystems and portable health monitoring due to its responsive power management approaches and structured data transmission, which improve the resiliency of the system. The system ensures the reliability of signals whilst lowering latency and data volume in comparison to conventional cloud-only systems. Limitations include the requirement for energy profiling, distinctive hardware miniaturizing, and sustained real-world validation. By integrating context-aware processing, flexible design, and effective communication, the Monit4Healthy system complements existing IoT health solutions and promotes better integration in clinical and smart city healthcare environments.

Keywords: IoT; connected health ecosystems; health monitoring; Monit4Healthy system; sensor fusion; real-time biomedical signal processing

1. Introduction

By enabling highly connected, real-time health monitoring ecosystems featuring the use of biomedical sensors, cutting-edge communication technologies, and complex data processing frameworks, the Internet of Things (IoT) has totally reshaped the world of healthcare. The growing demand for ongoing, remote patient monitoring, notably for the management of chronic diseases, rehabilitation from surgeries, and eldercare, is a major driver of this approach toward societal transformation [1]. Traditional healthcare

 https://doi.org/10.3390/s25072292

models are changing dramatically in their pursuit of data-centric, preventative, and highly customized medicine. Such models have previously hinged on established in-person consultations and reactive forms of therapy. Real-time patient monitoring is distinguished as an integral part of clinical decision making in this constantly evolving setting, facilitating preventative care and greatly boosting patient outcomes.

Patient care, disease prevention, and comprehensive health management are all influenced by a complex system of healthcare professionals, medical establishments, innovative technologies, and regulatory frameworks that jointly make up a health ecosystem. These ecosystems are typically confined to big healthcare institutions, notably clinics, hospitals, and specialized care centers, where physicians establish diagnoses and manage patients mostly through recurrent relationships with patients and subsequent data analysis [2]. The easy share of clinical information, the merging of interdisciplinary healthcare services, and the accessibility of medical infrastructure that enables both therapeutic and preventive procedures are all key to the effectiveness of such ecosystems. Concerns ranging from fragmented data, inadequate interoperability, and limited access to real-time patient information are prevalent in traditional health ecosystems and can compromise clinical decision making and interfere with the continuity of qualitative care.

By taking advantage of digital transformation, IoT-driven sensor networks, cloud-powered analytics, and artificial intelligence (AI) for providing real-time, data-driven healthcare services, a connected health ecosystem optimizes traditional healthcare setups. This approach allows for the uninterrupted gathering of data, secure transmission, and smart processing throughout different medical contexts by smoothly integrating medical devices, wearable sensors, remote monitoring systems, and electronic health records. In addition to strengthening patient monitoring, this integrated model supports more proactive, tailored, and reliable healthcare delivery. A connected health ecosystem facilitates the systematic detection of abnormalities, the predictive modeling of disease evolution, and continuous remote health monitoring and management, as opposed to traditional health systems where the observation of patients is intermittent and exclusive to hospitals and medical centers.

Through the application of biomedical sensors, edge computing, and wireless communication protocols to provide real-time, automated, and continuous health evaluations, the integration of IoT-enabled monitoring technologies has radically changed the medical field [3]. Such advances enhance patient monitoring capabilities while rendering it more feasible to identify health abnormalities at an early stage as it allows for the straightforward gathering, exchange, and analysis of physiological and physical data [4]. IoT-enabled systems feature the real-time tracking of key health parameters within medical centers, remote healthcare environments, and smart urban infrastructures, in contrast to traditional health monitoring solutions that primarily depend on random assessments conducted at clinical appointments. A more thorough patient monitoring and improved clinical decision support originated from the development of multi-sensor systems that integrate technologies like electrocardiography (ECG), oxygen saturation (SpO_2), electromyography (EMG), galvanic skin response (GSR), and motion-tracking technologies. Such systems have also significantly boosted the accuracy and extent of digital health assessments. Incorporating edge computing architectures with low-latency communication protocols maximizes data processing efficiency, reduces dependency on centralized cloud computing, and enables the prompt identification of abnormalities and system resilience. In line with rising demands for integrated healthcare ecosystems and efficient smart city integration, these improvements in technology have widened up the possibility for scalable, secure, and energy-efficient health monitoring solutions.

Multi-sensor IoT health monitoring solutions have lately progressed amid notable developments in wireless communication, sensor miniaturization, and real-time analytics. The timely detection of abnormalities has been enhanced by the merging of AI-driven health analytics with real-time physiological data collection, facilitating built-in alerts, predictive analytics, and the recognition of health fluctuations at the outset. By simply integrating IoT-enabled health solutions into telemedicine infrastructures, remote patient health monitoring systems most especially have radically shifted the way healthcare is delivered [5]. They make certain that real-time physiological data are adequately shared and processed in order to facilitate remote clinical evaluations.

Emerging IoT health ecosystems improve personalized care by fostering efficient interaction between patients, healthcare providers, and smart urban health infrastructure. As IoT-enabled technologies develop further, their impact transcends well beyond monitoring a single patient; they could ultimately be applied to guide health policy decisions, streamline hospital resource management, and even deliver population-level smart health oversight on an extensive basis.

IoT-enabled health monitoring is evolving quickly, but there are still a number of important challenges to overcome, especially when it comes to real-time data transmission, sensor communication, system scalability, and strict security standard compliance. For healthcare applications like movement analysis, neuromuscular evaluations, and cardiac analysis that need ultra-low latency and high-performance monitoring in order to provide reliable and helpful data to healthcare professionals, they have to be operated practically in real time. There are compromises among latency, bandwidth capacity, and energy efficiency brought about by frequently implemented wireless communication protocols such as Wi-Fi (Wireless Fidelity) and Bluetooth Low Energy (BLE) [6]. Although Wi-Fi allows for fast data transfer, in high-density healthcare environments, it requires significant amounts of power and can be vulnerable to network bottlenecks. Large-scale healthcare implementations could face limitations due to BLE's limited bandwidth and inadequate data throughput, despite its low-power operation optimization.

The implementation of low-power, long-range IoT communication technologies, including NB-IoT and Long-Range Wide-Area Network (LoRaWAN), which are increasingly endorsed for having the capacity for establishing scalable remote health monitoring infrastructure [7,8], is an alternative option. These protocols are especially appropriate for the continuous monitoring of patients with increased risk in residence-based care settings because they convey wide-area data transmission with low power consumption. Their slower transfer speeds restrict their ability to gather high-frequency biomedical parameters, exposing a basic compromise between real-time responsiveness and energy efficiency. Adaptive communication models that combine system performance, scalability, and real-time medical insights by boosting data transmission efficiency depending on particular healthcare solution demands need to be developed to tackle this issue.

Achieving flattened multi-sensor integration and interoperability is one of the key hurdles in IoT-based healthcare ecosystems. It remains difficult to manage data fusion, generate effective sensor communication, and maintain shared interoperability since plenty of the current health monitoring systems are based on fragmented and heterogeneous sensor architectures. Scalability and simple integration into current health information systems and urban IoT infrastructures are hampered by large variations in sensor approaches, operating frequencies, and proprietary communication protocols, which add complexity to comprehensive data standards. MQTT-based IoT messaging and HL7 Fast Healthcare Interoperability Resources (FHIRs) are two recent improvements in standardized frameworks that attempt to address these shortcomings in interoperability [9,10], but acceptance is still lagging because of ongoing standardization shortages, existing system limitations, and regulatory complexities.

As IoT-enabled healthcare solutions keep developing, data security and patient privacy become more important challenges transcending interoperability. End-to-end encryption, reliable authentication procedures, and strict compliance with healthcare data protection security like the General Data Protection Regulation (GDPR) and the Health Insurance Portability and Accountability Act (HIPAA) [11] are prerequisites for widespread IoT health solutions. In order to prevent data breaches, unauthorized access, and cyberattacks on critical healthcare infrastructure, robust cybersecurity measures must be implemented given the increasing number of connected medical devices distributing highly sensitive patient data. Scalable and secure data-sharing frameworks are becoming incrementally more significant as IoT health monitoring devices [12] connect more and more with smart city infrastructures. Decentralized health data management systems, encrypted data-sharing protocols, and secure cloud infrastructures must be implemented to tackle these challenges and protect patient data while facilitating real-time, data-driven healthcare delivery.

Recent research efforts have been focused on bolstering real-time processing frameworks, improving data transmission protocols, and expanding sensor-to-network communication in IoT healthcare settings [13–15] in order to mitigate such barriers. A key driver in increasing data processing performance, reducing network bottlenecks, and maintaining continuous health monitoring in insightful, application scenarios [16] is the widespread use of versatile IoT communication models that incorporate low-latency transmission protocols and edge computing architectures. The emergence of edge computing architectures in IoT-based healthcare solutions is among the most significant improvements in this domain. Edge computing renders it possible to preprocess data in real time at the sensor or gateway level, as opposed to conventional cloud-centric approaches that are frequently limited by latency and bandwidth issues. It lessens the computational load on centralized healthcare servers while improving transmission efficiency. These solutions might instantly recognize abnormalities, filter redundant data, and trigger alerts by integrating AI-driven signal processing capabilities. This boosts system responsiveness and supports medical decision making.

New opportunities for extensive, real-time health monitoring are being brought about by the convergence of IoT health monitoring with smart city infrastructures. It opens up new opportunities for public health monitoring, emergency response synchronization, and AI-enhanced demographic surveillance when IoT-enabled healthcare solutions are plugged into smart city ecosystems. Real-time predictive analytics for major health events and location-aware health services may be realized through the use of low-power, distributed health monitoring devices that proactively link with urban IoT infrastructures. Such developments strengthen the role of IoT in influencing healthcare's evolution by fostering the emergence of public health systems that are more robust and responsive.

This paper presents the Monit4Healthy system, an extensive platform that integrates multi-sensor fusion, optimized real-time biomedical signal processing, and effective IoT communication frameworks to improve connected health ecosystems in line with a rising need for proficient IoT-enabled health monitoring solutions. By tackling important concerns with sensor interoperability, low-latency data transmission, and secure health data exchange, the system contributes to developing healthcare infrastructures that are more savvy, versatile, and interconnected.

The Monit4Healthy system features a modular sensor fusion architecture to provide thorough physiological and biomechanical health evaluations, in contrast with other IoT health platforms that frequently focus on collecting single-sensor data. By ensuring consistent data integration, this system promotes both the accuracy and efficacy of health monitoring. By combining edge computing, secure communication protocols, and smart data transmission approaches, the system improves the scalability, efficiency, and reliability of real-time health monitoring solutions. Its structure endorses remote healthcare management and continuous patient monitoring, where all of the above are part of the larger context of next-generation IoT-based connected health ecosystems. The conceptual framework of the system may be customized to fit the health infrastructures of smart cities, enabling population-wide health analytics, emergency response coordination, and real-time health tracking. With the aforementioned approaches, the Monit4Healthy system accounts for the establishment of smart and data-driven healthcare solutions by demonstrating an affordable and reliable way of overcoming current bottlenecks in IoT healthcare solutions.

By merging multi-sensor fusion with a flexible and configurable architecture, the Monit4Healthy system enables a novel approach to IoT-enabled health monitoring that is appropriate for a range of healthcare settings. By providing context-aware physiological and biomechanical monitoring, the Monit4Healthy system helps to guarantee a better degree of accuracy and accessibility in both clinical and non-clinical environments, in contrast to traditional IoT health solutions that typically focus on collecting single-sensor data. Its focus on smart information interchange and real-time physiological signal processing facilitates better decision making in connected health ecosystems. The Monit4Healthy system's architecture takes into account real-world implementation concerns certifying that the system can be smoothly integrated into current digital health infrastructures with no mandating major architectural changes. In tackling current challenges with interoperability, data accuracy, and secure sensor communication, these characteristics showcase the Monit4Healthy system as a constructive tool for IoT-driven healthcare and boost the position of connected health ecosystems in next-generation healthcare solutions.

Many present solutions are still constrained by fragmented sensor architectures, a lack of real-time edge processing, and ineffective data transmission procedures, which impede scalability, accuracy, and adaptability in actual healthcare environments, even with notable advancements in IoT-based healthcare monitoring. These drawbacks underscore the urgent need for integrated solutions that provide context-aware processing, set up, low-latency communication, and multi-sensor data fusion across adaptable and interoperable technologies. The Monit4Healthy system was designed specifically to meet these gaps by providing a cohesive and flexible approach that overcomes the present drawbacks of IoT health monitoring technology.

Even though several IoT-enabled health monitoring systems have emerged lately, they often face significant shortcomings when it comes to integrating several sensors easily, processing data adaptively, and communicating effectively in real time when resources are limited. Particularly in wearable and portable health monitoring circumstances, high latency, inefficient bandwidth, and increased energy consumption are caused by the inflexible architectures found in several current digital health solutions, such as in cloud-dependent processing and continuous raw data streaming. An extensive number of IoT-based health solutions rely on single-sensor data collection processes and are not sufficiently versatile to provide scaled, context-aware health monitoring in a variety of settings. Section 2, which assesses the current situation of IoT-based health communication technologies and underlines the functional constraints of present systems, provides a detailed analysis of these shortcomings.

In order to ensure low-latency communication and interoperability with current digital health infrastructures, adaptive, modular IoT health systems that can perform multi-sensor fusion, structured data transmission, and edge-level signal processing are required. These challenges highlight a clearly defined research gap. As a way to bridge this gap, the digital health solutions must be able to minimize computational overhead and energy consumption while constantly adjusting gathering and transmission techniques in response to the real-time physiological parameters and limitations of the network.

In an attempt to swiftly tackle these operational and technological issues, the Monit4Healthy system was designed and developed. In particular, in environments that demand trustworthy remote management and compatibility with connected health ecosystems, such as smart city healthcare infrastructures, its modular architecture, context-aware signal processing, and structured, event-driven data transmission framework as a whole comprise an alternative and integrated way of advancing IoT-based health monitoring. Thus, the Monit4Healthy system is a base for future implementations in real-world healthcare and city environments, offering a scalable and adaptable solution to the present drawbacks of IoT health monitoring systems.

This paper's remaining sections are organized as follows: Section 2 provides a comprehensive review of IoT-based health communication technologies, focusing on advancements in sensor fusion and real-time biomedical signal processing. Section 3 details the Monit4Healthy system architecture, covering sensor integration, communication frameworks, and data acquisition processes. Current results are discussed in Section 4, where the system's sensor performance, data transmission efficiency, and overall functional reliability are evaluated. The Monit4Healthy system's broader impacts are covered in Section 5, which also examines its potential to integrate with smart city infrastructures. Lastly, Section 6 highlights the significance of IoT in forming next-generation healthcare solutions while summarizing significant accomplishments and outlining future research areas.

2. Literature Review

2.1. IoT Technologies in Health Monitoring

Proactive medical interventions, early disease identification, and ongoing remote patient surveillance are made possible by IoT-enabled health monitoring systems. Digital healthcare is changing as a result of the integration of wearable sensors, cloud-based platforms, and AI-driven analytics. By improving operational efficiency, giving medical personnel data-driven insights, and encouraging proactive, personalized treatment, the ongoing development of IoT technologies in healthcare has completely changed how patient care is delivered.

2.1.1. Key Advancements and Applications of IoT in Healthcare

IoT-enabled devices facilitate the 24/7 real-time health monitoring of vital signs like blood pressure, heart rate, and SpO_2. This enables more effective patient care by facilitating the early detection of medical conditions, improved chronic disease management, and timely alerts for healthcare professionals [17].

Advanced sensors are used by wearable devices like smartwatches, fitness trackers, and biosensors to monitor a variety of parameters, including heart rate and rhythm, ECG, sleep patterns (duration and quality of sleep), activity levels (steps, total movement, motion tracking), physiological signs of stress, etc. Algorithms are used to gather huge amounts of real-time health data [18]. Medical practitioners can take preventative measures for early warning systems for metabolic and cardiovascular disorders.

Remote diagnosis and seamless communication are made possible by integrating IoT devices with telemedicine platforms. This decreases needless hospital stays, increases access to healthcare in underserved and rural areas, and improves senior care through remote monitoring [19].

Large volumes of data are produced by IoT devices, and machine learning and AI are used to integrate and analyze these data. Insights from predictive analytics are utilized to prevent disease, provide personalized treatment plans, and make better operational decisions in hospitals [20].

Real-time patient, staff, and medical equipment tracking is made possible by IoT-enabled infrastructure, which speeds up emergency response times and optimizes resource use in hospital and smart home systems [21].

Key IoT applications in the healthcare sector include the following:

a. Chronic Disease Management—In order to effectively manage chronic conditions, IoT devices like blood pressure monitors, connected inhalers, and blood glucose monitors are essential [22]. This includes continuous blood pressure monitoring for patients with hypertension, smart inhalers for monitoring COPD, and real-time glucose tracking for managing diabetes [23].

b. Remote Patient Monitoring (RPM)—Heart rate, blood pressure, temperature, and other health data can be automatically collected through wearables and portable ECG monitors and sent to a software program where algorithms can evaluate the information to recommend treatments or generate alarms [24]. RPM makes it easier to track health in real time, which improves care for older people who live independently and enables preventive treatments for high-risk patients [25]. Healthcare professionals may monitor patient health in real time thanks to the continuous flow of data, which reduces the need for frequent in-person visits.

c. Preventive Healthcare—Fitness trackers and smartwatches aid in health maintenance by monitoring physical activity, sleep patterns, and stress levels through heart rate variability analysis [26]. These devices promote proactive well-being management and healthier lifestyle choices.

2.1.2. Relevance to Connected Ecosystems

IoT-enabled solutions facilitate smooth integration and real-time interaction between patients, healthcare providers, and medical devices across many healthcare stakeholders, laying the groundwork for connected health ecosystems [27].

Critical health and sensitive patient data are continuously gathered by a variety of medical devices and sent via the Internet for further analysis and decision making in an interconnected environment made possible by IoT technologies [28,29]. Real-time interactions are made possible through the seamless communication between these systems and devices inside the ecosystem, which also keeps patients, physicians, hospitals, and caregivers informed. Through the aggregation of data from several IoT devices, this integration promotes a linked healthcare network and offers comprehensive insights.

Cloud-based platforms and data integration technologies are used by connected health ecosystems to facilitate the gathering, secure storage, and analysis of patient data from IoT devices [30], improving decision making and enhancing collaboration across healthcare institutions.

Through the continuous analysis of patient data within the ecosystem and the provision of real-time health insights by means of wearables or applications, IoT-enabled monitoring enhances proactive and personalized care [31]. Furthermore, healthcare practitioners receive early warning alerts from AI-driven analytics, which helps avoid complications and decrease the number of hospitalizations.

Regardless of geographical location, connected ecosystems in healthcare bring together different stakeholders—hospitals, clinics, patients, and devices—into a cohesive framework that improves collaboration and provides treatment continuity [32]. For enhanced overall efficiency, hospitals deploy IoT systems for equipment optimization, resource management, and patient tracking.

These ecosystems also contribute to population health monitoring by detecting trends like disease outbreaks or high-risk patients [33], enabling targeted interventions through the utilization of data-driven insights [34].

Seniors can live independently with personalized care plans and remote monitoring thanks to the seamless integration of smart home health systems and linked wearables into connected ecosystems. Smart home health systems further integrate wearable devices, allowing seniors to maintain independent living with continuous health support [35].

2.2. Challenges in IoT-Based Health Monitoring

Despite IoT benefits for health monitoring, a number of challenges prevent its widespread adoption of IoT-enabled healthcare systems. These challenges span the technical, operational, and ethical dimensions [36]:

- *Data accuracy and reliability:* ensuring that IoT devices provide accurate and consistent health data for effective decision making;
- *Data security and privacy*: strong data encryption, authentication mechanisms, secure transmission in IoT health applications, and compliance with healthcare regulations like GDPR (EU) and HIPAA (USA) are required to protect sensitive patient information;
- *Infrastructure and connectivity:* reliable internet connections (such as Wi-Fi and 5G) and sufficient network capacity are essential for real-time data transmission;
- *Energy efficiency:* continuous data transmission in wearable devices must be optimized for low power consumption to extend battery life;
- *AI and machine learning integration*: implementing real-time anomaly detection and predictive analytics necessitates powerful computers and optimized software;
- *Interoperability and scalability:* the adoption of interoperability frameworks (such as FHIR) is necessary to ensure the seamless integration of various IoT devices and healthcare systems;
- *Ethical considerations*: addressing issues with patient autonomy, consent, and data ownership is critical for responsible IoT implementation in healthcare;
- *Cost and accessibility:* removing financial obstacles for smaller healthcare providers and guaranteeing fair access to IoT-enabled healthcare, especially in low-income or vulnerable regions, are important for avoiding a digital divide.

Resolving these issues is essential to guaranteeing the success of connected health ecosystems and requires continuous innovation and cooperation between producers, healthcare providers, and legislators.

2.3. Communication Technologies for IoT Healthcare

Efficient communication protocols are fundamental to IoT-driven healthcare systems, ensuring real-time monitoring, secure data transmission, and minimal latency for many IoT-driven healthcare applications in order to guarantee real-time monitoring and timely alerts. For example, rapid data transmission is necessary for emergency response apps, smart wearables, and remote patient monitoring systems to identify critical medical problems like hypoglycemia or cardiac arrhythmias.

2.3.1. Latency and Bandwidth Challenges

Significant difficulties arise from latency and bandwidth constraints, especially in settings with a high device density or unreliable network infrastructure. Many IoT healthcare applications require ultra-low latency to ensure timely alerts and immediate responses in many IoT health applications, like insulin administration systems, fall detection, and cardiac monitoring, to guarantee prompt reactions in emergency situations. The effectiveness of treatment and patient safety could be compromised by any delay in data transmission.

These issues are addressed by several wireless technologies with differing trade-offs:

- Wi-Fi's fast transmission rates make it ideal for sending massive amounts of medical data, including continuous vital sign monitoring or high-resolution imaging. It can be used for high-bandwidth medical image transfers and real-time video consultations, but its high-power consumption limits it from being used in battery-operated devices [37].
- BLE is intended for low-power consumption, making it perfect for wearable health devices and short-range communication between medical sensors and smartphones, yet its short range prevents widespread coverage in hospital settings or remote monitoring scenarios [38].
- LoRaWAN offers long-range connectivity with low power consumption, which makes it ideal for asset tracking in hospitals and rural healthcare applications for sending small health-related data packets over long distances. However, because of its low data rate, LoRaWAN is not appropriate for sending high-resolution medical data [39].

Communication bottlenecks in high-density environments or unreliable network infrastructure can hinder real-time patient monitoring. In order to maximize effectiveness, reliability, and real-time responsiveness in IoT-driven healthcare systems, a hybrid strategy combining many wireless technologies is needed to address these latency and bandwidth limitations.

2.3.2. Energy Constraints and the Need for Low-Power IoT Communication

With the use of wearable and implanted technology, remote health monitoring systems are essential to contemporary healthcare since they allow for continuous patient surveillance. Managing energy restrictions to increase device longevity and guarantee dependable performance is an essential challenge in these systems.

Power-Efficient Sensing and Transmission: Wireless Body Area Networks, which are made up of sensor nodes that run on batteries and gather and send physiological data, are frequently utilized in health monitoring. Since regular battery replacements are impractical, particularly for implantable devices, energy efficiency in these networks is essential. Research indicates that the transmission and processing of data uses a lot of energy. Low-power communication protocols such as BLE and LoRaWAN offer extended battery life while ensuring reliable data transmission.

Techniques that include eliminating redundant data at the sensor level, lowering the amount of transmitted data, and conserving energy have been suggested as ways for reducing this [40].

2.3.3. Advances in Low-Power Communication Protocols

The primary objective of communication protocol advancements has been to utilize less energy. Emerging protocols such as NB-IoT and Zigbee optimize data transmission efficiency while minimizing power consumption. Low-power communication protocols such as BLE and LoRaWAN offer extended battery life while ensuring reliable data transmission. The treNch protocol is capable of lowering power consumption by one to two orders of magnitude in comparison to various BLE modes in a variety of scenarios while maintaining an equivalent level of service quality [38,41].

2.3.4. Energy Harvesting Techniques

One possible solution to address energy limits is to incorporate energy harvesting techniques into IoT devices. Battery power can be substituted or supplemented by methods including solar, thermal, and kinetic energy harvesting, allowing devices to run for extended periods of time between charges [42–45].

A variety of strategies, such as energy harvesting methods, sophisticated low-power communication protocols, and power-efficient sensing, are used to address energy limits in remote health monitoring.

2.3.5. Cloud vs. Edge Computing in Healthcare IoT

In healthcare IoT systems, the choice between cloud and edge computing architectures significantly impacts performance, particularly concerning latency and bandwidth [46,47].

Cloud-Based IoT Platforms: Cloud computing has significant scalability. However, because data must be transmitted from the IoT devices to the cloud and back, this centralized solution may cause latency problems and network dependency. Even though cloud computing enables scalability, in time-sensitive healthcare applications where instantaneous data processing is essential, such delays could be critical [48].

Edge Computing Solutions: By processing data locally on embedded devices or nearby servers, edge computing overcomes these issues by lowering latency and bandwidth consumption. Real-time data analysis and decision making are made possible through this decentralized method, which is essential for applications like emergency response and patient monitoring [49].

2.4. Sensor Fusion and Data Processing in IoT Health Systems

In order to produce a more accurate and complete representation of a subject or environment, multi-sensor fusion enhances the accuracy and reliability of health monitoring by combining data from multiple sources such as motion sensors, biomedical sensors, and environmental sensors, producing insights faster and allowing for more sophisticated analysis.

2.4.1. Importance of Sensor Fusion for Accurate Health Monitoring

Sensor fusion enables a more comprehensive understanding of patient health, enabling activity monitoring through the integration [50] of multiple sensor modalities. For example, combining electrocardiography (ECG) with motion-tracking technologies enhances cardiac monitoring accuracy.

Some examples of multi-sensor fusion applications are provided below:

Gait Analysis: Combining wearable technology with data from vision-based systems, such as Kinect sensors, improves the comprehension of joint movements in gait analysis and results in more accurate evaluations of gait anomalies [51,52].

Vital Signs Monitoring: By combining information from many biomedical sensors, vital signs can be monitored more precisely, which enhances the identification of disorders including arrhythmias and sleep apnea [53,54].

Activity Recognition: Through activity tracking, multimodal fusion sensors in body-worn devices have enhanced physical activity monitoring, preserving or improving the quality of social and personal lives [55].

2.4.2. AI and Machine Learning for Predictive Analytics

Advanced AI-driven analytics improve healthcare decision making by detecting health anomalies, predicting disease progression, and offering personalized treatment recommendations.

2.4.3. Emerging Trends in Multi-Sensor Data Processing

Recent advancements in wearable technology and edge computing facilitate real-time data analysis while minimizing network dependency. AI-enhanced data filtering techniques enhance data accuracy and decrease redundant transmissions.

These advancements in sensor fusion technologies are essential for facilitating medical diagnosis and improving healthcare quality.

This literature review has examined key advances, challenges, and emerging trends in IoT-enabled healthcare monitoring. By leveraging multi-sensor fusion, AI-driven analytics, and efficient communication protocols, IoT-based health ecosystems have the potential to enhance real-time patient monitoring and personalized healthcare delivery. However, addressing challenges related to data security, interoperability, and energy efficiency remains critical for the widespread adoption of IoT in healthcare.

3. Materials and Methods

3.1. Monit4Healthy System Architecture and Design

The Monit4Healthy system was designed using a modular sensor fusion architecture that ensures scalability, interoperability, and real-time data acquisition. The system is structured across multiple levels. The architecture consists of five core layers:

- *Data Layer*: This layer involves the real-time collection of physiological and environmental data using advanced biomedical sensors integrated into multiple monitoring devices.
- *Communication Layer*: This layer facilitates efficient and secure data transmission. Wearable devices rely on BLE for short-range communication, while high-bandwidth data transfers occur via Wi-Fi to ensure seamless connectivity with cloud-based storage and processing systems.
- *Edge Processing Layer*: This layer is responsible for local data processing and initial analysis before transmitting physiological and biomechanical data to the cloud. This layer operates on a Raspberry Pi, which processes raw sensor signals from the Medical Blackbox, AcU Blackbox, EMG Blackbox, and Gaitband IoT-based devices (designed and developed for the Monit4Healthy system). Compared to a cloud-only setup, this approach improves latency, minimizes bandwidth usage by preprocessing data locally, and ensures real-time responsiveness. The implementation of edge computing in IoT systems introduces several significant advantages: (1) reduced latency, by processing data directly at the edge nodes, ensuring fast response times essential for real-time healthcare monitoring; (2) bandwidth efficiency, as only relevant, processed data are transmitted to cloud servers, reducing network congestion; (3) system autonomy and fault tolerance, allowing continued local operation even when cloud services are temporarily unavailable; and (4) enhanced privacy and security, as sensitive patient data are filtered and processed locally, limiting external exposure and improving compliance with data protection standards. Figure 1 highlights the data flow between edge and cloud layers [56].
- *Processing and Cloud Layer and Application Sublayer*: The Processing and Cloud Layer is responsible for data storage, structured representation, and alerting mechanisms. This layer aggregates sensor data transmitted from the Edge Processing Layer, organizing them for structured analysis and long-term storage. It includes interactive data processing tools, enabling customizable health metric representations and structured reports. A cloud-based alert and notification system operates within this layer, continuously analyzing incoming physiological and biomechanical data. It detects deviations from predefined health thresholds and triggers automated alerts through multi-channel notifications. Secure storage mechanisms are integrated to ensure data encryption, controlled access, and regulatory compliance for sensitive health information. The Application Sublayer acts as the central processing interface for managing sensor data, health alerts, and system configurations. This layer facilitates remote monitoring functionalities, allowing healthcare providers to manage real-time and

historical patient data. It is responsible for user authentication, system configuration, and data synchronization across cloud servers and edge devices. It also handles communication between medical professionals and patients, ensuring the seamless integration of monitoring data with personalized medical recommendations.

- *Visualization Layer*: This layer focuses on presenting real-time health insights in a user-friendly format. It provides interactive interfaces for different stakeholders, including patients, clinicians, and researchers, ensuring efficient access to health data. Features include dynamic dashboards, trend graphs, and anomaly detection alerts, ensuring that medical professionals and patients receive timely information.

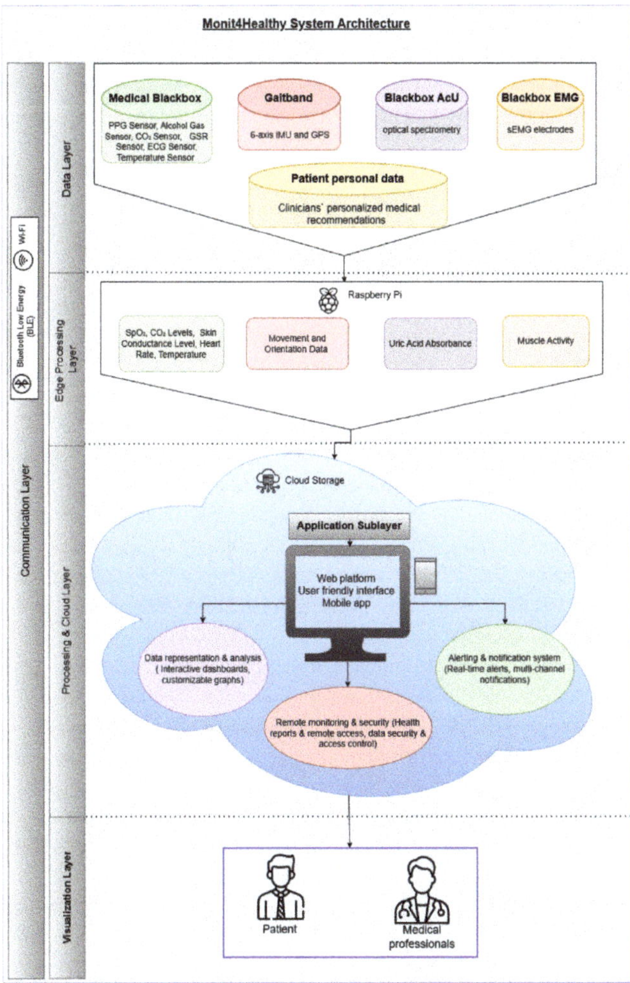

Figure 1. Overview of the Monit4Healthy system architecture, showing the interaction between data collection (wearable and portable medical devices), wireless communication (BLE and Wi-Fi modules), edge processing (Raspberry Pi), cloud storage and analytics (ICIPRO platform), and visualization layers (clinician dashboards and mobile interfaces). The diagram highlights the structured flow of biomedical data—acquired via multi-sensor fusion—transmitted securely from the Medical Blackbox and other devices to the cloud infrastructure for real-time monitoring, anomaly detection, and visualization within connected health ecosystems.

To guarantee organized aggregation as well as secure handling, real-time health data gathered from wearable devices are sequentially processed through many levels in the Monit4Healthy system. In addition to transmitting raw sensor data, the shift from edge computing to cloud infrastructure additionally involves preprocessed signals that have already been filtered and structured for additional analysis. The integrity and accuracy of physiological and biomechanical parameters are maintained while bandwidth consumption is reduced using this sequential method.

The structural organization and interaction between these layers are visually represented in Figure 1, which illustrates the Monit4Healthy system architecture. This figure provides a comprehensive overview of how data flow from the initial sensor measurements to real-time decision-making and visualization tools. It highlights the integration of hardware and software components, showcasing the seamless connectivity between IoT devices, cloud infrastructure, and end-user applications. This architecture was designed for ensuring the system's ability to deliver accurate, efficient, and scalable health monitoring solutions.

The Monit4Healthy system integrates a suite of specialized biomedical IoT-based devices, each designed for specific health monitoring applications:

- *Medical Blackbox*: A multifunctional enclosure that integrates various medical-grade sensors, designed to facilitate efficient data acquisition and processing. It captures and analyzes vital parameters such as heart rate, SpO_2, carbon dioxide levels, and GSR. This device integrates high-precision sensors for real-time physiological assessment, ensuring the accurate detection of health anomalies.
- *Gaitband*: A wearable device designed for real-time gait pattern monitoring, optimized for ergonomic comfort and sensor placement stability. Equipped with an LSM303DLHC magnetometer and accelerometer, the Gaitband continuously monitors gait patterns, posture stability, and movement dynamics. The SAM-M8Q GPS (Global Positioning System) module enhances location tracking and movement analysis, while the HM-10 Bluetooth module and ESP8266 (NodeMCU) facilitate secure wireless data transmission. By analyzing motion data, the device detects irregular walking patterns, potential fall risks, and neuromuscular impairments, providing valuable insights for mobility assessment and rehabilitation.
- *EMG Blackbox*: A specialized enclosure engineered for housing the EMG sensor array, ensuring both signal integrity and protection from external interference. Utilizes surface electromyography (sEMG) sensors for neuromuscular function assessment, measuring electrical muscle activity to monitor conditions such as muscle fatigue and neurodegenerative disorders.
- *AcU Blackbox*: A device that employs advanced optical spectrometry techniques for non-invasive uric acid monitoring in urine samples. The device provides real-time biochemical analysis, enabling the early detection of metabolic imbalances.
- *Patient personal data*: Based on personal medical history and previous measured parameters and information from the Monit4Healthy database, clinicians can issue personalized medical recommendations to the patients under their care, in order to communicate clinical or lifestyle actions that the patient should follow. Each recommendation comprises several attributes, such as the title, an optional reference to the medical device that the patient might use in order to complete the action, and the medical recommendation from the physician. The title serves as a brief summary of the action, while the description offers details and guidance for the patient. The medical device is optional, mentioned if the patient needs to use it in order to collect data for ongoing assessment or monitoring, e.g., the Gaitband, for movement analysis. It is a means of communication between clinicians and patients, allowing physicians

to remind patients of the necessary clinical or lifestyle actions they should undertake to improve their health. These actions may include tasks such as performing daily exercise, using a medical device to acquire specific data, or scheduling appointments with medical specialists. Each medical recommendation is uniquely associated with the medical professional and their assigned patient, and it is time-stamped and can be easily tracked. The recommendation's title, associated device, and description are all stored with structured data fields to allow easy retrieval and management.

Each device transmits data securely via the Monit4Healthy gateway (Raspberry Pi), which preprocesses and optimizes data for cloud storage.

3.2. Blackbox Devices' Concept and Development

The Blackbox devices within the Monit4Healthy system follow a modular sensor-fusion framework, integrating low-power microcontrollers (ESP32, STM32) for energy-efficient processing. Edge computing optimizes local data analysis, reducing bandwidth usage and improving response times. To ensure data security, encryption protocols and secure communication channels are implemented, aligning with regulatory standards.

For precise integration and manufacturability, Autodesk Fusion 360 [57,58] was chosen due to its all-in-one 3D modeling, parametric design, simulation, and real-time collaboration capabilities. This cloud-based platform enabled efficient transitions from concept to production.

Development began with manual measurements and hand sketches, later digitized in Autodesk Fusion 360 to ensure accuracy in geometric constraints, tolerances, and material properties. This structured workflow, applied to all Blackbox devices, minimized design inconsistencies and optimized fabrication processes.

The design process included defining functional specifications, developing 2D sketches, and transitioning to parametric 3D models with manufacturability optimizations (e.g., fileting, draft angles, sensor housings). High-fidelity simulations (FEA, thermal analysis) ensured structural robustness and hardware integration. Iterative refinements addressed discrepancies, optimizing material use and production feasibility.

Final designs were formalized in STL format for seamless fabrication, ensuring scalability and real-world deployment. The combination of manual sketching, digital modeling, and iterative validation enhanced reliability and performance, as illustrated in subsequent figures.

The finalized prototype of the EMG Blackbox is illustrated in Figure 2, showcasing its compact design, component layout, and integration of core hardware elements such as the Arduino Mega, NodeMCU ESP8266, and Li-Po battery. This configuration supports real-time data acquisition, wireless transmission, and portable biomedical monitoring capabilities.

Figure 2 illustrates the 3D extruded model derived from a 2D sketch in Autodesk Fusion 360. The enclosure is designed for housing electronic or medical sensors and features:

- Corner mounting holes for secure fastening.
- Side cutout ventilation and USB-C connectivity.
- A circular opening on the angled face, for sensor integration.
- *Geometric features*: Rectangular base with an inclined edge for structural compatibility; the chamfered surface enhances ergonomics and esthetics.
- *Mounting points*: Four precisely positioned screw holes ensure secure fastening and stability.
- *Recessed cavity*: Central section accommodates a display, sensor, or functional interface.
- *Manufacturing considerations*: Optimized for 3D printing, balancing material efficiency and structural integrity.

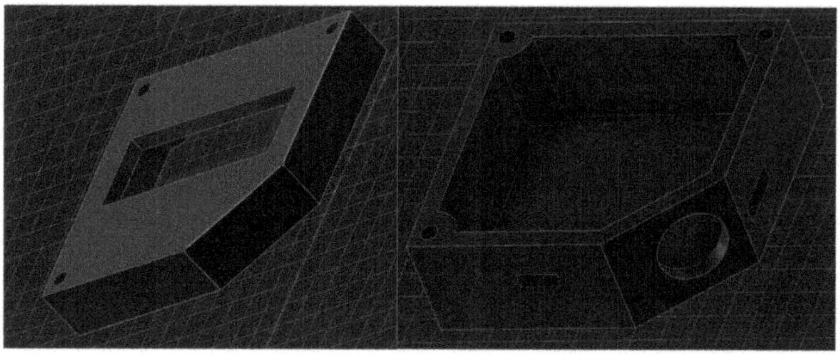

Figure 2. The final model of the EMG Blackbox. The final assembled prototype of the EMG Blackbox, showcasing its compact design for portable biomedical monitoring applications. The image highlights key components, including the embedded microcontroller board (Arduino Mega), Li-Po battery pack (7.4 V, 2500 mAh), wireless communication module (NodeMCU ESP8266), and integrated EMG signal acquisition sensors. Internal wiring and modular connectors are organized to support reliable signal acquisition and efficient power distribution. This configuration allows for real-time data processing and wireless transmission to edge computing nodes, ensuring robust operation in laboratory and home-care settings.

Figure 3 illustrates the fully developed 3D model of the Medical Blackbox, designed in Autodesk Fusion 360 for the Monit4Healthy system. The enclosure integrates structural, functional, and ergonomic optimizations for housing medical-grade electronics and sensors.

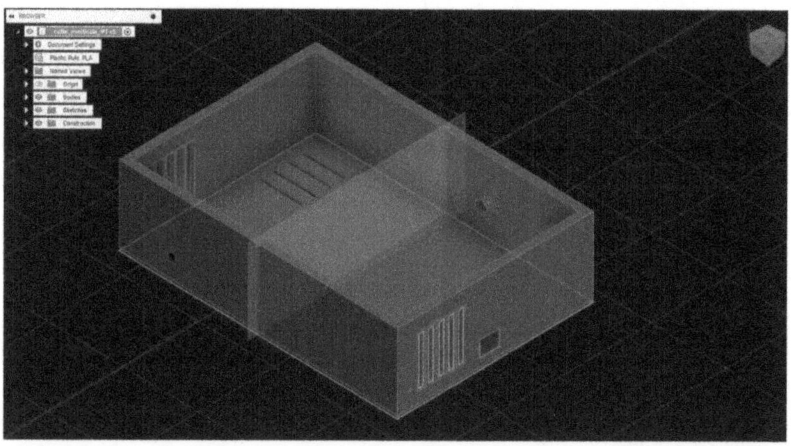

Figure 3. The final model of the Medical Blackbox. The finalized prototype of the Medical Blackbox device used in the Monit4Healthy system. The image shows the internal layout of key components, including the Arduino Mega microcontroller, signal conditioning circuits, NodeMCU ESP8266 for wireless communication, and the Li-Po battery pack. The device integrates multiple biomedical sensors (ECG, EMG, photoplethysmography (PPG), GSR) for real-time physiological data acquisition and processing. The compact and modular design supports portable operation in clinical and home environments, facilitating edge-level signal processing and secure data transmission to cloud infrastructure.

Technical and functional design considerations

- *Structural optimization*: Rigid, lightweight construction ensuring mechanical protection and spatial efficiency.
- *Ventilation system*: Engineered side slots for effective heat dissipation and sensor accuracy.

- *Custom interface openings*: Strategically placed cutouts for sensors, power, and data ports, minimizing interference.
- *Fastening and accessibility*: Predefined mounting holes and alignment guides facilitate secure assembly and maintenance.
- *Manufacturing readiness*: Optimized for 3D printing, ensuring precision, repeatability, and material efficiency.

Key design considerations

- *Dimensional accuracy:* Defined length/width parameters for precise fit and structural integrity.
- *Feature placement:* Central cutout for connectors, sensors, or ventilation.
- *Alignment and constraints:* Geometric symmetry ensures parametric adaptability.
- *Manufacturing preparation:* Sketch optimized for extrusion, fileting, and fastening integration.

Figure 4 presents the fully developed 3D model of the Medical Blackbox cover, designed for seamless integration with the enclosure. The model incorporates precisely positioned cutouts, fastening points, and structural reinforcements to enhance functionality and system performance.

- *Component-specific cutouts*: Geometrically optimized openings for sensor exposure, display integration, and thermal dissipation.
- *Precision fastening*: Predefined corner mounting holes ensure secure, vibration-resistant attachment while maintaining accessibility.
- *Multi-plane design*: Slanted and recessed surfaces enhance ergonomics, esthetics, and mechanical robustness.
- *Manufacturing and assembly readiness*: Optimized for additive/subtractive manufacturing, ensuring tight tolerance and reliable interlocking with system components.

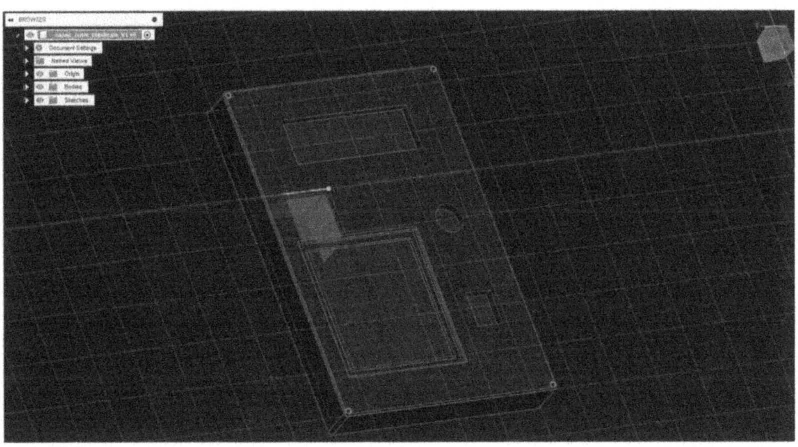

Figure 4. The finished model of the Medical Blackbox's lid. The finalized model of the Medical Blackbox's external casing, specifically its custom-designed lid. The enclosure was designed to ensure physical protection for internal electronic components, including the microcontroller, sensors, and communication modules, while maintaining accessibility for maintenance and upgrades. The lid incorporates ventilation slots for thermal management and designated ports for sensor connections, power input, and wireless communication interfaces, ensuring reliable and safe operation during portable health monitoring activities.

Data Transmission Framework

Through the use of BLE and Wi-Fi protocols, which both enable Transport Layer Security (TLS) 1.2/1.3 and Secure Socket Layer (SSL) for encrypted communication, ensuring authentication and integrity, the Monit4Healthy system enables secure information exchange. TLS guarantees all external data transmissions to cloud services, whereas BLE is employed for local, short-range biomedical data communication. In particular, the system uploads data to the ICIPRO cloud infrastructure—which is designed for the secure processing and storage of sensitive data—using TLS-encrypted communication. The ICIPRO infrastructure provides certified secure settings for digital health solutions. As part of its compliance to national cybersecurity regulations, ICIPRO uses TLS protocols to establish server-side encryption and enforces data isolation, access control constraints, and encrypted backups. The system uses an adaptive data packet framework, resulting in only pertinent and priority information for transmission, edge-level data preprocessing, and this lowers the total amount of raw signals, and dynamic protocol selection depending on the operational context to accomplish this.

In particular, Wi-Fi transmission facilitates the transfer of higher-resolution biomedical data to cloud platforms for additional analysis, whilst BLE allows for short-range, low-power communication among sensors and local processing units. This hybrid communication approach ensures low-latency data sharing, which is important for real-time health monitoring, while optimizing energy efficiency and network bandwidth consumption. The system constantly determines between BLE and Wi-Fi in order to improve data transmission by considering the trade-offs between data throughput, communication range, and power efficiency. Although BLE guarantees minimal power consumption for short-range exchanges, its data speeds are constrained, and it performs most effectively for sending low-volume, sensitive information. The reliable transmission of high-resolution biomedical data is made possible by Wi-Fi's greater bandwidth and range, but at the expense of higher power consumption.

The intrinsic interference mitigation approaches included into BLE and Wi-Fi protocols are used to lessen any signal interference in high-density environments. To further boost transmission accuracy under overloaded network contexts, future system versions might consider adaptive communication approaches.

The following sections present the key modules and protocols utilized within the system, emphasizing their role in enabling seamless and secure data exchange.

Data Flow Structure Within the Monit4Healthy System

Various sensors track vital parameters such as heart rate, SpO_2, temperature, and EMG signals. These sensors communicate through standardized wireless protocols, ensuring seamless data transmission.

The edge level of the Monit4Healthy system is where integrated feature extraction and multi-sensor data synchronization are carried out. Individual preprocessing stages (such as noise filtering, timestamp alignment, and feature extraction) are applied to physiological signals from ECG, EMG, PPG, and GSR sensors. The features extracted are subsequently categorized into integrated data packets for transmission and real-time analysis. At the microcontroller level, each signal is subjected to distinct filtering and amplification based on its frequency and amplitude features. This is complemented by analog-to-digital conversion and timestamp-based alignment. After being synchronized, the signals are sent to an edge computing unit for initial signal processing. Despite the current lack of a formal fusion approach (such as rule-based fusion), the system architecture enhances context awareness by supporting simultaneous data processing and displaying correlated physiological parameters in real time. Future stages of development will involve the integration of advanced sensor fusion algorithms to provide complex health metrics based on multi-sensor inputs.

The data flowchart in Figure 5 illustrates the process of multi-sensor data acquisition, synchronization, preprocessing, and transmission in the Monit4Healthy system. It represents the flow of data from initial acquisition through multi-sensor systems to final data transmission, ensuring that the physiological signals from the patient are accurately monitored, processed, and transmitted to healthcare professionals for further analysis and timely interventions. A detailed description of the data flow and processing workflow is presented as follows:

- *Data Layer—Multi-Sensor Data Acquisition:* The process starts with the Data Layer, where physiological signals are continuously collected by various sensors integrated within the Monit4Healthy system. The key sensors involved in data acquisition include the Medical Blackbox, Gaitband, Blackbox AcU, and EMG Blackbox.
- *Communication Layer—Data Transmission:* Once the data are collected by the sensors, they are transmitted through two communication channels:
  - BLE, used for short-range communication, ideal for wearable or portable sensors that require low-power consumption.
  - Wi-Fi, used for higher-bandwidth, long-range data transmission, enabling the system to send large datasets such as high-resolution ECG and other complex signals to the cloud for further processing.
- *Edge Processing Layer—Initial Signal Processing:* After the data are transmitted, they enter the Edge Processing Layer. Here, the raw signals undergo initial processing, including the following:
  - Signal Conditioning: The filtering and amplification of the data to remove noise and enhance signal quality.
  - Feature Extraction: Relevant features such as heart rate from ECG and PPG signals, muscle activity from EMG, and skin conductance levels from GSR are extracted for further analysis.
  - Analog-to-Digital Conversion: The analog signals from the sensors are converted to digital format to facilitate further computational processing.

  A Raspberry Pi or similar edge computing devices are used in this layer to perform the initial processing, reducing the amount of raw data sent to the cloud and ensuring that only essential information is transmitted, thus minimizing bandwidth usage and latency.
- *Data Structuring and Integration:* Once the feature extraction is performed, the processed data are structured and integrated for synchronization across multiple sensors. This step ensures that the data from all sensors are aligned in time by applying timestamps, facilitating multi-sensor data fusion and providing a more comprehensive evaluation of physiological conditions. The integration of data from various sensors such as ECG, EMG, and GSR enhances the system's ability to detect abnormalities and generate alerts when necessary.
- *Data Transmission to the Cloud:* After data fusion, the structured data are transmitted using the communication protocols (BLE or Wi-Fi) to the cloud server (ICIPRO). This enables the following:
  - Data Storage: The cloud serves as the repository for the processed data, ensuring they are stored securely and accessible for further analysis.
  - Data Synchronization: The real-time synchronization of the patient's health data across devices and systems allows healthcare professionals to access the most up-to-date information at any time.
- *Application Sublayer—Real-Time Monitoring and Alerts:* At this stage, the structured and synchronized data are accessible via an Application Sublayer, where healthcare professionals and patients can view the data in real time through interactive dashboards. These dashboards provide visualizations such as the following:
  - Trend graphs showing the patient's health metrics over time.

- ○ Alerts generated when the data deviate from predefined thresholds, indicating potential health risks. This allows for timely interventions, as healthcare providers can receive immediate alerts if a patient's vital signs indicate a need for attention.

- *Visualization Layer—Health Data Visualization:* Finally, the processed data are displayed in the Visualization Layer, providing clear and easy-to-interpret health metrics for both patients and medical professionals. This layer facilitates the following:
  - ○ The real-time visualization of key health parameters such as heart rate, SpO_2, gait patterns, and muscle activity.
  - ○ Alert notifications that inform healthcare providers about critical changes in the patient's condition.
  - ○ Access to multiple functionalities that provide real-time, updated data and information related to the patient's profile, based on the information stored in the system's database. These functionalities are considered and utilized as part of a personalized, proactive, and preventive health management support system for the patient.

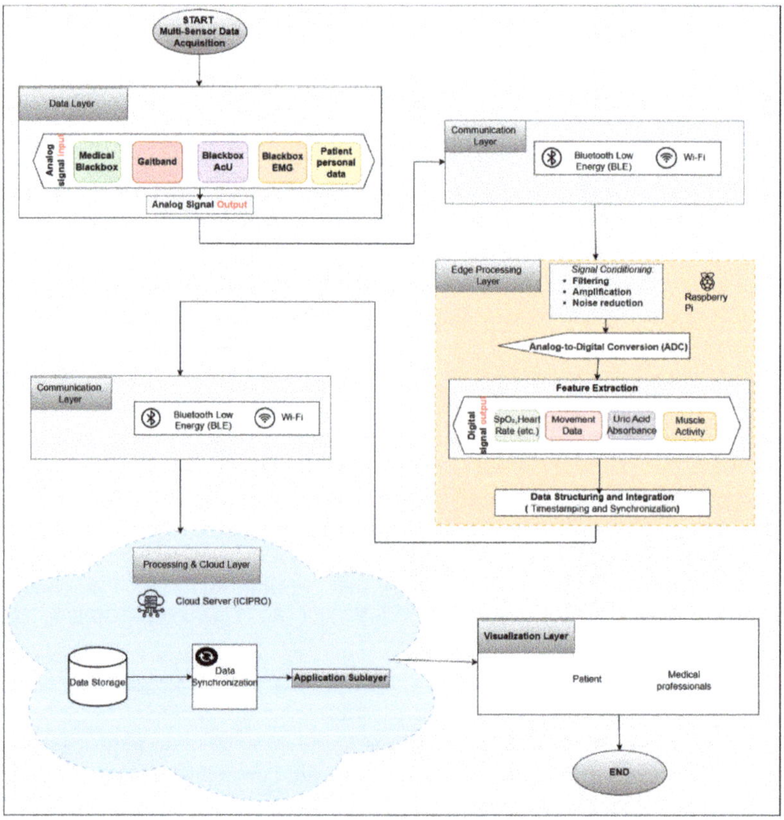

Figure 5. Multi-sensor data acquisition, synchronization, and structured transmission workflow in the Monit4Healthy system.

Communication Protocols and Optimization Strategies

Efficient data exchange is crucial for real-time health monitoring systems. Wi-Fi and Bluetooth serve as the primary communication protocols for data transmission in IoT-enabled medical devices. Wi-Fi provides high-speed, long-range connectivity, enabling continuous data synchronization with cloud storage. The Bluetooth module ensures real-time data transmission by establishing a seamless wireless connection, which is vital for telemetry and remote monitoring applications.

- *Bluetooth Module for Real-Time Data Transmission:* The Bluetooth module facilitates real-time data transmission, ensuring reliable connectivity. This feature is essential for applications requiring continuous data flow, such as telemetry and remote patient monitoring. Advanced BLE technology ensures low power consumption and reduced latency, making it suitable for wearable health applications. It employs adaptive frequency hopping (AFH) to minimize interference and optimize bandwidth utilization. BLE and Wi-Fi are widely adopted wireless communication technologies for data transmission, each optimized for different application scenarios. BLE is designed for low-power, short-range communication, making it ideal for battery-operated devices such as wearable sensors and IoT nodes. In contrast, Wi-Fi provides higher data throughput and extended coverage, supporting applications that require continuous, high-bandwidth connectivity. Compared to other protocols such as Zigbee and LoRaWan, BLE offers higher data rates but at the cost of a reduced range, while Wi-Fi consumes significantly more power but ensures robust data transmission over greater distances. Although this work does not explicitly analyze the trade-offs in terms of power efficiency, range, and data throughput, it focuses on leveraging both BLE and Wi-Fi to optimize data transmission for specific use cases. BLE has significantly advanced from BLE4 to BLE5, delivering notable enhancements in communication reliability, the data transfer rate, and network scalability—essential elements for contemporary IoT and sensor network applications. Both BLE4 and BLE5 are designed primarily for low-power use—an important feature for battery-powered IoT devices and sensor networks. BLE4, launched with Bluetooth 4.0, utilizes a frequency-hopping spread spectrum technique within the 2.4 GHz ISM band to reach data rates of up to 1 Mbps, featuring a design tailored for energy efficiency and low-power functionality via simplified connection and advertising protocols. Conversely, BLE5 brings substantial improvements, including the adoption of a coded physical layer that boosts range and reliability in environments with high interference, along with a doubling of the nominal data rate to 2 Mbps. BLE5 increases the advertising capability by allowing larger payloads, which enhances detailed device discovery and connectionless communications. These technological advancements not only boost overall network efficiency but also allow for the incorporation of diverse sensors—like GPS units, EMG devices, and environmental sensors such as the DS18B20 and MAX30100—into more intricate and resilient monitoring frameworks. Both BLE and Wi-Fi are prone to interference in high-density environments. This is due to the shared spectrum in the 2.4 GHz ISM band, as well as due to the overlapping channels. To avoid congested channels, BLE utilizes adaptive frequency hopping (AFH). However, AFH's effectiveness may diminish in dense environments, therefore leading to packet loss and reduced reliability. Wi-Fi networks experience co-channel interference when multiple access points operate on the same or adjacent channels, leading to slower data rates and increased latency, particularly in areas with many overlapping access points [59].
- *GPS Module:* The GPS module integrates a chip antenna and Qwiic cable connection, ensuring efficient hardware integration. It significantly reduces power consumption while enhancing performance and accuracy. The ability to receive signals from up

to four Global Navigation Satellite System constellations simultaneously improves positional accuracy, particularly in environments with limited sky visibility, such as urban canyons or dense foliage. Additionally, Assisted-GPS technology is used to reduce the time-to-first-fix, ensuring faster positioning data. A differential GPS can also be employed to enhance precision, reducing errors to sub-meter accuracy. Based on the u-blox MAX-M10 GNSS platform, the GPS module used in the Monit4Healthy system supports simultaneous reception from up to four GNSS constellations (GPS, Galileo, GLONASS, and BeiDou), improving positioning accuracy and signal availability even in difficult-to-reach places like urban canyons. For power-optimized performance, it makes use of the u-blox Super-E technology, which uses as low as 25 mW while in continuous tracking mode. The Qwiic interface facilitates communication over I^2C, allowing for effective data transmission with low latency and simplified wiring. In order to improve energy consumption in wearable health monitoring applications, the module incorporates automated low-power modes, such as power save and periodic tracking, and enables an Assisted Global Navigation Satellite System for an expedited time-to-first-fix. It has anti-jamming and anti-spoofing characteristics, which are essential for preserving data integrity in linked health ecosystems, and enables satellite-based augmentation systems for improved location precision. The MAX-M10 GPS module is a dependable and energy-efficient option for real-time geolocation in the Monit4Healthy system because of these features [60].

- *EMG Sensor (Myoware Sparkfun) for Muscle Signal Analysis:* This sensor measures EMG signals from muscle activity and features an integrated amplifier with an adjustable gain of up to 1000x. It includes a high-pass filter (0.5 Hz) to eliminate direct current offset and a low-pass filter (125 Hz) to reduce high-frequency noise. The sensor uses three disposable gel electrodes and outputs both raw and envelope signals, making it suitable for applications such as robotics and prosthetics. Signal conditioning is performed to reduce motion artifacts and interference from ambient electrical sources. A notch filter (50/60 Hz) is applied to mitigate power line interference. The Monit4Healthy system's integrated EMG module uses a multi-stage analog signal conditioning architecture to record and process surface electromyography data with high fidelity. In order to separate the physiological EMG frequencies while reducing low-frequency motion artifacts and high-frequency noise, the signal collection chain starts with an instrumentation amplifier set up for an initial gain of 200× and operating within a passband of 20 Hz to 498 Hz. Signal scaling according to electrode location and specific muscle activation levels is made possible through the following gain stage, which may be adjusted between 1× and 100×. A 3.6 Hz high-pass filter is incorporated into the module to remove DC offset, guaranteeing baseline stability and reducing signal drift over time. After amplification, the signal passes through an envelope detection stage and a full-wave rectification circuit, enabling the recording of both rectified and envelope outputs. A high-resolution analog-to-digital converter digitizes these analog outputs, guaranteeing precise EMG signal sampling for edge processing. In the Monit4Healthy Edge Processing Layer, adaptive feature extraction and threshold-based event detection are made possible through the architecture's support for the real-time monitoring of both raw and processed EMG data. With a low power consumption appropriate for wearable usage, its modular and adjustable architecture enables the EMG module to be tailored for a variety of health monitoring applications, such as neuromuscular evaluations and rehabilitation situations [61].
- *DS18B20 Digital Temperature Sensor:* The DS18B20 temperature sensor operates using a 1-wire interface, allowing multiple sensors to share a single data line. It provides accurate temperature readings with a configurable resolution (9–12 bits) and dig-

ital transmission to minimize errors compared to analog sensors, making it ideal for biomedical and environmental applications. The sensor includes an integrated temperature compensation feature, ensuring stable readings even under varying environmental conditions. It supports cyclic redundancy check for error detection, enhancing data integrity. The DS18B20 sensor was chosen for its networking advantages, using a 1-wire interface that allows multiple sensors to share a single data line, simplifying wiring in large networks. Data from the sensors were collected with a microcontroller and transmitted via a wireless connection to a cloud platform. The digital transmission of data reduces errors compared to analog sensors, providing accurate and reliable measurements over time [62].

- *MAX30100/MAX30102:* Pulse Oximetry and Heart Rate Sensors utilize photoplethysmography (PPG) to measure blood SpO_2 and heart rate. In our project, the integration of the MAX30102 biosensor with the ESP32 microcontroller utilizes the Inter-Integrated Circuit (I^2C) protocol to facilitate seamless data transfer, which is essential for monitoring key health parameters such as heart rate and SpO_2 [63]. The sensor includes infrared and red light emitters for dual-wavelength analysis, improving accuracy by compensating for motion artifacts. An integrated ambient light cancelation system enhances signal clarity. The MAX30100 sensor measures PPG signals to determine the SpO_2 levels and heart rate. Data are then sent wirelessly (via Bluetooth or Wi-Fi) to cloud storage, enabling real-time remote monitoring. This setup allows healthcare professionals to access patient data remotely and make timely interventions if necessary [64].

3.3. *Development Environment and Tools*

To ensure the seamless management of patient and medical professional data while maintaining secure accessibility, the Monit4Healthy system leverages robust web frameworks. The system integrates Django for backend processing, RESTful APIs for efficient data exchange, and secure authentication protocols for access control. Additionally, it supports medical professional management by handling medical results, tracking patient recommendation history, managing referrals, and overseeing the integration of patients with connected medical devices.

3.3.1. Software Frameworks and Programming Environments

The Monit4Healthy system consists of a backend application developed using Python 3.13.1 and the Django 5.1.5 framework.

Django Framework: A Secure Backend for Health Data Management

Django, a high-level Python web framework, was selected for the Monit4Healthy system due to its advanced security model, scalable architecture, and native support for seamless interoperability with IoT ecosystems. Its built-in ORM, middleware capabilities, and modular design facilitate the development of a highly maintainable and performant backend, optimized for handling concurrent requests and large-scale data processing. These features are essential in the context of web-based and IoT-driven healthcare applications, where stringent data security, efficient real-time communication, and system resilience are critical requirements. Django's architecture provides a robust and scalable framework ideally suited to the requirements of our project. A Django project encapsulates the entire web application, integrating all its components, while individual applications serve as modular units responsible for specific functionalities. The framework adheres to the Model–View–Template architectural pattern, ensuring a clear separation of concerns and promoting maintainability. The model layer defines the data schema and orchestrates database interactions, the view layer processes Hypertext Transfer Protocol requests and generates responses, and the template layer structures presentation logic, enabling the efficient rendering of response data in formats such as HTML or JSON. These characteristics make Django an optimal choice for developing a structured, maintainable, and high-performance web application [65].

Django's Object Relational Mapper (ORM) provides an abstraction layer that simplifies database interactions by enabling developers to define and manipulate database models using Python code instead of raw SQL (Structured Query Language). This approach enhances code readability, maintainability, and portability while ensuring database-agnostic development, thereby facilitating seamless integration with various relational database management systems [66].

Django provides native support for a wide range of database management systems (DBMS), including MariaDB, PostgreSQL, MySQL, SQLite, and Oracle [67], through dedicated database drivers that handle connections and execute operations specific to each database management system.

For this system, the MariaDB management system was selected as the primary database management system due to its optimized performance, scalability, and extensive compatibility with Django's ORM. It also enhances data storage and retrieval, ensuring fast and efficient access to patient records.

Data Storage and Management: MariaDB

This database provides the following:

- *SQL Support*: It facilitates the efficient and scalable querying, management, and manipulation of structured data.
- *Data Integrity and Security*: It implements strong encryption protocols and authentication mechanisms to protect sensitive health information, ensuring compliance with data protection regulations.

The key functionalities implemented within the platform include the following:

- *User Authentication and Role-Based Access Control*—Django's built-in authentication system ensures that access to sensitive patient data is restricted to authorized personnel, enforcing strict role-based access policies.
- *Patient Management*—The framework enables the efficient storage and retrieval of patient data, appointment scheduling, and real-time notifications, ensuring seamless coordination between healthcare providers and patients.
- *Medical Professional Management*—Django supports the management of medical results, tracking of patient recommendation histories, and coordination of referrals. Additionally, it facilitates the integration of connected medical devices, with automated alert mechanisms notifying healthcare professionals of critical updates, such as abnormal test results or significant changes in a patient's status.
- *Secure Data Handling*—The framework incorporates advanced security mechanisms, including built-in protection against SQL injection, cross-site scripting, and cross-site request forgery attacks, ensuring the integrity and confidentiality of medical data.

3.3.2. Hardware Components Used

The Monit4Healthy system integrates a selection of high-performance microcontrollers and sensor technologies to enable real-time health monitoring and environmental data acquisition. These components ensure efficient data processing, seamless connectivity, and high reliability, which are critical for IoT-driven healthcare applications.

Microcontrollers

- *ESP32*—Chosen for its dual-core architecture, integrated Wi-Fi and Bluetooth capabilities, and ultra-low power consumption, the ESP32 is particularly suited for remote patient monitoring systems that require real-time data transmission over wireless networks. Its deep-sleep functionality minimizes energy usage, ensuring extended operation in battery-powered medical devices [68].

- *STM32*—These are a family of ARM Cortex-based microcontrollers offering high computational efficiency and low power consumption, making them ideal for embedded medical applications requiring real-time processing. An STM32 microcontroller compresses the digital ECG signals using a basic matrix–vector multiplication technique based on compressed sensing [69].
- *Raspberry Pi*—Serving as an edge computing node, Raspberry Pi offers high processing power and versatile I/O capabilities, enabling the real-time preprocessing of biomedical signals, anomaly detection, and encrypted data transmission to cloud servers. This reduces network latency and offloads computational tasks from the central cloud infrastructure [70].
- **ATmega2560 Microcontroller**—Applications requiring extensive firmware storage and multitasking can benefit from the ATmega2560, an 8-bit AVR microcontroller with 256 KB of flash memory, 8 KB of SRAM, and 4 KB of EEPROM. Four UARTs, an SPI, I^2C, and sixteen 10-bit analog-to-digital converter channels are among the many peripherals it supports. The AT-mega2560, which runs at a clock speed of 16 MHz, provides dependable performance for activities involving real-time data collection and control while consuming little power, which is crucial for embedded health monitoring applications [71].
- **Arduino Mega Board**—With 54 digital input/output pins, 16 analog inputs, and various serial communication interfaces, the Arduino Mega board, which is based on the ATmega2560 microcontroller, has extensive I/O capabilities that enable integration with multiple biomedical sensors at once. The Arduino Mega is used in the Medical Blackbox device of the Monit4Healthy system to collect and interpret physiological signals in real time. Strong and scalable health monitoring systems are made possible by its dependable interrupt management, hardware timers, and interoperability with common sensor libraries, which provide effective and coordinated data collection from multi-sensor arrays [72]. Currently, the Medical Blackbox is powered via a regulated 5V DC input, compatible with external USB power adapters and battery packs, ensuring safe and portable operation during laboratory testing.
- *Biomedical Sensors*—Modules such as MAX30100, for heart rate and SpO_2 monitoring via PPG; AD8232, for real-time ECG signal acquisition; a custom EMG module, for electromyography signal capture, designed with adjustable gain, bandpass filtering, and analog signal conditioning tailored for muscle activity monitoring; a GSR sensor, for galvanic skin response measurement, used in stress and autonomic nervous system assessments; and a digital temperature sensor, for continuous body temperature monitoring for real-time ECG signal acquisition, enable continuous physiological monitoring, facilitating the early detection of health anomalies.

Hardware Integration and Connectivity

In addition to the microcontrollers outlined above, the Monit4Healthy system incorporates a range of hardware components that support efficient sensor integration, signal acquisition, and wireless communication.

To support seamless multi-sensor integration and real-time data acquisition, the Monit4Healthy system utilizes modular hardware assemblies designed for efficient sensor interconnection and reliable data transmission. Each Blackbox device incorporates standardized connectors (e.g., Qwiic interface for I^2C communication), allowing for rapid prototyping and scalability. Signal conditioning circuits, such as analog filters and amplifiers, are embedded within the hardware design to ensure high-fidelity signal acquisition and compatibility with microcontroller input ranges. Custom PCB layouts and compact enclosures, developed using parametric CAD tools, facilitate ergonomic deployment in clinical and remote settings. Wireless connectivity is achieved through integrated BLE and Wi-Fi modules, optimized for low-power consumption and robust communication, as detailed in Section 3.2.

3.3.3. Data Processing Frameworks

Edge and Fog Computing: Raspberry Pi as a Gateway

The Monit4Healthy system employs edge computing paradigms to enhance data processing efficiency and reduce latency. Raspberry Pi serves as the central gateway, collecting sensor data and transmitting it to cloud infrastructure via Wi-Fi. The Raspberry Pi 4 Model B, equipped with wireless LAN and Bluetooth connectivity, plays a pivotal role in health monitoring applications by aggregating sensor readings, performing preliminary data processing, and ensuring real-time monitoring. This setup enhances system scalability, efficiency, and responsiveness, thereby improving both the accuracy and accessibility of patient data. With low power consumption (~2.5 W during typical operation) and optimized data transmission, the Raspberry Pi 4 Model B ensures energy-efficient performance, making it well suited for continuous health monitoring scenarios. Its capability to process and filter data locally before transmission significantly reduces network congestion and minimizes latency [73].

Cloud Computing for Monit4Healthy System: ICIPRO IaaS

ICIPRO—a significant cloud-based initiative designed to modernize and improve public services in Romania by leveraging cloud computing technology—provides Infrastructure as a Service (IaaS), offering virtualized computing resources, storage, and network connectivity through its secure cloud infrastructure. The platform is built on high-performance hardware, including servers, routers, and firewalls, ensuring reliability and scalability for systems like the Monit4Healthy system, which runs on a virtual machine (VM). ICIPRO provides high availability and load balancing, reducing downtime and improving system resiliency. VMs in this cloud run in a redundant data center environment, minimizing the risk of failure [74].

Network Isolation and Security

ICIPRO implements Network Virtualization using Generic Routing Encapsulation, ensuring complete network isolation for the Monit4Healthy system from other cloud tenants. A total of 10 Gbps of connectivity between hardware components provides low-latency and high-speed data transfers, essential for real-time health monitoring applications. Built-in firewall rules and virtual network segmentation enhance data security and prevent unauthorized access. ICIPRO integrates a System Center Data Protection Manager to offer continuous data backup and disaster recovery for critical health monitoring data [74].

Figure 6 shows the ICIPRO dashboard controlling a virtual machine running Ubuntu, along with a terminal window—a Command Line Interface (CLI) running a Django development server using the "python3 manage.py runserver 0.0.0.0:8088" command—showing both system resource utilization and real-time CLIs. This architecture enhances predictive diagnostics and enables timely medical interventions, optimizing patient outcomes in a digitally connected healthcare ecosystem [75].

In addition to technical measures, data privacy and regulatory compliance are essential aspects of the Monit4Healthy system. The integration of cloud-based data transmission ensures low-latency, high-throughput, and encrypted data exchange, adhering to healthcare standards such as GDPR and, where applicable, HIPAA. Data privacy is maintained by implementing secure data transmission protocols, user authentication, and access control mechanisms within the system. The ICIPRO cloud infrastructure, where data are stored and processed, enforces strict security and privacy policies, including encrypted data transfer, controlled access, and secure storage, supporting compliance with GDPR and HIPAA principles related to health data confidentiality, integrity, and availability. In accordance with GDPR requirements, the system operates based on explicit informed consent from users, including healthcare professionals and patients, for data collection and processing. Consent procedures are designed to ensure transparency and user rights, including access, rectification, and data deletion upon request.

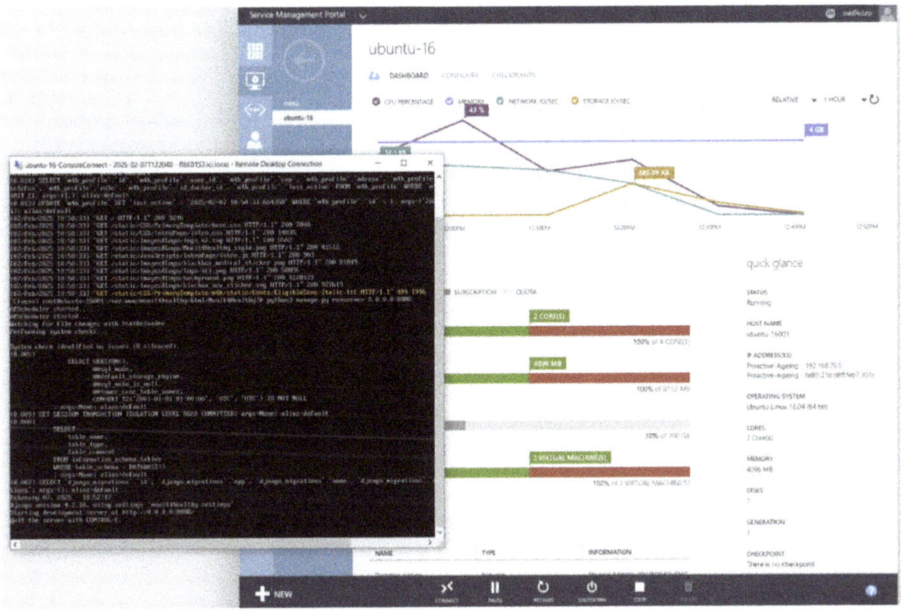

Figure 6. ICIPRO dashboard hosting the Monit4Healthy system's cloud backend. The virtual machine runs Ubuntu OS with a Django development server (python3 manage.py runserver 0.0.0.0:8088), enabling secure data handling and real-time system monitoring through resource usage metrics and CLI interactions.

Using Python, HTML, CSS, PHP, SQL, and JavaScript for embedded programming, along with ICIPRO for UI and cloud integration, enables scalable and efficient software frameworks for real-time data exchange in connected health systems.

Throughout its sensor network, the Monit4Healthy system uses an efficient data processing framework that maximizes computing efficiency and minimizes energy usage. The technology preprocesses raw physiological signals locally before sending only relevant, organized data to the cloud through the use of edge computing at the device level. This approach ensures real-time sensitivity without using a large amount of power by effectively reducing bandwidth usage, network congestion, and cloud computing requirements. The Monit4Healthy system successfully establishes a balance between dependable data transfer and energy economy by integrating low-power transmission protocols like BLE for short-range communication. Through these procedures, the system optimizes total power consumption, preserves device battery life, and addresses key issues in latency-sensitive biomedical applications, therefore mitigating the typical limits of IoT health monitoring.

4. Results

The implementation of the Monit4Healthy system has advanced to the laboratory level, which includes the manufacturing of Blackbox devices and the associated monitoring platform, following the design and conceptual development phase. At this point, the system's components—i.e., sensor-integrated Blackbox devices, data collection systems, early processing frameworks, and the platform—have been integrated and put through limited testing. Functional testing and validation, device calibration, and first performance evaluations were carried out in this phase, verifying that the basic architecture performs as planned prior to more extensive deployment considerations. The results, which demonstrate the system's capabilities in a laboratory setting and its current degree of development, are shown in the following subsections.

4.1. Selected Illustrations of Implemented Devices

4.1.1. The Medical Blackbox Device

The Medical Blackbox is a multi-sensor health monitoring device designed to measure key physiological parameters with high precision. The Medical Blackbox is designed as a stationary device intended for use in both clinical and home environments, where it can be deployed near the patient for real-time monitoring. While it is not wearable, the device's form factor and low-power operation support relocation between settings, ensuring flexibility in remote healthcare scenarios. The device features an intuitive LCD display and a user-friendly keypad for selecting different monitoring modes, ensuring accessibility across various healthcare environments, from clinical use to home settings (Figure 7). Built on the Monit4Healthy architecture, it integrates biosensors previously outlined in this study, including those for ECG, GSR, and SpO_2 measurements. Its design incorporates PPG sensors for enhanced heart rate and SpO_2 accuracy, as well as EMG signal acquisition capabilities, which will be further discussed in subsequent sections. Figure 7 showcases a view of the Medical Blackbox, with its internal structure and components.

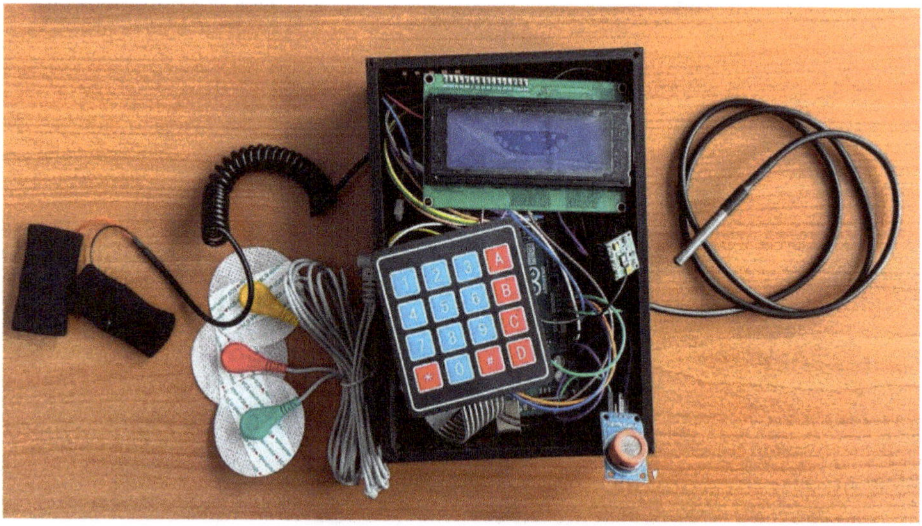

Figure 7. Internal structure and components the Medical Blackbox. The image shows the spatial arrangement of hardware components used in the Medical Blackbox, including the Arduino Mega board for multi-sensor signal acquisition, the ESP8266 module for wireless communication, and biomedical sensors for ECG and EMG monitoring. The power supply connection and sensor interface wiring are also visible, illustrating the system's design for real-time physiological data acquisition and transmission within connected health monitoring scenarios.

The integration of these sensors within the Medical Blackbox (Figure 8), in combination with the Arduino Mega microcontroller board, enables comprehensive and continuous data acquisition. Unlike other devices in the system that use low-power microcontrollers such as ESP32 and STM32, the Medical Blackbox employs an Arduino Mega, chosen for its versatile I/O capabilities and suitability for prototyping multi-sensor signal acquisition. To ensure high-fidelity signal capture, signal conditioning techniques—specifically filtering and amplification—are applied to the raw data streams, effectively enhancing signal quality and minimizing noise interference. The microcontroller unit (MCU) facilitates real-time processing by executing embedded algorithms that assess sensor outputs, enabling the detection of anomalies and supporting immediate alert generation and system responses. Data communication is established wirelessly via the NodeMCU ESP8266 module. The processed data are subsequently transmitted to a Raspberry Pi 4 Model B, functioning as a gateway device, which aggregates the data and forwards them to cloud-based infrastructures for advanced processing, storage, and visualization. Communication between the integrated sensors and the Arduino Mega is implemented using the I^2C protocol—a widely adopted serial communication interface recognized for its efficiency and reliability, requiring only two conductors for data exchange.

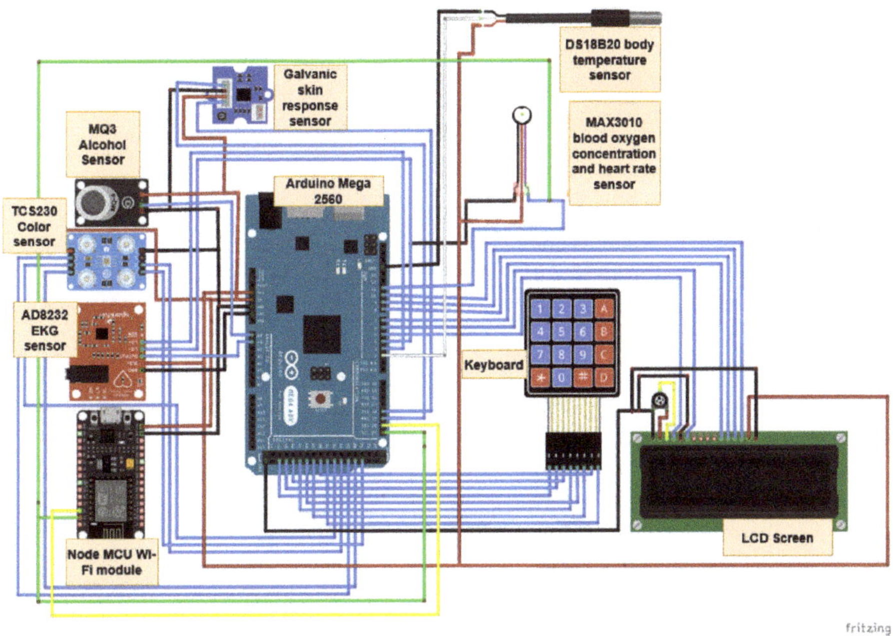

Figure 8. Electrical schematic of the Medical Blackbox, illustrating the integration of multiple biomedical sensors with the Arduino Mega microcontroller. The schematic was developed.

The Medical Blackbox is powered through a regulated external adapter compatible with standard 220V outlets, ensuring the stable and safe operation of all internal components as illustrated in the electrical schematic.

4.1.2. The EMG Blackbox Device

Geared toward monitoring patient neuromuscular activity, the EMG Blackbox employs high-sensitivity gel electrodes to record muscle electrical activity with minimal interference. To improve the accuracy of the EMG signal, the device incorporates noise-filtering techniques for eliminating artifacts. Preliminary laboratory tests indicate that the EMG

Blackbox is highly effective in capturing muscle activity patterns with high precision and minimal latency, therefore allowing healthcare specialists to assess the evolution of muscle function, and detect early signs of neuromuscular deterioration, ensuring a proactive approach to patient monitoring. A frontal view of the device, as well as its internal structure, can be observed in Figure 9.

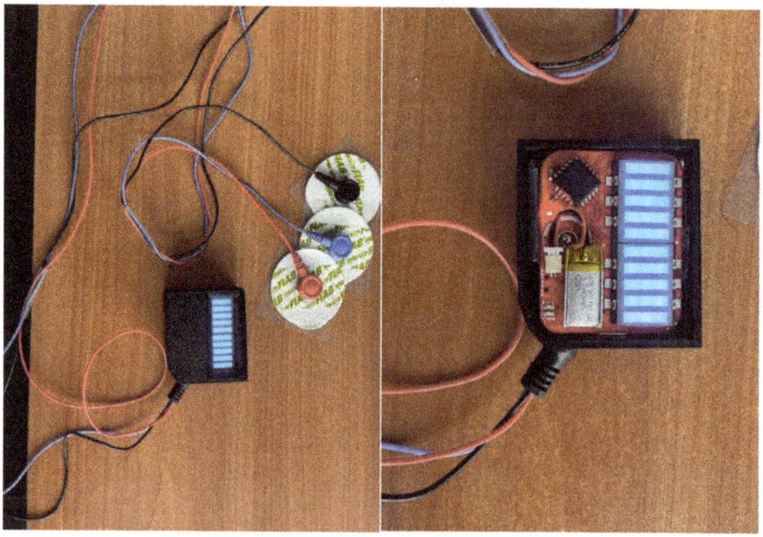

(a) (b)

Figure 9. The EMG Blackbox: (**a**) frontal view showing the compact enclosure housing the data acquisition unit; (**b**) internal structure illustrating the integration of biomedical sensors, signal conditioning circuitry, and microcontroller components used for real-time EMG signal processing and wireless data transmission.

The EMG Blackbox device is designed for capturing and analyzing EMG signals in real time. It integrates multiple technological components to ensure the high-fidelity acquisition and wireless transmission of muscle activity data. The core sensing component consists of surface EMG electrodes, which record the electrical activity of the target muscles. The MyoWare 2.0 Wireless Shield processes the EMG signals, while the MyoWare 2.0 LED Shield visually indicates muscle contraction intensity. For wireless communication, the system employs the ESPRESSIF ESP32-WROOM-32E module, which facilitates secure Wi-Fi connectivity to the local network and platform. Power is supplied either via a USB-C interface or through a rechargeable battery, ensuring operational flexibility. The system is configured for straightforward deployment: the EMG sensors are attached to the specific muscle groups, and the ESP32 microcontroller is programmed to collect, process, and transmit the acquired signals. Data integration into the Monit4Healthy platform is accomplished via API configuration, enabling comprehensive clinical analysis and patient management based on EMG data.

The EMG Blackbox is built using off-the-shelf components, selected specifically for their reliability, accessibility, and ease of integration in prototyping and research environments. The device utilizes commercially available modules such as the MyoWare 2.0 EMG sensor and the ESPRESSIF ESP32-WROOM-32E Wi-Fi module, interconnected via standard breadboard configurations. Schematics (presented in Figure 10) were designed using Fritzing software 1.0.4, to clearly illustrate the hardware integration and signal flow. While the current implementation leverages off-the-shelf components to expedite development and testing, future iterations of the EMG Blackbox will focus on designing a custom PCB, which will allow for miniaturization, improved signal integrity, and enhanced suitability for clinical use cases.

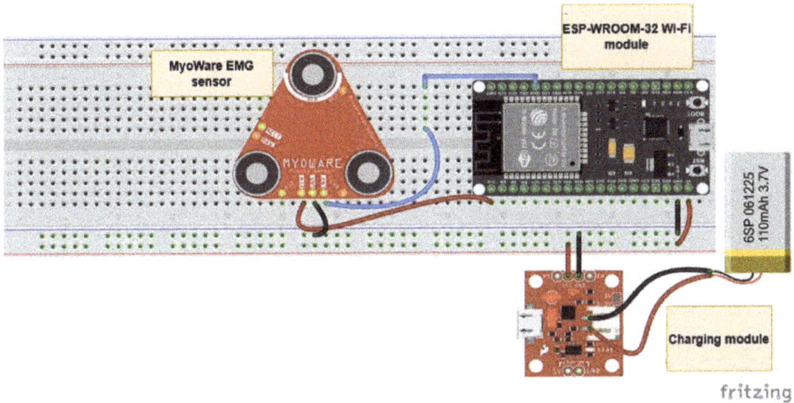

Figure 10. Electrical schematic of the EMG Blackbox device developed using Fritzing software. The schematic illustrates the integration of a MyoWare EMG sensor with the ESPRESSIF ESP32-WROOM-32E Wi-Fi module for wireless data transmission. Power supply is provided through a rechargeable 3.7V Li-Po battery, managed by a dedicated charging module.

4.2. *Monit4Healthy Platform Development*

The Monit4Healthy system continuously collects, processes, and visualizes real-time health data, permitting patients to monitor their vital signs, movement stability, and neuromuscular data through an interactive interface. Concomitantly, medical professionals can access detailed patient profiles to visualize historical data and analyze long-term health evolution, with the platform allowing clinicians to continuously monitor a patient and perform risk assessment. Automated alerts notify users of abnormal values or events, contributing to timely interventions.

4.2.1. Monit4Healthy Platform Overview

The Monit4Healthy platform is designed to support two primary user roles—patients and medical professionals—each with specific, tailored functionalities. Following authentication, users are directed to their personalized dashboards, displaying real-time data. The patient dashboard presents the main health parameters of the patient, which include the pulse, SpO_2, temperature, movement stability, and fall detection status, while the medical professional's dashboard provides a detailed overview of patient data, allowing them to monitor patient health data and respond promptly to urgent events.

The platform allows the analysis of historical data and graphical representations of medical information, intended to offer a clearer visualization of the data, and ultimately aid in proactive risk management. The platform's primary purpose is to integrate health information into a user-friendly interface, allowing for the monitoring of metrics, managing medical recommendations, and addressing potential health concerns.

The Monit4Healthy platform is designed with a Romanian-language interface, in order to support the accessibility of its primary users. To make navigation intuitive for both patients and healthcare professionals, all system functionalities, including dashboards, medical data visualizations, device management, and recommendations, are presented in Romanian. Throughout this section, the platform interface is illustrated alongside translated Romanian key terms to facilitate a better understanding of the interface.

4.2.2. Patient Interface

The Monit4Healthy patient dashboard provides an overview of the most recent medical data collected, allowing the patient to be informed and engaged in their health management. The dashboard displays physiological and biomechanical data, within measuring dials, including "Puls (Pulse)", "SpO_2", "Temperatură (Body Temperature)", "Alcool Expirat (Alcohol Concentration)", "Unghi Postural (Postural Angle)", "Deplasare Laterală (Lateral Displacement)", and "Detecție Cădere (Fall Detection)". These parameters are continuously acquired from the medical devices and updated in real-time, therefore allowing patients to track changes in their vital signs and mobility status.

Additionally, as illustrated in Figure 11, to the right of the page, the dashboard displays "Ultima Recomandare Medicală", which is the most recent medical recommendation provided by their assigned clinician. This functionality enables efficient communication between the healthcare specialist and the patients in their care, supporting adherence to medical advice.

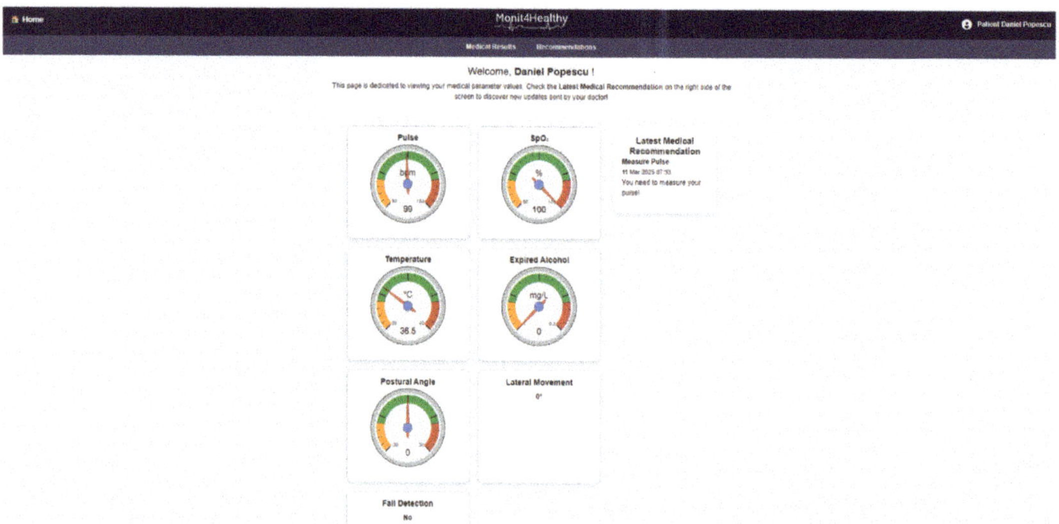

Figure 11. The Monit4Healthy patient dashboard. The dashboard interface designed for patient use provides real-time visualization of key health parameters acquired via the Monit4Healthy system. It displays physiological data such as ECG, EMG, heart rate, and temperature in both numeric and graphical formats, allowing users to track their current status and detect anomalies. The interface also includes trend graphs for historical data analysis and interactive alerts for significant deviations from normal values. This component ensures user-friendly access to health information, supporting timely awareness and personalized health management.

Amidst the key sections of the platform are "Rezultate medicale (Medical Results)" and "Recomandări (Recommendations)", both of which are designed to allow patient access to health information. The former is dedicated to providing a detailed view of the patient's most recent medical alerts ("Ultimele Alerte") and recorded physiological parameters ("Rezultate"), as well as a list of devices that have been assigned to them for data collection ("Dispozitive Asociate"). As shown in Figure 12, a detailed log of historical health data permits the patient to analyze detected anomalies over time, while the overview of medical devices assigned offers transparency regarding the sources of the collected data and reinforces the patient's awareness of their monitoring regimen. The section offers a retrospective analysis of health trends, therefore offering patients the opportunity to identify patterns in their physiological data, potentially assisting in recognizing factors that contribute to fluctuations in their health status.

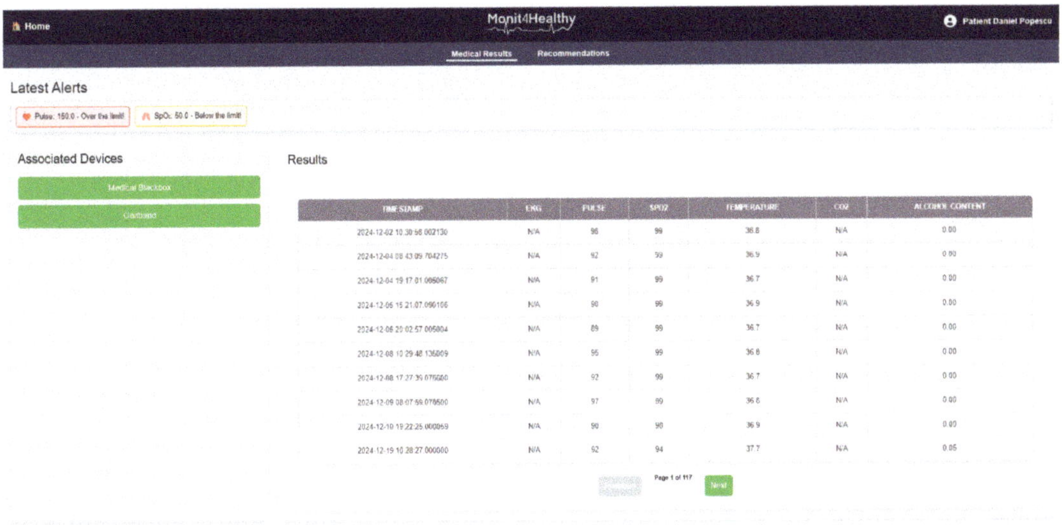

Figure 12. The Medical Results section in the patient's platform interface. This section of the Monit4Healthy patient platform enables users to review detailed medical evaluations and historical health data. It consolidates diagnostic results, laboratory findings, and physician observations in a structured format. The interface supports secure access to medical documents, visual summaries of health trends, and export options for sharing data with healthcare professionals. This functionality enhances patient engagement and facilitates continuity of care through transparent access to individualized medical records.

The "Recomandări (Recommendations)" section provides a chronological view of all medical recommendations that the patient has previously received from their healthcare provider. The "Recomandările Mele (My Recommendations)" table maintains an organized history of all past clinical guidance, mentioning the date, a summary of the recommendation ("Titlu"), a detailed description of the clinical advice ("Recomandare"), the device with which the patient should monitor their vitals ("Dispozitiv, dacă este specificat"), and the medical professional that issued the recommendation. The section supports the patient in monitoring their own adherence over time and encourages patient compliance with prescribed recommendations. Thus, the platform ensures transparent access to medical data and contributes to a patient-centered approach to digital health monitoring. Figure 13 presents the "Recomandări" (Recommendations) page from the patient interface.

4.2.3. Medical Professional Interface

Designed to facilitate real-time patient management and health monitoring, the Monit4Healthy clinician dashboard serves as a main control panel through which healthcare professionals can oversee patient data, manage device assignments, and promptly respond to sensor-triggered alerts issued by the devices within the Monit4Healthy system. The clinician dashboard provides an exhaustive overview of all monitored patients, ensuring that the medical professionals have instant access to patient health information.

A centralized table "Ultimii pacienți monitorizați (Last Monitored Patients)" presents the key patient details, including the first and last name of the patient, their e-mail, and personal identification code ("CNP"), alongside the assigned medical devices ("Dispozitive asociate (Associated Devices)"), offering a clear understanding of the monitoring capabilities available for each patient.

Figure 13. The Recommendations section in the patient platform interface. This section provides personalized health advice, lifestyle recommendations, and system-generated alerts based on the analysis of physiological data collected by Monit4Healthy devices. It offers users tailored guidance aimed at improving health outcomes, supporting self-management, and fostering proactive engagement in personal health monitoring.

On the right panel of the interface, a series of statistics are displayed within cards to assist the clinician in resource allocation and patient monitoring.

The "Numărul de pacienți monitorizați curent (Number of currently monitored patients)", "Dispozitive nealocate (Number of unassigned devices)", "Pacienți Monitorizați (Monitored Patients)", and "Alerte noi (New Alerts)" are meant for trend analysis in order to detect potential deteriorations in the patients' conditions. As Figure 14 illustrates, at the bottom of the dashboard, the "Istoricul Alertelor (Alert History)" section provides a chronological view of past alerts, allowing medical professionals to track variations in patient health status.

The Monit4Healthy platform offers an efficient patient management interface, allowing clinicians to associate and dissociate medical devices as needed. This section includes a structured table "Pacienți și Dispozitive (Patients and Devices)" that lists the patients and their associated devices, for the clear visualization of ongoing monitoring activities. Interactive features such as a dropdown menu "Alege dispozitiv (Choose a device)" and the "Atribuie Dispozitiv (Assign the device)" button allow the medical professional to select and assign an available medical device to a patient. Once a selection has been made, the platform ensures integration with the patient's records. Similarly, if a device is no longer relevant for monitoring a certain patient's health status, it can be dissociated from the patient profile, by pressing the "Șterge Dispozitiv (Remove device)" button, therefore allowing for the optimal distribution of the devices among the monitored patients. Additionally, the list of patients under the clinician's care can be modified dynamically, providing flexibility for managing assigned patients, with the medical professionals being able to eliminate a patient with "Șterge Pacient (Remove a Patient)" from their patient list.

As shown in Figure 15, this section offers an overview of the assigned patients, therefore ensuring optimal device allocation and resource usage.

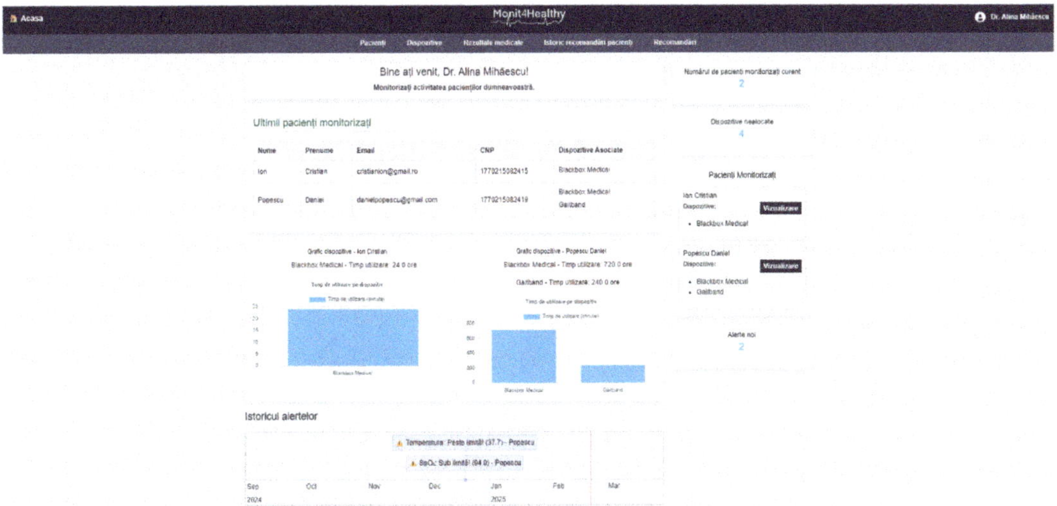

Figure 14. The Monit4Healthy medical professional dashboard. This dashboard provides healthcare professionals with real-time access to patient data, including vital signs, monitoring trends, and alert notifications. It enables efficient patient management through data visualization tools, customizable reports, and secure access to longitudinal health records, supporting informed clinical decision making and timely interventions.

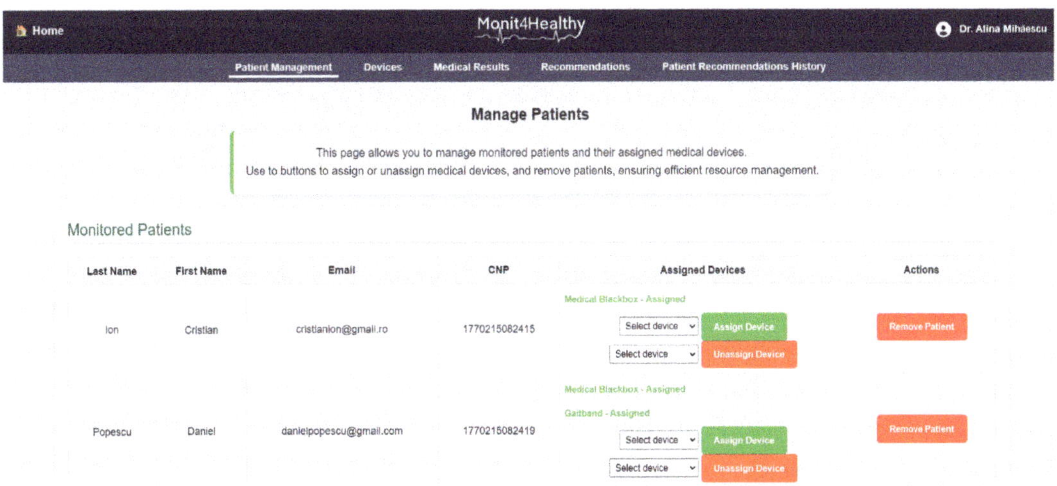

Figure 15. The Patients section from the medical professional's interface. This section allows clinicians to view and manage their assigned patients, providing a summary of each patient's monitoring status, recent alerts, and key health parameters. The interface supports efficient navigation through patient profiles, enabling streamlined access to individualized health data and historical records for continuous care management.

The Devices section provides an overview of medical device assignments, with an integrated calendar facilitating a clear visualization of assignment history. This functionality ensures that all device assignments are properly documented. The section includes two tables: "Pacienți Monitorizați (Monitored Patients)" detailing monitored patients along with their assigned devices, mapping the patients to the physiological parameters tracked; and "Dispozitive (Devices)", acting as an inventory of all medical devices within the system. This offers medical professionals an overview of available resources and contributes to optimizing device allocations. The Devices page from the clinician's interface can be viewed in Figure 16.

Figure 16. The Devices section from the medical professional's interface. This section displays all active Monit4Healthy devices assigned to patients, including their operational status, device identifiers, and recent data transmission logs. Medical professionals can monitor the connectivity and functionality of each device in real time, ensuring seamless data acquisition and prompt identification of technical issues.

The Medical Results section of the Monit4Healthy platform serves as an interface for accessing and analyzing patient data collected from multiple connected devices. This section integrates real-time data visualization, filtering mechanisms, and customizable graphical representations, offering flexible access to medical professionals to patient metrics.

The interface offers a structured and easily navigable layout, with a left-side panel displaying a list of all monitored patients alongside their relevant information such as names, assigned medical devices, and available data records. Upon selecting a patient, the most relevant medical results are displayed, organized into structured tables, each corresponding to a different monitoring device, as illustrated in Figure 17. The system allows for a detailed examination of patient data, with timestamps ensuring a clear chronological perspective on health trends. To further strengthen accessibility, the platform includes sorting and filtering options, allowing clinicians to refine the dataset based on specific timeframes, or parameters of interest.

A core feature of this section is the ability to graphically represent medical data by clicking the "Generează grafic (Generate a graph)" button, providing a more intuitive way to interpret the evolution of patient health status. Users can select specific physiological parameters for visualization, generating dynamic graphs that highlight fluctuations over time, aiding in the early detection of health deterioration or improvement.

- The History of Medical Recommendations section of the Monit4Healthy platform is designed to facilitate the review of past clinical guidance provided to patients. This section provides an interactive interface incorporating a dropdown menu, allowing clinicians to select a specific patient. Upon selection, the platform dynamically displays all previously issued recommendations for the specific patient. This historical overview permits the medical professional to identify trends in patient response to recommendations.

The Recommendations section allows clinicians to issue new medical recommendations. To the left of the interface, the section features an input form where physicians are required to select a patient ("Selectează Pacient—Select a pacient") and, optionally, an associated medical device ("Selectează Dispozitiv, opțional"—Select a device, optional), before submitting a new recommendation. On the right side of the interface, once a patient is selected, their clinical details are displayed, including the "Nume, Prenume, Vârstă (Patient first and last name and age)", "Patologie (Associated pathologies)", and "Ultimele date înregistrate (Most recently recorded data)".

Figure 17. The Medical Results section from the medical professional's interface. This section provides healthcare professionals with detailed visualizations of biomedical data collected from patients, including charts, anomaly indicators, and historical trends. It supports clinical decision making by offering real-time access to physiological parameters, ensuring comprehensive and efficient patient monitoring.

As illustrated in Figure 18, the input form requires a "Title (Titlu)", providing a summary of its purpose, and a dedicated section for detailed clinical guidance ("Recomandare"), specifying the parameters to monitor. This functionality guarantees that clinical advice is well documented, as well as systematically recorded, therefore assuring clinical workflow efficiency and improved patient engagement and adherence to medical advice.

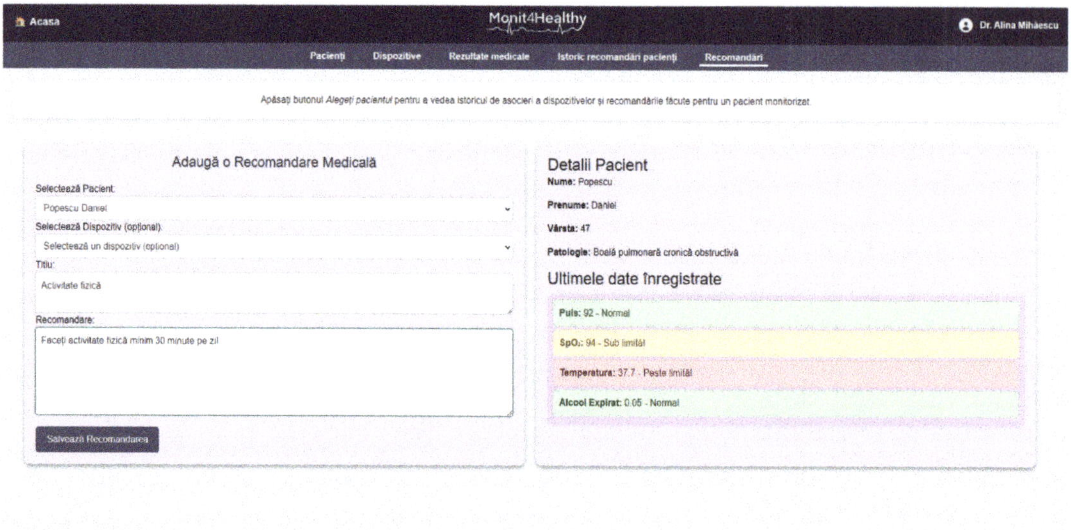

Figure 18. The Recommendations section from the medical professional's interface. This section allows medical professionals to provide personalized health recommendations based on patients' monitored data. It facilitates the creation, management, and communication of tailored advice, supporting proactive care and fostering continuous engagement between clinicians and patients.

4.2.4. Real-Time Data Transmission and Alerts

A fundamental aspect of the Monit4Healthy platform is its real-time data acquisition and visualization, with continuous patient health metric monitoring. As soon as a connected medical device receives new information, the data are transmitted to the platform, processed, and presented on the user's dashboard with minimal latency. To ensure patient safety, the system includes an alert mechanism to notify users of deviations from predefined physiological thresholds, with the system instantly generating visible alerts for both the patient and the physician.

4.2.5. Platform Authentication and User Management

The Monit4Healthy platform utilizes a secure authentication framework to ensure role-based access control. The system differentiates between the two primary user roles, each granted distinct access permissions. During account creation, patients register by providing personal details, while medical professionals must provide a unique medical identification code ("*Cod Parafă*") to verify their professional credentials. This verification step ensures that only licensed healthcare professionals can access patient data, therefore assuring platform security and regulatory compliance. Figure 19 showcases the Monit4Healthy system authentication page, with the selected medical professional role, in filling out the identification code in order to create a new account.

4.3. Development Phase Observations

Preliminary laboratory tests have indicated that the Monit4Healthy platform offers a user-friendly experience, being highly intuitive and with a structured, easily navigable interface for both patients and healthcare professionals. System responsiveness was assessed, with multiple devices transmitting data concurrently. The platform demonstrated minimal latency, with real-time updates appearing nearly instantly in the platform upon data acquisition.

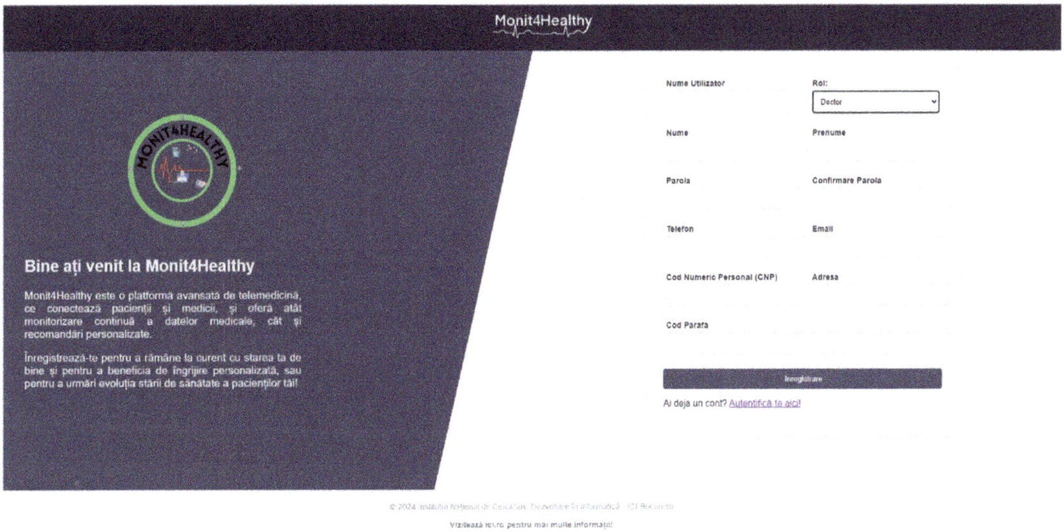

Figure 19. The Authentication page in the Monit4Healthy platform. This page ensures secure access to the platform through user authentication. It supports login for both patients and healthcare professionals, maintaining data confidentiality and system integrity by restricting access to authorized users only.

The accuracy, transmission efficiency, and reliability of the sensor data acquisition was tested, with the medical devices demonstrating a robust integration within the platform, ensuring efficient data flow from the sensors to the cloud and to the user interface. With the use of Raspberry Pi as a gateway, the process of data transmission was optimized for minimal latency and bandwidth efficiency.

To ensure data accuracy in the Monit4Healthy system, all sensors used for biomedical signal acquisition—specifically ECG, EMG, PPG, GSR, and body temperature sensors—underwent calibration and validation procedures during laboratory testing. Calibration was conducted using reference measurements obtained from standardized commercial devices and controlled environmental conditions, with sensor outputs assessed to confirm compliance with tolerances specified for healthcare applications. Where applicable, factory calibration parameters were reviewed and verified. For example, the PPG sensor (MAX30102) used for heart rate and SpO_2 monitoring was factory-calibrated and provided an SpO_2 accuracy of ±2% (within a 70–100% range) and heart rate accuracy of ±3 BPM, as specified by the manufacturer. Similarly, the temperature sensor (DS18B20) offers an accuracy of ±0.5 °C within the range of −10 °C to +85 °C and was validated under laboratory conditions. For the ECG and EMG modules, signal accuracy was evaluated using laboratory-grade signal simulators with standardized output parameters. These simulators generated reference waveforms with amplitude ranges between 0.5 mV and 4 mV, enabling the verification of acquisition accuracy and confirming appropriate filtering for clinical relevance. Although the system is currently in the final development and validation phase, with laboratory testing confirming stable and reliable sensor performance, comprehensive clinical accuracy benchmarking against certified medical devices is scheduled as part of the real-world deployment. These activities are detailed in the implementation roadmap and will involve comparative evaluations under clinical conditions to confirm sensor precision and consistency in real-world scenarios.

As previously mentioned, among the evaluated aspects during the development phase was the performance and placement optimization of the Gaitband. While initially placed on the ankle, laboratory tests revealed that resting positions were frequently misinterpreted as falls. Upon detecting changes in orientation, such as leg repositioning or shifting weight, the accelerometer and gyroscope triggered false alerts. The most reliable and stable alternative was placing the device on the chest. This reduced false positive alerts and improved the accuracy of the balance assessment. GPS tracking was also incorporated in order to detect the location of the patient upon detecting a fall event.

As the system is still in its final stages of development, the experimental evaluation of the Monit4Healthy system so far has shown tangible gains in communication efficiency, confirming the effective application of structured data processing and communication techniques. The outcomes demonstrate that the system consistently gathers and processes biomedical signals while maintaining steady and dependable data exchange in a laboratory environment. Performance assessments further show that the system successfully reduces transmission inconsistencies, preserving an ongoing communication flow under various operational conditions. In enhancing the dependability and continuity of real-time data exchange, the Monit4Healthy system has the potential to overcome important challenges in connected health ecosystems, which is in line with the goals of next-generation IoT health monitoring systems.

A preliminary laboratory evaluation of the system's reliability focused on evaluating fault tolerance, connection resilience, and signal consistency throughout extended usage. No significant issues were found over multiple continuous 8 h testing sessions, and under controlled conditions, the system maintained consistent wireless connectivity with a packet loss rate of less than 1%. Real-time data transmission was uninterrupted, and sensor signal integrity was maintained. Although these results show dependable short-term performance, further research to guarantee long-term dependability in various healthcare contexts will still need to include large-scale deployment measurements and prolonged runtime evaluations (such as the Mean Time Between Failures).

These findings demonstrate that the Monit4Healthy system, in its current laboratory configuration, achieves stable performance and efficient resource utilization under typical monitoring conditions, validating its readiness for further clinical validation and optimization.

5. Discussion

By incorporating multi-sensor fusion, real-time biomedical signal processing, and efficient data transmission approaches, the Monit4Healthy system enables a cohesive structure for IoT-enabled health monitoring. The system addresses challenges related to sensor interoperability, reliable information exchange, and real-time monitoring features in order to address meaningful challenges in IoT-driven healthcare. Its multi-layered architecture allows highly effective data gathering, edge processing, and secure communication protocols, providing accurate physiological and biomechanical data analysis. Its usefulness across connected health ecosystems is sustained by laboratory validation, which demonstrates its capacity to provide responsive signal processing, continuous monitoring, and comprehensive health data management.

Main contributions and foreseen performance impact

The Monit4Healthy system's *structured data transmission process* has been designed as one of the most important factors that aims to decrease energy consumption and network congestion while improving real-time biomedical signal processing. Future assessments will be carried out to establish the extent of these expected improvements. Only relevant and processed health parameters are transmitted owing to the system's edge-level data preprocessing, responsive data selection, and event-driven alert mechanism. It is a better alternative for typical IoT-based health monitoring systems as it successfully minimizes data redundancy while maintaining high data integrity and real-time responsiveness.

Preprocessing based on edge computing is a key element of this approach, since it considerably decreases the amount of raw health data sent to the cloud. In order to make sure that only the most relevant health parameters can be processed into optimum datasets, physiological signals collected through sensors like ECG, EMG, PPG, and GSR are subject to initial filtering, feature extraction, and aggregation at the data gathering stage. By reducing used bandwidth in the same time with preserving signal reliability, this process allows for smooth data transmission while not burdening computational resources.

The Monit4Healthy system updates its communication protocols according to network scenarios and the data level of importance to further enhance structured data transmission. Wi-Fi transmission is used for the high-resolution transmission of health data when mandatory, while BLE is applied for short-range, energy-efficient data sharing. In order to facilitate the real-time detection of abnormalities and multi-sensor interaction, the system likewise makes sure that delivered data are structured with timestamps, sensor identifiers, and abnormal flags.

The Monit4Healthy system's event-driven trigger alert system, which evaluates measured parameters according to clinical relevance, is a prominent part of its structured transmission process. Instead of continuously transmitting gathered data, the system assesses patterns in medical standards and ultimately triggers an alert when particular limits are surpassed. For instance, typical transmission cycles are overridden by a persistently aberrant ECG rhythm, guaranteeing timely response. In contrast, data are aggregated and sent at predetermined intervals when physiological values remain within normal limits. This minimizes excessive network congestion while providing medical professionals with thorough and organized insights.

Maintaining low-latency, energy-efficient, and high-integrity transmission of data while preserving real-time responsiveness is a significant challenge in IoT-enabled health monitoring. The Monit4Healthy system was designed with a hierarchical communication approach that incorporates edge computing to minimize needless data transmission. In contrast with typical IoT health monitoring solutions that depend on constant raw data streaming to cloud-based infrastructures, this approach is expected to improve power efficiency and network resource allocation, and future evaluations will assess the performance of Monit4Healthy under real medical conditions.

Prior to being sent to an ICIPRO cloud-based server, collected data first go through edge-level filtering, feature extraction, and structuring. While maintaining data integrity and system responsiveness, this approach successfully lowers network congestion, maximizes bandwidth usage, and supports appropriate real-time monitoring.

The Monit4Healthy system's fusion of edge and cloud computing is strengthened by the fact that edge-level processing reduces the need for constant raw data streaming to cloud servers by enabling real-time filtering, feature extraction, and data structuring. This approach significantly lowers the amount of data transmitted, which consequently decreases network congestion and improves reaction time for critical alerts compared to conventional cloud-only architectures. The system can handle time-sensitive tasks at the edges, like identifying anomalies and providing event-driven alerts, using a Raspberry Pi as a local processing unit. Solely necessary structured data for long-term storage and analysis are offloaded to the ICIPRO cloud infrastructure.

To illustrate the benefits of integrating edge computing in the Monit4Healthy system, a comparative analysis was conducted under laboratory conditions, focusing on latency and bandwidth usage in two configurations: edge–cloud hybrid (Monit4Healthy system) and cloud-only. In both scenarios, physiological data from ECG and PPG sensors were collected and processed. Edge-level preprocessing, performed on Raspberry Pi, reduced the volume of transmitted data by approximately 70%, as only filtered and feature-extracted data packets were sent to the cloud, compared to raw signal transmission in the cloud-only setup. For a 60 s monitoring session with ECG sampled at 250 Hz and PPG at 100 Hz (16-bit resolution), cloud-only transmission required approximately 0.040 MB per minute, while edge–cloud processing reduced this to around 0.012 MB/min. Latency measurements, defined as the time from data acquisition to cloud reception and processing, were reduced from an average of 350 ms (cloud-only) to 120 ms (edge–cloud), representing a 65% latency improvement. This efficiency supports scalable deployment in environments with constrained network resources and enables real-time responsiveness, crucial in health monitoring applications. While these results are preliminary and limited to controlled laboratory testing, they validate Monit4Healthy's design objectives and establish a technical foundation for future benchmarking in clinical environments.

The Medical Blackbox device is powered by an external adapter in the current lab setting so that it can operate continuously. The proposed design takes into account the overall utilization of the active biomedical sensors, ESP8266 module, and Arduino Mega. Although these findings are suggestive, a thorough energy determination evaluation will be conducted to evaluate runtime variability in various operating scenarios.

A more thorough quantitative comparison between edge-enabled and cloud-only configurations—measuring specific metrics like transmission delay, data volume, and processing time—is planned in the system's deployment roadmap, despite early laboratory evaluations confirming that edge computing contributes to lower latency and optimized bandwidth usage. In order to compare the system's performance in real-life scenarios and to further assess the technological advantages brought about by the edge computing architecture, this future analysis will incorporate simulation-based and physical testing.

The Monit4Healthy system optimizes its flexibility by setting priorities and organizing the transmission of data at the edge level so that measured parameters undergo processing with the lowest possible level of latency. The system optimizes network use while keeping high-accuracy biomedical data collected by dynamically changing data transmission priorities based on sensor type and real-time health parameters. These features define the Monit4Healthy system as a scalable option for healthcare and remote patient monitoring solutions, wherein energy-efficient, low-latency functioning is required for the continuous tracking of the health status.

The Monit4Healthy system's *multi-sensor fusion-based architecture* faces the drawbacks of single-sensor monitoring systems by enabling thorough health evaluations. The system increases the accuracy of diagnosis through linking many health information by means of the integration of ECG, EMG, PPG, and GSR sensors. Its flexible and modular architecture reinforces its application within connected healthcare ecosystems by facilitating adoption throughout a variety of healthcare environments, from clinical settings to home-based monitoring.

The Monit4Healthy system represents *an opportunity for integration with smart city infrastructures* beyond its main capabilities. The system may assist and provide support to digital health analytics platforms, emergency response coordination, and real-time health surveillance by integrating into already-existing urban IoT infrastructures. Large-scale urban health monitoring solutions may benefit from its scalable architecture and acquired data transmission structure through easy interoperability with healthcare services at the city level and AI-driven analytics, leading to the improvement of population-wide health assessment and resource management. In enabling predictive disease modeling and enhancing the use of healthcare resources, this potential integration fosters proactive health monitoring enlarged to the population level. The Monit4Healthy system seems to be appropriate for usage in assisted living environments, such as chronic disease management within smart healthcare regions, and elderly care, due to its secure data transmission capabilities and real-time biomedical sensor acquisition.

Although Monit4Healthy is primarily designed for remote and individualized health monitoring, the structured biomedical data it collects and processes can contribute to broader smart city healthcare ecosystems. In integrating anonymized health trends into municipal health platforms, the system can support data-driven decision making for resource allocation, urban health planning, and the early detection of community health risks. The potential interoperability of Monit4Healthy with digital health infrastructures enables its integration with emergency response systems, allowing healthcare providers to access real-time patient data when needed. This capability underlines the role of the Monit4Healthy system as a complementary component within connected health ecosystems, contributing to the development of more responsive and data-informed urban healthcare solutions.

To make sure that its sensor fusion approaches, data transmission protocols, and real-time signal processing features operate as designed under simulated healthcare circumstances, the Monit4Healthy system has undergone functional testing and validation in a controlled laboratory context. To verify that the system maintains high signal accuracy and reliable network transmission under various test situations, the testing process focused on assessing data reliability, sensor responsiveness, and connectivity stability. The system was evaluated using simulated physiological data inputs that emulated changing health monitoring circumstances in order to gauge real-time performance.

Estimated performance contributions of the Monit4Healthy system

With the goal of enhancing real-time biomedical monitoring, the Monit4Healthy system implements structured data transmission, edge computing abilities, and optimized sensor communication. The architectural design and laboratory testing and validation results highlight improvements in key performance aspects, such as latency reduction, bandwidth efficiency, and energy optimization, although formal large-scale benchmarking remains an area for future research.

- *Latency optimization*: The minimization of transmission delays is accomplished by employing local edge-level preprocessing, therefore reducing the volume of raw biomedical signals being delivered to cloud infrastructures. By substituting the mere reliance on cloud-based processing with real-time physiological filtering and feature extraction occurring at the acquisition level, the system ensures that it is clinically relevant and only transmits well structured data. This approach is expected to reduce latency in scenarios of high-priority health monitoring, such as in detecting arrhythmias based on ECG data, when event-triggered alerts are dire.
- *Bandwidth utilization efficiency:* The Monit4Healthy system implements adaptive data selection mechanisms, with the purpose of improving network efficiency. To reduce redundant data transmissions, the system replaces the approach of continuously streaming high-frequency signals with periodic reporting and event-triggered alerts. Given that network congestion and resource allocations need to be optimized when designing remote monitoring systems, this design is specifically efficient. The system's dual-mode communication approach—using BLE for short-range sensor connectivity and Wi-Fi for higher-resolution data transmission when necessary—allows scalable and energy-efficient data exchange.
- *Energy consumption considerations:* The system improves energy efficiency by implementing a structured power management strategy, minimizing cloud reliance by passing computational tasks to local processing units. To optimize sensor-device interactions, the system utilizes BLE-based communications protocols, which have established low-power characteristics. Further quantitative analysis is required in

order to validate the impact of these features in continuous monitoring scenarios, although they are anticipated to enhance energy-efficient operations. The reduction in continuous high-power transmissions, combined with the event-driven adaptive communication is anticipated to lengthen the lifespan of wearable and portable monitoring units. Despite these projections, battery consumption metrics need prolonged real-world evaluation. To additionally improve energy efficiency in extended health monitoring scenarios, future research could focus on duty-cycling mechanisms and machine learning-driven power scheduling.

Limitations and challenges:

- Real-world clinical assessments are still unexplored, despite laboratory testing confirming the system's fundamental functioning. Future research is required to evaluate the Monit4Healthy system's effectiveness in dynamic real-world settings, especially where network unpredictability, patient mobility, and telemedicine integration are involved.
- Another challenge is scalability, especially when it comes to easy integration with current telemedicine solutions and healthcare IT infrastructures. Even though the system's modular architecture allows for flexibility, further testing is necessary to guarantee that it can handle extensive medical applications.
- Measurements for the long-term power usage of continuous monitoring have not yet been assessed apart from controlled laboratory settings. Although edge computing decreases redundant data transmission, energy efficiency is expected to be boosted using adaptive power management approaches, especially for wearable and portable applications; future experimental validations need to be performed to identify the practical advantages of these approaches in real-word settings.

Future work involves further research in several important directions in order to improve the Monit4Healthy system's practical relevance:

- Carrying out extensive experiments to assess system performance in genuine clinical and digital health settings.
- Implementing a full English language option to improve usability for international users.
- Connecting EHRs and other type of healthcare solutions to provide simple interoperability and data sharing.
- Investigating hybrid IoT communication models that provide high-bandwidth, low-latency medical data transmission by integrating 5G-enabled frameworks.
- Future developments focusing on implementing advanced multi-sensor fusion algorithms, enabling the generation of composite health indicators by correlating synchronized ECG, EMG, PPG, and GSR data.
- Developing predictive analytics powered by AI to improve the automated identification of abnormalities and individualized health monitoring.
- To optimize network efficiency, current implementations are targeted toward structured data transmission. Despite ongoing approaches, future research could employ adaptive power management techniques, such as dynamic sensor sampling rate adjustments, to allow the system to reduce data acquisition frequency, and duty-cycling mechanisms, to power down non-essential components during periods of inactivity. To further lengthen the battery life, energy-aware task scheduling could be explored as well.
- Future experimental validations will quantitatively analyze Monit4Healthy's energy consumption across various operating conditions, assessing the real impact of edge computing and adaptive data transmission on power efficiency.
- Integration with IoT-specific protocols like MQTT is being considered for future development phases, particularly to improve scalability and interoperability in smart

city ecosystems or more intricate, dispersed healthcare infrastructures. This will further align the Monit4Healthy system with IoT communication standards by enabling efficient device-to-device and device-to-cloud connection.
- The potential benefits of edge computing will be addressed in more depth. This consists of real-world and simulation-based comparisons of edge-enabled and cloud-only setups, with an emphasis on important performance indicators including power consumption, processing time, data volume, and transmission latency. The findings from these investigations will allow for the additional optimization of the Monit4Healthy system's edge-cloud integration strategy and quantitatively validate the architectural benefits of the system across various healthcare situations.
- Comprehensive clinical accuracy benchmarking against certified medical equipment is scheduled during real-world deployment phases to validate sensor performance under operational healthcare conditions.
- Adaptive communication strategies that dynamically optimize protocol selection and channel allocation in real time, aiming to enhance transmission reliability and energy efficiency in high-density and variable network environments, will be taken into consideration.
- Miniaturizing the Medical Blackbox by designing a custom PCB to replace the current prototyping board aims to optimize the device's size, improve power efficiency, and enhance hardware reliability for long-term deployment in both clinical and home-based settings.
- Future research will involve thorough power profiling under a range of operating situations, such as continuous monitoring, event-driven alerts, and inactivity states, in order to comprehensively assess the energy efficiency of the Monit4Healthy system. All system components' current drawing and consumption trends will be tracked using standardized energy measuring instruments. We will analyze metrics like overall battery duration under particular tasks, energy per transmitted data unit (mWh/MB), and average energy consumption (mW). These parameters will corroborate the system's energy optimization assertions and direct future hardware and software improvements by enabling comparisons with other IoT-based health monitoring solutions. Future analyses will measure energy consumption per unit of processed biomedical data (e.g., milliwatts per kilobyte transmitted) and energy per monitoring session (e.g., Wh per 1 h operation under standard use settings) in order to assess the system's energy efficiency. The baseline measurements from cloud-only monitoring systems with comparable sensor configurations but no edge-level preprocessing or structured data transfer will be compared to these measurements. The energy-saving effects of the Monit4Healthy system's edge computing and adaptive transmission approaches will be measurable through this comparison.

Through the implementation of these improvements, the Monit4Healthy system will become more fitting for long-term deployment in remote health monitoring and wearable medical devices, thus guaranteeing sustained operation in resource-constrained environments.

Roadmap for real-world deployment

To ensure a structured transition of the Monit4Healthy system from laboratory validation to real-world healthcare environments, the following roadmap outlines the key phases of deployment, each with specific objectives, methodologies, and performance targets:

- Phase 1—Pilot Trials in Controlled Clinical Settings
    - *Objective*: Evaluate reliability, usability, and accuracy under supervision of medical professionals.

- *Activities*: Deploy Monit4Healthy units in medical units for short-term trials; monitor ECG, EMG, and PPG signal acquisition; assess edge–cloud performance in operational conditions.
- *Metrics*: User feedback, real-time alert responsiveness, signal integrity, data transmission stability, and power consumption profiles.
- *Duration*: 3–6 months.
- *Deliverables*: Technical report on system performance; validated integration with EHR platforms and telemedicine systems.

- Phase 2—Multi-Site and Home-Based Testing
  - *Objective*: Assess scalability, adaptability, and long-term reliability in diverse healthcare scenarios.
  - *Activities*: Deploy in multiple sites including home-care environments; focus on patient mobility, variable network conditions, and extended monitoring periods.
  - *Metrics*: Energy consumption trends, battery life in real conditions, network adaptability, anomaly detection rates, data privacy compliance.
  - *Duration*: 6–9 months.
  - *Deliverables*: Comparative performance analysis vs. cloud-only systems; refinement of adaptive power management; user satisfaction studies.
- Phase 3—System Optimization and Hardware Refinement
  - *Objective*: Miniaturize hardware components, improve energy efficiency, and enhance user ergonomics.
  - *Activities*: Design and fabricate custom PCBs for Medical Blackbox and EMG devices; integrate energy-aware scheduling; implement secure boot and firmware updates.
  - *Metrics*: Reduction in form factor, increased battery efficiency (target >12 h), and hardware validation.
  - *Duration*: 6 months (overlapping with Phase 2).
  - Deliverables: Optimized prototypes; industrial design for wearable/portable use; regulatory compliance planning (e.g., CE, FDA).
- Phase 4—Integration into Smart Health Ecosystems
  - *Objective*: Align Monit4Healthy system with smart city infrastructure.
  - *Activities*: Pilot integration with municipal health dashboards, emergency services, and AI-powered health trend analytics; implementation of MQTT for scalable IoT communication.
  - *Metrics*: Interoperability benchmarks, real-time data aggregation at population level.
  - *Duration*: 9–12 months.
  - *Deliverables*: Full-scale deployment proposal; partnership reports; scalability studies.

Together, these perspectives and planned developments position the Monit4Healthy system as a scalable, adaptive, and forward-looking solution, capable of enhancing real-time health monitoring and advancing the integration of IoT technologies in connected healthcare environments, including future deployment within smart city infrastructures.

6. Conclusions

Through the integration of multi-sensor fusion, IoT-enabled monitoring technologies, and appropriate data transmission frameworks, the Monit4Healthy system enables connected health ecosystems. The system tackles important issues in IoT-driven healthcare by enhancing sensor interoperability, real-time health monitoring, and secure data exchange. This allows for effective, scalable, and flexible health data processing.

The Monit4Healthy system improves biomedical signal integrity, network performance, and energy efficiency using edge computing and the structured transmission of information approaches. Thorough health evaluations are supported by its modular sensor fusion architecture, which enables easy integration into clinical, telemedicine, and smart city infrastructures. By means of the flexible prioritization of real-time health data and the decrease in needless network loads, the system promises high-reliability, low-latency medical monitoring. By enabling interoperability with digital health solutions, and the potential integration into smart city infrastructures, the Monit4Healthy system brings an opportunity to extend connected health ecosystems beyond conventional healthcare environments. To further consolidate its standing as a next-generation IoT-based health monitoring solution, future work and research will be centered around scalability, AI-driven analytics, and expanded implementations in current healthcare environments.

Author Contributions: Conceptualization, M.I. and V.-Ș.C.; methodology V.-Ș.C. and M.I.; software, V.-Ș.C., A.-G.M., M.-C.P. and O.B.; validation, V.-Ș.C., M.I. and A.A.; formal analysis, A.A.; investigation, A.-M.G.; writing—original draft preparation, M.I., A.-G.M., M.-C.P., A.A., O.B. and A.-M.G.; writing—review and editing, M.I., A.-G.M., M.-C.P., A.A., O.B. and A.-M.G.; visualization, V.-Ș.C.; supervision, A.A. and V.-Ș.C.; project administration, M.I. All authors have read and agreed to the published version of the manuscript.

Funding: This research received no external funding.

Institutional Review Board Statement: Not applicable.

Informed Consent Statement: Not applicable.

Data Availability Statement: Data are contained within the article.

Conflicts of Interest: The authors declare no conflicts of interest.

Abbreviations

The following abbreviations are used in this manuscript:

AFH	Adaptive Frequency Hopping;
AI	Artificial Intelligence;
BLE	Bluetooth Low Energy;
CAD	Computer-Aided Design;
CLI	Command Line Interface;
ECG	Electrocardiography;
EMG	Electromyography;
FHIR	Fast Healthcare Interoperability Resources;
GDPR	General Data Protection Regulation;
GPS	Global Positioning System;
GSR	Galvanic Skin Response;
HIPAA	Health Insurance Portability and Accountability Act;
I^2C	Inter-Integrated Circuit;
IaaS	Infrastructure as a Service;
IoT	Internet of Things;
LoRaWAN	Long-Range Wide-Area Network;
ORM	Object Relational Mapper;
MCU	Microcontroller Unit;
PPG	Photoplethysmogram;
RPM	Remote Patient Monitoring;
SpO_2	Oxygen Saturation;

SQL	Structured Query Language;
SSL	Secure Sockets Layer;
TLS	Transport Layer Security;
VM	Virtual Machine;
Wi-Fi	Wireless Fidelity.

References

1. Sarkar, M.; Lee, T.-H.; Sahoo, P.K. Smart Healthcare: Exploring the Internet of Medical Things with Ambient Intelligence. *Electronics* **2024**, *13*, 2309. [CrossRef]
2. Copenhagen: WHO Regional Office for Europe Building a Public Health Innovation Ecosystem in the WHO European, Region: Meeting Report. 2024. Available online: https://www.who.int/europe/publications/i/item/WHO-EURO-2024-10102-49874-74930 (accessed on 21 March 2025).
3. Băjenaru, L.; Marinescu, I.A.; Tomescu, M.; Drăghici, R. Evaluarea Satisfacției Persoanelor Vârstnice În Utilizarea de Tehnologii Inteligente Asistive: VINCI Studiu de Caz. *Rev. Română De Informatică Și Autom.* **2022**, *32*, 19–32. [CrossRef]
4. Kouchaki, S.; Ding, X.; Sanei, S. AI- and IoT-Enabled Solutions for Healthcare. *Sensors* **2024**, *24*, 2607. [CrossRef]
5. AlAdwani, H.; Zahra. A Systematic Review of IoT Integration on Health Monitoring System. *Int. J. Eng. Manag. Res.* **2023**, *13*, 50–59. [CrossRef]
6. Koulouras, G.; Katsoulis, S.; Zantalis, F. Evolution of Bluetooth Technology: BLE in the IoT Ecosystem. *Sensors* **2025**, *25*, 996. [CrossRef]
7. Safi, H.; Jehangiri, A.I.; Ahmad, Z.; Ala'anzy, M.A.; Alramli, O.I.; Algarni, A. Design and Evaluation of a Low-Power Wide-Area Network (LPWAN)-Based Emergency Response System for Individuals with Special Needs in Smart Buildings. *Sensors* **2024**, *24*, 3433. [CrossRef]
8. Durand, T.G.; Visagie, L.; Booysen, M.J. Evaluation of Next-generation Low-power Communication Technology to Replace GSM in IoT-applications. *IET Commun.* **2019**, *13*, 2533–2540. [CrossRef]
9. Ait Abdelouahid, R.; Debauche, O.; Mahmoudi, S.; Marzak, A. Literature Review: Clinical Data Interoperability Models. *Information* **2023**, *14*, 364. [CrossRef]
10. Azzedin, F.; Alhazmi, T. Secure Data Distribution Architecture in IoT Using MQTT. *Appl. Sci.* **2023**, *13*, 2515. [CrossRef]
11. Shahid, J.; Ahmad, R.; Kiani, A.K.; Ahmad, T.; Saeed, S.; Almuhaideb, A.M. Data Protection and Privacy of the Internet of Healthcare Things (IoHTs). *Appl. Sci.* **2022**, *12*, 1927. [CrossRef]
12. Popoola, O.; Rodrigues, M.; Marchang, J.; Shenfield, A.; Ikpehai, A.; Popoola, J. A Critical Literature Review of Security and Privacy in Smart Home Healthcare Schemes Adopting IoT & Blockchain: Problems, Challenges and Solutions. *Blockchain Res. Appl.* **2024**, *5*, 100178. [CrossRef]
13. da Costa, T.P.; da Costa, D.M.B.; Murphy, F. A Systematic Review of Real-Time Data Monitoring and Its Potential Application to Support Dynamic Life Cycle Inventories. *Environ. Impact Assess. Rev.* **2024**, *105*, 107416. [CrossRef]
14. Al Khatib, I.; Shamayleh, A.; Ndiaye, M. Healthcare and the Internet of Medical Things: Applications, Trends, Key Challenges, and Proposed Resolutions. *Informatics* **2024**, *11*, 47. [CrossRef]
15. Rupanetti, D.; Kaabouch, N. Combining Edge Computing-Assisted Internet of Things Security with Artificial Intelligence: Applications, Challenges, and Opportunities. *Appl. Sci.* **2024**, *14*, 7104. [CrossRef]
16. Bhuiyan, M.N.; Rahman, M.M.; Billah, M.M.; Saha, D. Internet of Things (IoT): A Review of Its Enabling Technologies in Healthcare Applications, Standards Protocols, Security, and Market Opportunities. *IEEE Internet Things J.* **2021**, *8*, 10474–10498. [CrossRef]
17. Wu, X.; Liu, C.; Wang, L.; Bilal, M. Internet of Things-Enabled Real-Time Health Monitoring System Using Deep Learning. *Neural Comput. Appl.* **2023**, *35*, 14565–14576. [CrossRef]
18. Su, M.; Hua, J.; Sun, X.; Liu, Z.; Shi, Y.; Pan, L. Wireless Wearable Devices and Recent Applications in Health Monitoring and Clinical Diagnosis. *Biomed. Mater. Devices* **2024**, *2*, 669–694. [CrossRef]
19. Omaghomi, T.T.; Elufioye, O.A.; Akomolafe, O.; Anyanwu, E.C.; Odilibe, I.P. A Comprehensive Review of Telemedicine Technologies: Past, Present, and Future Prospects. *Int. Med. Sci. Res. J.* **2024**, *4*, 183–193. [CrossRef]
20. Buleje, I.; Siu, V.S.; Hsieh, K.Y.; Hinds, N.; Dang, B.; Bilal, E.; Nguyen, T.; Lee, E.E.; Depp, C.A.; Rogers, J.L. A Versatile Data Fabric for Advanced IoT-Based Remote Health Monitoring. In Proceedings of the 2023 IEEE International Conference on Digital Health (ICDH), Chicago, IL, USA, 2–8 July 2023; IEEE: Piscataway, NJ, USA, 2023; pp. 88–90.
21. Laska, B.; Xi, P.; Valdés, J.J.; Wallace, B.; Goubran, R. Zero-Shot Multi-Task Cough Sound Analysis with Speech Foundation Model Embeddings. In Proceedings of the 2024 IEEE International Symposium on Medical Measurements and Applications (MeMeA), Eindhoven, The Netherlands, 26–28 June 2024; IEEE: Piscataway, NJ, USA, 2024; pp. 1–6.
22. Dadkhah, M.; Mehraeen, M.; Rahimnia, F.; Kimiafar, K. Use of Internet of Things for Chronic Disease Management. *J. Med. Signals Sens.* **2021**, *11*, 138–157. [CrossRef]

23. Talpur, M.S.H.; Abro, A.A.; Ebrahim, M.; Kandhro, I.A.; Manickam, S.; Laghari, S.U.A.; Dandoush, A.; Uddin, M. Illuminating Healthcare Management: A Comprehensive Review of IoT-Enabled Chronic Disease Monitoring. *IEEE Access* **2024**, *12*, 48189–48209. [CrossRef]
24. de Farias, F.A.C.; Dagostini, C.M.; de Assunção Bicca, Y.; Falavigna, V.F.; Falavigna, A. Remote Patient Monitoring: A Systematic Review. *Telemed. e-Health* **2020**, *26*, 576–583. [CrossRef] [PubMed]
25. Ianculescu, M.; Paraschiv, E.-A.; Alexandru, A. Addressing Mild Cognitive Impairment and Boosting Wellness for the Elderly through Personalized Remote Monitoring. *Healthcare* **2022**, *10*, 1214. [CrossRef] [PubMed]
26. De Sario Velasquez, G.D.; Borna, S.; Maniaci, M.J.; Coffey, J.D.; Haider, C.R.; Demaerschalk, B.M.; Forte, A.J. Economic Perspective of the Use of Wearables in Health Care: A Systematic Review. *Mayo Clin. Proc. Digit. Health* **2024**, *2*, 299–317. [CrossRef]
27. Lekwijit, S.; Terwiesch, C.; Asch, D.A.; Volpp, K.G. Evaluating the Efficacy of Connected Healthcare: An Empirical Examination of Patient Engagement Approaches and Their Impact on Readmission. *Manag. Sci.* **2024**, *70*, 3417–3446. [CrossRef]
28. Abdulmalek, S.; Nasir, A.; Jabbar, W.A.; Almuhaya, M.A.M.; Bairagi, A.K.; Khan, M.A.-M.; Kee, S.-H. IoT-Based Healthcare-Monitoring System towards Improving Quality of Life: A Review. *Healthcare* **2022**, *10*, 1993. [CrossRef]
29. Hashim, H.; Salihudin, S.F.B.; Saad, P.S.M. Development of IoT Based Healthcare Monitoring System. In Proceedings of the 2022 IEEE International Conference in Power Engineering Application (ICPEA), Shah Alam, Malaysia, 7–8 March 2022; IEEE: Piscataway, NJ, USA, 2022; pp. 1–5.
30. Dankan Gowda, V.; Chaithra, S.M.; Gujar, S.S.; Firoz Shaikh, S.; Ingole, B.S.; Sudhakar Reddy, N. Scalable AI Solutions for IoT-Based Healthcare Systems Using Cloud Platforms. In Proceedings of the 2024 8th International Conference on I-SMAC (IoT in Social, Mobile, Analytics and Cloud) (I-SMAC), Kirtipur, Nepal, 3–5 October 2024; IEEE: Piscataway, NJ, USA, 2024; pp. 156–162.
31. Martins, A.; Londral, A.; Nunes, I.L.; Lapão, L.V. Unlocking Human-like Conversations: Scoping Review of Automation Techniques for Personalized Healthcare Interventions Using Conversational Agents. *Int. J. Med. Inform.* **2024**, *185*, 105385. [CrossRef]
32. Navaz, A.N.; Serhani, M.A.; El Kassabi, H.T.; Al-Qirim, N.; Ismail, H. Trends, Technologies, and Key Challenges in Smart and Connected Healthcare. *IEEE Access* **2021**, *9*, 74044–74067. [CrossRef]
33. Ramachandran, G. Creating a Digital Spine for Connected Healthcare Ecosystem: Digitalization for an Effective Supply Chain. In *Resilient Health*; Academic Press: Cambridge, MA, USA, 2024.
34. Rehman, A.; Naz, S.; Razzak, I. Leveraging Big Data Analytics in Healthcare Enhancement: Trends, Challenges and Opportunities. *Multimed. Syst.* **2022**, *28*, 1339–1371. [CrossRef]
35. Tian, Y.J.; Felber, N.A.; Pageau, F.; Schwab, D.R.; Wangmo, T. Benefits and Barriers Associated with the Use of Smart Home Health Technologies in the Care of Older Persons: A Systematic Review. *BMC Geriatr.* **2024**, *24*, 152. [CrossRef]
36. Mahmmod, B.M.; Naser, M.A.; Al-Sudani, A.H.S.; Alsabah, M.; Mohammed, H.J.; Alaskar, H.; Almarshad, F.; Hussain, A.; Abdulhussain, S.H. Patient Monitoring System Based on Internet of Things: A Review and Related Challenges with Open Research Issues. *IEEE Access* **2024**, *12*, 132444–132479. [CrossRef]
37. Islam, S.M.R.; Lloret, J.; Zikria, Y. Bin Internet of Things (IoT)-Based Wireless Health: Enabling Technologies and Applications. *Electronics* **2021**, *10*, 148. [CrossRef]
38. Amezquita Garcia, J.A.; Bravo Zanoguera, M.E.; Murrieta-Rico, F.N. Advances and Classification of Autonomous Systems in Biomedical Devices: Integration of Energy Harvesting and Ultra-Low Power Consumption. *Electronics* **2025**, *14*, 144. [CrossRef]
39. Al-Sofi, S.J.; Atroshey, S.M.S.; Ali, I.A. IEEE 802.15.6 and LoRaWAN for WBAN in Healthcare: A Comparative Study on Communication Efficiency and Energy Optimization. *Computers* **2024**, *13*, 313. [CrossRef]
40. Jain, S.; Jain, P.; Upadhyay, P.K.; Moualeu, J.M.; Srivastava, A. An Energy Efficient Health Monitoring Approach with Wireless Body Area Networks. *ACM Trans. Comput. Healthc.* **2022**, *3*, 1–22. [CrossRef]
41. Moreno-Cruz, F.; Toral-López, V.; Escobar-Molero, A.; Ruíz, V.U.; Rivadeneyra, A.; Morales, D.P. TreNch: Ultra-Low Power Wireless Communication Protocol for IoT and Energy Harvesting. *Sensors* **2020**, *20*, 6156. [CrossRef]
42. Rabah, H.; Albasha, L.; Mir, H.; Quadir, N.; Abbas, S.Z. High-Efficiency, Low-Loss, and Wideband 5.8 GHz Energy Harvester Designed Using TSMC 65 nm Process for IoT Self-Powered Nodes. *Energies* **2025**, *18*, 862. [CrossRef]
43. Tohidinejad, Z.; Danyali, S.; Valizadeh, M.; Seepold, R.; TaheriNejad, N.; Haghi, M. Designing a Hybrid Energy-Efficient Harvesting System for Head- or Wrist-Worn Healthcare Wearable Devices. *Sensors* **2024**, *24*, 5219. [CrossRef]
44. Famitafreshi, G.; Afaqui, M.S.; Melià-Seguí, J. A Comprehensive Review on Energy Harvesting Integration in IoT Systems from MAC Layer Perspective: Challenges and Opportunities. *Sensors* **2021**, *21*, 3097. [CrossRef]
45. Dizdarević, J.; Blažević, D.; Grunewald, M.; Jukan, A. An Edge/Cloud Continuum with Wearable Kinetic Energy Harvesting IoT Devices in Remote Areas. In Proceedings of the ICC 2024—IEEE International Conference on Communications, Denver, CO, USA, 9–13 June 2024; IEEE: Piscataway, NJ, USA, 2024; pp. 879–884.
46. Kurduban, V. Available online: https://Www.Digi.Com/Blog/Post/Edge-Computing-vs-Cloud-Computing (accessed on 22 March 2025).

47. Sarangi, A.K.; Mohapatra, A.G.; Mishra, T.C.; Keswani, B. Healthcare 4.0: A Voyage of Fog Computing with IOT, Cloud Computing, Big Data, and Machine Learning. In *Fog Computing for Healthcare 4.0 Environments*; Springer: Cham, Switzerland, 2021; pp. 177–210.
48. Al-Marsy, A.; Chaudhary, P.; Rodger, J.A. A Model for Examining Challenges and Opportunities in Use of Cloud Computing for Health Information Systems. *Appl. Syst. Innov.* **2021**, *4*, 15. [CrossRef]
49. Awad, A.I.; Fouda, M.M.; Khashaba, M.M.; Mohamed, E.R.; Hosny, K.M. Utilization of Mobile Edge Computing on the Internet of Medical Things: A Survey. *ICT Express* **2023**, *9*, 473–485. [CrossRef]
50. Issa, M.E.; Helmi, A.M.; Al-Qaness, M.A.A.; Dahou, A.; Abd Elaziz, M.; Damaševičius, R. Human Activity Recognition Based on Embedded Sensor Data Fusion for the Internet of Healthcare Things. *Healthcare* **2022**, *10*, 1084. [CrossRef]
51. Bijalwan, V.; Semwal, V.B.; Mandal, T.K. Fusion of Multi-Sensor-Based Biomechanical Gait Analysis Using Vision and Wearable Sensor. *IEEE Sens. J.* **2021**, *21*, 14213–14220. [CrossRef]
52. Salis, F.; Bertuletti, S.; Scott, K.; Caruso, M.; Bonci, T.; Buckley, E.; Della Croce, U.; Mazza, C.; Cereatti, A. A Wearable Multi-Sensor System for Real World Gait Analysis. In Proceedings of the 2021 43rd Annual International Conference of the IEEE Engineering in Medicine & Biology Society (EMBC), Mexico, 1–5 November 2021; IEEE: Piscataway, NJ, USA, 2021; pp. 7020–7023.
53. Qiu, S.; Zhao, H.; Jiang, N.; Wang, Z.; Liu, L.; An, Y.; Zhao, H.; Miao, X.; Liu, R.; Fortino, G. Multi-Sensor Information Fusion Based on Machine Learning for Real Applications in Human Activity Recognition: State-of-the-Art and Research Challenges. *Inf. Fusion* **2022**, *80*, 241–265. [CrossRef]
54. John, A.; Cardiff, B.; John, D. A Review on Multisensor Data Fusion for Wearable Health Monitoring. *arXiv* **2024**, arXiv:2412.05895.
55. Teran-Pineda, D.; Thurnhofer-Hemsi, K.; Dominguez, E. Analysis and Recognition of Human Gait Activity Based on Multimodal Sensors. *Mathematics* **2023**, *11*, 1538. [CrossRef]
56. Zhou, S.; Sun, J.; Xu, K.; Wang, G. AI-Driven Data Processing and Decision Optimization in IoT through Edge Computing and Cloud Architecture. *Preprints* **2024**, 2024100736. [CrossRef]
57. Still, T. Autodesk Fusion and Cadence: The Power of an Integrated Electromechanical Design Workflow. Available online: https://www.autodesk.com/products/fusion-360/blog/autodesk-fusion-cadence-integrated-workflow-electromechanical-design/ (accessed on 24 March 2025).
58. Machine, M.F.G. The Advantages of CAD Software and Why Every Engineer Should Use It. Available online: https://shop.machinemfg.com/the-advantages-of-cad-software-and-why-every-engineer-should-use-it/ (accessed on 24 March 2025).
59. Chang, C.-S.; Wu, T.-H.; Wu, Y.-C.; Han, C.-C. Bluetooth-Based Healthcare Information and Medical Resource Management System. *Sensors* **2023**, *23*, 5389. [CrossRef]
60. Specht, C. Maritime DGPS System Positioning Accuracy as a Function of the HDOP in the Context of Hydrographic Survey Performance. *Remote Sens.* **2022**, *15*, 10. [CrossRef]
61. Lee, K.H.; Min, J.Y.; Byun, S. Electromyogram-Based Classification of Hand and Finger Gestures Using Artificial Neural Networks. *Sensors* **2021**, *22*, 225. [CrossRef]
62. Elyounsi, A.; Kalashnikov, A.N. Evaluating Suitability of a DS18B20 Temperature Sensor for Use in an Accurate Air Temperature Distribution Measurement Network. In Proceedings of the 8th International Electronic Conference on Sensors and Applications, Online, 1–15 November 2021; MDPI: Basel, Switzerland, 2021; p. 56.
63. Uddin, R.; Koo, I. Real-Time Remote Patient Monitoring: A Review of Biosensors Integrated with Multi-Hop IoT Systems via Cloud Connectivity. *Appl. Sci.* **2024**, *14*, 1876. [CrossRef]
64. Heaney, J.; Buick, J.; Hadi, M.U.; Soin, N. Internet of Things-Based ECG and Vitals Healthcare Monitoring System. *Micromachines* **2022**, *13*, 2153. [CrossRef] [PubMed]
65. Divyashree, N.; Prasad, N.K.S. Design and Development of We-CDSS Using Django Framework: Conducing Predictive and Prescriptive Analytics for Coronary Artery Disease. *IEEE Access* **2022**, *10*, 119575–119592. [CrossRef]
66. Alimbayeva, Z.; Alimbayev, C.; Ozhikenov, K.; Bayanbay, N.; Zhikenova, A. Wearable ECG Device and Machine Learning for Heart Monitoring. *Sensors* **2024**, *24*, 4201. [CrossRef]
67. Duart, X.; Quiles, E.; Suay, F.; Chio, N.; García, E.; Morant, F. Evaluating the Effect of Stimuli Color and Frequency on SSVEP. *Sensors* **2020**, *21*, 117. [CrossRef] [PubMed]
68. Ibáñez Castillo, M. Wireless Based Wearable Patient Health Monitoring System Using ESP32. Bachelor's Thesis, Universitat Politècnica de Catalunya, Barcelona, Spain, 2024.
69. Cheng, Y.; Ye, Y.; Hou, M.; He, W.; Li, Y.; Deng, X. A Fast and Robust Non-Sparse Signal Recovery Algorithm for Wearable ECG Telemonitoring Using ADMM-Based Block Sparse Bayesian Learning. *Sensors* **2018**, *18*, 2021. [CrossRef]
70. Suresha, P.B.; Hegde, C.; Jiang, Z.; Clifford, G.D. An Edge Computing and Ambient Data Capture System for Clinical and Home Environments. *Sensors* **2022**, *22*, 2511. [CrossRef]
71. Bossard, A. Memory Optimisation on AVR Microcontrollers for IoT Devices' Minimalistic Displays. *Chips* **2022**, *1*, 2–13. [CrossRef]
72. Ehrmann, G.; Blachowicz, T.; Homburg, S.V.; Ehrmann, A. Measuring Biosignals with Single Circuit Boards. *Bioengineering* **2022**, *9*, 84. [CrossRef]

73. Mathe, S.E.; Kondaveeti, H.K.; Vappangi, S.; Vanambathina, S.D.; Kumaravelu, N.K. A Comprehensive Review on Ap plications of Raspberry Pi. *Comput. Sci. Rev.* **2024**, *52*, 100636. [CrossRef]
74. ICI Bucharest ICIPRO Cloud Computing Services. Available online: https://www.icipro.ro/ (accessed on 25 March 2025).
75. Nancy, A.A.; Ravindran, D.; Raj Vincent, P.M.D.; Srinivasan, K.; Gutierrez Reina, D. IoT-Cloud-Based Smart Healthcare Monitoring System for Heart Disease Prediction via Deep Learning. *Electronics* **2022**, *11*, 2292. [CrossRef]

sensors

MDPI

Article

A Mold Damage Monitoring Algorithm for Power Metallurgy Molding Machines Using Bidirectional Long Short-Term Memory on an Internet of Things Platform †

Hao-Pu Lin [1], Yuan-Chieh Chen [2], Chin-Chuan Han [2,*], Yu-Chi Wu [3] and Jin-Yuan Lin [3]

[1] Ph. D. Program in Material and Chemical Engineering, National United University, MiaoLi 360302, Taiwan; d1012005@o365.nuu.edu.tw
[2] Department of Computer Science and Information Engineering, National United University, MiaoLi 360302, Taiwan; tryit320495@gmail.com
[3] Department of Electrical Engineering, National United University, MiaoLi 360302, Taiwan; ycwu@nuu.edu.tw (Y.-C.W.); yuan@nuu.edu.tw (J.-Y.L.)
* Correspondence: cchan@nuu.edu.tw; Tel.: +886-920528098

† This article is an extended version of the conference paper: Lin, H.P.; Chen, Y.C.; Han, C.C.; Wu, Y.C.; Chang, C.S.; Lin, J.Y. Mold Damage Monitoring for Power Metallurgy Molding Machines Using Deep Learning Methods. In Proceedings of the 10th IIAE International Conference on Intelligent Systems and Image Processing, Beppu, Japan, 4–8 September 2023.

Highlights:

What are the main findings?

- Damage to the mold will be reflected in the vibration.
- The vibration caused by the damaged mold is very small.

What are the implications of the main finding?

- Bidirectional LSTM can be used to determine the mold status through vibration.
- The accuracy highly depends on the captured data, with a high sampling rate.

Abstract: In this paper, an analysis and monitoring algorithm is proposed for mold health evaluation using vibration data. Two inertial measurement units (IMUs) and an embedded system are first used to acquire vibration data from a powder metallurgy molding machine. These data are collected on an Internet of Things (IoT) platform using the Message Queueing Telemetry Transport (MQTT) protocol. For data analysis, the vibration signal on the Z axis is segmented to label the contact section of the upper and middle molds, and the corresponding vibration data of the stamping friction on the X, Y, and Z axes are extracted. Using only historical vibration data from normal stamping, a Bidirectional Long Short-Term Memory (Bi-LSTM) model with an attention mechanism is trained to predict normal stamping vibrations several minutes in advance. By comparing the predicted stamping vibrations with the observed data at the current time, the mean square errors (MSEs) are calculated to evaluate the health status of the mold. Several ablation experiments were conducted to assess the performance of the trained model. The average MSE values for normal samples and abnormal samples were smaller than 0.5 and larger than 1.0, respectively. The experimental results confirm that the trained prediction model and evaluation indicators can effectively notify operators in advance. An early warning system using vibration data for mold damage was successfully implemented, enhancing predictive maintenance.

Keywords: intelligence system; vibration data; Internet of Things (IoT); deep learning; inertial measurement unit (IMU); mean square error

1. Introduction

In response to the demands of Industry 4.0, particularly the shift towards intelligent automation, traditional industries have begun transforming their production processes by incorporating automation, networked systems, data analytics, and intelligent monitoring. This transformation aims to reduce labor dependency, enhance production efficiency, and mitigate operational risks for personnel. In Taiwan, the powder metallurgy industry has been active for more than 40 years. The manufacturing process in this industry involves feeding raw metal powders into a mold, stamping them under a certain pressure using hydraulic machines, forming products of the required shape and size, sintering them at high temperatures, and then performing post-processing, such as reshaping and anti-rust treatment. Traditional mold health monitoring methods rely primarily on manual inspection and periodic maintenance schedules. However, manual inspections are subjective, time-consuming, and highly dependent on operator expertise, making them prone to human error. Additionally, simple threshold-based monitoring systems can only detect severe failures but lack the capability to identify early-stage mold degradation. These methods often result in unexpected mold failures, leading to production downtime and financial losses.

Anomaly detection in vibration data has garnered significant attention due to its critical role in fault diagnosis and structural health monitoring using machine learning and deep learning techniques. While conventional machine learning models, such as support vector machines (SVMs) and decision trees, have been applied for fault diagnosis, they often require handcrafted feature extraction and may struggle with complex, high-dimensional data. For instance, Torres et al. [1] proposed an Isolation Forest algorithm for vibration data analysis to detect anomalies in induction motors within a predictive maintenance context. Adaptive training methods for vibration-based anomaly detection have proven effective in early diagnostics for wind turbine condition monitoring, thereby reducing maintenance costs [2]. Additionally, vibration-based anomaly detection frameworks have been extensively applied to rolling element bearings, successfully identifying distinct fault modes and contributing to the development of robust fault diagnosis systems [3]. In all approaches, the vibration data are first transformed into the frequency domain for feature extraction, and posterior probabilities are calculated for classification using traditional machine learning methods.

Recently, many deep learning-based methods have been proposed for anomaly detection using time-series vibration data across various applications. These methods can be categorized into several network architectures, such as convolutional neural network (CNN) + deep neural network (DNN) [4], Long Short-Term Memory (LSTM) + autoencoder [5–8], feed-forward NN + SVM [9], LSTM + SVM [10], CNN + LSTM [11,12], bi-directional Long Short-Term Memory (Bi-LSTM) [13–19], and others. They have been applied to domains such as mechanical and industrial systems, energy and power systems, civil and structural monitoring, aerospace, and more, as shown in Table 1.

Since vibration data are usually acquired in high dimensions, Deac et al. [4] designed a CNN for dimensionality reduction and a DNN for classification in motor bearing condition assessment for fault diagnosis. LSTM networks are well suited for handling time-series data due to their ability to capture and retain long-term dependencies. LSTM-based autoencoders have demonstrated promising results in detecting anomalies in vertical carousel storage systems, particularly by improving data preprocessing to enhance accuracy [5]. Additionally, low-sampling-rate vibration data have been explored using artificial intelligence techniques to address cost and data efficiency challenges, which is particularly valuable for structural fault detection [6]. This LSTM plus autoencoder mechanism is also designed for identifying abnormal patterns in rotary machinery [7] and wind turbines [8], respectively. The integration of turbine data and vibration signals into unified models has further improved the predictive accuracy of wind turbine component failures based on data from the Supervisory Control and Data Acquisition (SCADA) system [9]. In their approaches, a feed-forward neural network (NN) was trained to predict future patterns, and a one-class SVM classifier was subsequently trained for anomaly detection. Similarly, Vos et al. [10] replaced the feed-forward NN module with a deep LSTM network for anomaly detection on datasets of helicopter gearboxes and real helicopter flight tests. Other CNN-LSTM-based network architectures have been proposed by Nguyen-Da et al. [11] and Zhang and Zhou [12]. Hybrid models that combine CNNs and LSTM architectures are proven to be effective in capturing both temporal and spatial features in vibration data for structural health monitoring. Their methods were applied to the monitoring of industrial diesel generators [11] and suspension bridges [12], respectively.

Table 1. Anomaly detection using deep learning methods and vibration data across various applications.

Network Architectures	Applications
CNN + DNN	motor bearing [4]
LSTM + Autoencoder	carousel storage system [5], motor bearing [6], rotary machinery [7], wind turbine [8]
feed-forward NN + SVM	generator bearing of wind turbine [9]
LSTM + SVM	helicopter gearboxes and real helicopter flight tests [10]
CNN + LSTM	diesel generator [11], suspension bridge [12]
Bi-directional LSTM (Bi-LSTM)	bearing of rotating machinery [13], rotating machinery [14], rolling bearing [15], weather forecast [16], wheelset rearing in high-speed trains [17], unmanned aircraft [18], mold damage monitoring [19]
TCN + autoencoder	construction vibration monitoring of buildings [20]

Compared to LSTM, Bi-LSTM captures both forward and backward temporal dependencies. Several Bi-LSTM layers are often stacked to capture more complex patterns and dependencies in sequential data for enhancing model performance. For example, Lee et al. [13] proposed a stacked convolutional Bi-LSTM model for bearing fault diagnosis in rotating machinery. Patra et al. [14] further compared the performance of Bi-LSTM and Gated Recurrent Units (GRU) in anomaly detection for rotating machinery. They showed that Bi-LSTM outperforms GRU in modeling long-term vibration patterns. Similarly, Qiu et al. [15] designed a modified Bi-LSTM network to process rolling bearing vibration data using only long-term memory. In other applications of stacked Bi-LSTM networks, Sharma and Sharma [16] applied it to weather forecasting, while Zhu et al. [17] used it for fault diagnosis of wheelset bearings in high-speed trains.

All used input features, i.e., time-series data and anomaly detection, can be executed in real time. Basically, time-series patterns are processed to extracted the discriminant features between normal or abnormal classes, and the convolutional networks, such as CNN or autoencoder, are further adopted for classification. Most of their approaches are implemented on the Internet of Things (IoT)-driven industrial monitoring. In other applications of anomaly detection using vibration data, vibration analysis is crucial in the aerospace sector for detecting flight anomalies in unmanned aircraft, ensuring operational safety [19]. Vibration data have also been analyzed on datasets of helicopter gearboxes and real helicopter flight tests [10]. Additionally, the temporal convolutional network (TCN) combined with an autoencoder and integrated with cloud computing has enabled real-time anomaly detection in construction vibration monitoring [20]. From the above literature, existing deep learning models for anomaly detection in industrial applications primarily focus on rolling bearings and rotating machinery, with limited research on powder metallurgy mold health monitoring [19].

Molds must be installed by experienced experts before production. During operation, a molding machine performs tens or even hundreds of thousands of stamping cycles for a specific product. However, the timing of mold failure is unpredictable, making it difficult to collect labeled training samples for supervised learning. To tackle this issue, the proposed method compares vibration data collected immediately after mold installation with subsequent vibration data to identify deviations in magnitude. The greater the deviation, the further the mold's health status deviates from its original condition. Once the deviation surpasses a predefined threshold, an early warning system alerts operators to halt production and inspect the mold to prevent severe damage.

The contributions of the proposed method are summarized as follows:

- The IMU-based devices are embedded in an Internet of Things (IoT) platform [21]. This platform ensures scalability and adaptability to different molding machines, allowing manufacturers to apply it across various machine settings with minimal modifications.
- A vibration-based predictive framework is developed to enable early fault warning before visible damage occurs, thereby reducing unexpected downtime.
- An unsupervised deep learning approach is adopted for anomaly detection, where the model is trained using only historical normal vibration samples; the labels of samples are unnecessary.
- The model achieves high anomaly detection accuracy, effectively distinguishing between normal and abnormal conditions based on various metric criteria.
- A real-time monitoring and early warning system is implemented, integrating IoT-based sensing to provide proactive alerts for preventive maintenance.

The originality of this study is briefly described as follows. As we know, each forming machine is expensive and has been in operation for more than a decade in Taiwanese factories. Most old machines lack digital and network capabilities and do not have any sensors. Currently, mold health monitoring relies on manual inspection of product dimensions, as mentioned above. Modifying these machines by embedding sensors is costly, and manufacturers—primarily small and medium-sized enterprises—are unwilling to discard old machines and purchase new, network-enabled equipment. To address this issue, this study establishes an IoT platform. We designed an IMU-based sensor device that can be externally attached to forming machines to collect vibration data from the molds without modifying any part of the forming machine's system. This data acquisition device is cost-effective, with two devices installed on each forming machine. A single NodeMCU is responsible for transmitting the vibration data, with a total cost of less than USD 25. The data are sent to the broker on the IoT platform for further analysis. Combined with the subsequent deep learning algorithm, the system can provide early warnings and reduce the likelihood of mold damage. Moreover, the IoT-based environment developed in this study can be scaled to accommodate dozens of forming machines within a factory without requiring complex network configurations. It also supports the flexible addition of new sensors, offering high scalability. Additionally, to the best of our knowledge, no related studies have applied vibration data and deep learning methods to mold damage detection in the powder metallurgy industry. Our research lab is the first to utilize these approaches for mold health monitoring. By combining deep learning, IoT-based sensing, and real-time anomaly detection, this study contributes to the advancement of intelligent predictive maintenance strategies, enhancing manufacturing efficiency and reducing operational risks.

The rest of the paper is organized as follows: The proposed method, including the data acquiring device, signal segmentation, deep network architecture, and the design of warning rules, is described in Section 2. In addition, the ablation experiments conducted to demonstrate the validity of the proposed method are described in Section 3. Finally, the discussions and conclusions are provided in Sections 4 and 5, respectively.

2. Methods

Molds have a finite lifespan, typically lasting for tens or hundreds of thousands of stamping cycles. Once the mold surpasses its operational lifespan or sustains damage, it affects the product's shape due to the extremely small tolerance range (approximately 0.05 mm to 0.1 mm). Figure 1 illustrates a damaged mold with a horizontal crevice, which is highlighted by a red circle. Mold degradation occurs progressively and is challenging to detect through visual inspection. Furthermore, producing extra spare parts for molds is not feasible due to their high costs. As a result, when a mold is damaged, immediate replacement is often impossible, leading to production delays and significant financial losses. Our observations indicate that subtle variations in vibration patterns occur during the stamping process before mold failure. However, these changes are imperceptible to the human eye, and conventional molding machines lack the capability to detect them or provide early warning signs. To address this challenge, this study proposes a machine learning-based approach to monitor mold health by capturing vibration data generated during stamping. Specifically, three-axis inertial measurement units (IMUs) are installed at the upper and middle mold positions to collect vibration data, which are then used to develop a predictive model for mold health estimation.

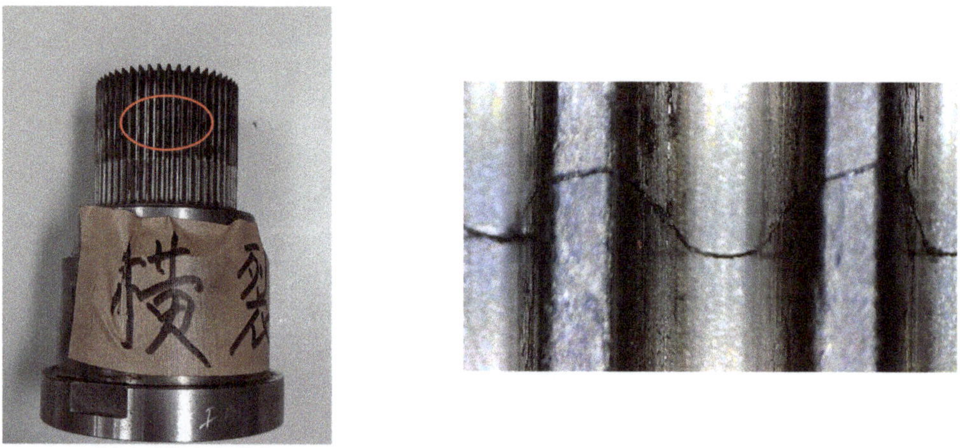

(a) (b)

Figure 1. The damaged mold. (**a**) A horizontal crevice (Chinese words) can be seen in the red circle; (**b**) A close-up image of the crevice.

The proposed method consists of the following four stages to establish an anomaly detection model trained with vibration data from healthy molds: (1) a vibration data acquisition device, (2) signal segmentation and feature extraction, (3) deep learning-based regression model training, and (4) anomaly monitoring. The flowchart of the proposed method, as shown in Figure 2, is briefly described as follows: First, IMU-based devices are mounted on the upper and middle molds to acquire vibration data from three-axis acceleration at a 500 Hz sampling rate during the stamping process. These data are collected on an IoT platform using the Message Queueing Telemetry Transport (MQTT) protocol.

Since the upper mold regularly moves up and down, the corresponding signals on the Z axis are segmented to identify the contact section of the two molds based on the detected peaks. Next, feature vectors are extracted from the accelerations of the two IMUs on the X, Y, and Z axes according to the segmented contact section. A Bi-LSTM-based deep model is then trained to construct a regression model that predicts future vibration features using historical data from a healthy mold. The mold status is determined by comparing the predicted features with the currently observed features. If the difference exceeds a predefined threshold, an alarm is triggered. Each stage is described in detail in the following subsections.

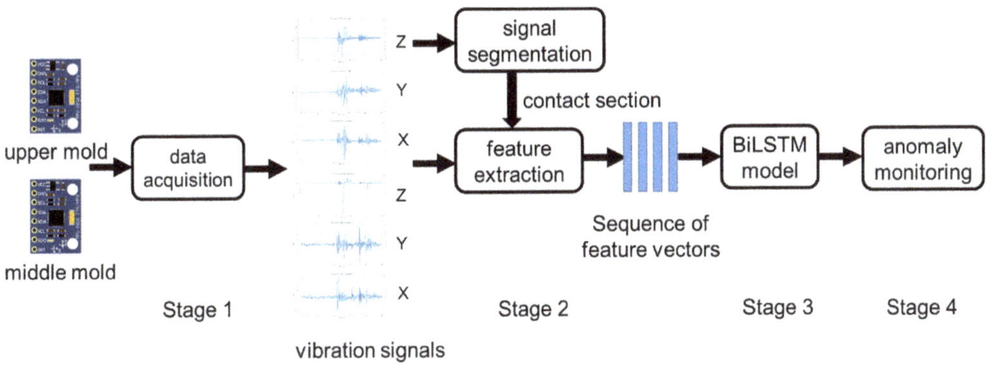

Figure 2. A diagram of the proposed method.

2.1. Vibration Data Acquisition Device

In this study, an inertial measurement unit (IMU, e.g., MPU6050 [22]) and a NodeMCU comprise vibration data acquisition devices, as shown in Figure 3a. Although MPU6050 is a multi-data sensor, e.g., detecting three-axis acceleration, three-axis rotation speed, and temperature, only the accelerations of the X, Y, and Z axes were acquired in a 500 Hz sampling rate in this study. Peripheral passive components were required to connect the IMU module, as shown on the right-hand side of Figure 3. Pins Vcc, GND, SCL, SDA, and AD0 were used. Here, pins Vcc and GND were connected to a 3.3 V power supply; pins SCL and SDA were used for the I^2C communication protocol; and pin AD0 identified the I^2C address that allows multiple sensors to be connected on the same bus. In order to control the accelerometer chip and transmit the collected vibration data through the network, NodeMCU, a Micro Control Unit (MCU) based on the core module ESP-12E, as shown on the left-hand side of Figure 3a, initialized sensor MPU6050, set the sampling rate, and performed the communication in the I^2C protocol. Due to the communication characteristics, two sensors can be connected by connecting their SDA and SCL together on the same bus topology. The address, either 0x68 or 0x69, identified the specified sensor for data transmission, as shown on the right-hand side of Figure 3b. NodeMCU was responsible for connecting the input/output pins to the sensors, and the power supply was stepped down to supply ESP-12E via AMS1117. For the sake of easy installation, the connection ports were designed in the USB Type-C ports. Type-C cables are easy to obtain and install by non-professionals. In the case of faults or cable disconnections, on-site personnel can quickly troubleshoot and restore the monitoring system. The vibration acquisition devices were installed at the positions of the upper and middle molds on the molding machine, as shown in Figure 4.

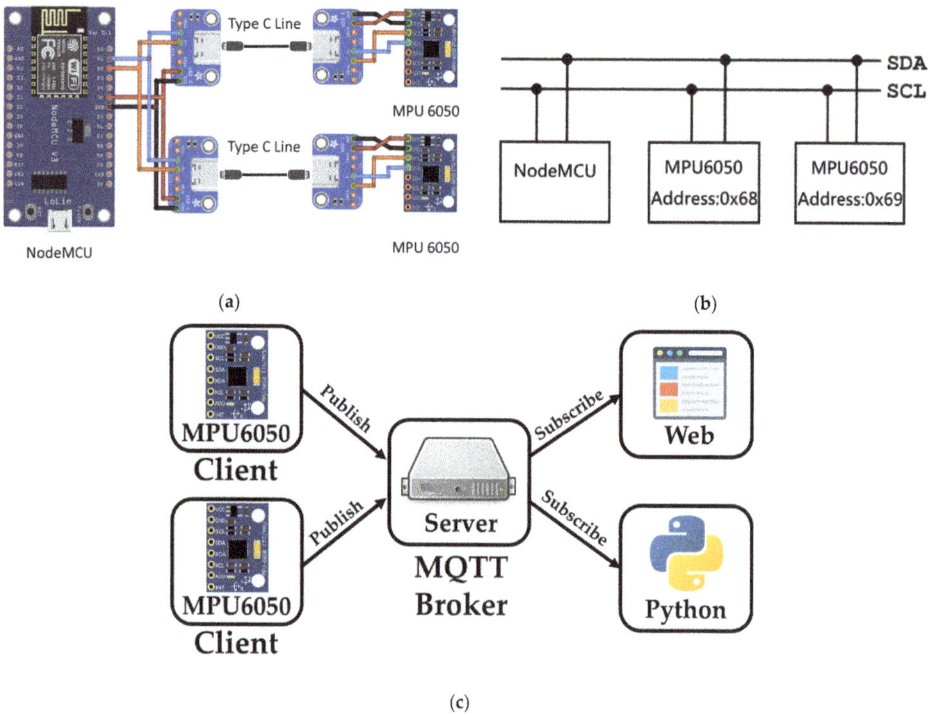

Figure 3. The vibration data acquisition devices and data collection platform. (**a**) Data acquisition hardware; (**b**) Data bus; (**c**) Platform based on the MQTT protocol.

Figure 4. The mounted vibration data acquisition devices; red circles depict the positions of the upper and middle molds. (**a**) Front view of the molding machine; (**b**) Close-up shot of the upper mold.

The IoT platform [21] operates based on three primary roles: the broker, the publisher, and the subscriber. The publisher sends messages to the broker, attaching specific topic tags. The subscriber, in turn, subscribes to topics of interest from the broker. When the publisher transmits new data associated with a specific topic, the broker forwards it to the relevant subscribers based on their requests. This intermediary approach eliminates the need for devices to directly exchange IP addresses, simplifying network configurations and reducing setup complexity. Additionally, the system is highly scalable, allowing for the seamless addition of new sensors as required, thereby ensuring flexibility for future expansion. This IoT-based data collection platform was constructed as shown in Figure 3c using interconnecting sensors, embedded devices, and servers [23]. It was designed to be a data collection, analysis, and monitoring system for reducing manual monitoring, enabling automated judgment and issuing early warning during the entire operation process. In order to achieve stable data transmission for multiple molding machines and future expandability, the Message Queueing Telemetry Transport (MQTT) protocol [24,25] was adopted for collecting the vibration data from IMUs mounted on the upper and middle molds.

2.2. Signal Segmentation and Feature Extraction

The structure of a mold is composed of three parts: the upper, middle, and lower parts. After the mold is filled with powder, a molding machine sequentially performs the following actions of (1) the upper mold moving down, (2) the upper mold/middle mold contacting and stamping, (3) the upper mold moving up, and (4) the extra powder pushing out, as shown in Figure 5. These actions are repeated and are known as a stamping cycle. During the stamping cycle of the molding machine, high pressure is only applied during the contact of the mold, i.e., the upper mold contacts the middle mold. Some signals in a cycle do not contain any discriminant features because they contain the data purely resulting from the movement of the mold. Thus, we must identify the corresponding signals in the stamping operations to analyze the vibration data caused by the increased friction force on the mold. The vibration signal section in which the upper mold contacts with the middle mold is found, the mold is stamped, and the mold is demolded. A graphical illustration is given to show the positions of the upper and middle molds during a stamping cycle (Figure 5). The signals at four positions on the Z axis are acquired and drawn by the rectangles; these are shown in zones 1 to 4 in Figure 6, in which zone 2 is the section used for feature extraction. The stamping cycles are regularly and repeatedly operated throughout the entire production process. The Z axis signals of zones 1 and 3 represent the highest positions of the upper mold. Therefore, the peaks of zones 1 and 3 on the Z axis signals should be detected first. Four steps are executed to detect the peak position in order to identify the contact section on the original Z axis signals. Next, the feature vectors are extracted from the signals on the X, Y, and Z axes according to the segmented contact section, i.e., zone 2.

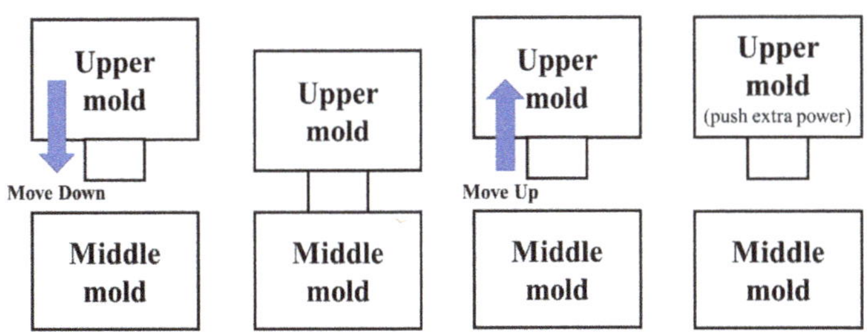

Figure 5. The stamping cycle of the upper mold.

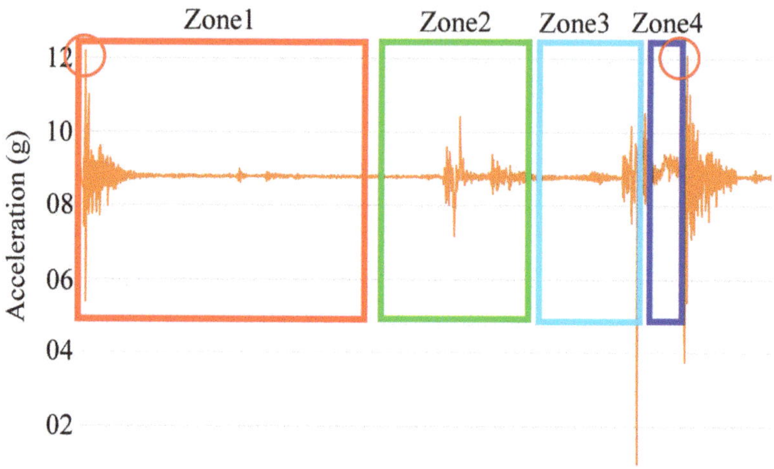

Figure 6. The Z axis vibration signal in a stamping cycle. Two peaks are detected in Step 3 as drawn in the two red circles.

1. Signal smoothing: The original signals are smoothed for noise removal using a one-by-three convolutional filter. This operation smooths the signal and effectively modifies the pulses with very large abnormal amplitudes.
2. Stamping segment identification: The mean μ and standard deviation σ of the Z axis vibration data are first calculated. The outliers whose values are out of the range [$\mu - 4\sigma$, $\mu + 4\sigma$] are identified. If the outliers are located within an interval, e.g., 4000 sampling points in 8 s, this interval is regarded as the normal stamping process because of the movement of the upper mold. The signals within the interval are subsequently segmented and the peak positions are identified. On the other hand, the molding machine is considered to be in a state of downtime if there are no outliers within the interval.
3. Peak detection: The local maximum value is found to be the vertex in the outlier section in step (b). These are the peaks drawn in the two red circles in Figure 6, which are the highest point of the upper mold during a stamping operation.
4. Feature extraction: Two vibration data acquisition devices are independently mounted at positions on the upper and middle molds. Due to the asynchronous sampling rate of the upper mold's hydraulic system and sensors, there may be a few milliseconds of difference within the mechanical stamping cycle, and a fixed sampling rate cannot be used. The length between peaks still has a difference of 1–5 data points. The vibration data of the X, Y, and Z axes from two IMUs are extracted to be used as the features for abnormal detection. Since molding machines are repeatedly operated at a regular speed, the contact section—the green box area (zone 2) in Figure 6—is obtained with a fixed offset from the peak position. The local minimal peak in the segment is next identified on the Z axis signal, and 500 points are taken from this point in the forward and backward direction. The corresponding 3000 acceleration vibration values on the X, Y, and Z axes are extracted to be the feature vector, i.e., the green box area in Figure 6 and six feature vectors of two IMUs (mounted on the upper and middle molds), as shown in Figure 7. The vibration data of the stamping friction between the two molds are effectively extracted for the mold's health evaluation during the upper and middle mold contact process.

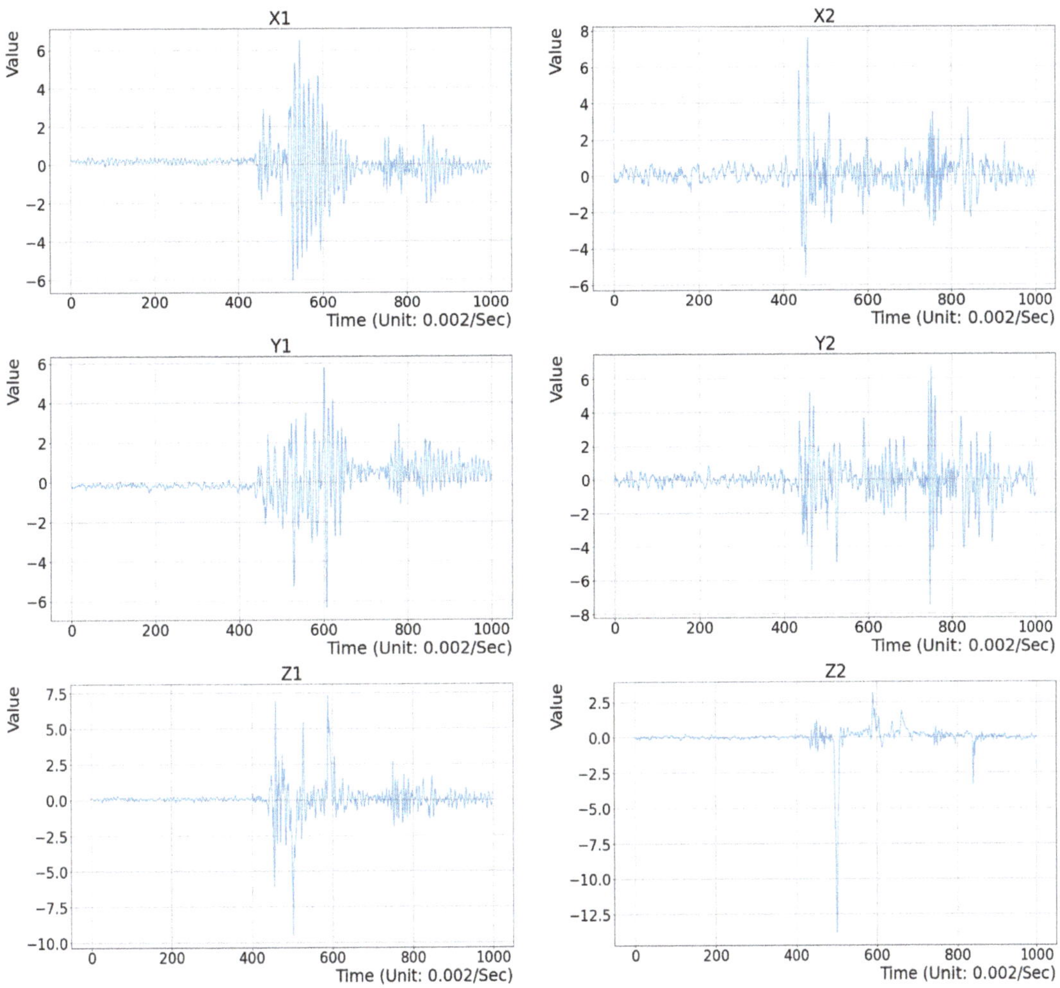

(**a**) (**b**)

Figure 7. The extracted vibration signals on the X, Y, and Z axes for feature extraction. (**a**) The signals of the IMU on the upper mold; (**b**) The signals of the IMU on the middle mold.

2.3. Training of Deep Regression Model

Since molds are instantaneously and randomly damaged and the vibration data on mold damage are challenging to collect, supervised machine learning techniques cannot be used to train the two-class classification model, i.e., the normal and abnormal classes. Therefore, only normal data are collected to train a deep regression model using unsupervised learning methods.

We designed a mold prediction model using the training vectors extracted from periodic signals in this study. If a CNN (convolutional neural network) or an attention mechanism is adopted as the training model, too many parameters need to be trained and a significant amount of training time is needed. It is difficult to implement the model in single-board computers or edge computing devices for real-time detection. In addition, there are many types of molding machine molds, and it is impossible to replicate the technology for bearing mechanical life detection [26]. Therefore, we refer to the approach in papers [17,18] and design a network architecture for bearing damage analysis.

In this study, a deep learning regression model was trained to later predict the vibration data. As mentioned in the previous section, the original Z axis data were segmented to identify the segments of feature vectors, and the feature vectors were extracted from the corresponding segmented signal of the X, Y, and Z axes. The input vectors consist of a sequence of temporal patterns. This regression model on temporal patterns is used to analyze sequences of historical patterns for predicting future patterns. Bi-LSTM is adopted to enhance the model's ability to learn from time-series data. Compared to traditional LSTM, Bi-LSTM captures both forward and backward temporal dependencies, improving the accuracy of anomaly detection. Additionally, a Multi-Head Attention (MHA) mechanism is incorporated into the regression model to strengthen its capacity to capture long-range temporal dependencies. MHA can simultaneously focus on critical features across different time steps, mitigating the vanishing gradient problem and enhancing the model's ability to recognize anomalous patterns. After training the model, if the prediction closely matches the current observation, the mold is classified as healthy; otherwise, it is classified as abnormal.

In this study, a deep network architecture was used for seismic data regression prediction, as shown in Figure 8 and referred to in [17,18]; the network architecture sequentially comprised one input layer, one Bi-LSTM layer with an attention mechanism, one normalization layer, one feed forward layer, one normalization layer, one Bi-LSTM layer, and one full connected output layer. This network model was a regression model and the scenario for its input and output data is described as follows.

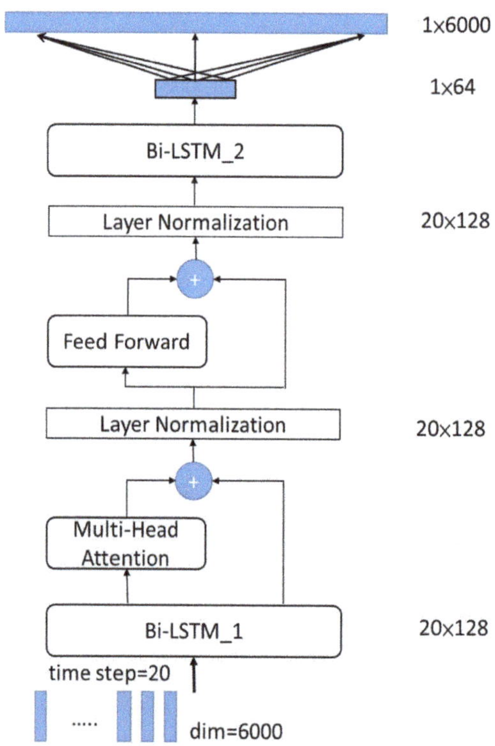

Figure 8. The network architecture of Bi-LSTM with an attention mechanism when parameters of 'feature dimension' and 'time step' are 6000 and 20, respectively. Here, symbol '+' represents an addition operation.

- When the feature dimension and time step parameters are set to 6000 and 20, respectively, a data array of 6000 by 20 is input into the "InputLayer", representing a sequence of extracted vibration data from two IMUs located on the contact sections of the upper and middle molds over the past 20 stamping cycles.
- The first "Bi-LSTM" layer was configured with 128 units to capture bidirectional dependencies in the time-series through forward and backward computations. This layer outputs a sequence of features for each time step and enhances the representation of temporal dependencies. Additionally, the dropout rate is set to be 0.2 in this module.
- An attention mechanism was adopted to model global dependencies among high-dimensional features. Four attention heads were utilized to extract long-range interactions between features in the 'Multi-Head Attention' layer. Similarly, the dropout rate is set to 0.2 again.
- Additionally, residual connections from the outputs of the "Multi-Head Attention" and "Bi-LSTM_1" layers, along with layer normalization, are applied to improve training stability and convergence.
- The "Feed Forward" layer was composed of two fully connected layers with a dimension of 128 and ReLU activation for further enhancing the non-linear representation capacity.
- Furthermore, residual connections and layer normalization were adopted to improve training stability and convergence.
- The output of the feed forward layer was processed by the second "Bi-LSTM_2" layer with 64 units. This layer further compressed the feature dimensions and generated the final sequential representation. Here, the dropout rate is set to 0.2.
- A fully connected layer maps the final representation to a 6000-dimensional output, predicting the corresponding features for the next time step.

This deep learning-based regression model effectively captures long-term dependencies in vibration signals and provides a predictive mechanism for mold health assessment, enabling early detection of potential mold degradation.

2.4. Anomaly Monitoring

Considering the current feature vector of vibration data V_t at time t and the historical vibration feature vectors $V_{t-s}, V_{t-s+2}, \ldots, V_{t-1}$, here, the parameter of time step in the Bi-LSTM model was set to be s. The parameters were input into the designed regression model, and a predicted feature vector $\hat{V}_t$ was output, as shown in Figure 9. During the training process, a large amount vector sequences were collected for training, and the loss function was defined as the mean square error (MSE) between the predicted vector $\hat{V}_t$ and the observed feature vectors V_t. The regression network can be considered as a prediction model of vibration data after p time steps, and we hope the errors between vectors $\hat{V}_t$ and V_t (e.g., MSE $\left(\hat{V}_t - V_t\right)$) are as small as possible when the mold is operated in its healthy status. Otherwise, the vibration data V_t will be in an abnormal state. To design the warning rule, the input historical feature vectors $V_{t-s}, V_{t-s+1}, \ldots, V_{t-1}$ were input into the model, and the predicted output $\hat{V}_{t-1}$ was compared with the current vibration data V_t; when MSE $\left(\hat{V}_{t-1} - V_t\right) < t$ is smaller than a threshold t, no alarm was given because of the normal operation, and vice versa. An alarm should be issued due to its abnormal occurrence.

After training the model, it was evaluated using various evaluation metrics, such as MSE, Mean Absolute Error (MAE), and Coefficient of Determination (R-squared). Once the model has been trained and evaluated using the MSE metric criterion, as shown in Figure 10, it can be used for mold health status prediction. During the operation of the stamping process, if the model detects an abnormality in the vibration signal, it can alert the operators to check the mold and prevent mold damage.

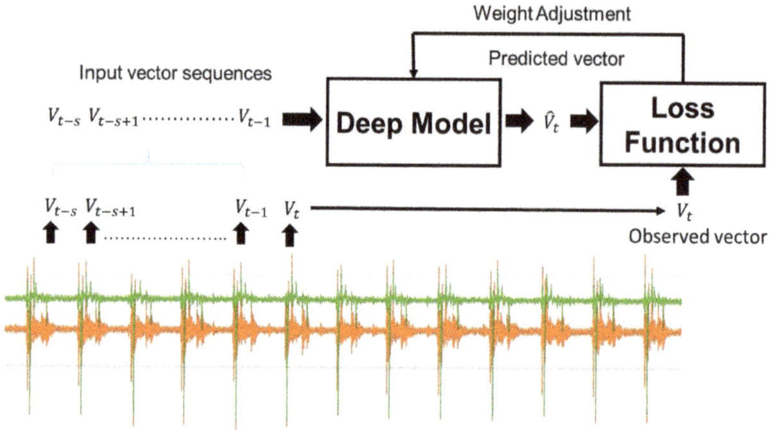

Figure 9. The training process of the deep model.

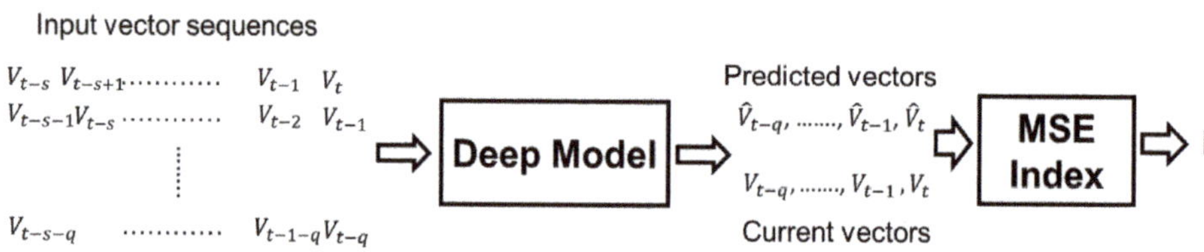

Figure 10. The prediction process of the deep model.

3. Experimental Results

To assess the effectiveness of the proposed method, several experiments were conducted. First, three toy examples were designed to illustrate the difference between the observed vibration data and the predicted data generated by the deep regression model within a single stamping cycle, as shown in Figures 11–13. Here, the configurations of the illustrated examples are set with a time step of 20, an input feature dimension of 6000, and the MSE metric as the criterion. As described in the previous section, historical vibration data from $t-20$ to $t-1$ were input into the regression model to predict the current vibration data $\hat{V}_t$ at time t. Here, feature vectors $V_{t-20}, V_{t-19}, \ldots, V_{t-1}$ with a time-step parameter of 20 were extracted from approximately 150 s of past data. The MSE value between the predicted vector $\hat{V}_t$ and the observed vector V_t along the X, Y, and Z axes was computed to evaluate prediction accuracy. Figure 11 presents the evaluation results for the training dataset (i.e., the inside test), where the differences between the observed vibration data and the model's predictions were compared. The average MSE value for the X, Y, and Z axes was 0.1985, indicating minimal discrepancies. To highlight these differences, the range from dimension 400 to 700 in Figure 11 shows that the variations were not significant. Next, Figure 12 illustrates the evaluation results for vibration data collected during a normal stamping cycle (i.e., the outside test), yielding an MSE value of 0.2551. The observed deviations were also minor, particularly within the time window 400 to 740, as shown in Figure 12. Finally, Figure 13 demonstrates the evaluation results for an abnormal stamping cycle, where the MSE value increased substantially to 1.1076. Notably, significant discrepancies were observed along the X and Y axes, highlighting the model's sensitivity to abnormal vibration patterns. This MSE value far exceeded those obtained for normal stamping cycles in Figures 11 and 12. These results confirm that the trained model accurately predicts the expected vibration data during normal stamping operations, while effectively distinguishing abnormal conditions.

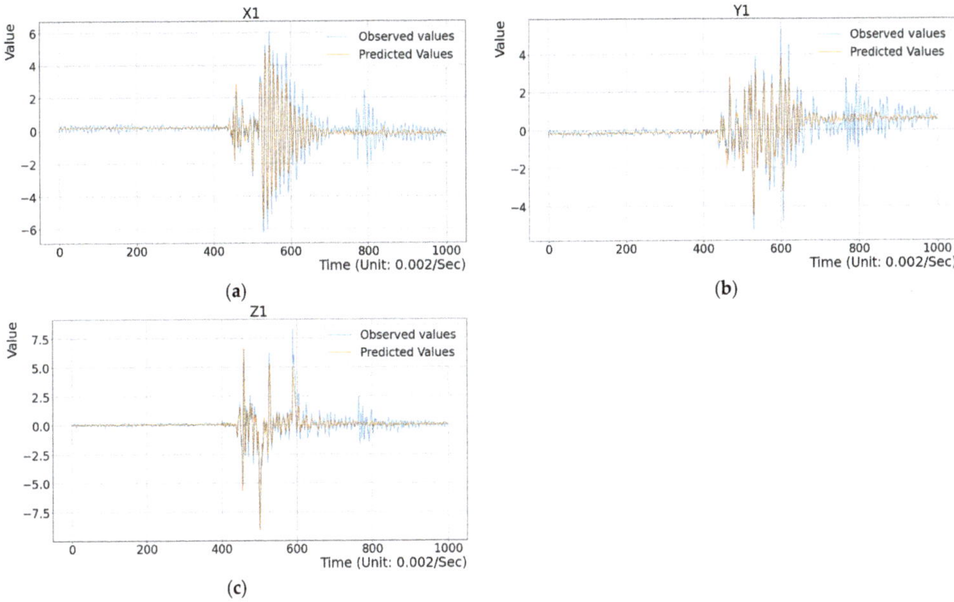

Figure 11. A comparison between the predicted and observed data with an MSE value = 0.1985 in the training dataset (inside test). (**a**) The data on the X axis; (**b**) The data on the Y axis; (**c**) The data on the Z axis.

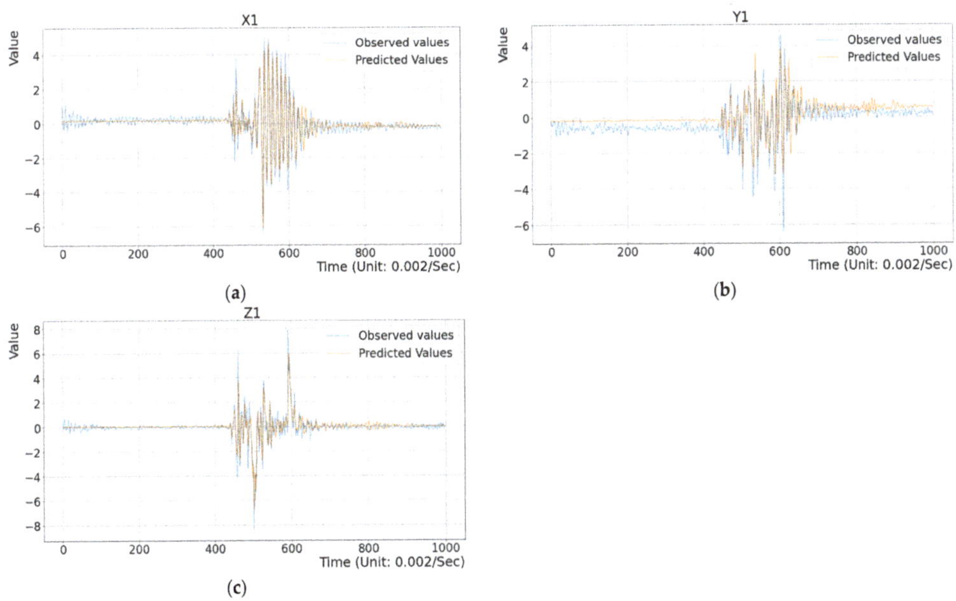

Figure 12. A comparison between the predicted and the observed data under the healthy condition with an MSE value = 0.2551 (outside test). (**a**) The data on the X axis; (**b**) The data on the Y axis; (**c**) The data on the Z axis.

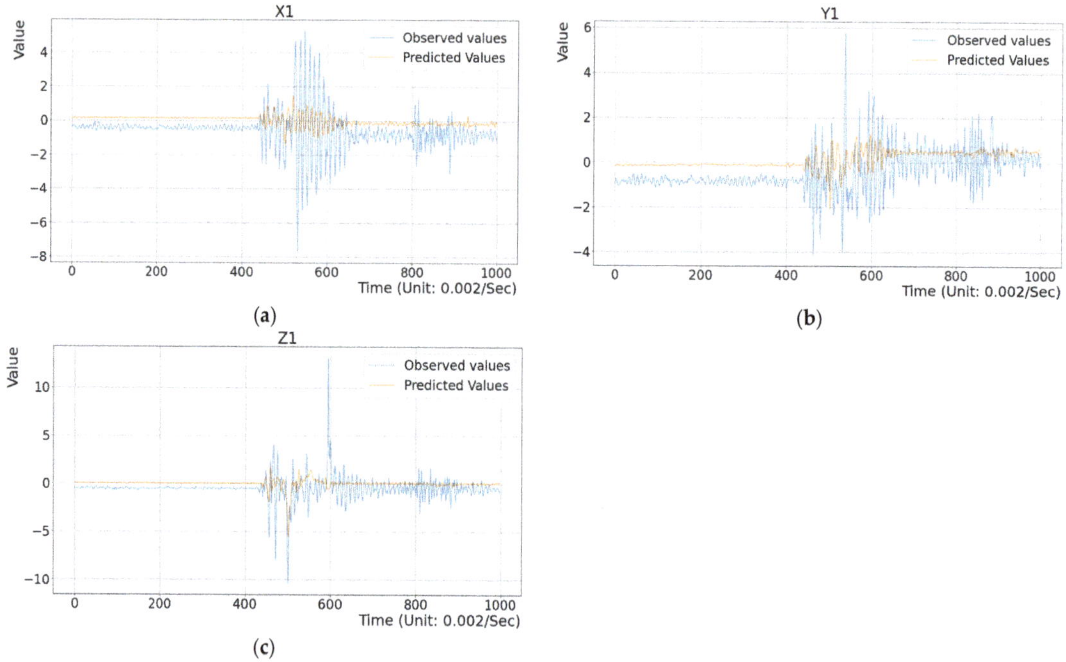

Figure 13. A comparison between the predicted and observed data in the damaged condition with an MSE value = 1.1076. (**a**) The data on the X axis; (**b**) The data on the Y axis; (**c**) The data on the Z axis.

To further evaluate the model's performance, a dataset was constructed from stamping process data, including nine hours of vibration data under healthy mold conditions and two hours of data preceding mold replacement. The healthy mold data were divided into the following two mutually exclusive subsets: the first seven hours for training and the remaining two hours for testing. During the training process, 80% of the samples were used for model weight optimization, while 20% were allocated for validation. Following the segmentation and feature extraction steps described in the previous section, a total of 2383 feature vectors were obtained for training, each with a dimensionality of 6000 (comprising 1000 data points per X-Y-Z axis from two IMUs). Additionally, the other two-hour testing dataset containing 720 feature vectors was collected under normal conditions. The collected vibration data typically exhibit a fixed periodic pattern. However, temporary pauses occur when operators temporarily halt the molding machine to inspect and verify whether the stamped products meet quality standards, introducing variations in the number of feature vectors. Furthermore, 547 feature vectors were collected during the two hours preceding mold failure and replacement. This dataset was used to assess the model's effectiveness in detecting abnormal conditions before mold failure.

Moreover, ablation experiments were conducted to demonstrate the effectiveness and robustness of the proposed method, as shown in Table 2. Table 2 presents the predictive error metrics of the model under different conditions, including mean square error (MSE), Mean Absolute Error (MAE), Coefficient of Determination (R^2), and Mean Absolute Percentage Error (MAPE). The MSE quantifies the mean of the squared differences between predicted and actual values, where lower values indicate a higher prediction accuracy. The MAE represents the average absolute deviation between predicted and actual values, offering robustness against extreme errors. The R^2 score in our model reflects the mold's health status; a higher R^2 value indicates that the mold is in a healthier condition, with its vibration patterns closely following the expected ones. Conversely, a lower R^2 value suggests that the mold is deviating from its normal operational state, indicating progressive wear and increased likelihood of failure. Additionally, the MAPE expresses the prediction error as a percentage of actual values, facilitating the comparison of errors across different scales. In addition, two parameters of input patterns, feature dimension and time step, are also varied in the experiments for both healthy and abnormal statuses. The input feature dimensions are set to be 6000, 4800, 3600, 2400, and 1200. This means that 1000, 800, 600, 400, and 200 vibration data points are extracted from each axis of the IMU.

Table 2. The results of ablation experiments using the Bi-LSTM module.

Status, Time Step / Metric, Dimension		Healthy 10	Healthy 20	Healthy 30	Abnormal 10	Abnormal 20	Abnormal 30
MSE	6000	0.4924	0.4710	0.4777	1.2870	1.3161	1.2443
	4800	0.4687	0.4686	0.4658	1.2343	1.2078	1.3077
	3600	0.4532	0.4382	0.4286	1.2381	1.2639	1.2280
	2400	0.4158	0.3988	0.4355	1.2918	1.2390	1.2386
	1200	0.4445	0.4463	0.4303	1.2322	1.1716	1.2633
MAE	6000	0.4175	0.4113	0.4137	0.7406	0.7472	0.7309
	4800	0.4080	0.4081	0.4068	0.7258	0.7207	0.7409
	3600	0.3916	0.3886	0.3863	0.7116	0.7238	0.7125
	2400	0.3844	0.3801	0.3983	0.7229	0.7124	0.7089
	1200	0.4154	0.4143	0.4063	0.7437	0.7209	0.7564
R^2	6000	−0.9626	−0.9682	−0.9740	−6.7291	−7.0028	−7.2110
	4800	−0.9085	−0.9489	−0.9607	−6.7984	−7.0538	−7.2994
	3600	−0.9259	−0.9449	−0.9279	−6.6974	−6.9014	−7.0825
	2400	−0.9469	−0.8846	−0.9521	−6.1434	−6.2602	−6.4312
	1200	−0.8180	−0.7781	−0.7324	−4.4574	−4.4492	−4.9106
MAPE (%)	6000	248.24	255.55	256.44	252.56	257.75	228.05
	4800	254.00	266.60	267.79	251.29	246.61	262.76
	3600	384.02	379.21	372.47	398.72	438.60	414.73
	2400	409.71	371.38	360.61	371.88	335.70	316.06
	1200	343.15	343.18	317.13	465.26	435.89	482.07

The results indicate that as the feature dimension decreases, the metrics MSE and MAE consistently decline under the "Healthy" condition. For example, when the time step is 20, the MSE decreases from 0.4710 to 0.4463, and the MAE decreases from 0.4113 to 0.4143. Additionally, under the "Abnormal" condition, these errors are significantly higher across all three metrics. This trend suggests that the model effectively distinguishes between normal and abnormal conditions, as greater prediction deviations under the abnormal condition indicate that the model significantly differentiates the observed data from the model's predictions. Furthermore, R^2 values are used to evaluate the fitness between the observed data and the model's predictions, where higher values indicate a better model fit. The results show notably lower R^2 values under the "Abnormal" condition, ranging from approximately −4.4492 to −7.0028. In contrast, the "Healthy" condition exhibits significantly higher R^2 values, ranging from −0.9682 to −0.7781. This indicates that the model effectively differentiates between normal and abnormal conditions rather than merely fitting all data points. However, in this study, MAPE does not demonstrate strong discriminative ability between healthy and abnormal states compared to MSE or MAE. The MAPE metric is highly sensitive to extreme values and small actual values. It is also undefined for zero actual values and becomes problematic when actual values are close to zero. Based on the observations in Figures 11–13 and the statistical data of actual values, more than 75% of the actuals are zero or near zero.

On the other hand, the choice of feature dimension may influence prediction accuracy. Higher-dimensional data may contain information beyond the critical section of compression, potentially affecting the model's performance. However, the results are not conclusive. In summary, the errors before replacing the damaged mold are much higher than those observed under normal stamping operations across various parameters.

To clearly visualize the sequential evolution of MSE values, we graphically represent the MSE values of the healthy mold (Figure 14a) and the damaged mold (Figure 14b) over time. We observed that most MSE values in Figure 14a are below 0.75, with occasional slight deviations exceeding 1.0. However, in Figure 14b, there is one point where the MSE value exceeds 12 (indicated by a red circle), which is suspected to correspond to the moment when the mold was damaged. Additionally, the frequency of MSE values exceeding 0.75 in Figure 14b is much higher than in Figure 14a. From Figure 14b, there is a significant change in MSE values before the mold is damaged, indicating a gradual decline in the prediction accuracy. To design warning rules with a more effective performance and fewer false alarms, we applied a minimal filter of size 1 by 3, selecting the minimal MSE values within three cycles in both Figure 14a,b. The filtered results are displayed in Figure 14c,d. While some sporadic false alarms are present, their frequency in Figure 14c is lower than that in Figure 14d. Based on Figure 14c,d, the warning conditions are designed to trigger an alert and notify operators when the filtered MSE value exceeds 0.75.

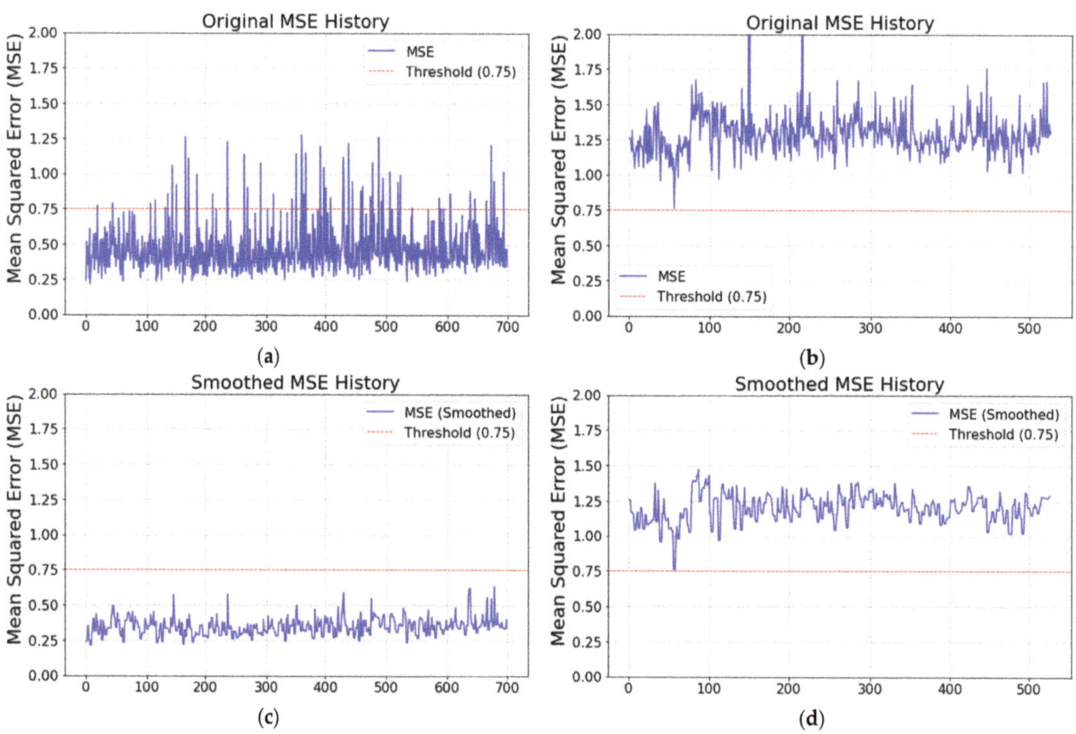

Figure 14. The MSE values between the predicted vectors and observed vectors. (**a**) The MSE values in the health status over two hours; (**b**) The MSE values in the damage status over two hours; (**c**) The MSE values of (**a**) after filtering with a 1 × 3 window size; (**d**) The MSE values of (**b**) after filtering with a 1 × 3 window size.

The deep learning architecture, inspired by [17,18], is designed to train a prediction model for the current observed pattern based on historical data from normal stamping cycles. According to the results in [17,18], Bi-LSTM extracts more discriminative features from input patterns over a longer range in both directions. As we know, there are also many deep learning models available for time-series pattern prediction. To evaluate their effectiveness, we modified the network in Figure 8 by replacing the Bi-LSTM module with two widely used models: LSTM [27] and GRU [28]. The unit sizes in all three modules are set as similarly as possible to ensure a fair comparison. Additionally, all parameters, including feature dimensionality, time step, and distance metrics, are kept identical to those in Table 2. The results for the LSTM and GRU modules are presented in Tables 3 and 4, respectively. Here, the MAPE metric is excluded due to its limited discriminating ability, as discussed previously. Overall, all three modules effectively evaluate the healthy status of the mold, as shown in Tables 2–4. To further analyze their performance, we design a simple separation index. As we know, the error values between predictions and observations should be as small as possible for a healthy mold and as large as possible for an abnormal state. The greater the difference in errors between the healthy and abnormal states, the better the model's distinguishing capability. For example, the errors of Bi-LSTM, LSTM, and GRU under healthy and abnormal conditions are (0.4710, 1.3161), (0.4669, 1.2281), and (0.4728, 1.2703), respectively, when using the MSE metric, with a feature dimension of 6000 and a time step of 20. The corresponding separation degrees are 0.8451, 0.7612, and 0.7975, respectively, by calculating the distance under the healthy and abnormal conditions. These results indicate that the Bi-LSTM module achieves better separation than the other two models. The separation index is set to one if the Bi-LSTM module performs better; otherwise, it is set to zero. All separation indexes are computed to compare the Bi-LSTM module with LSTM and GRU, as shown in Table 5. From this table, we can see that Bi-LSTM is more effective than LSTM and GRU, particularly when using the R^2 metric.

On the other hand, the execution time for each prediction is analyzed as follows: First, the workspace for the prediction model is set up on a general personal computer equipped with an Intel Core i9-11900 processor, 64 GB of RAM, and Windows 11 operating system. Due to the initialization overhead of the GPU workspace, the prediction time for the first sample exceeds 1 s. After that, the execution times of each prediction for modules Bi-LSTM, LSTM and GRU are reduced to 0.0866, 0.0892, and 0.0873 s, with the standard deviation 0.0345, 0.0541, and 0.0477, respectively.

Table 3. The results of ablation experiments using the LSTM module.

Metric, Dimension \ Status, Time Step		Healthy 10	Healthy 20	Healthy 30	Abnormal 10	Abnormal 20	Abnormal 30
MSE	6000	0.4677	0.4669	0.4646	1.2393	1.2281	1.2859
	4800	0.4592	0.4529	0.4629	1.2193	1.1516	1.2238
	3600	0.4231	0.4165	0.4173	1.2091	1.1842	1.2120
	2400	0.3803	0.3798	0.3738	1.2487	1.2325	1.2420
	1200	0.3698	0.3698	0.3703	1.2020	1.2020	1.1951
MAE	6000	0.4103	0.4099	0.4105	0.7278	0.7279	0.7423
	4800	0.4046	0.4042	0.4085	0.7233	0.7060	0.7218
	3600	0.3872	0.3801	0.3839	0.7050	0.7015	0.7102
	2400	0.3726	0.3738	0.3724	0.7100	0.7064	0.7142
	1200	0.3813	0.3813	0.3806	0.7286	0.7286	0.7276
R^2	6000	−0.9274	−0.9452	−0.9528	−6.6675	−6.8689	−7.1395
	4800	−0.9266	−0.9294	−0.9607	−6.8421	−6.9253	−7.2647
	3600	−0.9001	−0.8985	−0.9244	−6.5607	−6.8211	−7.0496
	2400	−0.8593	−0.8719	−0.8566	−6.0252	−6.2414	−6.5422
	1200	−0.6363	−0.6363	−0.6349	−4.3503	−4.3503	−4.6468

Table 4. The results of ablation experiments using the GRU module.

Metric, Dimension \ Status, Time Step		Healthy 10	Healthy 20	Healthy 30	Abnormal 10	Abnormal 20	Abnormal 30
MSE	6000	0.4667	0.4728	0.4641	1.2971	1.2703	1.2699
	4800	0.4639	0.4572	0.4601	1.2290	1.1947	1.2060
	3600	0.4370	0.4425	0.4217	1.2378	1.2179	1.2088
	2400	0.3898	0.4069	0.4026	1.2254	1.2071	1.2274
	1200	0.4113	0.4242	0.4163	1.2761	1.2056	1.1605
MAE	6000	0.4108	0.4120	0.4110	0.7421	0.7367	0.7374
	4800	0.4060	0.4054	0.4060	0.7222	0.7160	0.7212
	3600	0.3886	0.3937	0.3850	0.7138	0.7112	0.7121
	2400	0.3754	0.3854	0.3835	0.7086	0.7019	0.7113
	1200	0.4025	0.4100	0.4052	0.7549	0.7294	0.7141
R^2	6000	−0.9649	−0.9621	−0.9724	−6.7546	−6.9372	−7.1268
	4800	−0.9418	−0.9613	−0.9676	−6.7956	−6.9987	−7.2716
	3600	−0.9436	−0.9591	−0.9344	−6.6405	−6.9001	−7.1006
	2400	−0.8935	−0.9235	−0.9108	−6.0634	−6.2334	−6.4430
	1200	−0.7150	−0.7236	−0.7117	−4.4663	−4.3689	−4.5842

Table 5. The separation indexes of Bi-LSTM module compared with modules LSTM and GRU.

Metric, Dimension \ Status, Time Step		Healthy 10	Healthy 20	Healthy 30	Abnormal 10	Abnormal 20	Abnormal 30
MSE	6000	1	1	0	0	1	0
	4800	1	1	1	1	1	1
	3600	0	1	1	0	1	1
	2400	1	0	0	1	1	0
	1200	0	0	1	0	0	1
MAE	6000	1	1	0	0	1	0
	4800	0	1	1	1	1	1
	3600	1	1	0	0	1	0
	2400	1	0	0	1	1	0
	1200	0	0	1	0	0	1
R^2	6000	1	1	1	0	1	1
	4800	0	1	1	1	1	1
	3600	1	1	1	1	1	0
	2400	1	1	0	1	1	0
	1200	0	0	1	0	1	1

4. Discussion

In this study, we have identified the critical vibration patterns that are associated with mold damage in powder metallurgy molding machines. One notable observation was the presence of a small secondary wave pattern in the Z axis acceleration during the downward motion of a damaged upper mold. This anomaly indicates the section of mold contact. By carefully analyzing the experimental results in Table 2, several key findings of this research are as follows:

- Effective Early Fault Detection: By acquiring vibration data along the X, Y, and Z axes using IMU sensors and estimating a prediction function through a deep learning model, the system successfully detects early signs of mold deterioration (see Figure 14). The results indicate that abnormal stamping cycles exhibit significantly higher error values compared to normal cycles across various metrics in Table 2, demonstrating the model's ability to differentiate between healthy and damaged molds.
- High Prediction Accuracy and Fewer False Alarm: The system's MSE value for normal stamping data remains as low as 0.6442, while for abnormal conditions, the MSE

increases to 1.1344, further validating the effectiveness of the predictive approach. Additionally, the MAE value for normal samples is 0.3548, whereas for damaged molds, it rises to 0.7406, confirming that the system effectively detects anomalies.

- Robustness of Trained Model: From the ablation experiments, the trained model demonstrated a strong robustness to various parameters, such as feature dimension, metric criteria, and time step. These parameters had minimal impact on prediction accuracy, indicating the model's robust design.
- Robust and Scalable Monitoring: The model was trained and evaluated over a long period, using 2383 feature vectors from healthy mold operations and 547 feature vectors collected from the two hours preceding mold failure, demonstrating its robustness in identifying deviations before mold breakdown. Furthermore, the system adapts to different molding machines and can be extended by adjusting feature extraction methods and training datasets.
- Real-time Monitoring Feasibility: The model's inference time per sample needs 0.0866 s after initialization and optimization, demonstrating its capability for real-time deployment in industries.

5. Conclusions

This study developed a deep learning-based mold health monitoring system using vibration data to detect mold damage in powder metallurgy molding machines. An IMU plus NodeMCU module was designed for data acquisition, capturing vibration signals along the X, Y, and Z axes on an IoT platform. For data analysis, Z axis vibration data were used to mark the contact sections of the upper and middle molds, while friction vibration signals on the X, Y, and Z axes were captured during the stamping process. A Bi-LSTM regression model with an attention mechanism was trained using only historical normal vibration data to predict future vibration patterns. From the experimental results, the error values for normal and abnormal samples are as follows: less than 0.5 and greater than 1.0 on the MSE metric, less than 0.45 and greater than 0.70 on the MAE metric, and greater than −1.0 and less than −4.0 on the R^2 metric, respectively. Observing Table 2, we find that the errors are small under healthy conditions, whereas the errors are large under abnormal conditions and are clearly separated. The threshold value for issuing an alarm, such as the middle value, can be easily determined. This early warning system effectively reduces the risk of mold damage and decreases machine downtime.

This research establishes a foundation for intelligent predictive maintenance in industrial manufacturing, enhancing efficiency and reducing production losses. Despite these promising findings, this study highlights several areas for future exploration. Firstly, while the current dataset effectively demonstrated the feasibility of mold health monitoring, expanding the dataset to include more examples of mold damage under varying operational conditions is crucial. This expansion would enhance the model's generalizability and its ability to handle the diverse scenarios encountered in real-world production. Second, we believe that increasing the sampling rate from 500 Hz to 1 KHz would be highly beneficial for segmenting the Z axis signal. This enhancement would allow for the more precise detection of the contact section between the upper and middle molds. The corresponding extracted features from this section could be used to train the prediction model more effectively, further improving the accuracy of anomaly detection. Next, edge computing could be explored for real-time on-site processing, reducing reliance on centralized computing. Finally, the development of an adaptive early warning system, fine-tuned to specific mold types and operational environments, would be a valuable extension of this research. Such systems could optimize maintenance schedules, minimize downtime, and further validate the practical application of this technology in industrial applications.

6. Patents

Based on this work, a patent application (# 112132226) has been filed to the Intellectual Property Office, Ministry of Economic Affairs, Taiwan, R.O.C.

Author Contributions: Conceptualization: C.-C.H., Y.-C.W. and J.-Y.L.; methodology: C.-C.H., Y.-C.W. and J.-Y.L.; hardware: H.-P.L.; software: H.-P.L., Y.-C.C. and C.-C.H.; validation: H.-P.L., Y.-C.C. and C.-C.H.; formal analysis: C.-C.H., H.-P.L. and Y.-C.W.; investigation: H.-P.L., C.-C.H., Y.-C.W. and Y.-C.C.; resources: C.-C.H., Y.-C.W. and J.-Y.L.; data curation: H.-P.L. and Y.-C.W.; writing—original draft preparation: H.-P.L., C.-C.H. and Y.-C.C.; writing—review and editing: H.-P.L., C.-C.H. and Y.-C.W.; supervision: C.-C.H., Y.-C.W. and J.-Y.L.; project administration: C.-C.H., Y.-C.W. and J.-Y.L.; funding acquisition: C.-C.H., Y.-C.W. and J.-Y.L. All authors have read and agreed to the published version of the manuscript.

Funding: This study was partially supported by the National Science and Technology Council (No. 111-2221-E-239-026) and the industry–academia collaboration project of Chao-Chuan Industrial Co., Ltd.

Institutional Review Board Statement: Not applicable.

Informed Consent Statement: Not applicable.

Data Availability Statement: The datasets presented in this article are not readily available because the data are part of an ongoing study.

Acknowledgments: This article is a revised and expanded version of a paper entitled "Mold Damage Monitoring for Power Metallurgy Molding Machines Using Deep Learning Methods", which was presented at the 10th IIAE International Conference on Intelligent Systems and Image Processing, Beppu, Japan, held from 4 September 2023 to 8 September 2023.

Conflicts of Interest: The authors declare no conflicts of interest. The funders had no role in the design of the study; in the collection, analyses, or interpretation of data; in the writing of the manuscript; or in the decision to publish the results.

Abbreviations

The following abbreviations are used in this manuscript:

IMU	inertial measurement unit
IoT	Internet of Things
MCU	Micro Control Unit
MSE	mean square error
MAE	Mean Absolute Error
MQTT	Message Queueing Telemetry Transport
I^2C	Inter-Integrated Circuit
LSTM	Long Short-Term Memory
CNN	convolutional neural network

References

1. Torres, P.; Ramalho, A.; Correia, L. Automatic Anomaly Detection in Vibration Analysis Based on Machine Learning Algorithms. In *Proceedings of the International Conference Innovation in Mechatronics Engineering and Advanced Manufacturing*; Springer: Cham, Switzerland, 2022.
2. Hasegawa, T.; Ogata, J.; Murakawa, M. Adaptive training of vibration-based anomaly detector for wind turbine condition monitoring. *Int. J. Progn. Health Manag.* **2017**, *8*, 1–12.
3. Zhang, B.; Georgoulas, G.; Orchard, M. Rolling element bearing feature extraction and anomaly detection based on vibration monitoring. *IEEE Trans. Control Autom.* **2008**, *34*, 923–935.
4. Deac, C.; Deac, G.C.; Parpală, R.C.; Popa, C.L. Vibration anomaly detection using deep neural network and convolutional neural network. *Int. J. Mech. Optim.* **2021**, *11*, 19–28. [CrossRef]

5. Do, J.S.; Kareem, A.B.; Hur, J.W. LSTM-autoencoder for vibration anomaly detection in vertical carousel storage and retrieval system (VCSRS). *Sensors* **2023**, *23*, 1009. [CrossRef] [PubMed]
6. Jung, Y.; Park, E.G.; Jeong, S.H.; Kim, J.H. AI-based anomaly detection techniques for structural fault diagnosis using low-sampling-rate vibration data. *Aerospace* **2024**, *11*, 509. [CrossRef]
7. Li, J.; Wang, Y. A deep learning approach for anomaly detection based on SAE and LSTM in mechanical equipment. *Int. J. Adv. Manuf. Technol.* **2019**, *105*, 4421–4433.
8. Lee, Y.; Park, C.; Kim, N.; Ahn, J.; Jeong, J. LSTM-autoencoder based anomaly detection using vibration data of wind turbines. *Sensors* **2024**, *24*, 2833. [CrossRef] [PubMed]
9. Turnbull, A.; Carroll, J.; McDonald, A. Combining SCADA and vibration data into a single anomaly detection model to predict wind turbine component failure. *Wind Energy* **2021**, *24*, 123–135. [CrossRef]
10. Vos, K.; Peng, Z.; Jenkins, C.; Shahriar, M.R. Vibration-based anomaly detection using LSTM/SVM approaches. *Mech. Syst. Signal Process.* **2022**, *165*, 108752.
11. Nguyen-Da, T.; Nguyen-Thanh, P.; Cho, M.Y. Real-time AIoT anomaly detection for industrial diesel generator based on an efficient deep learning CNN-LSTM in Industry 4.0. *Internet Things* **2024**, *27*, 101280.
12. Zhang, X.; Zhou, W. Structural vibration data anomaly detection based on multiple feature information using CNN-LSTM Model. *Struct. Control Health Monit.* **2023**, *2023*, 3906180.
13. Lee, K.; Kim, J.K.; Kim, J.; Hur, K. Stacked Convolutional Bidirectional LSTM Recurrent Neural Network for Bearing Anomaly Detection in Rotating Machinery Diagnostics. In Proceedings of the 2018 1st IEEE International Conference on Knowledge Innovation and Invention (ICKII), Jeju, Republic of Korea, 23–27 July 2018; pp. 98–101.
14. Patra, K.; Sethi, R.N.; Behera, D.K. Anomaly detection in rotating machinery using autoencoders based on bidirectional LSTM and GRU neural networks. *Turk. J. Electr. Eng. Comput. Sci.* **2022**, *30*, 1256–1272. [CrossRef]
15. Qiu, D.; Liu, Z.; Zhou, Y.; Shi, J. Modified Bi-directional LSTM Neural Networks for Rolling Bearing Fault Diagnosis. In Proceedings of the 2019 IEEE International Conference on Communications, Shanghai, China, 20–24 May 2019; pp. 1–6.
16. Sharma, U.; Sharma, C. Deep Learning Based Prediction of Weather Using Hybrid Stacked Bi-Long Short Term Memory. In Proceedings of the 12th International Conference on Cloud Computing, Data Science and Engineering, Noida, India, 27–28 January 2022; pp. 422–427.
17. Zhu, Z.; Wang, H.; Liu, Z.; Meng, S. Fault Diagnosis of Wheelset Bearings Using Deep Bidirectional Long Short Term Memory Network. In Proceedings of the 2019 Prognostics and System Health Management Conference, Qingdao, China, 25–27 October 2019; pp. 1–7.
18. Bektash, O.M.; Cour-Harbo, A. Vibration Analysis for Anomaly Detection in Unmanned Aircraft. In Proceedings of the 2020 Annual Conference of the Prognostics and Health Management Society (PHM), Virtual, 9–13 November 2020.
19. Lin, H.P.; Chen, Y.C.; Han, C.C.; Wu, Y.C.; Chang, C.S.; Lin, J.Y. Mold Damage Monitoring for Power Metallurgy Molding Machines Using Deep Learning Methods. In Proceedings of the 10th IIAE International Conference on Intelligent Systems and Image Processing, Beppu, Japan, 4–8 September 2023.
20. Meng, Q.; Zhu, S. Anomaly detection for construction vibration signals using unsupervised deep learning and cloud computing. *Adv. Eng. Inform.* **2023**, *55*, 101907.
21. Su, X.; Li, P.; Riekki, J.; Liu, X.; Kiljander, J.; Soininen, J.-P.; Prehofer, C.; Flores, H.; Li, Y. Distribution of Semantic Reasoning on the Edge of Internet of Things. In Proceedings of the 2018 IEEE International Conference on Pervasive Computing and Communications, Athens, Greece, 19–23 March 2018; pp. 1–9.
22. Introduction to MPU6050. Available online: https://reurl.cc/5MNxVn (accessed on 15 February 2023).
23. Tatyasaheb, D.N.; Kumar, B. Implementation and Comparison of MQTT Protocol to Check the Drawbacks for Future Enhancement. In Proceedings of the 2021 International Conference on Computing, Communication and Green Engineering, Pune, India, 23–25 September 2021; pp. 1–6.
24. Hunkeler, U.; Truong, H.L.; Stanford-Clark, A. MQTT-S—A Publish/Subscribe Protocol for Wireless Sensor Networks. In Proceedings of the 3rd International Conference on Communication Systems Software and Middleware and Workshops, Bangalore, India, 6–10 January 2008; pp. 791–798.
25. Kashyap, M.; Sharma, V.; Verma, G. Implementation and Analysis of IoT Based Automation Using MQTT. In Proceedings of the IEEE 4th International Conference on Computing, Power and Communication Technologies, Kuala Lumpur, Malaysia, 24–26 September 2021; pp. 1–5.
26. Li, W.; Liu, D. An Approach of Rolling Bearing Remaining Useful Life Prediction Based on Adaptive Ensemble Model. In Proceedings of the 2022 International Conference Sensing, Measurement & Data Analytics in the Era of Artificial Intelligence, Harbin, China, 30 November–2 December 2022; pp. 1–5.

27. Hochreiter, S.; Schmidhuber, J. Long short term memory. *Neural Comput.* **1997**, *9*, 1735–1780. [PubMed]
28. Cho, K.; Merriënboer, B.; Gulcehre, C.; Bahdanau, D.; Bougares, F.; Schwenk, H.; Bengio, Y. Learning Phrase Representations Using RNN Encoder–Decoder for Statistical Machine Translation. In Proceedings of the 2014 Conference on Empirical Methods in Natural Language Processing (EMNLP), Doha, Qatar, 25–29 October 2014; pp. 1724–1734.

sensors

MDPI

Article

Development of an IMU-Based Post-Stroke Gait Data Acquisition and Analysis System for the Gait Assessment and Intervention Tool

Yu-Chi Wu [1,*], Yu-Jung Huang [1], Chin-Chuan Han [2], Yuan-Yang Cheng [3] and Chao-Shu Chang [4]

[1] Department of Electrical Engineering, National United University, Miaoli 36003, Taiwan; yujungzip@gmail.com
[2] Department of Computer Science and Information Engineering, National United University, Miaoli 36003, Taiwan; cchan@nuu.edu.tw
[3] Department of Physical Medicine and Rehabilitation, Taichung Veterans General Hospital, Taichung City 40705, Taiwan; rifampin@gmail.com
[4] Department of Information Management, National United University, Miaoli 36003, Taiwan; cschang@nuu.edu.tw
* Correspondence: ycwu@nuu.edu.tw; Tel.: +886-939967722

Abstract: Stroke is the fifth leading cause of death in Taiwan. In the process of stroke treatment, rehabilitation for gait recovery is one of the most critical aspects of treatment. The Gait Assessment and Intervention Tool (G.A.I.T.) is currently used in clinical practice to assess the gait recovery level; however, G.A.I.T. heavily depends on physician training and clinical judgment. With the advancement of technology, today's small, lightweight inertial measurement unit (IMU) wearable sensors are rapidly revolutionizing gait assessment and may be incorporated into routine clinical practice. In this paper, we developed a gait data acquisition and analysis system based on IMU wearable devices, proposed a simple yet accurate calibration process to reduce the IMU drifting errors, designed a machine learning algorithm to obtain real-time coordinates from IMU data, computed gait parameters, and derived a formula for G.A.I.T. scores with significant correlation with the physician's observational scores.

Keywords: post-stroke gait assessment; inertial measurement unit; machine learning; gait assessment and intervention tool

1. Introduction

Stroke is a major global health issue, affecting 15 million people annually and causing significant mortality and disability [1,2]. In Taiwan, it is the fifth leading cause of death, a primary cause of adult disability, and a significant burden on healthcare resources. Rehabilitation is key to stroke treatment. Based on the literature [3], the initial three months following a stroke are often referred to as the golden period for rehabilitation. During this crucial time, actively participating in a variety of therapeutic exercises can significantly reduce the physical impairments caused by the stroke. These impairments can affect stroke patients to varying degrees, depending on the severity and location of the brain injury, such as minor difficulties with coordination or balance for mild impairment, the difficulty of walking without assistance or using mobility aids for moderate impairment, and complete inability to walk without extensive support or assistive devices for severe impairment. Rehabilitation efforts during this golden period may enable some individuals to regain normalcy and lead independent lives. During the rehabilitation, clinicians use standardized stroke scales to quantify damage, including neurological deficits and gait abnormalities. Post-stroke patients often exhibit altered gait patterns with deviations in spatiotemporal characteristics [4], including reduced walking speed, shorter strides, increased energy expenditure, and asymmetry with uneven step lengths, stance times, and swing times between the affected and unaffected sides, impairing mobility and quality of life. While many regain walking ability, their gait remains biomechanically different from healthy individuals. For example, the right foot's step length and stride length may differ from the step length and stride length of the left foot for a post-stroke patient. However, the step length and stride length of both feet may be similar for a healthy individual. Analysis of the time–distance parameters in patients who have had strokes also reveals that they walk more slowly than healthy age-matched subjects, and the amplitude of peak values of lower extremity joints is reduced for patients who have had strokes [5]. Proper treatment and therapy can help optimize mobility.

Identifying clinical features related to post-stroke walking ability is crucial for developing effective gait training programs. Gait recovery has consistently been recognized as a primary rehabilitation goal, especially since impairments in muscle strength, motor function, and balance have been observed to be highly correlated with walking ability. The strength of affected hip flexors, ankle plantar flexors, knee extensors, and unaffected knee flexors and ankle plantar flexors is moderately to highly correlated with walking and stair-climbing speed ($r = 0.5\sim0.8$) [6,7] (moderate correlation: $r = 0.5\sim0.7$, indicating a noticeable relationship between two variables; high correlation: $r = 0.7$ to 0.9, suggesting a much stronger association where the variables are closely aligned). The Fugl-Meyer assessment of affected lower limb motor function significantly correlates with gait speed in mild to moderate stroke patients ($r\sim0.6$) [8]. Balance function measured by the Berg Balance Scale is highly correlated with functional walking distance [6-minute walk test (6 MWT) and 12-minute walk test (12 MWT)] at self-selected speeds ($r = 0.78$–0.80) [9].

In clinical gait assessment, a person's walking "ability" and "how" they walk are relevant. A stroke patient's walking ability is a function of stroke severity, typically based on two main aspects: how far a person can walk and their endurance. These are usually assessed using 3-minute, 6-minute, or 10-minute walk tests [1]. Functional Ambulation Category (FAC), Short Physical Performance Battery (SPPB), and/or Motor Assessment Scale (MAS) can also be used for further evaluation. On the other hand, assessing gait quality or "how" one walks involves studying gait patterns and specific gait characteristics. Historically, gait characteristics were more often studied in research settings rather than clinical environments. Still, instrumented gait analysis, including kinematic and kinetic assessments, can be used today to help clinicians establish patient-specific quantitative functional mobility benchmarks and goals.

In the early stages of post-stroke recovery, doctors perform qualitative observations, also known as visual gait analysis (using the naked eye or video), to assess gait performance and functional improvement [10]. Current clinical assessment methods based on visual observation largely rely on training and clinical judgment. Despite being scrutinized for inter-observer variability [11], "observational gait analysis" methods remain popular among clinicians due to their simplicity, availability, and low cost [12]. However, these qualitative analyses' validity, reliability, specificity, and responsiveness [13,14] have been questioned [11]. Overall, subjective observational gait scales may be useful for preliminary assessment of some spatiotemporal and/or kinematic gait parameters but are insufficient for analyzing the multifaceted nature of gait variability and complexity (e.g., motor and balance parameters). The timing for more complex gait analysis in stroke patients may depend on stroke severity and other factors such as fatigue, instability, pain, and persistent poor walking patterns despite adequate muscle activity during passive examination [1]. Nishi et al. [15] highlighted the importance of evaluating gait speed adjustability and instability in free-living environments to understand how individuals with post-stroke dysfunction navigate diverse contexts. They studied 50 stroke survivors wearing accelerometers on their lumbar spine (L5) for 24 h in real-life conditions. Directed acyclic graphs (DAGs) were generated using gait parameters derived from the acceleration data, and clustering analyses identified three distinct patterns based on four key variables: Ashman's D (step velocity), Fugl-Meyer Assessment, step time asymmetry, and step length. The findings underscore the diversity in gait control strategies among stroke survivors, suggesting the need for tailored rehabilitation approaches to address individual variations in gait patterns. Faisal and Deen [16] introduced a simplified gait index based on critical parameters—walking speed, maximum knee flexion angle, stride length, and stance-swing phase ratio—to assess overall gait quality. They used a gait dataset of 120 healthy individuals from two gait datasets [17,18] and found the healthy gait index range (0.5–0.67). A support vector machine algorithm was applied to validate this index, achieving approximately 95% classification accuracy. The study also confirmed the index's reliability and effectiveness by comparing it with other published datasets, which aligned with the predictions.

Since the 1990s, instrumented gait analysis has become the standard in research, providing an accurate and reliable biomechanical gait assessment method, including key parameter data (spatiotemporal data measurements and kinematic and kinetic measurements) [5,19,20]. However, gait laboratories typically include large, cumbersome, and expensive equipment such as motion capture systems (Vicon, Motion Analysis Inc., Qualisys, OptiTrack, etc.), force plates (Bertec, Kistler, Noraxon, etc.), instrumented walkways (Tekscan, GaitRite, etc.), balance platforms, and other instruments considered important for specific research applications (including post-stroke gait assessment) [1]. Among these devices, motion analysis systems (kinematic systems) are used to obtain the magnitude and timing of individual joint motion. For this purpose, the position and orientation of the limb segments and the join angles are recorded, using electrogoniometers or video cameras. The force platform or plate (kinetic system) provides a full three-dimensional description of the average ground reaction force vector beneath the foot, including three components of force (vertical, lateral (side-to-side), and fore-aft), the two coordinates of the center of pressure, and the moments about the x, y, and z axes. The construction and configuration of these laboratories are specific, aiming to conduct accurate tests, for example an area at least 9 m wide, 11 m long, with 3–5 m high ceilings (higher for stair climbing and motion applications) and special flooring. Therefore, it is not surprising that these setups, possibly only found in laboratory environments, are uncommon in clinical settings due to construction challenges and costs, as well as the lack of appropriately skilled clinicians to operate and manage/interpret the large amounts of data generated from various sensing technologies used in post-stroke gait rehabilitation.

With technological advancements, today's small, lightweight wearable IMU sensors are rapidly revolutionizing gait assessment in research environments and have the potential to be incorporated into routine clinical practice. These sensors can generally be fixed to specific body parts using straps and, with enhanced wireless communication capabilities, can detect real-time synchronized information from multiple devices worn on limb joints during walking activities, thereby monitoring movement changes and assessing gait performance [21]. They provide researchers with new opportunities for continuous gait recording, offering dynamic methods for quantifying gait over time and providing real-time feedback to patients and clinicians. Their lightweight, portability, and low-cost offer research potential for data collection outside the laboratory and in natural environments (clinics, sports fields, etc.). Moreover, these sensors can easily synchronize with other physiological measurement devices (such as EMG, ECG, and EEG), providing valuable multifactorial continuous data on subjects/patients in various environments, better reflecting the true living experiences and functional walking conditions in community settings [22].

IMUs, through IC (integrated circuits) processing and MEMS (micro-electromechanical systems) technology, are typically just a few millimeters in size, like 2–5 mm per side, and have functions such as accelerometers, gyroscopes, and sometimes magnetometers, capable of measuring and reporting information about an object's speed, acceleration, vector, and gravity. In principle, IMUs derive a series of data calculations based on Newton's mechanics formulas. Therefore, given known physical mass and force, applying the $F = M \times a$ rule allows for calculating acceleration values. Using advanced mathematical concepts, integrating acceleration yields speed, and similarly, integrating speed yields displacement results in corresponding three-axis coordinates. Thus, once the basic signals are detected by the accelerometer and appropriate algorithms are selected for filtering and analysis, we can practically capture relevant gait variables during the walking period, even covering estimated values such as step frequency and stride length [21].

With the advent of small, lightweight wearable measurement devices, significant progress has been made in processing gait data. Artificial intelligence (AI), including machine learning (ML), provides data analysis and model building, effectively performing recognition and prediction [23–27]. Reference [26] used dynamic gait data, employing SVM (support vector machine), RF (random forest), and ANN (artificial neural network) to classify young adults, older adults, and elderly patients, using KPCA (kernel principal component analysis) for data dimensionality reduction in the SVM classifier. Reference [24] used SVM, ANN, and RBFNN (radial basis function neural networks) classifiers to distinguish different walking conditions in hemiplegic patients, finding that a SVM had a higher recognition accuracy. Reference [23] used a general regression neural network (GRNN) as a decision tool, and reference [25] used an optimized ANN to identify hemiplegic stroke and able-bodied ambulation. Reference [28] described the characteristics of hemiplegic gait using a PCA (principal component analysis) of trunk movement (TM) and gait event (GE) parameters from IMUs. Reference [29] analyzed different IMU-based gait parameters proposed in the literature to assess frailty status (robust, pre-frail, or frail) or fall risk, showing that the most important parameters for frailty diagnosis were double support time, gait speed, stride time, step count, and steps per day or walking percentage per day. In the case of fall risk detection, the most relevant parameters were those related to trunk stability or motion. Luo et al. [30] proposed a gait analysis algorithm designed for heel-mounted IMUs to accurately assess heel dynamics and compute a comprehensive range of spatiotemporal gait parameters and parameters related to symmetry and variability. Three IMUs were mounted on a shoe: one on the instep, one in the center of the insole with a cavity, and one on the heel. Experiments included straight walking and daily activity simulations, with an optical motion capture (OMC) system as a reference. Results showed strong correlations ($r > 0.9$) and high reliability for all parameters (minimum intraclass correlation coefficient of 0.921). Lin et al. [31] addressed the growing issue of unbalanced walking among older adults by developing a deep learning (DL) model to improve balance assessment methods. The proposed model combines a convolutional neural network (CNN) with a long short-term memory (LSTM) network to predict scores for three balance scales: Berg Balance Scale (BBS), Timed Up and Go (TUG), and Single-Leg Stance (SLS). Gait data from 15-meter walks were collected using seven IMUs (on the lumbar spine, right upper arm, left upper arm, right thigh, left thigh, right calf, and left calf), with the CNN-LSTM model achieving accurate predictions for BBS and TUG scores. BBS and TUG assessments require only two IMUs placed on the thighs, producing mean absolute errors (MAEs) of 1.2562 and 1.4016 s, respectively. While the SLS showed an average MAE of 20.8886. Testing also demonstrated that using just one IMU on the right calf for BBS and TUG yielded comparable results, with MAEs of 1.4334 and 1.5229 s. Rokhmanova et al. [32] explored the impact of kinematic errors on learning therapeutic gait using vibrotactile feedback for knee osteoarthritis patients. The accuracy of inertial measurement units (IMUs) was compared to marker-based motion capture data, revealing that higher tracking errors resulted in slower learning rates and more incorrect responses to feedback. Participants with lower tracking errors showed better retention of learned gait in an outdoor setting compared to the control group, while those with higher errors did not. Tracking errors were correlated with foot size and angle magnitude, indicating potential bias. Guo et al. [33] introduced a temporal convolutional network (TCN)-based approach for robust and automatic gait phase recognition across various scenarios and walking speeds. Data from real-world overground walking experiments and public treadmill datasets were used to validate the method. In the overground walking experiment, two IMU sensors mounted on the shank with velcro belts were used to obtain gait data. A force plate synchronized with a motion capture system (Mocap) was used as the gold standard in the gait analysis. The TCN approach achieved 97% accuracy in gait-phase estimation, outperforming six machine learning models (K-nearest neighbor, naïve Bayes, support vector machine, decision tree, logistic regression, and multilayer perception) and two neural network models (fully convolutional network and long short-term memory).

In summary, the effectiveness of gait recovery in post-stroke rehabilitation is crucial. Collecting gait parameters during rehabilitation and daily life can help medical teams establish patient-specific quantitative functional mobility benchmarks and goals. Lightweight wearable devices can assist patients in daily rehabilitation and walking training at home, which is meaningful for clinical treatment.

In this paper, we used wearable devices to collect six-axis data on stroke patients' walking gait (XYZ three-axis linear accelerations and XYZ three-axis angular velocities). We also designed a calibration process and a machine learning model to obtain the required gait parameters and proposed a formula for assessing the G.A.I.T. score. These scores and the doctor's observational scores were evaluated through statistical analysis to find their correlations. The contributions of this work are summarized as follows.

- Development of an IMU-Based Gait Data Acquisition System: Proposed a wearable device based on inertial measurement units (IMUs) that can collect real-time gait data from stroke patients, overcoming the limitations of traditional gait assessment tools;
- Simple and Accurate Calibration Process: Designed a straightforward calibration method to mitigate drift errors in IMU data, ensuring the accuracy and reliability of the measurements;
- Application of Convolutional Neural Networks (CNNs): Utilized a CNN model for gait data analysis, enabling accurate prediction of gait parameters and providing real-time feedback;
- Analysis of Eight Stages of the Gait Cycle: Clearly defined the eight stages of the gait cycle and developed a scoring formula based on these stages, enhancing the objectivity of gait assessment;
- High Correlations with Physician Observations: Demonstrated significant correlations between G.A.I.T. scores calculated from IMU data and those assessed by physicians, validating the system's effectiveness;
- Clinical Application Potential: Provided a tool for gait training in daily life, assisting healthcare teams in establishing patient-specific functional mobility benchmarks, thereby improving recovery outcomes for stroke patients;

- Advancement of Gait Assessment Technology: Promoted the application of gait assessment technology in clinical practice through small, lightweight wearable devices, offering new directions for future research and practice.

2. Materials and Methods

In gait data collection, it is necessary to identify the various phases and parameters of the human gait cycle (GC), as shown in Figure 1 [34]. Eight states are required to complete a gait cycle: heel strike (HS) (initial contact), loading response, mid-stance, terminal stance, pre-swing, toe-off (initial swing), mid-swing, and terminal swing.

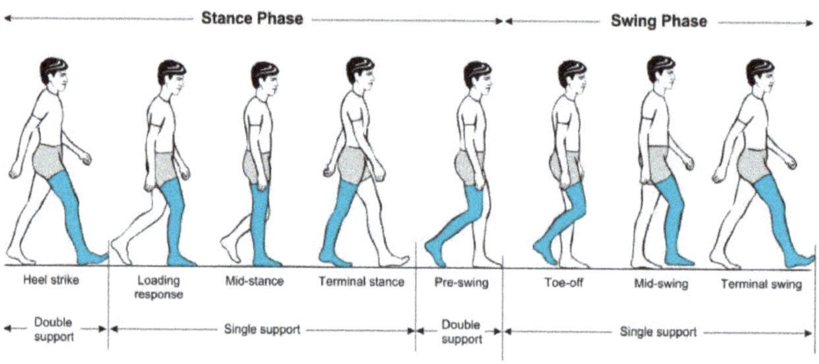

Figure 1. Gait cycle (GC) [34].

2.1. IMU Sensor

This paper used NGIMU [35] as the data acquisition device to facilitate gait data collection and ensure data stability. The dimensions of the NGIMU are 56 × 39 × 18 mm, and it weighs 46 g. The appearance is shown in the left image of Figure 2. The right image of Figure 2 is a schematic diagram of the nine-axis sensor coordinate system, where the x-axis points forward, the y-axis points to the left, and the z-axis points upward (towards the sky). The definition of rotational angular velocity is based on the right-hand rule, where the thumb points along the axis, and the direction of the curled fingers defines the positive rotational velocity.

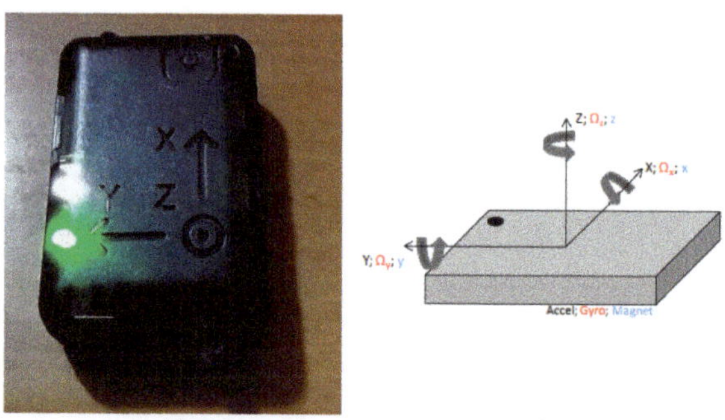

(a) (b)

Figure 2. (**a**) NGIMU; (**b**) schematic diagram of the nine-axis sensor coordinate system.

NGIMU has a built-in Attitude and Heading Reference System (AHRS) algorithm, which calculates and provides data types and related static accuracy values. The data types include quaternion, rotation matrix, Euler angles, linear acceleration, and earth acceleration.

NGIMU measurement data can be transmitted via a USB wired connection or WiFi/UDP wireless connection.

To accurately measure the height of the shoe sole from the ground during actual walking, we do not attach the NGIMU to the ankle. Instead, we designed a wearable fixture for the heel of the shoe, placing the NGIMU sensor in a fixed position, as shown in Figure 3. This avoids the height error caused by attaching it to the ankle and provides a position and angle closer to the sole. Figure 4 shows the 3D-printed fixture for placing the NGIMU (for both left and right feet).

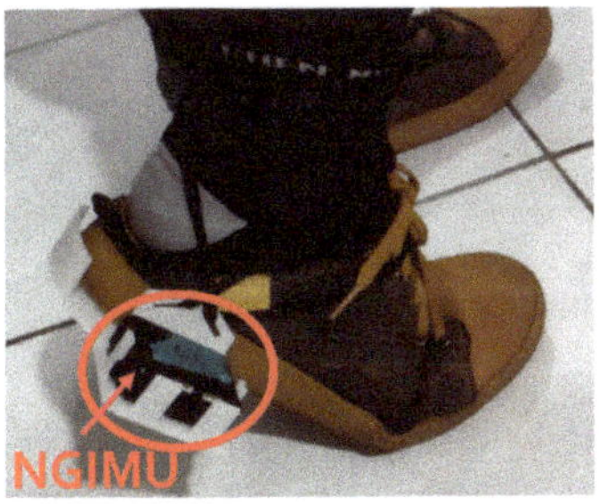

Figure 3. NGIMU in the fixture at the heel of the shoe.

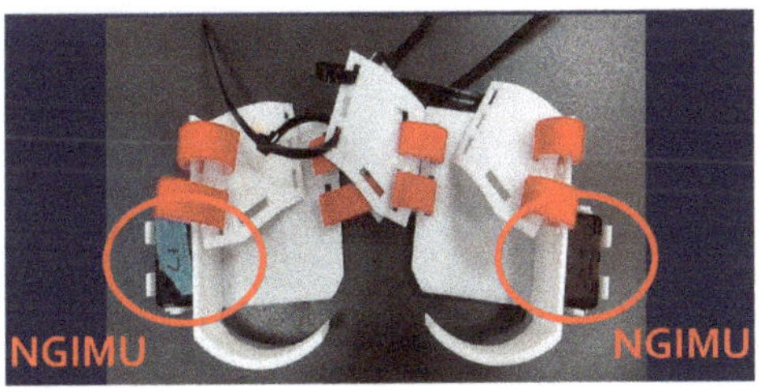

Figure 4. 3D-printed fixtures for placing the NGIMU.

2.2. Data Analysis Procedure

Figure 5 illustrates the data analysis process. Using WiFi, the data measured by the NGIMU (six-axis data and Euler angles) is saved to a computer in Excel format. These data are then converted into XYZ spatial coordinate distance values, with the x-axis representing forward distance, the Y-axis representing left (+) and right (−) deviation during forward movement, and the Z-axis representing heel height. The calibration procedure is used to calibrate the XYZ distance measurements. The calibrated values can determine the occurrence points of the GC's eight states and serve as labels for supervised learning, training a model to map the original six-axis data to XYZ spatial distance values. This paper uses a convolutional neural network (CNN) model to train the data.

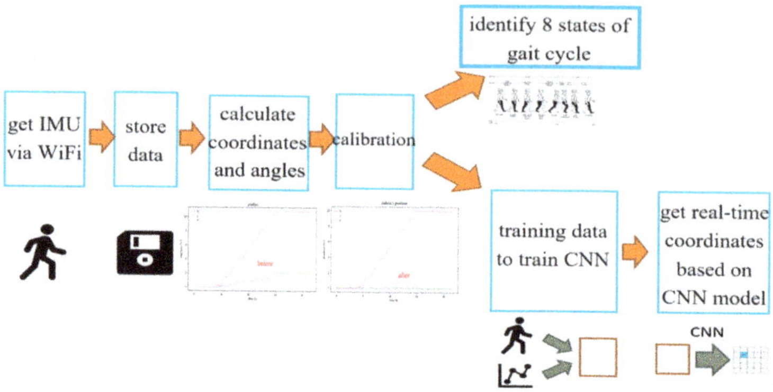

Figure 5. Data analysis process.

2.3. Three-Axis XYZ Distance Calibration

Although the NGIMU has good static angular velocity accuracy, it was found that the calculated spatial XYZ three-axis distance values have cumulative integration errors, resulting in data drift. Therefore, calibration is necessary. Typically, precise calibration requires expensive optical motion capture (OMoCap) systems, which need multiple optical cameras and a setup space. However, due to budget constraints and practical site space, a simple yet relatively accurate calibration procedure is designed to be suitable for gait data. The procedure is as follows:

1. First, fix the walking x-axis linear distance (e.g., ten meters for post-stroke patient to walk), mark the starting and ending lines, and ideally draw a straight line between the starting and ending lines. The subject should walk between the starting and ending points;
2. X-axis distance calibration: The predetermined x-axis length value (e.g., 10 m) calibrates the calculated x distance based on Equation (1) as shown in Figure 6.

$$X_{cali} = \frac{10}{Xend} \times X_{meas} \quad (1)$$

   where X_{meas} is the measured x-axis distance obtained from IMU, $Xend$ is the measured x-axis distance at the end point, and X_{cali} is the calibrated x-axis distance;
3. Y-axis distance calibration: Set the Y distance to zero when both feet stand at the starting point. Assume the Y distance at the end of the walk is approximately the same as at the starting point (since walking along the x-axis line), and calibrate the Y-distance accordingly based on Equation (2) as shown in Figure 7.

$$Y_{cali} = Y_{meas} - (Yend - Ystart) \quad (2)$$

   where Y_{meas} is the measured Y-axis distance obtained from IMU, $Yend$ and $Ystart$ are the measured Y-axis distances at the end and starting points, respectively, and Y_{cali} is the calibrated Y-axis distance. Since the walking lane has starting and end lines, it is reasonable to assume the patient begins and finishes walking with both feet standing at the starting and end line. The measured $Yend$ and $Ystart$ are supposed to be the same, and the difference between these two values can be used to calibrate the measured Y-axis distance;
4. Z-axis distance calibration: Since walking is on a flat floor, the Z-distance when the foot is fully on the ground (stationary) is regarded as zero to calibrate the z-distance based on Equation (3) as shown in Figure 8.

$$Z_{cali} = Z_{meas} - h \quad (3)$$

where Z_{meas} is the measured Z-axis distance obtained from IMU, h is the measured Z-axis distance at double limb support state, and Z_{cali} is the calibrated Z-axis distance. Since the Z-axis distance (foot height) is supposed to be zero at the double limb support state, the difference between the measured Z-axis distance at this state h and zero can be used to calibrate the measured Z-axis distance.

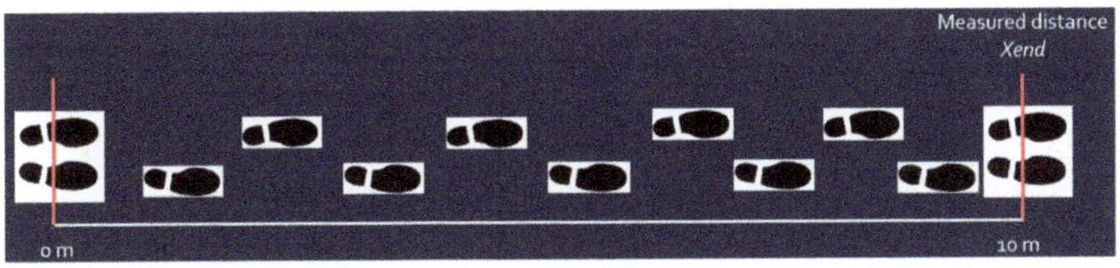

Figure 6. X-distance calibration.

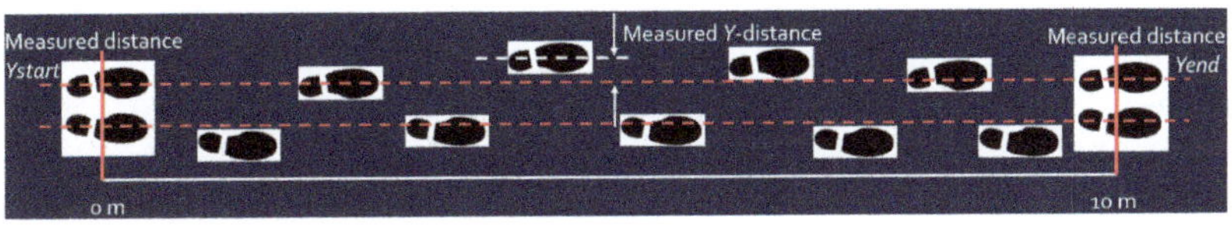

Figure 7. Y-distance calibration.

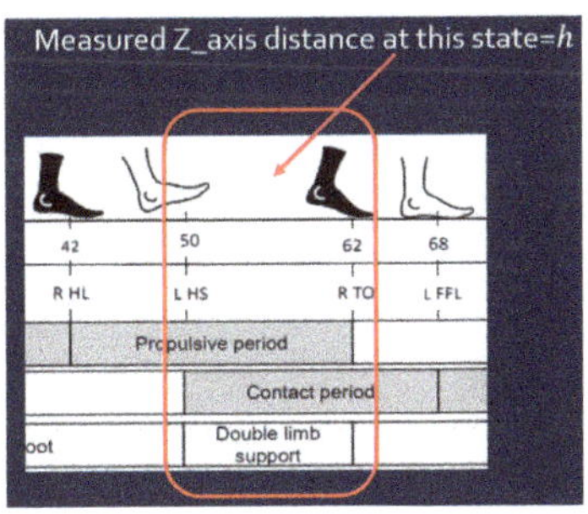

Figure 8. Z-distance calibration.

2.4. Determining the Eight States of the Gait Cycle

To identify the gait cycle, data from both feet are required. The analysis of the corresponding movement relationships is as follows (with the right foot as the primary test foot and the left foot as the secondary test foot):

1. Initial Contact (Heel Strike): The movement begins with the heel of the primary test foot touching the ground (start of first double support);
2. Loading Response: The end of the first double support with the heel of the secondary test foot at its highest point under the condition that the X-distance of the secondary test foot is almost not changing;
3. Mid-stance: The secondary test foot reaches the body's center point (the secondary test foot position surpasses the primary test foot);
4. Terminal Stance: The heel of the secondary test foot touches the ground (start of the second double support);
5. Pre-swing: The end of the second double support phase with the heel of the primary test foot at its highest point under the condition that the X distance of the primary test foot is almost not changing;

6. Initial Swing (Toe-off): The highest point of the primary test foot (maximum knee flexion);
7. Mid-swing: The primary test foot is parallel to the floor;
8. Terminal Swing: The heel of the primary test foot touches the ground.

Based on the collected six-axis posture data, the criteria for determining each gait phase are as follows:

1. Initial Contact: The maximum point of the right pitch angle;
2. Loading Response: The minimum point of the left pitch angle;
3. Mid-stance: The left x-axis position surpasses the right x-axis position;
4. Terminal Stance: The maximum point of the left pitch angle;
5. Pre-swing: The minimum point of the right pitch angle;
6. Initial Swing: The maximum point of the right Z-axis height;
7. Mid-swing: The right pitch angle returns to zero;
8. Terminal Swing: The maximum point of the right pitch angle.

2.5. Convolutional Neural Network (CNN)

Convolutional Neural Networks (CNNs) are primarily used for processing and analyzing data with a grid-like structure, such as images and videos. The uniqueness of CNNs lies in their use of convolutional layers, pooling layers, and flattening layers to automatically learn features from the input data. Figure 9 shows the schematic diagram of the CNN model used in this work. The model consists of 11 layers with a kernel size of 41. The data are processed with convolutions and pooling before being upscaled to predict coordinate positions. Each training data set has a length of 1392 points, with each point containing six-axis data (XYZ-axis linear acceleration values and XYZ-axis angular velocity). The output data are 3D coordinate positions (XYZ three-dimensional position coordinates).

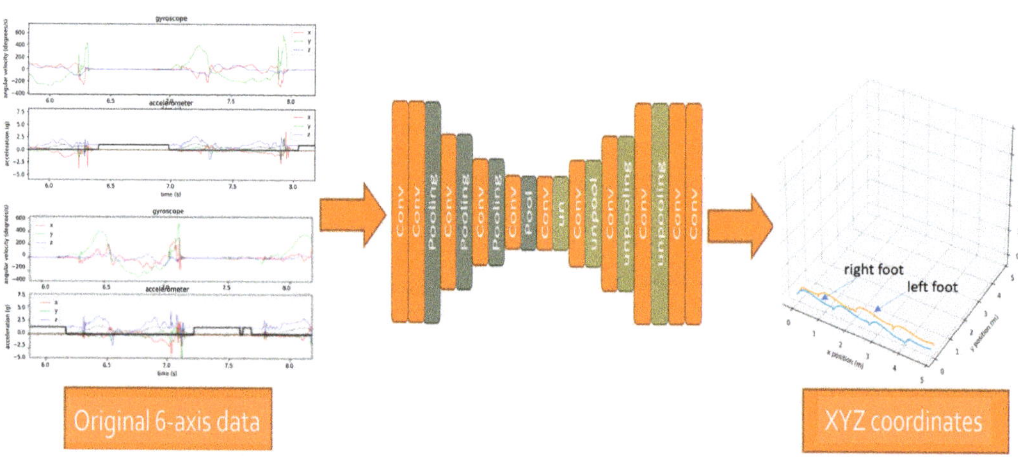

Figure 9. CNN for XYZ coordinates.

2.6. Gait Assessment and Intervention Tool (G.A.I.T.)

To accurately assess the gait of a post-stroke patient, we need gait coordination measurement and force production, including gait kinematics and gait kinetics, customarily measured using motion capture and force plate systems in well-equipped gait research laboratories. However, in clinical practice and certain gait research settings, the access to such technology is often unavailable. Additionally, the time constraints of clinical practice make it impractical to use this instrumented technology in its current form due to the labor-intensive process of data collection and analysis. However, observational gait coordination measures can be employed in these cases [36]. The Gait Assessment and Intervention Tool (G.A.I.T.) is a 31-item measure designed to evaluate the coordinated movement components of gait and identify associated gait deficits. It offers several advantages, including a comprehensive, objective-based scoring method and the ability to incrementally measure improvement within specific items. An inexperienced clinician, after receiving training, achieved an inter-rater reliability with an experienced rater of an intraclass correlation coefficient (ICC) equal to 0.99 [10]. The G.A.I.T. effectively distinguished between two gait training interventions, demonstrating an additional benefit of functional electrical stimulation (FES) compared to similar comprehensive gait training.

In this paper, we used simple IMU-based wearable devices to collect gait data of post-stroke patients and derive G.A.I.T. scores based on the GC's state parameters. These scores are compared with those evaluated by the patients' doctor. The parameters of GC's states used for G.A.I.T. score are the pitch angles of the left terminal stance, right terminal stance, left loading stance, and right pre-swing and the times associated with these states, as shown in Figure 10 where the limped foot of the patient is left foot. The following variables are defined and used for calculating G.A.I.T. score based on Equations (4)–(20).

Maximum pitch angle of left foot step i: $P_{L,max,i}$,
Minimum pitch angle of left foot step i: $P_{L,min,i}$,
Maximum pitch angle of right foot step i: $P_{R,max,i}$,
Minimum pitch angle of right foot step i: $P_{R,min,i}$,

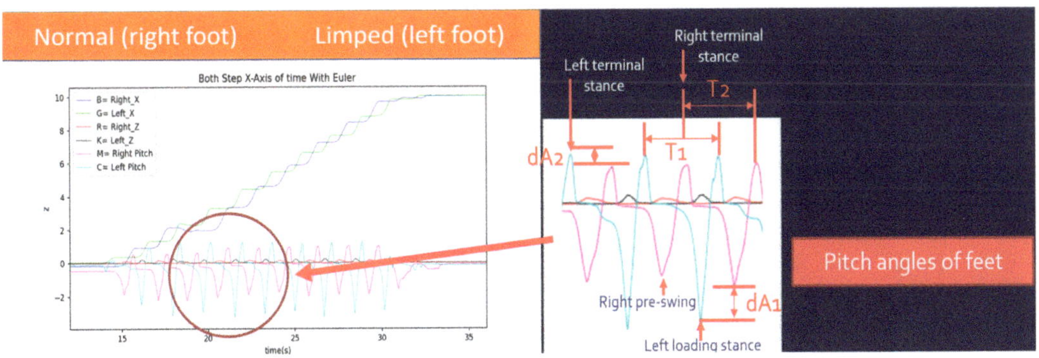

Figure 10. Parameters of GC's states for G.A.I.T.

$$\text{Average of the maximum pitch angle of left foot over all steps}: \overline{P_{L,max,i}} = e, \tag{4}$$

$$\text{Average of the minimum pitch angle of left foot over all steps}: \overline{P_{L,min,i}} = f, \tag{5}$$

$$\text{Average of the maximum pitch angle of right foot over all steps}: \overline{P_{R,max,i}} = g, \tag{6}$$

$$\text{Average of the minimum pitch angle of right foot over all steps}: \overline{P_{R,min,i}} = h, \tag{7}$$

$$\max(P_{L,max,i}) = c, \tag{8}$$

$$\min(P_{L,min,i}) = d, \tag{9}$$

$$\max(P_{R,max,i}) = k, \tag{10}$$

$$\min(P_{R,min,i}) = m, \tag{11}$$

$$|P_{L,max,i} - P_{R,max,i}| = \mathrm{dA}_{2,i} \tag{12}$$

$$|P_{L,min,i} - P_{R,min,i}| = \mathrm{dA}_{1,i} \tag{13}$$

$$\text{Average of } \mathrm{dA}_{2,i} = a \tag{14}$$

$$\text{Average of } \mathrm{dA}_{1,i} = b \tag{15}$$

$$Arange_L = (e - f) \tag{16}$$

$$Arange_R = (g - h) \tag{17}$$

$$p = |\overline{\mathrm{T1}_i} - \overline{\mathrm{T2}_i}| \tag{18}$$

$$q = \frac{(\overline{\mathrm{T1}_i} + \overline{\mathrm{T2}_i})}{2} \tag{19}$$

$$\text{G.A.I.T.Score} = \frac{a + b + (c-e) + (k-g) + (f-d) + (h-m)}{\min(Arange_L,\ Arange_R)} \times \left(\frac{p}{q} + 1\right) \tag{20}$$

The first term on the right-hand side in Equation (20) deals with spatial factors of the gait, while the second term on the right-hand side copes with temporal factors. The value of $\mathrm{dA}_{2,i}$ is the difference between the maximum pitch angles of left and right feet for step i, and a is the average of $\mathrm{dA}_{2,i}$. Therefore, the larger the value of a, the more asymmetric both feet are and, therefore, the higher the G.A.I.T. score should be. The same logic applies to variables b and $\mathrm{dA}_{1,i}$. The value of c represents the maximum pitch angle of the left foot over all steps, and the value of e represents the average of the maximum pitch angles of the left foot over all steps. The difference between these two, $(c - e)$, represents the walking consistency of the left foot. Therefore, the larger the value of $(c - e)$, the less consistency the left foot is and the higher the G.A.I.T. score should be. The same logic applies to $(k - g)$. The minimum pitch angles can also be used to evaluate the walking consistency; therefore, $(f - d)$ and $(h - m)$ are used in Equation (20). The numerator of the first term on the right-hand side in Equation (20) may vary from person to person; the denominator of the first term on the right-hand side in Equation (20) is used to remove this factor. The smaller value from $Arange_L$ and $Arange_R$ is chosen as the denominator. $\overline{\mathrm{T1}_i}$ and $\overline{\mathrm{T2}_i}$ represent the average times of the strides of left foot and right foot, respectively. The value of p is the difference between $\overline{\mathrm{T1}_i}$ and $\overline{\mathrm{T2}_i}$. p is then divided by q, the average of $\overline{\mathrm{T1}_i}$ and $\overline{\mathrm{T2}_i}$. The smaller the value of p/q, the more consistent both feet are. The spatial term (the first term) on the right-hand side is adjusted using the second term (temporal factor p/q plus 1) on the right-hand side.

3. Results

The NGIMU measures data from both feet, sampling every 5 ms at a rate of 200 Hz and recording the position and angle of the feet. The stored data include time. The first batch of data was collected in a laboratory environment, with the walking distance along the x-axis set to 5 m. The dataset consists of 120 entries, with three participants (two males and one female, aged 19–21, with an average age of 20). After randomization, 80% of the data, 96 entries, was used as training data, and the remaining 24 entries were used as test data. Each data entry has a length of 1920 points (sampling frequency of 200 Hz, sampling time of about 10 s, including stationary time before and after walking).

The second batch of data were collected at Taichung Veterans General Hospital, Taiwan, with the walking distance along the x-axis set to 10 m. Currently, 26 data entries have been collected successfully. The collected data are stored in a local database and converted to CSV file format at the end of sampling. The files are divided into four parts, each containing six-axis acceleration data (excluding magnetic angle data due to their slow sampling rate) and Euler angle data for the left and right feet. The accelerometer data format is arranged as follows: the first column is time, the second to fourth columns are angular velocity, and the fifth to seventh columns are acceleration. The Euler angle data format is arranged with the first column being time, the second column being roll angle, the third column being pitch angle, and the fourth column being yaw angle.

3.1. XYZ Distance Calibration

After the data are collected and stored as a CSV file, they are calibrated using the proposed simple method. The Y-axis and Z-axis are adjusted based on the x-axis values, and the x-axis position is calibrated by setting the final distance to eliminate cumulative errors from integration. The trends of the XYZ distances before and after calibration are shown in Figure 11. The left figure shows that the calculated X distance from the built-in AHRS is not exactly 10 m due to integration errors. The shifts of Y-distance are also noticed. The Z-distance (height) gradually increases due to integration errors. The right figure shows the results after calibration, eliminating all the aforementioned phenomena. Figure 12 shows detailed comparisons for Z- and Y-distances before and after calibration. The upper figure of Figure 12 shows the calibration results of the Z-distance (height), where each stationary point is flat on the ground after calibration. The lower figure of Figure 12 shows the calibration results of the Y-distance, eliminating the difference in walking direction caused by the different foot angles before and during walking.

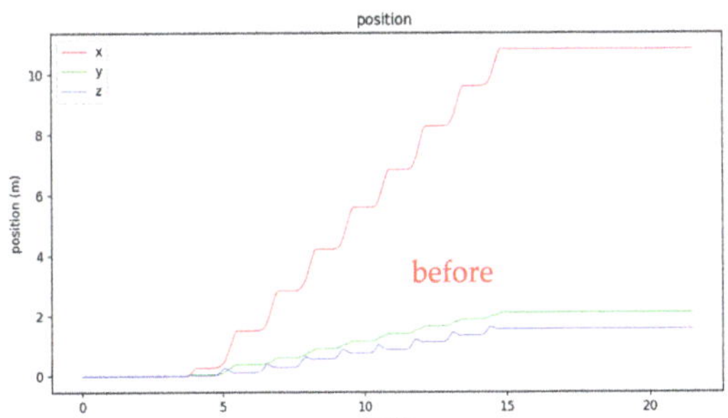

(a)

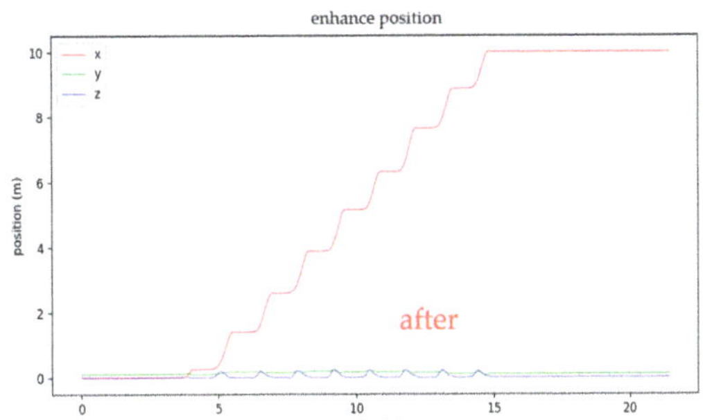

(b)

Figure 11. XYZ distance calibration: (**a**) before and (**b**) after calibration.

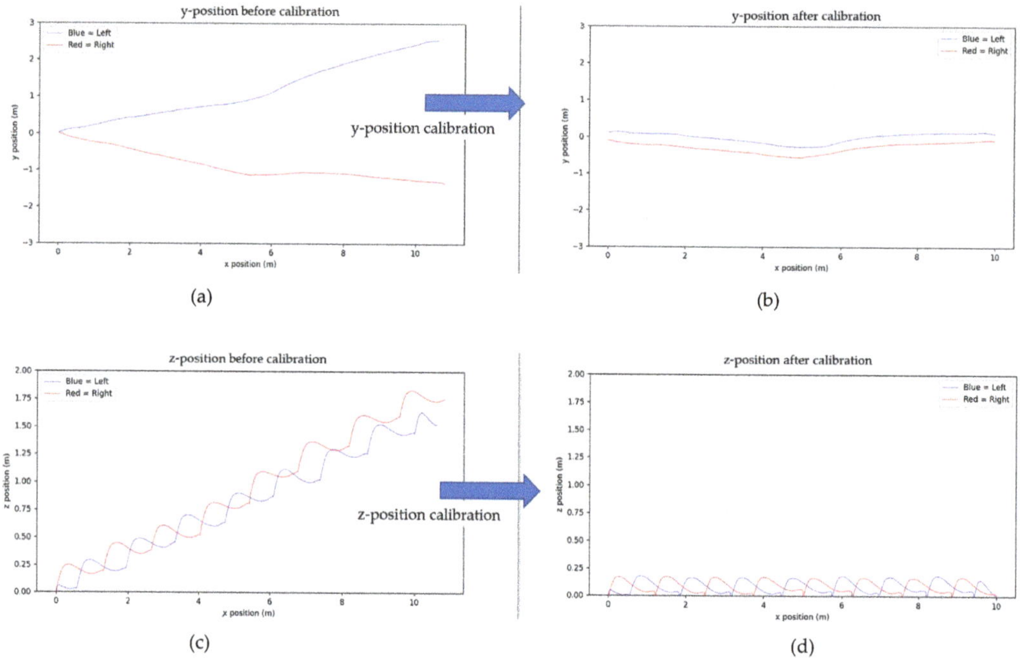

Figure 12. Close look of x- and z-positions before and after calibration. (**a**) y-position before calibration; (**b**) y-position after calibration; (**c**) z-position before calibration; (**d**) z-position after calibration.

3.2. Eight States of Gait Cycle

Based on the criteria for determining the GC's eight states, the corresponding results are shown in Figure 13. In the top graph of Figure 13b, blue represents the distance of the right foot along the x-axis (Right X), green represents the distance of the left foot along the x-axis (Left X), red represents the height of the right foot along the Z-axis (Right Z), black represents the height of the left foot along the Z-axis (Left Z), magenta represents the pitch angle of the right foot (right pitch), and cyan represents the pitch angle of the left foot (left pitch). The numbers within the solid circles indicate the time points of the eight states. The corresponding XYZ accelerations and angular velocities are shown in the bottom graph of Figure 13b. Figure 13a shows the successfully annotated time points of the eight states in each gait cycle. In this graph, the red solid/dash lines represent the primary test foot (right foot), and the blue solid/dash lines represent the secondary test foot (left foot). Therefore, red and blue solid lines illustrate the X-position and the Z-axis height, and the dashed red and blue lines represent the Euler angles (values are proportionally reduced). The green segments represent the eight states in the gait cycle (the eighth state is the starting point of the next cycle, so there are seven segments).

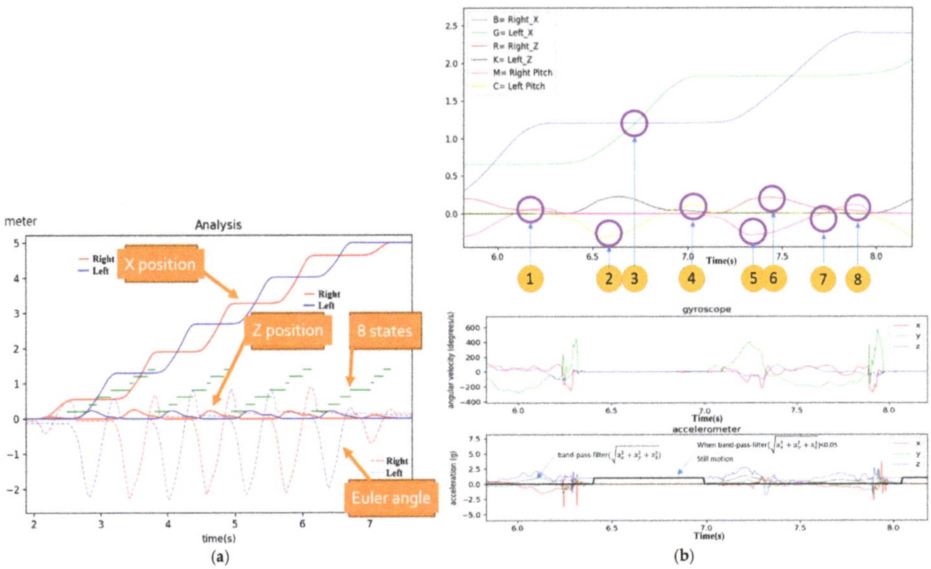

Figure 13. Determining the GC's eight states. (**a**) Annotated time points of the eight states in each gait cycle; (**b**) top: x distance, z height, and pitch angle and bottom: gyroscope and accelerometer data.

3.3. Convolutional Neural Network (CNN)

The 1D CNN is written in Python (version 3.8) and executed on the PyCharm (version 2024.1.6) platform using Pytorch (version 2.1.1+cu118). PySide2 (version 5.15.2.1) is chosen for GUI (Graphic User Interface). PyCharm is the development platform because it has better support for PySide2. PySide2 is similar to PyQt, making the development process equally simple. Additionally, PySide2 follows the LGPL license, allowing commercial use without being restricted by sharing regulations. Pytorch, an open-source AI development package since 2017, is known for its ease of use, quick model building, and excellent support for GPU computation, making it one of the most popular AI development tools in academia.

The computer specifications for running the program are an Intel i7-12700H/2.30 GHz CPU (Santa Clara, CA, USA) and a Nvidia GeForce RTX 3060 GPU (Santa Clara, CA, USA). Training the 11-layer 1D CNN model with 96 data entries takes approximately 3 s per epoch. If only the CPU is used for training, it takes about 60–120 s. Figure 14 shows the training and validation loss (mean squared error, MSE) curve, and Figure 15 shows the prediction results after training with the epoch set at 500. The root of the MSE of the test data (24 data entries) is 5.42 cm, which is accurate enough for engineering purposes.

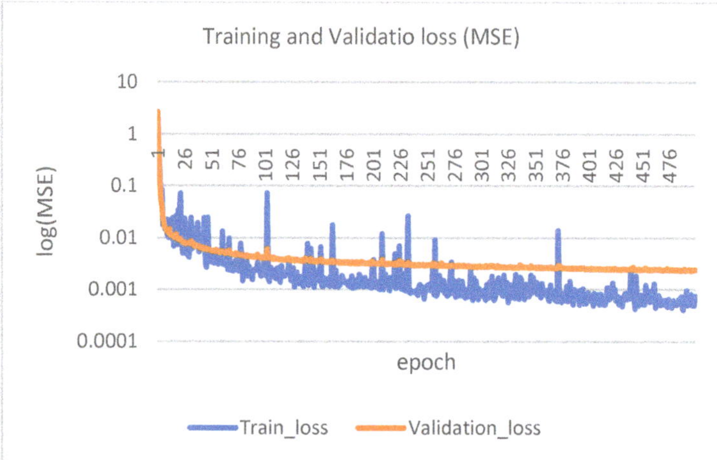

Figure 14. The training and validation loss (MSE).

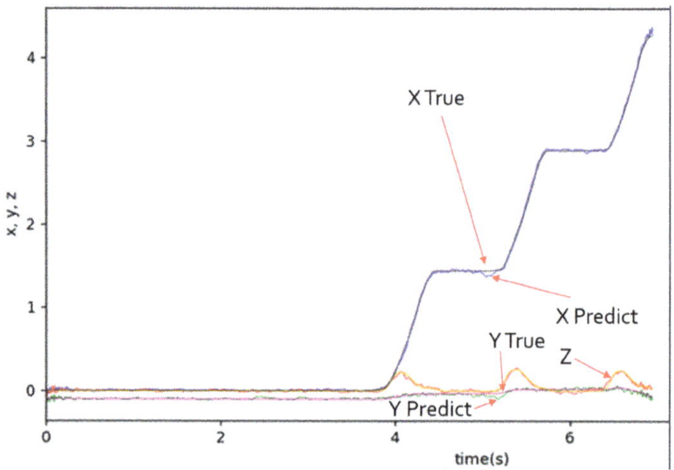

Figure 15. The prediction results after training.

3.4. G.A.I.T. Score

Based on 26 data entries collected from the Taichung Veterans General Hospital (TVGH), Taiwan, Equation (20) calculates the G.A.I.T. scores for these data entries. Then, the calculated scores are compared with those evaluated by the director of the Rehabilitation Department of the TVGH. Table 1 lists these scores. Table 2 shows the mean and standard deviations of these two sets of scores. Table 3 shows the Pearson correlation of these two data sets with 0.752, which is significant at the 0.01 level (two-tailed). Table 4 shows the Spearman's rho = 0.933 of these two data sets, which is also significant at the 0.01 level (two-tailed).

Table 1. G.A.I.T scores.

Subject #	Score by Doctor	Score by IMU
1	0	0.2732
2	4	0.7906
3	5	0.7266
4	3	0.3859
5	3	0.3620
6	5	0.5506
7	10	1.1644
8	6	1.1418
9	3	0.5846
10	3	0.4490
11	9	4.0168
12	9	2.3792
13	3	0.7701
14	0	0.3567
15	0	0.1239
16	0	0.2664
17	4	0.4509
18	0	0.1600
19	3	0.3820
20	0	0.2640
21	5	0.7700
22	0	0.2070
23	7	0.8810
24	3	0.4000
25	0	0.1640
26	0	0.1860

Table 2. Means and standard deviations of two sets of scores.

	Mean	Std. Deviation
Score by doctor	3.35	3.032
Score by IMU	0.7174	0.83552

Table 3. Pearson Correlation.

		Score by Doctor	Score by IMU
Score by doctor	Pearson Correlation	1.000	0.752
	Sig. (2-tailed)		0.000
Score by IMU	Pearson Correlation	0.752	1.000
	Sig. (2-tailed)	0.000	

Table 4. Spearman's rho.

			Score by Doctor	Score by IMU
Spearman's rho	Score by doctor	Correlation Coefficient	1.000	0.933
		Sig. (2-tailed)		0.000
	Score by IMU	Correlation Coefficient	0.933	1.000
		Sig. (2-tailed)	0.000	

4. Discussion

The advantages of using NGIMU in this work include its excellent static angular velocity accuracy, which allows for more accurate distance measurements. It has built-in

WiFi wireless transmission, making installation easy and requiring only a WiFi adapter without additional hardware. This feature makes collecting patients' data convenient in the hospital environment. Additionally, NGIMU offers a user-friendly setup environment, making data collection more manageable for users.

NGIMU can measure nine-axis data using a gyroscope, accelerometer, and magnetometer; however, only six-axis data from NGIMU were used in this work. The sampling rates of the gyroscope and accelerometer both can reach 400 Hz, but the sampling rate of magnetic angles is only 20 Hz. We use 200 Hz as the sampling rate for measuring gait data, which is fast enough for walking movement.

Although NGIMU provides good static angular velocity and has a built-in AHRS algorithm, the accumulated error in calculating XYZ coordinates still exists, which can be observed in the left diagram of Figure 11. The diagram shows that the final X-distance exceeds 10 m, the final Y-distance is apart from the walking straight line by 2 m, and the Z-distances (height) of the GC's double support are not zero (unrealistic floating above the floor). These errors make the gait data not practical to use. Therefore, a calibration process becomes essential. In this paper, a simple yet accurate enough calibration is proposed by using the special characteristics of 10 m linear walking. The calibration outcomes are satisfactory, which can be observed from the right diagram in Figures 11 and 12. The final X-distance is at 10 m, the Y-distance is close to the walking straight line, and the Z-distance (height) is periodic and is zero at double support states.

By observing Figure 15, the predicted output results of the CNN model still exhibit some irregularities. We can further use a low-pass filter to filter the prediction results. In this paper, a filter based on the Slide Window method averages the data by extending 10 points to the left and right, effectively removing irregular waves. The filtered results are shown in Figure 16.

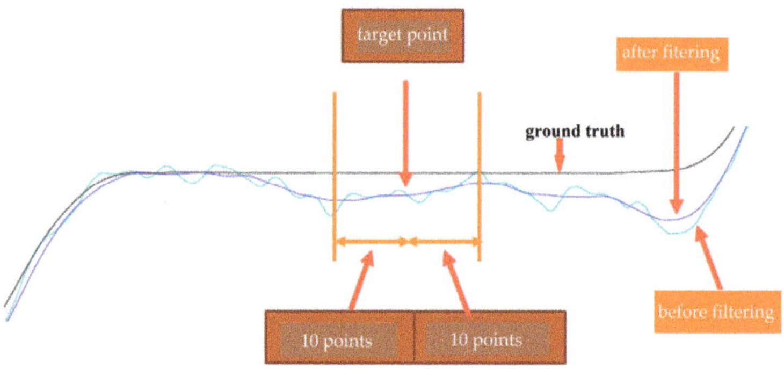

Figure 16. The filtered results.

The input data for the CNN training set consists of six-axis data (linear accelerations: a_x, a_y, a_z and angular velocities: ω_x, ω_y, ω_z) from the left and right feet, with each foot trained separately. Each foot has six features. The output labels of the training set are the calibrated XYZ-distances. The input layer size is (6, 1392). The kernel size is 41. The input data are first processed through a 1D CNN, followed by four 1D CNN and pooling layers, with the filter size growing from 64 to 1024, doubling at each layer. Then, it goes through four 1D CNN and unpooling layers, with the filter size decreasing from 1024 to 64, halving at each layer. Finally, a 1D CNN layer processes the output into XYZ coordinates of 1392 points.

Based on the walking distance–time curve of the X-distance, each step of a single foot's walk can be segmented. Then, gait parameters such as step length, stride length, foot angle, walking speed, and the time difference between the left and right foot landing in the gait cycle can be calculated:

1. Gait Cycle Time: The time between two consecutive landings of the same foot;
2. One Step Move Time: The time between alternating landings of the left and right feet;
3. Gait Cycle Length: The step length between two consecutive landings of the same foot;
4. One Step Move Length: The distance between alternating landings of the left and right feet;
5. Angle At Stop: The foot angle when a single foot stops during a step.

Based on the steps of the object, we can use the form of (standard deviation, mean), (standard deviation, mean) to present the above parameters. The first set of data represents the standard deviation and mean of the left foot, and the second set represents the standard deviation and mean of the right foot. Taking one healthy participant and one post-stroke patient as examples, Table 5 shows a healthy person's gait parameters, where the standard deviations are very small, and the average values of the left and right feet are not significantly different. Table 6 shows the gait parameters of a patient, where the standard deviations of the One Step Move Time for the left foot (0.075) and the right foot (0.115), the standard deviations of the Gait Cycle Length for the left foot (0.138) and the right foot (0.079), and the means of the One Step Move Length for the left foot (0.153) and the right foot (0.477) can be seen, and the standard deviations and mean of the Angle At Stop for the left foot (3.354, −11.985) and the right foot (0.545, −2.184) show significant differences. These parameters may provide some insights for scoring the G.A.I.T. in the future.

Table 5. Gait parameters of a healthy person.

	Left Foot (std. dev., Mean)	Right Foot (std. dev., Mean)
Gait Cycle Time	(0.08744, 1.35300)	(0.03945, 1.34750)
One Step Move Time	(0.04715, 0.61750)	(0.01795, 0.71833)
Gait Cycle Length	(0.05711, 1.20995)	(0.02178, 1.33088)
One Step Move Length	(0.10636, 0.58199)	(0.10360, 0.66631)
Angle At Stop	(0.53145, 2.75407)	(0.94228, −1.17001)

Table 6. Gait parameters of a post-stroke patient.

	Left Foot (std. dev., Mean)	Right Foot (std. dev., Mean)
Gait Cycle Time	(0.12664, 1.85375)	(0.11065, 1.83958)
One Step Move Time	(0.07505, 0.86346)	(0.11515, 0.97821)
Gait Cycle Length	(0.13804, 0.69266)	(0.07912, 0.63808)
One Step Move Length	(0.18417, 0.15301)	(0.19998, 0.47711)
Angle At Stop	(3.35361, −11.98528)	(0.54522, −2.18445)

In the current work, we have derived a formula to score the G.A.I.T. based on the observations and the results from the gait parameters and GC's states. We found the pitch angles of the left terminal stance, right terminal stance, left loading stance, and right pre-swing, and the times associated with these states are related to the G.A.I.T. score. Score 0 of G.A.I.T. represents normal walking. Therefore, the smaller the p-value (similar to the Gait Cycle Time) of Equation (18) is the less the difference between the walking of the left and right feet is. The closer the two variables of c in Equation (8) and e in Equation (4) are the more consistent the left foot is walking. Similarly, the same logic applies to variables k in Equation (10) and g in Equation (6), f in Equation (5) and d in Equation (9), and h in Equation (7) and m in Equation (11). The larger the value of a in Equation (14) is the more unbalanced the two feet are. The same logic applies to the value of b in Equation (15). The q and $\min(Arange_L, Arange_R)$ in Equation (20) are used as the bases to normalize the calculated score. Based on statistical analysis, the calculated scores and the doctor's scores are significantly correlated with Pearson correlation = 0.752 and Spearman's rho = 0.933 at the 0.001 level (two-tailed).

5. Conclusions

We have successfully developed and tested an IMU-based gait data acquisition system that is easy to install in a clinic and hospital and effective for post-stroke patients. Our approach includes a straightforward yet precise calibration process to mitigate IMU drifting errors and a CNN model to obtain real-time coordinates of the subject's feet. The calibration process, tailored to the unique characteristics of 10 m linear walking, has yielded satisfactory results. The CNN model accurately predicts the coordinates in real time using six-axis data from each foot. By processing XYZ distances and pitch angles, we can easily determine the GC's eight states and extract key parameters to calculate G.A.I.T. scores using our proposed formula. Statistical analysis reveals a significant correlation between the calculated G.A.I.T. scores and the doctor's observational scores, with Pearson correlation at 0.752 and Spearman's rho at 0.933, both significant at the 0.001 level (two-tailed). These findings underscore the reliability and potential of our system in clinical settings.

6. Patents

Based on this work, a patent, application # 113214518, has been filed in the Intellectual Property Office, Ministry of Economic Affairs, Taiwan, R.O.C.

Author Contributions: Conceptualization, Y.-C.W., C.-C.H. and Y.-Y.C.; methodology, Y.-C.W., Y.-J.H. and C.-C.H.; software, Y.-J.H., C.-S.C. and Y.-C.W.; validation, Y.-C.W., C.-C.H., Y.-Y.C., C.-S.C. and Y.-Y.C.; formal analysis, Y.-J.H., Y.-C.W. and Y.-Y.C.; investigation, Y.-C.W., C.-S.C. and Y.-Y.C.; resources, Y.-C.W. and Y.-Y.C.; data curation, Y.-C.W., Y.-J.H. and Y.-Y.C.; writing—original draft preparation, Y.-J.H. and Y.-C.W.; writing—review and editing, Y.-C.W.; visualization, Y.-J.H. and Y.-C.W.; supervision, Y.-C.W., C.-C.H., Y.-Y.C. and C.-S.C.; project administration, Y.-C.W., C.-C.H. and Y.-Y.C.; funding acquisition, Y.-C.W. and Y.-Y.C. All authors have read and agreed to the published version of the manuscript.

Funding: This research was funded by the Taichung Veterans General Hospital and the National United University, Taiwan, grant number TCVGH-NUU1128902.

Institutional Review Board Statement: The study was conducted in accordance with the Declaration of Helsinki and approved by the Institutional Review Board of Tai-Chung Veterans General Hospital (TCVGH-IRB No. CF22222A, approved on 20 June 2022).

Informed Consent Statement: Informed consent was obtained from all subjects involved in the study.

Data Availability Statement: The datasets presented in this article are not readily available because the data are part of an ongoing study.

Conflicts of Interest: The authors declare no conflicts of interest. The funders had no role in the design of this study; in the collection, analyses, or interpretation of data; in the writing of this manuscript; or in the decision to publish the results.

References

1. Mohan, D.M.; Khandoker, A.H.; Wasti, S.A.; Alali, S.I.I.I.; Jelinek, H.F.; Khalaf, K. Assessment Methods of Post-stoke Gait: A Scoping Review of Technology-Driven Approaches to Gait Characterization and Analysis. *Front. Neurol.* **2021**, *22*, 650024.
2. Feigin, V.L.; Brainin, M.; Norrving, B.; Martins, S.O.; Pandian, J.; Lindsay, P.; Grupper, M.F.; Rautalin, I. World Stroke Organization: Global Stroke Fact Sheet 2025. *Int. J. Stroke* **2025**, *20*, 132–144. [PubMed]

3. Chan, C. Golden Recovery Period. Available online: https://www.strokefund.org.hk/en/golden-recovery-period/?form=MG0AV3 (accessed on 3 March 2025).
4. Ofran, Y.; Karniel, N.; Tsenter, J.; Schwartz, I.; Protnoy, S. Functional Gait Measurement Prediction by Spatiotemporal and Gait Symmetry in Individuals Post Stroke. *J. Dev. Phys. Disabil.* **2019**, *31*, 611–622.
5. Nadeau, S.; Betschart, M.; Bethoux, F. Gait Analysis for Poststroke Rehabilitation: The Relevance of Biomechanical Analysis and the Impact of Gait Speed. *Phys. Med. Rehabil. Clin.* **2013**, *24*, 265–276.
6. Hsu, A.L.; Tang, P.F.; Jan, M.H. Analysis of Impairments Influencing Gait Velocity and Asymmetry of Hemiplegic Patients after Mild to Moderate Stroke. *Arch. Phys. Med. Rehabil.* **2003**, *84*, 1185–1193.
7. Kim, C.M.; Eng, J.J. The Relationship of Lower-extremity Muscle Torque to Locomotor Performance in People with Stroke. *Phys. Ther.* **2003**, *83*, 49–57. [PubMed]
8. Nadeau, S.; Arsenault, A.B.; Gravel, D.; Bourbonnais, D. Analysis of the Clinical Factors Determining Natural and Maximal Gait Speeds in Adults with A Stroke1. *Am. J. Phys. Med. Rehabil.* **1999**, *78*, 123–130. [CrossRef]
9. Eng, J.J.; Chu, K.S.; Dawson, A.S.; Kim, C.M. Hepburn KE. Functional Walk Tests in Individuals with Stroke: Relation to Perceived Exertion and Myocardial Exertion. *Stroke* **2002**, *33*, 756–761.
10. Daly, J.; Nethery, J.; McCabe, J.; Brenner, I.; Rogers, J.; Gansen, J.; Butler, K.; Burdsall, R.; Roenigk, K.; Holcomb, J. Development and Testing of the Gait Assessment and Intervention Tool (GAIT): A Measure of Coordinated Gait Components. *J. Neurosci. Methods* **2009**, *178*, 334–339.
11. Toro, B.; Nester, C.; Farren, P. A Review of Observational Gait Assessment in Clinical Practice. *Physiother. Theory Pract.* **2003**, *19*, 137–149.
12. Wallmann, H.W. Introduction to Observational Gait Analysis. *Home Health Care Manag. Pract.* **2009**, *22*, 66–68.
13. Ferrarello, F.; Bianchi, V.A.M.; Baccini, M.; Rubbieri, G.; Mossello, E.; Cavallini, M.C.; Marchionni, N.; Di Bari, M. Tools for Observational Gait Analysis in Patients with Stroke: A systematic review. *Phys. Ther.* **2013**, *93*, 1673–1685. [PubMed]
14. Turani, N.; Kemiksizoglu, A.; Karatas, M.; Özker, R. Assessment of Hemiplegic Gait using the Wisconsin Gait Scale. *Scand. J. Caring Sci.* **2004**, *18*, 103–108. [PubMed]
15. Nishi, Y.; Iluno, K.; Takamura, Y.; Minamikawa, Y.; Moriloa, S. Modeling the Heterogeneity of Post-Stroke Gait Control in Free-Living Environments Using a Personalized Causal Network. *IEEE Trans. Neural Syst. Rehabil. Eng.* **2024**, *32*, 3522–3530.
16. Faisal, A.I.; Deen, M.J. Systematic Development of a Simple Human Gait Index. *IEEE Rev. Biomed. Eng.* **2024**, *17*, 229–242.
17. Faisal, A.I.; Majumder, S.; Scott, R.; Mondal, T.; Cowan, D.; Deen, M.J. A Simple, Low-cost Multi-sensor-Based Smart Wearable Knee Monitoring System. *IEEE Sens. J.* **2021**, *21*, 8253–8266.
18. Lencioni, T.; Carpinella, I.; Rabuffetti, M.; Marzegan, A.; Ferrarin, M. Human Kinematic, Kinetic and EMG Data During Different Walking and Stair Ascending and Descending Tasks. *Sci. Data* **2019**, *6*, 309.
19. Nadeau, S.; Duclos, C.; Bouyer, L.; Richards, C.L. Guiding task-oriented gait training after stroke or spinal cord injury (SCI) by means of a biomechanical gait analysis. In *Progress in Brain Research*; Green, A.M., Chapman, C.E., Kalaskav, J., Lepore, F., Eds.; Elsevier: Amsterdam, The Netherlands, 2011; pp. 161–180.
20. Harris, G.F.; Benson, L.; Riedel, S.; Acharaya, K.; Matesi, D.; Millar, E.A. Biomechanical Assessment Techniques in Rehabilitation. In Proceedings of the 1990 IEEE Colloquium in South America, Argentina, Brazil, Chile, Uruguay, 31 August–15 September 1990; pp. 31–34.
21. Chen, C.J. Gait Analysis—Comparison of Wearable and Non-Wearable Device Systems. Available online: https://mag.longgood.com.tw/ (accessed on 30 September 2022).
22. Tunca, C.; Pehlivan, N.; Ak, N.; Arnrich, B.; Salur, G.; Ersoy, C. Inertial Sensor-Based Robust Gait Analysis in Non-Hospital Settings for Neurological Disorders. *Sensors* **2017**, *17*, 825. [CrossRef]
23. Lee, H.; Guan, L.; Burne, J.A. Human Gait and Posture Analysis for Diagnosing Neurological Disorders. In Proceedings of the 2000 IEEE International Conference on Image Processing, Vancouver, BC, Canada, 10–13 September 2000; Volume 2, pp. 435–438.
24. Lau, H.; Tong, K.; Zhu, H. Support Vector Machine for Classification of Walking Conditions of Persons after Stroke with Dropped Foot. *Hum. Mov. Sci.* **2009**, *28*, 504–514.
25. Scheffer, C.; Cloete, T. Inertial Motion Capture in Conjunction with an Artificial Neural Network Can Differentiate the Gait Patterns of Hemiparetic Stroke Patients Compared with Able-bodied Counterparts. *Comp. Methods Biomech. Biomed. Eng.* **2012**, *15*, 285–294. [CrossRef]
26. Zhou, Y.; Romijnders, R.; Hansen, C.; van Campen, J.; Maetzler, W.; Hortobágyi, T.; Lamoth, C.J.C. The Detection of Age Groups by Dynamic Gait Outcomes Using Machine Learning Approaches. *Sci. Rep.* **2020**, *10*, 4426.
27. Iosa, M.; Capodaglio, E.; Pelà, S.; Persechino, B.; Morone, G.; Antonucci, G.; Paolucci, S.; Panigazzi, M. Artificial Neural Network Analyzing Wearable Device Gait Data for Identifying Patients with Stroke Unable to Return to Work. *Front. Neurol.* **2021**, *12*, 561.
28. Seo, J.W.; Kim, S.G.; Kim, J.I.; Ku, B.; Kim, K.; Lee, S.; Kim, J.U. Principal Characteristics of Affected and Unaffected Side Trunk Movement and Gait Event Parameters during Hemiplegic Stroke Gait with IMU Sensor. *Sensors* **2020**, *20*, 7338. [CrossRef]

29. Anwary, A.R.; Yu, H.; Vassallo, M. An Automatic Gait Feature Extraction Method for Identifying Gait Asymmetry Using Wearable Sensors. *Sensors* **2018**, *18*, 676. [CrossRef]
30. Luo, B.; Wang, Z.; Wang, D.; Chen, L.; Ma, X. Algorithm for Gait Parameters Estimation Based on Heel-Mounted Inertial Sensors. *IEEE Sens. J.* **2024**, *24*, 24723–24736.
31. Lin, C.H.; Peng, C.W.; Lee, I.J.; Wu, C.M.; Lin, B.S. Analysis of Gait Data Based on Deep Learning to Predict Multiple Balance Scale Scores. *IEEE Sens. J.* **2024**, *24*, 920–930.
32. Rokhmanova, N.; Pearl, O.; Kuchenbecker, K.J.; Halilaj, E. IMU-Based Kinematics Estimation Accuracy Affects Gait Retraining Using Vibrotactile Cues. *IEEE Trans. Neural Syst. Rehabil. Eng.* **2024**, *32*, 1005–1012. [PubMed]
33. Guo, Y.; Hutabarat, Y.; Owaki, D.; Hayashibe, M. Speed-Variable Gait Phase Estimation During Ambulation via Temporal Convolutional Network. *IEEE Sens. J.* **2024**, *24*, 5224–5236.
34. Sole Motion. Available online: https://solemotion.com.au/the-video-gait-analysis/ (accessed on 11 January 2025).
35. NGIMU. Available online: https://x-io.co.uk/ngimu/ (accessed on 11 January 2025).
36. Daly, J.J.; McCabe, J.P.; Gor-García-Fogeda, M.D.; Nethery, J.C. Update on an Observational, Clinically Useful Gait Coordination Measure: The Gait Assessment and Intervention Tool (G.A.I.T.). *Brain Sci.* **2022**, *12*, 1104. [CrossRef]

sensors

MDPI

Article

An Innovative IoT and Edge Intelligence Framework for Monitoring Elderly People Using Anomaly Detection on Data from Non-Wearable Sensors

Amir Ali [1], Teodoro Montanaro [1], Ilaria Sergi [1], Simone Carrisi [2], Daniele Galli [2], Cosimo Distante [1,3] and Luigi Patrono [1,*]

1 Department of Engineering for Innovation, Università del Salento, 73100 Lecce, Italy; amir.ali@unisalento.it (A.A.); teodoro.montanaro@unisalento.it (T.M.); ilaria.sergi@unisalento.it (I.S.); cosimo.distante@cnr.it (C.D.)

2 Gematica s.r.l., 73100 Lecce, Italy; s.carrisi@gematica.com (S.C.); d.galli@gematica.com (D.G.)

3 Institute of Applied Sciences and Intelligent Systems of CNR, 73100 Lecce, Italy

* Correspondence: luigi.patrono@unisalento.it

Abstract: The aging global population requires innovative remote monitoring systems to assist doctors and caregivers in assessing the health of elderly patients. Doctors often lack access to continuous behavioral data, making it difficult to detect deviations from normal patterns when elderly patients arrive for a consultation. Without historical insights into common behaviors and potential anomalies detected with unobtrusive techniques (e.g., non-wearable devices), timely and informed medical interventions become challenging. To address this, we propose an edge-based Internet of Things (IoT) framework that enables real-time monitoring and anomaly detection using non-wearable sensors to assist doctors and caregivers in assessing the health of elderly patients. By processing data locally, the system minimizes privacy concerns and ensures immediate data availability, allowing healthcare professionals to detect unusual behavioral patterns early. The system employs advanced machine learning (ML) models to identify deviations that may indicate potential health risks. A prototype of our system has been developed to test its feasibility and demonstrate, through the application of two of the most frequently used ML models, i.e., isolation forest and Long Short-Term Memory (LSTM) networks, that it can provide scalability, efficiency, and reliability in the context of elderly care. Further, the provided dashboard enables caregivers and healthcare professionals to access real-time alerts and longitudinal trends, facilitating proactive interventions. The proposed approach improves healthcare responsiveness by providing instant insights into patient behavior, facilitating more accurate diagnoses and interventions. This study lays the groundwork for future advancements in the field and offers valuable insights for the research community to harness the full potential of combining edge computing, artificial intelligence (AI), and the IoT in elderly care.

Keywords: IoT; healthcare; elderly; edge computing; anomaly detection; real-time health monitoring; non-wearable

1. Introduction

As the global population continues to age, with projections indicating that the number of people aged 65 and over will nearly double by 2050 [1], the need for innovative elderly care solutions is becoming increasingly important. Ensuring the safety and well-being of elderly individuals living independently is paramount, as this demographic shift poses significant challenges to healthcare systems worldwide [2]. Effective monitoring systems are crucial for detecting early signs of potential health issues and enabling timely interventions to prevent serious incidents, thereby reducing the burden on healthcare resources. In this context, the absence of continuous behavioral data significantly impedes the ability of doctors to detect deviations from normal patterns when elderly patients arrive for a consultation, making informed medical interventions challenging [3].

However, existing monitoring systems, predominantly relying on cloud computing for data analysis and storage [4,5], face significant limitations. These systems offer scalability but introduce challenges such as latency in response times, which can be detrimental in situations where immediate intervention is necessary [6]. Furthermore, privacy concerns are significantly heightened with cloud computing [7], as the transmission and storage of sensitive health data offsite pose risks of unauthorized access and breaches. Moreover, the current landscape of elderly monitoring solutions often incorporates wearable devices to track health metrics and activity levels. Despite their potential, these devices frequently encounter barriers related to user compliance and comfort. Elderly individuals may neglect to wear such devices consistently due to discomfort or forgetfulness, leading to gaps in monitoring and data collection, which undermine the reliability of health assessments [8].

Addressing these multifaceted challenges, this paper presents an Internet of Things (IoT) system that utilizes a combination of the IoT, AI, and edge computing techniques for monitoring and supporting elderly adults and their caregivers in smart home environments. Unlike traditional approaches, our system leverages non-wearable sensors integrated throughout the home environment to ensure unobtrusive surveillance without requiring any active effort from elderly individuals, thereby circumventing the compliance issues associated with wearables. We focus on developing an edge-based framework that processes data locally within the smart home environment. This approach enhances data privacy and security and minimizes latency, allowing for immediate analysis and response. A prototype of our system has been developed to test its feasibility and demonstrate that it can provide scalability, efficiency, and reliability in the context of elderly care. The developed prototype utilizes existing datasets to experiment with how anomaly detection techniques can help doctors in their daily routines through a dashboard that displays the collected data and predictions. By employing two of the most frequently used ML models for anomaly detection, i.e., isolation forest and LSTM networks [9,10], our experiments effectively identify behavioral deviations that may indicate potential health risks.

The selection of the two models was guided by their characteristics: the isolation forest algorithm is an advanced ML technique known for its effectiveness in identifying outliers within intricate, multidimensional datasets. It excels at detecting subtle yet critical anomalies that conventional techniques frequently miss. Additionally, we incorporate an LSTM network, which is known for its ability to recognize temporal dependencies in sequential data. By analyzing patterns over time, LSTM networks enable the prediction of health issues based on identified anomalies. Moreover, a real-time monitoring dashboard accessible to caregivers and medical professionals provides immediate alerts and comprehensive updates about the patient's status. This dashboard facilitates prompt and informed decision-making, improving response times and ultimately enhancing the quality of care for elderly people.

This work provides the foundation for future developments in the field and provides the research community with a valuable basis for fully leveraging the advantages of combining edge computing, AI, the IoT, and ML.

The main contributions of this work are as follows:

- We propose a non-wearable sensor-based IoT system in smart homes, eliminating the need for wearable devices and ensuring continuous and uninterrupted data collection.

- We design an edge-centric architecture that processes data locally within smart home environments, reducing latency and enhancing security while safeguarding patient privacy.
- We introduce an advanced anomaly detection approach for temporal analysis to identify subtle health risk indicators before they become critical issues.
- We engineer a real-time monitoring dashboard that provides immediate alerts and detailed health status updates, facilitating rapid and informed decision-making.
- We demonstrate the scalability and reliability of a functional prototype in real-world elderly care scenarios through extensive testing performed using isolation forest for spatial analysis and LSTM networks applied to existing datasets.

The remainder of this paper is organized as follows. Section 2 analyzes related works, while Section 3 presents the proposed architecture. Section 4 discusses the implementation details that could guide future or similar works. Section 5 reports the details of the experiments, and Section 6 discusses the results obtained from validating the proposed system. Finally, Section 7 concludes this paper and discusses future work.

2. Related Works

The field of smart home technology for elderly care has advanced considerably in recent decades, especially in systems that monitor daily activities to identify health risks. Traditional methods often rely on wearable devices to collect data and cloud AI-based solutions for their processing. Such solutions usually face challenges related to data privacy, due to the risks of sharing data over the internet; latency, due to the time needed to transfer data to the cloud and wait for results; and user compliance issues, due to the unwillingness of elderly individuals to wear devices. In response, recent innovations in non-wearable sensor networks integrated with edge computing frameworks offer enhanced privacy and real-time data processing. However, many of these systems do not focus on utilizing anomaly detection techniques applied to non-wearable data for health monitoring, which is the main focus of our research. Various studies have already been proposed in the literature that utilize ML models to process non-wearable data acquired in the home environment for health monitoring.

For instance, in the study by Paudel et al. [11], the authors explored anomaly detection to monitor the activities of elderly individuals in smart homes using a graph-based approach. They utilized the Kyoto dataset from the Center for Advanced Studies in Adaptive Systems (CASAS) [12] program, mapping daily routines via non-wearable sensors like motion, temperature, and door sensors. By examining activation frequency, duration, and spatial patterns, they identified deviations from normal behavior. Despite providing detailed insights, the study faced limitations, including cloud processing burdens, privacy concerns, and scalability issues due to complex graph structures and extensive sensor data. Our approach tackles these challenges by using lightweight algorithms like isolation forest for efficient anomaly detection without complex graphs, thereby enhancing system performance and scalability. In addition, through the utilization of edge computing, our work addresses the privacy and latency issues experienced by other authors.

Another interesting study was conducted by Shahid et al. [13]. It explored the use of anomaly detection algorithms in elderly care using non-wearable IoT devices in single-resident apartments. The study tracked kitchen and bathroom visits to establish behavioral norms, detect anomalies from deviations, and notify caregivers via SMS. Despite its successes, cloud processing delays hindered real-time analysis, and reliance on long-term norms reduced responsiveness to sudden activity changes. In contrast, our approach enables the use of advanced ML techniques, such as the isolation forest and LSTM models, to enhance adaptability and enable real-time detection of both static outliers and dynamic changes. Furthermore, using edge computing ensures responsive, privacy-preserving elderly monitoring and health risk detection. Gupta et al. [14] employed an LSTM-based neural network for fall detection in smart homes. However, its narrow focus limited health monitoring, and its reliance on cloud-based infrastructure introduced latency. Our work extends beyond fall detection by integrating sensors for sleep, motor function, and social interaction (i.e., periods of reduced engagement and low interaction with others that may signal a need for intervention), which are important indicators of cognitive health. Edge computing ensures local processing, thereby reducing latency and enhancing privacy. This enables real-time anomaly detection for comprehensive and prompt elderly care, unlike Gupta's fall-focused model, offering an expanded perspective on elderly monitoring.

Novak et al. [15] proposed an anomaly detection approach for elderly care in smart homes using non-wearable sensors like PIR and pressure sensors. Their methodology combines Self-Organizing Maps (SOMs), an unsupervised learning method for clustering and visualization on the cloud [16], and Markov models [17] to predict subsequent activities. By analyzing activities, location, timing, and duration, it establishes normal behavior and identifies deviations, which are crucial for monitoring the well-being of elderly residents. However, while the approach leverages the non-intrusive nature of sensors, it faces challenges in real-time anomaly detection and has limitations in the dataset's applicability for elderly monitoring. In comparison, our system is designed to utilize various ML algorithms on the edge and collect data from various devices in real time.

Pérez-Toro et al. [18] advanced Alzheimer's assessment using ForestNet, a hybrid ML framework combining decision trees and neural networks to analyze acoustic features. It achieved high recall in patient classification while identifying key diagnostic markers like rhythm and voice rates. Its interpretability provides clinicians with clear and actionable insights, bridging the gap between complex ML models and clinical practice. The study faced limitations, such as balancing interpretability with high predictive accuracy. However, although the study aimed to ensure privacy when sending data to the cloud, it did not utilize an edge system. Therefore, the system proposed in this paper could enhance reliability through edge computing and offer greater flexibility by enabling the integration of additional sensors and devices for real-time data collection.

Susnea et al. [19] used non-wearable sensors, such as Passive Infrared (PIR) motion detectors and magnetic door contacts, to monitor the activities of elderly individuals living alone. These sensors are integrated into a smart home environment to create activity maps that reflect spatial distribution and intensity. Their method uses encoded sensors to detect deviations from behavioral routines tested on the CASAS, HH126, and Kasteren House C datasets. The approach highlights the system's low cost, scalability, adaptability, minimal expert knowledge required for setup, and robust privacy protection; however, it is limited by the need for cloud infrastructure to process the data. Our system enhances this approach by leveraging edge computing and integrating additional sensors and devices to collect real-time data. Furthermore, it tests different ML techniques and incorporates previously unreported performance evaluations.

Sadhu et al. [20] demonstrated the potential of using ML applications on edge and fog computing architectures for remote assessment of Parkinson's disease. Their study uses smart gloves with flex sensors and an inertial measurement unit to capture hand movements, which are processed by ML classifiers for accurate, low-latency Parkinsonian movement classification. The edge–fog approach effectively addresses the latency and privacy concerns often associated with cloud-based telehealth systems, marking a significant advancement in real-time motor symptom monitoring. However, their approach remains limited to Parkinsonian symptom tracking, lacking a broader focus on cognitive health assessment. In addition, the use of an intrusive wearable device limits the effectiveness of the proposed solution.

Furthermore, Mokhtari et al. [21] proposed a hierarchical federated learning-based anomaly detection method for smart healthcare. The integration of edge computing with wearable and medical devices was explored to enhance data privacy and reduce latency in healthcare applications. While their research highlighted the advantages of using edge computing for anomaly detection, it was limited to the use of wearable devices. Our work extends these concepts specifically to non-wearable sensors within smart home environments, focusing on elderly care. The publication inspired our work, particularly in utilizing edge-based processing, enabling the efficient and secure detection of potential health anomalies and meeting the growing need for decentralized healthcare monitoring solutions.

Another interesting study that focused on the application of edge computing for human activity recognition was conducted by Singh et al. [22]. Their work presented a Human Activity Recognition (HAR) system employing a Convolutional Neural Network (CNN) designed for edge devices. Their approach employs wearable sensors, which efficiently process sensor data locally on edge devices, significantly reducing latency and enhancing data privacy. However, the utilization of wearable devices limited the effectiveness of the performed estimations.

Another notable study was conducted by Wang et al. [23]. After conducting a comprehensive analysis of the state of the art, demonstrating the need for frameworks to facilitate the application of ML, the IoT, and edge computing to the healthcare sector, the authors proposed an edge-based framework to process data from wearable sensors, ensuring timely detection of health anomalies and reducing latency in critical healthcare applications. The lack of support for non-wearable devices and the exclusive use of deep learning techniques for ML analysis limits the applicability of their framework to scenarios involving only wearable devices. However, it motivated us to incorporate support for non-wearable devices and apply lightweight anomaly detection techniques directly on the edge to detect anomalous behaviors in elderly patients.

The study conducted by Rancea et al. [24] further motivated our work. The authors presented an overview of the current landscape of edge computing intersecting with the healthcare domain, highlighting its role in providing innovative solutions in the context of the IoT revolution. The study discussed the critical role that edge-based systems play in the healthcare sector and underscored the need for real-time data processing and rapid decision-making in AI-optimized edge computing-based healthcare systems. In addition, it gave an overview of the advantages that solutions similar to the one proposed in this paper can offer in monitoring patients and providing prompt notifications to provide doctors with the necessary information for making well-informed decisions.

Additionally, Liu et al. [25] acknowledged the broader landscape of edge computing in healthcare by demonstrating significant improvements in energy efficiency within edge computing frameworks. Their work pursued a green computing design and formulated an optimization problem that minimizes the long-term energy consumption of the proposed network, reduces computing delays, and minimizes data requirements. A combination of their solution with the one proposed in this paper could significantly reduce computing delays and data requirements while preserving privacy and related characteristics that we demonstrated to be highly important in the healthcare domain.

Finally, it is worth mentioning the review presented by Punugoti et al. [26], who explored healthcare IoT and edge computing's benefits and drawbacks. Through a literature review, the authors demonstrated that edge computing can improve healthcare delivery, costs, and outcomes. In addition, they evaluated edge computing's enabling technologies, applications, and obstacles in healthcare and discussed possible solutions. As highlighted by the authors, existing research lacks solid empirical evaluations of proposed solutions. Furthermore, the proposed solutions do not seem ready to integrate real-time monitoring and ML techniques for data from non-wearable devices, as demonstrated in this paper. Therefore, the review further motivated our work and informed our suggestions for promising directions for future work, which will be discussed in the conclusions.

3. System Architecture

This research introduces a novel edge-based Internet of Things (IoT) framework that enables real-time monitoring and anomaly detection using non-wearable sensors to assist doctors and caregivers in assessing the health of elderly patients. Specifically, it leverages edge computing supported by cloud computing to locally process data using anomaly detection techniques and provide innovative services (e.g., a dashboard for doctors and an application for caregivers).

The envisioned system architecture, as shown in Figure 1, is divided into two main blocks: an edge block and a cloud block, each tailored to specific functionalities within the healthcare monitoring system. The edge block handles local data acquisition, preprocessing, and analysis, ensuring minimal latency and data privacy. Meanwhile, the cloud block provides long-term storage, trend analysis, and user interfaces for stakeholders such as doctors and caregivers. This modular and scalable architecture is designed to adapt to evolving technologies and growing requirements in elderly care. By leveraging local processing to reduce latency and cloud computing for storage and analytics, the system ensures timely, secure, and effective monitoring. The following subsections detail the components and roles of each block.

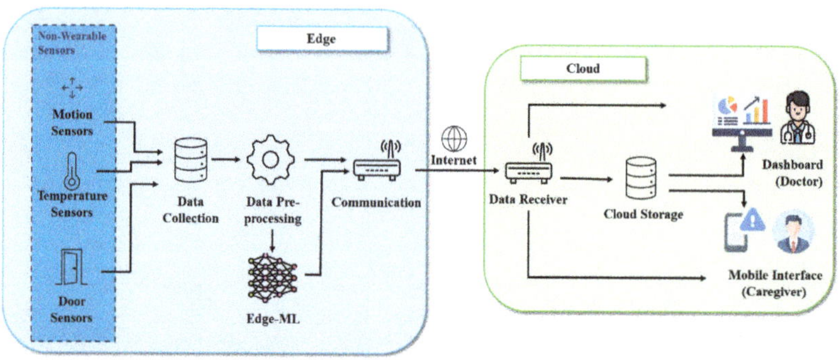

Figure 1. System architecture.

3.1. Edge Block

The edge block is the cornerstone of real-time operations in the system. It processes data locally, enabling swift anomaly detection and immediate responses while ensuring data privacy by reducing the need to transmit raw data over external networks. This modular block allows seamless integration of additional sensors, algorithms, or components without significant architectural changes. Therefore, the edge block is responsible for all operations that must be performed on the raw data, including processing. Local processing of data can guarantee precise predictions (which may only be possible with the collected raw data) and personalization, which are necessary for contacting the cloud block to display alerts or inferences to doctors and caregivers. This block consists of various modules: "non-wearable sensors" (such as motion sensors, temperature sensors, and door sensors), "data collection", "data preprocessing", "edge-ML", and "communication" modules, all operating at the edge of the network.

3.1.1. Non-Wearable Sensors

A home environment is typically equipped with various non-wearable sensors, including environmental sensors, location tracking systems, temperature monitors, motion detectors, and activity monitoring systems. These sensors continuously collect data on daily activities and health parameters. Our architecture leverages these sensors to collect environmental data, enabling continuous, non-intrusive monitoring tailored for real-time health and activity analysis. The use of environmental sensors provides a reliable and unobtrusive alternative to wearable devices, eliminating the dependency on user compliance. Wearables often require charging, regular use, and proper maintenance, which may not be practical for elderly individuals or those with physical or cognitive limitations. By focusing on environmental data, our system ensures continuous monitoring while reducing the burden on the user. Environmental sensors capture data passively and continuously, offering a broader perspective compared to wearable devices. While wearables primarily provide individual metrics like heart rate or step count, environmental sensors monitor the entire living environment. For instance, a motion sensor can detect whether an individual has not moved from one room to another for an unusually long time, which may indicate a potential problem. Similarly, door sensors can track exits to ensure safety, particularly for individuals with dementia who may wander. Additionally, temperature sensors can identify unsafe environmental conditions, such as a dangerously cold room. These capabilities make non-wearable sensors a holistic and practical solution for comprehensive monitoring, especially in sensitive applications like elderly care. The flexibility of our architecture lies in its modular design, which allows seamless integration of additional devices as they become available, whether real or simulated. Real devices installed in a home can send live data directly to the system, while simulated devices can be utilized during testing or in scenarios where physical installation is not feasible. This adaptability ensures that the system remains scalable, future-proof, and capable of incorporating emerging technologies.

3.1.2. Data Collection

The data collected by the sensors are first aggregated through the data collection module, which ensures that all relevant information is captured and prepared for further processing. This module acts as the initial interface between the sensors and the rest of the system, gathering data from various sources such as motion, temperature, door, and environmental sensors. To support future adaptability, the data collection module is designed as a stand-alone sub-framework capable of integrating new bundles at runtime (e.g., by integrating and adapting existing frameworks such as the Domotic OSGi Framework (DOG) [27] as an external module or by developing an innovative internal module) to handle a variety of communication protocols (such as MQTT, HTTP, or CoAP) and data formats (such as JSON, XML, or binary formats). Its intrinsic modularity ensures compatibility with both existing and new devices. This capability allows it to seamlessly integrate with sensors using different standards and technologies. The module also performs preprocessing tasks, including filtering noise, standardizing data formats, and applying basic validation to ensure that only accurate and meaningful data are forwarded to the next stage. By supporting multiple protocols and formats, the data collection module is a versatile and future-proof component, enabling the system to scale and adapt to emerging technologies and new devices as they become available.

3.1.3. Data Preprocessing

The data preprocessing module is a critical component designed to preprocess incoming sensor data in real time, ensuring it is structured and ready for further analysis. This module is responsible for filtering out noise, normalizing data from multiple sources, and extracting essential features to facilitate accurate anomaly detection. Its primary role is to transform raw, unprocessed data into a clean and interpretable format for subsequent machine learning analysis. This module is designed to handle high volumes of data efficiently, supporting the real-time nature of the system. It is scalable to accommodate additional sensors and adaptable to evolving requirements, ensuring compatibility with diverse data formats and structures. By optimizing data quality and reducing computational complexity, the data preprocessing module ensures that downstream modules, such as the "edge-ML" and the modules in the cloud, can operate effectively. The advantages of this module include its high throughput, reliability, and ability to process data without introducing latency. These characteristics make it a foundational part of the architecture, supporting the goal of the system of real-time, unobtrusive elderly care monitoring by enabling seamless integration of environmental sensor data, rapid preprocessing, and continuous anomaly detection.

3.1.4. Edge-ML

The edge-ML module is designed to analyze preprocessed data to detect patterns and deviations from normal behavior. Its primary role is to identify potential anomalies in the data, which may indicate health risks or unusual activity patterns in the monitored environment. By processing data outputs from the data preprocessing module, the edge-ML module evaluates the resident's behavioral trends and flags irregularities for further action. In addition to detecting anomalies, it generates real-time notifications based on identified irregularities or critical events. These notifications are routed through the communication module to the cloud infrastructure and are displayed on the monitoring "dashboard" and "mobile interface", ensuring that caregivers and healthcare providers are promptly informed of potential issues. This integration enables timely alerts and actionable insights to be delivered to caregivers and healthcare providers, facilitating prompt intervention when necessary. The design of this module ensures that it is adaptable to various types of data and capable of handling complex patterns, supporting the system's overall goal and enhancing decision-making for elderly care.

3.1.5. Communication

The communication module facilitates the flow of data between local processing units and external systems in the cloud block. It ensures that only relevant and processed data are securely transmitted to the cloud storage, reducing the exposure of sensitive raw data over external networks. In addition to data transmission, the communication module is responsible for routing real-time notifications generated by the edge-ML module to the dashboard and the mobile interface. This guarantees that caregivers and healthcare providers are promptly informed of detected anomalies or critical health events, enabling timely intervention. By efficiently managing both data and notifications, the communication module acts as a critical hub for real-time and secure communication within the system.

3.2. Cloud Block

The cloud block is responsible for providing long-term storage, trend analysis, and user interfaces for stakeholders such as doctors and caregivers. It is responsible for all operations that do not require access to the raw data. The cloud block displays alerts and inferred information communicated by the edge block after the processing of the collected raw data. This block consists of different modules: "data receiver", "cloud storage", "dashboard", and "mobile interface" modules, all operating in the cloud.

3.2.1. Data Receiver

The data receiver is the first module of the cloud block and acts as the gateway for data transitioning from the edge to the cloud. It ensures the secure, efficient, and reliable reception of data streams transmitted over the network. Upon receiving the data, the data receiver performs integrity and validation checks to verify that the data have not been corrupted or altered during transmission. These checks are essential for maintaining the reliability and trustworthiness of the data, which serve as the foundation for subsequent analyses and healthcare interventions. Beyond validation, the data receiver is responsible for sorting and filtering incoming data based on predefined parameters, such as data type, source, and priority. This organization optimizes the data management pipeline and ensures that only relevant and actionable information is processed or stored. The data receiver operates within an event-driven architecture, enabling the distribution of data to various cloud-based services and applications dynamically. This dynamic distribution is crucial for triggering immediate responses, such as sending alerts to caregivers or updating the dashboard in real time. By supporting a service-oriented approach, the data receiver enables different components within the cloud infrastructure to subscribe to specific data streams. This ensures the efficient dissemination of actionable insights to all relevant stakeholders, transforming raw data into meaningful outputs that drive healthcare monitoring and interventions.

3.2.2. Cloud Storage

The cloud storage is designed to store processed data for long-term archival and analysis. Its primary role is to maintain an organized repository of health and activity data for trend analysis and historical reviews. Unlike the real-time operations occurring on the edge, cloud storage supports non-time-sensitive functions, such as generating insights from long-term data and supporting data visualization tools. The data stored in this module can be accessed by the dashboard, where caregivers and healthcare providers review trends and patterns to make data-driven decisions. By offloading archival functions to the cloud, the system ensures scalability and supports future expansion to accommodate increasing data volumes.

3.2.3. Dashboard (Doctor) and Mobile Interface (Caregiver)

The dashboard and the mobile interface work together to provide comprehensive and accessible monitoring for both healthcare providers and caregivers. The dashboard, designed for doctors, offers a detailed and data-rich platform to track the resident's health and activity patterns in real time. It displays anomaly alerts generated by the edge-ML module, such as irregular activity patterns, prolonged inactivity, or unsafe environmental conditions, along with contextual details like timing and severity. In addition to these alerts, the dashboard presents health status metrics to monitor cognitive decline over time. For the experiments described in the next section, these metrics include motor function, evaluated using sitting-to-walking ratios; sleep quality, analyzed through wake-up patterns and disturbances; and social interactions, assessed based on home occupancy trends and room transitions, with periods of reduced engagement and low interactions with others potentially signaling a need for intervention. The dashboard's ability to visualize historical data allows healthcare providers to identify trends, evaluate long-term cognitive changes, and adjust care plans proactively. The mobile interface, on the other hand, is tailored for caregivers and focuses on simplicity and immediacy. It delivers real-time notifications about detected anomalies, such as inactivity, unusual motion, or hazardous environmental conditions, ensuring that caregivers are promptly informed and able to respond quickly. While it also provides health status metrics, these are presented in a more simplified format compared to the dashboard. In our scenario, as described in the next sections, caregivers can check the current sitting-to-walking ratio for mobility insights, recent sleep disturbances for assessing rest patterns, and updates on home occupancy trends for understanding social interaction. Designed for on-the-go access, the mobile app allows caregivers to stay connected and informed regardless of their location, prioritizing quick actions and immediate interventions.

4. Implementation

This section provides a comprehensive exploration of the methodologies used to develop a cutting-edge IoT system implementing the defined system architecture. This includes details of the various modules in the edge and cloud blocks, such as the communication, data preprocessing, edge-ML, and cloud storage blocks, and their integration with the dashboard and the mobile interface.

To validate the feasibility of advanced analytics within the designed edge computing framework, we decided to concentrate most of the effort in this research paper on only some of the architecture blocks. Specifically, we concentrated on (a) the modules responsible for the ML algorithms and models utilized at the edge, such as the data preprocessing module and the edge-ML module, and (b) the cloud block implementing cloud storage and the dashboard. Therefore, we decided to utilize existing datasets to simulate the presence of real-time sensors that send data to the system while leaving the implementation of other blocks (e.g., mobile interface) for future work.

This preliminary phase is crucial for assessing the practicality of real-time anomaly detection and serves as a foundational step toward realizing the full potential of the proposed architecture. To ensure scalability, the system is designed to allow easy addition of more computational nodes at the edge.

4.1. Data Collection and Preprocessing

The data collection and preprocessing modules ensure that data are prepared for effective anomaly detection and ML applications. To simulate real-time sensor data, the TM029 dataset from CASAS [12] was chosen. This dataset comprehensively represents daily activities in a single-resident smart home environment, making it ideal for analyzing behavioral patterns and detecting anomalies. This dataset includes readings from the following sensors:

- Motion Sensors (ZigBee Round Motion Detectors WMS10-C-ZP [28]): These sensors capture activity levels and mobility, which are essential for understanding movement patterns and energy expenditure.
- Door Sensors (ZigBee Magnetic Contact/Temperature Sensor WCS10A-2-ZP [28]): These sensors monitor room transitions and occupancy, providing insights into room usage and time spent in different areas of the home.
- Temperature Sensors (ZigBee Magnetic Contact/Temperature Sensor WCS10A-2-ZP [28]): These sensors add contextual information about environmental conditions that may impact comfort, health, and overall activity.

The raw sensor data required extensive preprocessing to ensure that it was suitable for analysis and model training. Missing values in the dataset were addressed using forward-filling techniques [29], a method that maintained temporal continuity and prevented disruptions in downstream analysis, which is especially critical for time-series data. Forward-filling replaces missing data points with the most recent valid observation, ensuring that the integrity of time-series patterns is preserved. Several key features were derived from the preprocessed data to capture meaningful activity and behavioral trends. These included sitting duration, representing the time spent seated during the day; walking duration, which measured the total time spent moving within the home; wake-up count, capturing the frequency of awakenings during sleep periods; out-of-home duration, reflecting the time spent outside the home; and sleep duration, indicating total sleep time as a measure of rest quality and overall health. Table 1 summarizes these parameters and their derivations.

Table 1. Summary of original and derived parameters.

Original Parameter	Derived Parameter	Description
Sensor ID	Sitting Duration	Time spent seated during the day
Timestamp	Walking Duration	Total time spent moving within the home
Motion Sensor Readings	Wake-Up Count	Frequency of awakenings during sleep periods
Door Sensor Activity	Out-of-Home Duration	Time spent outside the home
Environmental Sensors (e.g., temperature)	Sleep Duration	Total sleep time as a measure of rest quality and overall health

These features were carefully selected for their relevance in identifying behavioral anomalies and potential health risks. To ensure consistency across all features, min-max scaling was applied [30]. This normalization technique transforms data into a range suitable for ML models, ensuring uniformity and comparability between features. The preprocessing tasks were implemented using Python [31], leveraging well-established libraries to ensure efficiency and accuracy. The "pandas" library [32] was utilized for efficient data manipulation and preprocessing, providing robust capabilities for handling structured datasets. Additionally, the NumPy library [33] was employed for numerical computations and array operations, enabling seamless transformations and calculations during the preprocessing phase. These tools ensured that the raw sensor data were efficiently transformed into a structured format suitable for analysis and ML applications.

4.2. Technology Integration and System Configuration

The core infrastructure of our IoT system is built upon a carefully selected technology stack, comprising Node.js [34,35] and MySQL [36,37], chosen for their proven reliability and suitability for real-time applications in elderly care monitoring. This combination ensures efficient data handling, robust storage, and scalability, critical for the system's performance in real-world conditions. Node.js [34] is at the heart of our data processing operations due to its highly efficient, non-blocking, event-driven architecture. This technology is particularly well suited for applications that require real-time capabilities, as it can handle numerous data streams concurrently without delays. In our system, Node.js [34] is a central component of the data processing layer, orchestrating data flow from various sensors scattered throughout the care environment. Positioned within our IoT infrastructure, Node.js [34] manages and processes numerous data streams concurrently without delays, ensuring that all data are processed swiftly and accurately. This rapid processing capability is crucial for the timely detection of anomalies and potential emergencies, enabling immediate response and intervention. Node.js [34] also facilitates the development of scalable network applications. Its lightweight runtime supports our system's requirement for high throughput and data-intensive handling, which are essential for monitoring the nuanced behaviors of elderly residents. Node.js [34] also facilitates the integration of real-time data processing with other system components, such as the database management system (MySQL [37]) and the user interface, thereby enhancing the overall responsiveness and efficiency of our elderly care monitoring solution. Furthermore, Node.js [34] integrates seamlessly with Express.js [38], which serves as the framework for incoming and outgoing data. This setup manages data flow from various sensors and data processing units, ensuring a smooth and scalable communication pipeline. Complementing Node.js [34], MySQL [37] provides robust data management capabilities. Chosen for its reliability and efficiency, MySQL stores and manages the vast quantities of data generated by the IoT sensors. It ensures efficient data retrieval with features like indexing, which speeds up the search process, and query optimization, reducing the time to execute database commands. MySQL handles structured data storage with high transactional requirements in our deployment, ensuring data integrity and security. The advanced features of MySQL, such as partitioning, are utilized to manage large tables and complex queries more efficiently. This functionality is particularly important given the continuous data influx from our monitoring devices. The modular nature of our system allows the easy integration or substitution of the MySQL database with other similar solutions, like NoSQL databases. However, the choice was guided by the requirements of our experiment: While semi-structured data from sensors could potentially be managed using no-SQL databases, we opted for MySQL due to its robust transactional support, which is crucial for ensuring data integrity and consistency in our anomaly detection processes. Additionally, MySQL's advanced querying capabilities allow for the precise and complex data manipulations necessary for the sophisticated analysis models our system employs, which are integral to accurately processing structured sensor data streams.

In addition, to ensure seamless communication between the sensors and the data processing unit, the system employs MQTT (Message Queuing Telemetry Transport), one of the most frequently used publish/subscribe protocols for the IoT [39]. MQTT was chosen for its widespread adoption in healthcare IoT solutions, where its lightweight protocol facilitates efficient real-time communication, which is crucial for the rapid transmission of sensor data to the system. This capability allows the system to detect and process anomalies, ensuring timely intervention. The choice of MQTT over alternatives such as CoAP was guided by its documented advantages in similar applications across the healthcare sector, providing the necessary reliability and efficiency for our system's needs [40].

The integration of Node.js [34], Express.js [38], MySQL [37], and MQTT ensures that the system operates efficiently, handling the demands of real-time data processing, secure storage, and high-speed communication. This technological foundation is designed to scale with future enhancements, such as adding more sensors or supporting additional users, ensuring that the system remains robust and reliable under varying operational loads.

4.3. Edge-ML Module: Machine Learning Models and Anomaly Detection Techniques

The anomaly detection component utilizes two machine learning algorithms: isolation forest [41] and Long Short-Term Memory (LSTM) [14]. These models were chosen among the most frequently used anomaly detection algorithms for their ability to complement each other in detecting both isolated anomalies and time-dependent patterns in behavioral data. Isolation forest was selected for its ability to effectively identify anomalies in high-dimensional data. The algorithm works by isolating data points through random partitioning, where anomalies are easier to isolate due to their distinct characteristics. Isolation forest is well suited for detecting outliers in datasets with multiple features, such as those related to motion, door, and temperature sensor data. It was chosen because of its efficiency in handling large datasets and its ability to detect anomalies without relying on prior knowledge of the data distribution, making it ideal for real-time applications in a smart home environment. The isolation forest model was configured with 100 trees (n_estimators = 100), using a contamination factor of 0.1, and its training was executed efficiently within a few seconds (an average of 4 s).

In addition, Long Short-Term Memory (LSTM) networks were employed to capture temporal dependencies in the data. LSTM networks are specifically designed to work with sequential data, which is crucial for identifying anomalies over time. Recent studies, such as [42], have demonstrated that LSTM-based models outperform traditional statistical methods in anomaly detection, particularly in healthcare applications, due to their ability to effectively model both short-term and long-term dependencies. In our case, behavioral patterns, such as changes in walking duration, sitting duration, or wake-up count, are best understood in a temporal context. The LSTM model was chosen because it excels at modeling time-series data, making it well suited for predicting future behavioral states and detecting deviations from expected patterns. The LSTM model was configured with two layers, each with 50 units, and was trained over 20 epochs with a batch size of 32. An early stopping mechanism was employed after five epochs without improvement to prevent overfitting. The entire training process was optimized to be completed in a few minutes (an average of 8 min).

Both models were trained on an 80/20 train-test split of the dataset, which means that 80% of the data were used for training the models and the remaining 20% for testing, ensuring robust evaluation. Early stopping (a technique used during training) was employed to prevent overfitting. This method monitors the model's performance on a validation set during training and stops the training process once the model performance ceases to improve, thus ensuring that the model generalizes well to new, unseen data. Input data for both models were normalized using min-max scaling to ensure consistency across features. Min-max scaling is a preprocessing technique that transforms the features into a fixed range, typically 0 to 1, by subtracting the minimum value of each feature and then dividing by its range, thereby enhancing predictive accuracy.

The isolation forest model is expected to effectively flag irregularities in individual data points, such as sudden changes in sitting or walking durations. Meanwhile, the LSTM network was designed to predict future activity sequences, with anomalies flagged when the predictions deviate significantly from actual behavior. This configuration allows the system to identify both isolated incidents of anomalous behavior and broader shifts in activity patterns that may indicate cognitive or physical decline.

To enhance the understanding of the anomaly detection pipeline, Figure 2 illustrates the data flow from sensor collection to preprocessing, feature engineering, machine learning, and visualization. The diagram provides a concise visual representation of how the isolation forest and LSTM models are integrated within the system, offering a clear view of the relationships and processing steps. As already stated, in our experiments, we used the CASAS dataset, in which data collection involved continuous monitoring through non-wearable sensors deployed in a smart home environment. These sensors gathered information on motion activity, room occupancy, temperature variations, and door usage, providing a rich dataset for behavioral pattern analysis. The preprocessing stage ensures data quality and consistency by applying noise filtering, missing value interpolation, and min-max normalization, which standardizes feature values for uniform input to the models. Following preprocessing, feature engineering extracts key metrics that aid in anomaly detection. This includes statistical features, such as mean, variance, and standard deviation; temporal features, such as activity duration and frequency of movement; and event-based features that track sensor activations over time. These engineered features serve as inputs to the anomaly detection models, each employing a distinct approach to identifying irregularities. The isolation forest model assigns an anomaly score to each data point based on how easily it can be isolated, effectively detecting sudden behavioral deviations like prolonged inactivity or unusually frequent movements. In contrast, the LSTM model leverages time-series predictions, learning behavioral patterns from historical data and identifying anomalies when actual behaviors significantly deviate from predicted sequences. The final phase of the pipeline is visualization, where the detected anomalies are represented through time-series plots, heatmaps, and behavioral trend graphs. These visual outputs enable medical professionals and caregivers to interpret changes in behavior over time, allowing for early intervention in case of potential health risks.

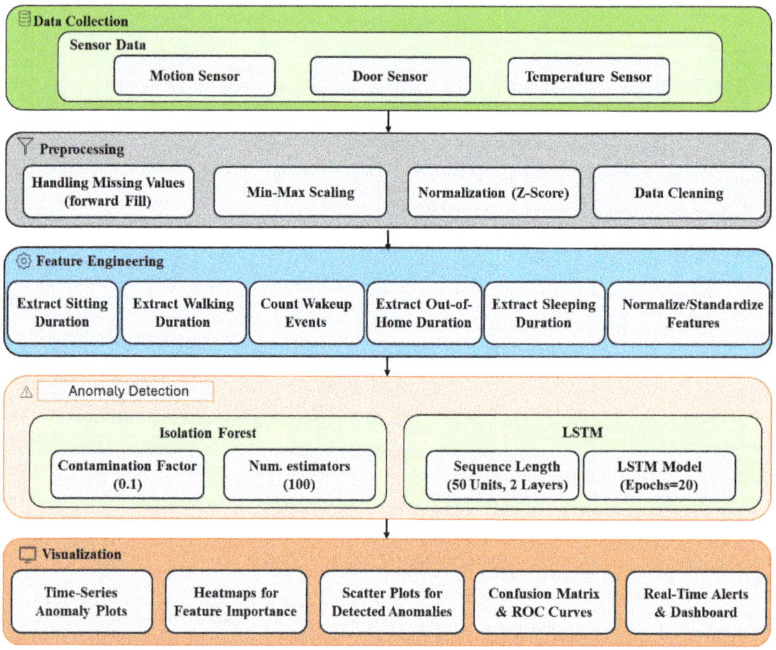

Figure 2. Data flow diagram illustrating the steps involved in anomaly detection, including sensor data collection, data preprocessing, feature engineering, machine learning, and visualization.

4.4. Dashboard Development and Integration

Our system features an interactive dashboard developed using React.js [43] to provide caregivers and doctors with real-time insights into the health and activities of monitored individuals. React.js was selected for its robust capability to handle dynamic updates and efficiently render complex interfaces, which are crucial for displaying live data streams from IoT sensors. This dashboard displays real-time data and alerts, enabling caregivers to promptly detect and address critical situations, thereby enhancing their response capabilities. Data management and routing are handled through Node.js [34] and Express.js [38], which efficiently manage data transmission between the IoT sensors and the dashboard. This setup ensures that data handling is both secure and robust, maintaining system integrity and responsiveness under various operational loads.

We prioritized user accessibility in the design of the dashboard to ensure that it remains intuitive and easy to navigate. This approach allows users of all technical skill levels to utilize the dashboard to make informed decisions quickly and effectively.

Figure 3 illustrates the developed interface of the monitoring dashboard, showcasing its capability to display live data and real-time alerts. The dashboard is designed not only to display health data but also to enable caregivers to immediately react to anomalies and emergencies, thus supporting proactive patient care. The architecture supports dynamic updates and real-time interactions without performance degradation, ensuring scalability. Prior to deployment, the system undergoes rigorous integration testing to confirm that all components function cohesively. This testing includes functional verification to ensure that each module operates correctly and interacts appropriately with other system parts, as well as performance assessments under simulated peak and normal conditions. These developments underscore the crucial role of the dashboard in our system, providing a reliable and effective tool for real-time health monitoring and emergency management in elderly care settings.

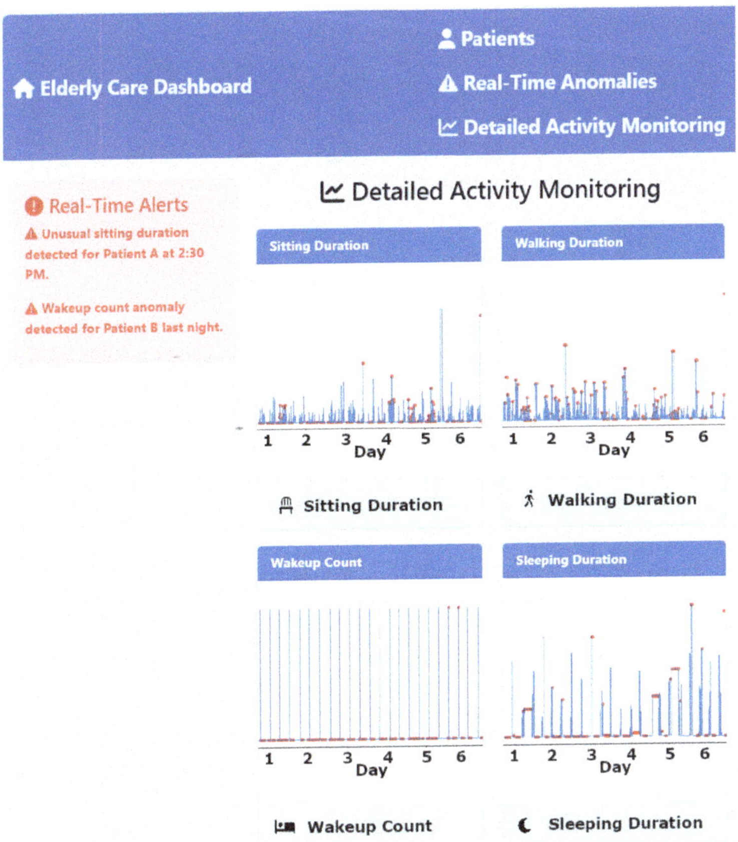

Figure 3. Real-time monitoring dashboard.

5. Experiments

This section rigorously examines the experiments conducted to test the implemented components of the IoT system designed to support elderly individuals and their caregivers through edge computing. The experiments demonstrate the system's real-time operational capabilities, its efficacy in anomaly detection, and the functionality of the user interface, which aids caregivers in making informed decisions.

5.1. Detailed Experimental Setup

Our experimental setup utilizes the advanced Raspberry Pi 4 Model B [44], an embedded low-cost computer for prototype experiments [45], as an edge device. It was selected for its robust processing power, which is essential for real-time data analytics and critical for immediate patient care responses at the edge. This edge computing device, equipped with a Quad-core Cortex-A72 (ARM v8) SoC at 1.5 GHz, 4 GB of RAM, and a 32 GB MicroSD card for data storage, manages extensive data processing on-site, minimizing latency crucial for timely interventions. The actual experimental data used in our study were sourced from the TM029 dataset provided by the CASAS project. This dataset includes comprehensive sensor data previously collected in real environments designed to monitor elderly individuals. The data encompass activities like cooking, eating, and sleeping, along with sensor activations such as motion detection, door usage, and temperature readings. To simulate a real-time operational environment, we employed a script that mimicked the presence of real devices. This script retrieves data from the dataset and sends it to our system for processing. The TM029 dataset was collected using non-wearable sensors, allowing us to test our system's data processing and analysis capabilities under realistic conditions. This approach ensures that our experimental validation reflects the practical application of the technology in real-world settings without the necessity for the physical deployment of sensors during the initial testing phase.

5.1.1. Data Selection and Utilization

The TM029 dataset from the CASAS [12] project, designed to monitor elderly individuals, provided comprehensive sensor data for this study. The dataset includes labeled activities, such as cooking, eating, and sleeping, along with raw sensor activations, such as motion detection, door usage, and temperature readings. These labels enable the system to differentiate between normal and anomalous behaviors.

We conducted a series of preprocessing steps to prepare the dataset for analysis. First, data cleaning was performed using custom scripts in Python to remove outliers and erroneous entries, ensuring data integrity. Subsequently, all numerical features were scaled using min-max normalization, implemented via the Scikit-Learn library, to maintain uniformity across data features. Lastly, timestamp alignment was achieved using the pandas library, which converted all time data into a consistent datetime format, facilitating accurate temporal analysis. This preprocessing ensured that the dataset was ready for feature extraction and machine learning model training.

5.1.2. Feature Engineering for Anomaly Detection

Strategic feature engineering is crucial for enhancing the anomaly detection capabilities of our system, mainly because it fine-tunes the input data for optimal relevance and significance, directly influencing model performance. We developed critical features that track and analyze daily routines and any deviations therein, which are vital for understanding patient behavior and potential health issues. For instance, sitting duration is calculated by aggregating periods when the subject is stationary, which helps identify sedentary behavior patterns critical in elderly care. Walking duration is monitored through motion sensors, providing a quantitative measure of physical activity essential for assessing mobility levels. Similarly, wake-up count is a crucial feature derived from movement data collected during typical sleeping hours, serving as a crucial indicator of sleep quality and disturbances. Out-of-home duration is another significant metric, tracked through door sensor activations to gauge the time spent outside the home, reflecting the individual's social activity or potential isolation. Lastly, sleep duration is estimated from periods of inactivity during night hours, providing critical insights into the subject's sleep patterns and overall rest quality. These features were specifically chosen for their strong predictive value regarding health concerns, such as cognitive decline, physical inactivity, and sleep disorders, which are prevalent issues in elderly care. To ensure the effectiveness of these features, each one undergoes a rigorous normalization process to maintain uniformity across different types of measurements before integration into our anomaly detection models. The models use these standardized features to establish behavioral baselines and identify deviations that may indicate potential health problems. This meticulous feature engineering process enables our anomaly detection models to respond effectively and accurately reflect real-world conditions. This capability significantly improves the system's utility in detecting and addressing emergent health issues, thereby enhancing the overall efficacy of patient care management.

5.1.3. Model Performance Evaluation

Following the extensive preparation and feature engineering detailed in the previous subsections, we rigorously evaluated the performance of our anomaly detection models using a set of standardized metrics crucial for confirming their effectiveness and reliability. Accuracy, which measures the proportion of correctly identified anomalies, provides an overarching view of the model's correctness. High accuracy indicates that our system is reliable and can be trusted to make critical decisions in real-time applications [46]. Precision is the ratio of true positive anomalies to all positive predictions, highlighting the model's ability to avoid false positives. Precision is particularly important in elderly care, where unnecessary alerts can lead to alarm fatigue among caregivers. Recall (sensitivity) reflects the model's ability to identify actual anomalies, emphasizing its detection capabilities. Ensuring high recall is crucial for the safety and well-being of elderly individuals, as missed detections could have serious consequences. The F1 score, the harmonic mean of precision and recall, offers a balanced measure of the model's performance and is particularly valuable when dealing with imbalanced data [47] typical of anomaly detection scenarios in smart homes. This ensures that both precision and recall are reasonably high, confirming the balanced sensitivity and specificity of the model. By rigorously applying these metrics, we fine-tune the models to optimize performance and reliability, ensuring that the system operates effectively under the diverse conditions found in elderly care environments. This rigorous application of standardized evaluation metrics helps adjust our models to meet operational requirements for real-time anomaly detection, ultimately enhancing the overall efficacy of patient care management.

6. Results and Discussion

This section elaborates on the findings from the deployment of our system, which leverages ML models such as isolation forest and LSTM to form a robust framework for supporting elderly care settings. These findings illustrate how the system's capabilities can significantly impact elderly care through the early warning of potential health issues to doctors and caregivers. As discussed in the previous section, we aimed to demonstrate the system's feasibility and effectiveness. Scripts were utilized to feed the system real data simulating input from IoT environmental devices. Therefore, the different components of the system acted as expected: they preprocessed the data, applied ML anomaly detection techniques to predict anomalies in the received data, transmitted the results to the cloud block, and displayed them on the dashboard. This section showcases and discusses the different aspects involved in the process.

6.1. Integrative Cognitive Health Assessment

The dashboard is a crucial component of our system. It synthesizes data from sleep quality, motor function, and social interaction analyses to comprehensively assess cognitive health. By integrating and analyzing trends and patterns across these key health domains, the dashboard offers a multifaceted solution that enhances the accuracy and timeliness of detecting potential cognitive decline. Therefore, it forms the basis for evaluating the effectiveness and reliability of our system. The specific results displayed on the dashboard, which can help doctors and caregivers in their daily activities, are detailed below. Then, in the subsequent sections, we present the results in terms of performance and accuracy, obtained from the different components that operated in the background to generate each result shown on the dashboard.

Sleep Quality Analysis: Sleep quality is a vital indicator of an individual's overall health and cognitive function. Our system uses motion sensors, as used for the CASAS TM029 dataset [12], to monitor various sleep parameters, such as duration, frequency of awakenings, and signs of restlessness. These sensors passively track movements during sleep to detect disruptions and quantify sleep quality without requiring active participation from the individual. The LSTM model analyzes these data to identify deviations from established sleep norms, which are indicative of potential sleep disorders or broader health issues that may impact cognitive function. This analysis helps caregivers customize interventions to improve sleep quality and, by extension, cognitive health.

Figure 4 illustrates specific instances of sleep disturbances over a selected period, highlighting patterns such as frequent awakenings or unusually short sleep durations, which may suggest underlying health conditions like insomnia.

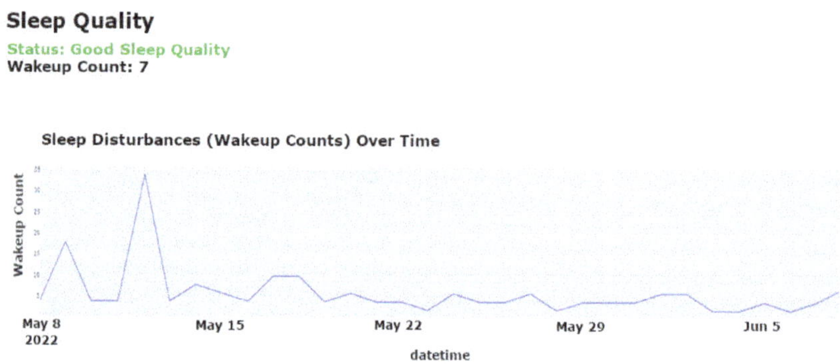

Figure 4. Sleep disturbances.

Figure 5 shows the evolution of sleep quality over time, visualizing how sleep patterns respond to interventions or changes in health status, helping caregivers assess the effectiveness of treatments or identify deteriorating conditions.

Motor Function Monitoring: Motor function provides insights into physical capabilities and potential cognitive issues. Motor function monitoring in our system is based on motion sensors and door sensors, as used in the CASAS TM029 dataset [12]. Motion sensors detect walking activity and periods of inactivity, while door sensors track entry and exit patterns. By analyzing the ratio of time spent sitting to walking using data collected from motion sensors, our isolation forest model identifies anomalies in these patterns that may indicate declining mobility, fatigue, or potential cognitive issues. Unusual increases in sedentariness or reductions in mobility may signal physical conditions affecting cognitive status, prompting further assessment and targeted interventions to enhance mobility and cognitive engagement.

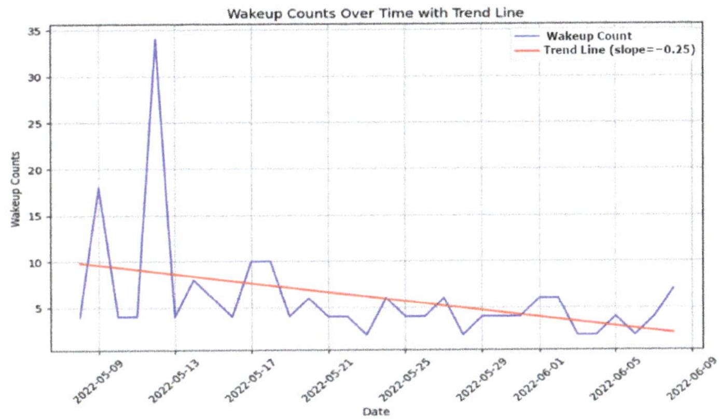

Figure 5. Sleep disturbance trend line.

Figure 6 depicts daily ratios of sitting to walking, clearly showing periods of reduced mobility or increased inactivity, which may indicate potential health risks or the need for targeted intervention.

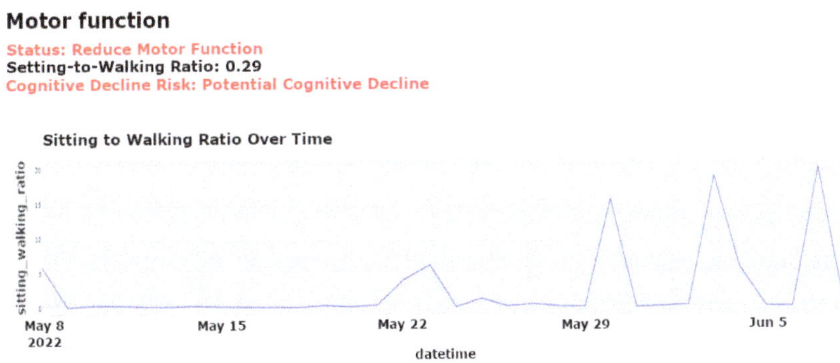

Figure 6. Motor function.

Figure 7 provides a longitudinal analysis of motor function, highlighting trends that may require medical attention or lifestyle adjustments, such as increasing physical therapy sessions or adjusting daily routines to enhance mobility.

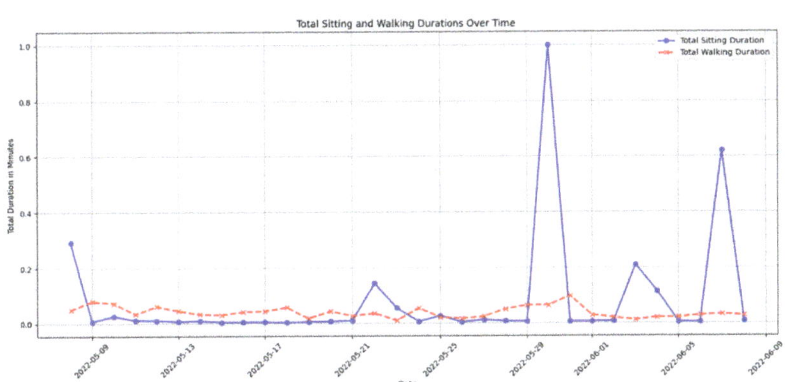

Figure 7. Sitting and walking trend line.

Social Interactions Evaluation: This feature tracks the frequency and quality of an individual's interactions with other people both within and outside their living environment. By analyzing variations in engagement patterns, the system detects potential social withdrawal, which may be an early indicator of cognitive or emotional decline.

Figure 8 illustrates the frequency of social interactions over time, identifying periods of reduced engagement that may signal a need for intervention.

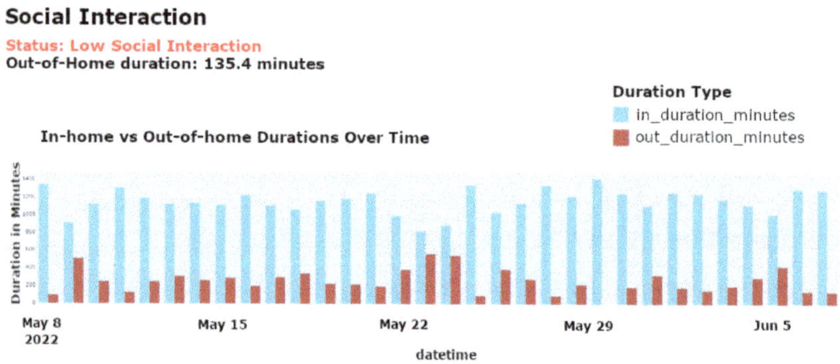

Figure 8. Visualization of social interaction frequency, highlighting periods of reduced engagement that may indicate cognitive or emotional changes.

Figure 9 compares home versus out-of-home activities, highlighting any significant changes in social behavior or potential withdrawal from usual activities, which are essential for assessing social habits and detecting early signs of depression that may indicate reduced social participation. Identifying these trends is crucial for initiating timely social stimulation interventions, which support cognitive well-being and mental health.

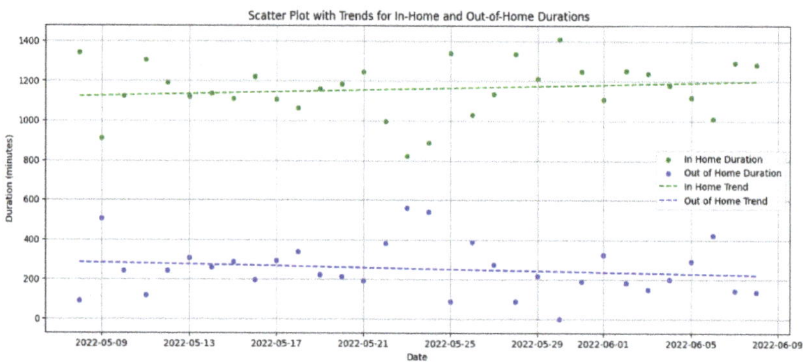

Figure 9. Comparison of in-home and out-of-home social activities, emphasizing behavioral shifts that may suggest social withdrawal or early cognitive decline.

Practical Benefits of Anomaly Detection

Our anomaly detection system supports caregivers and healthcare workers by closely monitoring key health indicators, such as sleep quality, motor function, and social interactions. By identifying anomalies in these areas, caregivers can take timely action to prevent health deterioration and improve elderly care outcomes. For example, our system detected decreased nocturnal mobility and increased awakenings in an 82-year-old male showing signs of cognitive decline. This triggered a telemedicine consultation, where caregivers adjusted his sleep environment and medication, leading to marked improvements in sleep quality and cognitive function. Similarly, the system flagged a decline in mobility in an

85-year-old female, prompting early intervention with physical therapy, which prevented further deterioration. Additionally, our system detected social withdrawal in a 78-year-old male based on reduced social interactions, allowing caregivers to introduce targeted social engagement activities, thereby improving the patient's overall well-being. These real-world scenarios demonstrate the system's effectiveness in transforming reactive healthcare into a proactive model, enabling caregivers to respond before serious health conditions develop.

Comprehensive Cognitive Health Dashboard

Our dashboard synthesizes data from the aforementioned analyses to provide an integrative view of cognitive health. By correlating sleep quality, motor function, and social interaction data, the dashboard identifies patterns that might not be apparent from individual measurements, ensuring a holistic approach to cognitive health assessment. The described dashboard provides a visual summary of the key metrics it monitors. Based on the aggregated data from the monitored metrics, it illustrates an individual's overall health status, with specific emphasis on the risk of cognitive decline.

6.2. Machine Learning: Anomaly Detection and Model Performance

As discussed, the proposed system employs advanced machine learning models, including isolation forest and LSTM, to effectively detect and analyze anomalies. The performance of these models is critically evaluated by their ability to identify deviations from normal behavior patterns, which are essential for preemptive healthcare interventions. Continuous tuning of the model parameters based on real-time data enhances detection accuracy, particularly in distinguishing between usual behavioral variability and genuine anomalies that require intervention. A critical demonstration of the efficacy of the dashboard was observed when it promptly issued an alert for an unusual sitting duration noted in an elderly patient. This alert enabled caregivers to quickly assess the situation and intervene appropriately, thereby showcasing the ability of the system to facilitate rapid response strategies and contribute to proactive patient management. The isolation forest model is used to identify anomalies in the daily activity patterns of elderly individuals. It effectively detects deviations from normal patterns by applying unsupervised learning techniques. This model isolates anomalies by randomly partitioning the dataset and typically requires fewer splits to isolate anomalous observations from normal instances, which are more abundant and thus deeper in the tree structure. The evaluation of our proposed edge-based anomaly detection system involved analyzing the daily activities of a single resident in a smart home using the CASAS [12] dataset. The key results obtained are shown in Figures 10 and 11.

As previously discussed, the system processes sensor data in real time, ensuring that anomalies are promptly detected and reported. This real-time processing capability is critical for ensuring timely intervention, especially in the context of elderly care. As shown in the previous subsection, machine learning algorithms were applied to various activities, including sleep duration, sitting duration, walking duration, and out-of-home duration, over a one-month period. The results highlight the system's ability to detect anomalies across these activities. The top graph in Figure 10 shows the daily sleep duration, with red markers indicating anomalies. These anomalies represent deviations from the typical sleep pattern. Days with significantly reduced sleep durations could indicate interrupted sleep or early waking times, which may be symptoms of sleep disorders or disturbances in the living environment. Consistent anomalies in sleep duration require attention as they could indicate underlying health issues. It is essential for caregivers or medical professionals to investigate these irregularities further to provide appropriate interventions and ensure the well-being of elderly people. Therefore, our system is able to generate real-time notifications based on identified irregularities or critical events, ensuring that caregivers and healthcare providers are promptly informed of potential issues. The bottom graph displays the daily sitting duration, with red markers indicating anomalies, i.e., prolonged periods of inactivity. Such patterns could be concerning, particularly for elderly individuals at risk of health complications associated with sedentary behavior. Extended periods of sitting without breaks can contribute to several health issues, including cardiovascular disease and increased risk of falls due to muscle weakening. Addressing these anomalies with scheduled activities or reminders to move could improve health outcomes.

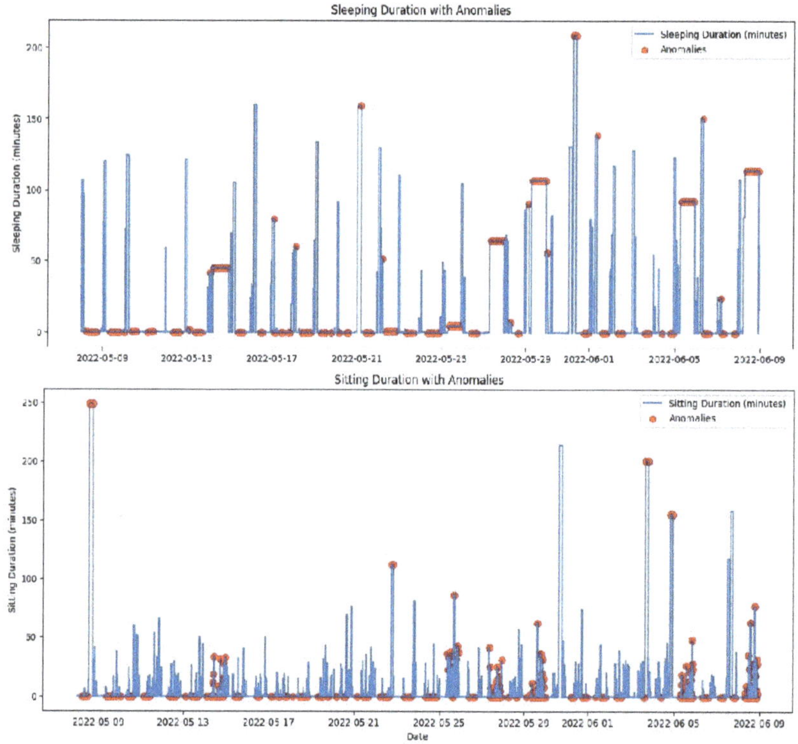

Figure 10. Detected anomalies, part 1.

The top graph in Figure 11 shows the daily walking duration, with red markers indicating anomalies. Deviations from the norm could suggest mobility issues or irregular daily activities. For instance, days with low values could point to reduced mobility. Monitoring walking duration is crucial for assessing elderly adults' physical activity levels and mobility. Anomalies might prompt assessments such as physical therapy evaluations or adjustments in daily mobility aids. The bottom graph illustrates out-of-home durations, with anomalies indicated by red markers. Deviations in these patterns could signify unexpected absences or changes in the usual routines of the individual, such as not leaving the house at all on days they normally would or being away for unusually long periods. Anomalies in out-of-home duration could raise safety concerns or highlight changes in social activities. The LSTM (Long Short-Term Memory) model was employed to analyze temporal dependencies within the activity data, enhancing the detection and understanding of anomalies in the behavior of elderly individuals. The model's capability to process time-series data makes it particularly well suited for monitoring systems in smart homes, where activities and their patterns over time are crucial for accurate anomaly detection.

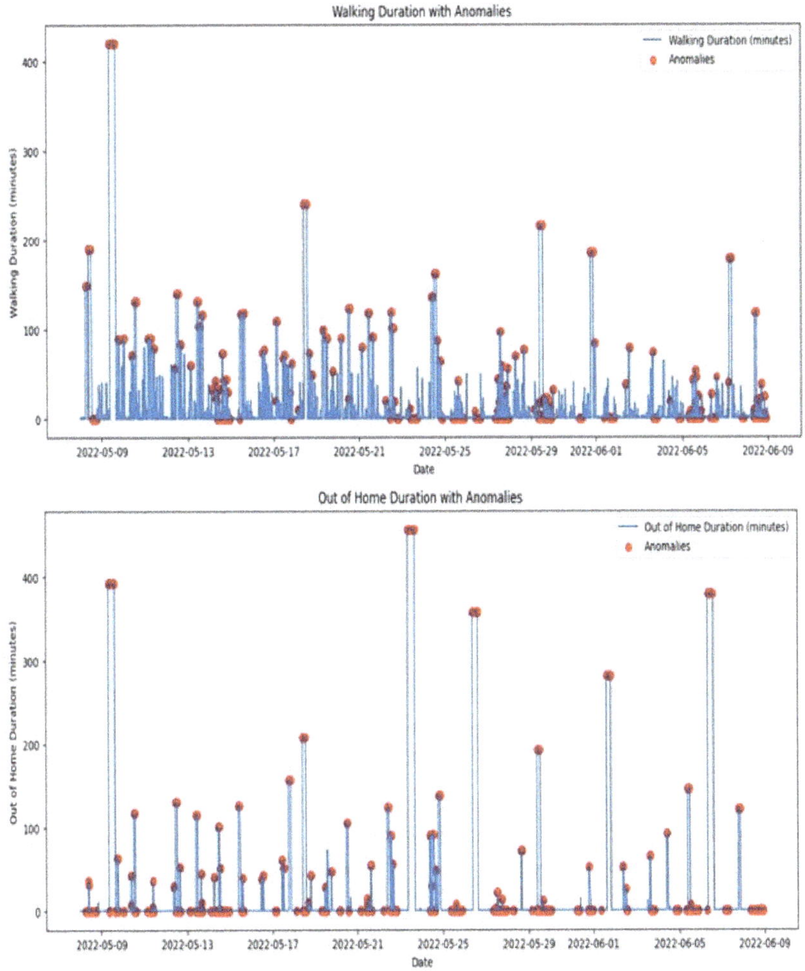

Figure 11. Detected anomalies, part 2.

Figure 12 shows the reconstruction error over time for the Long Short-Term Memory (LSTM) network, which is used to detect anomalies in activity data. The reconstruction error is calculated as the difference between the predicted and actual values. In the figure, significant spikes in the reconstruction error indicate instances where the observed activity deviated from the expected normal behavior, leading to anomaly detection. The threshold for anomaly detection was set at a reconstruction error of 0.25, determined through cross-validation to minimize false positives while ensuring sensitivity to potential health risks. Several instances exceeded this threshold during the observation period, with the highest reconstruction error reaching 0.35, signaling unusual behavior that warranted further investigation. Early stopping was applied during the training process after five epochs of no improvement in validation loss, ensuring that the model generalized effectively to new data.

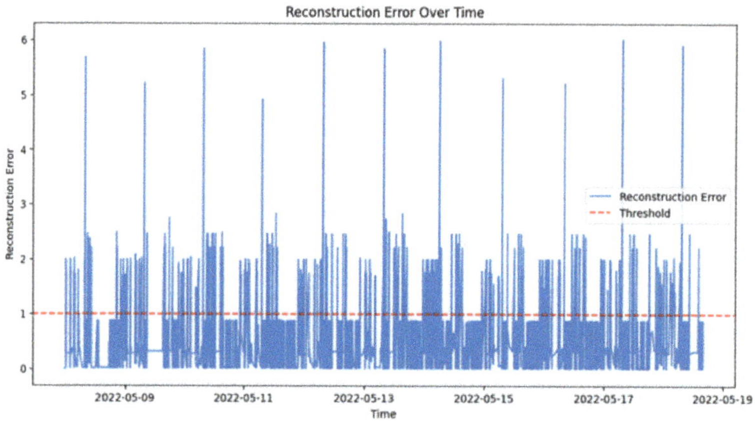

Figure 12. Reconstruction error over time.

The performance of the proposed model is shown in Figure 13, detailed through a confusion matrix, demonstrating strong predictive accuracy, with 37,557 true positives and 228,252 true negatives, reflecting the effectiveness of the model in accurately identifying both anomalies and normal behavior. It also records 1813 false positives and 3528 false negatives, indicating some areas for improvement. The model achieved an overall accuracy of 92.29%, a precision of 82.36%, a recall of 91.5%, and an F1 score of 86.6%. These metrics highlight the capability of the model to reliably detect most anomalies with a high degree of precision; however, the presence of false negatives suggests the potential for further refinement. Enhancements could include more nuanced feature engineering, training dataset expansion, or anomaly classification threshold adjustments. Continuous model evaluation and updates are crucial to adapt to evolving patterns of normal and anomalous behavior, ensuring that the model remains effective over time.

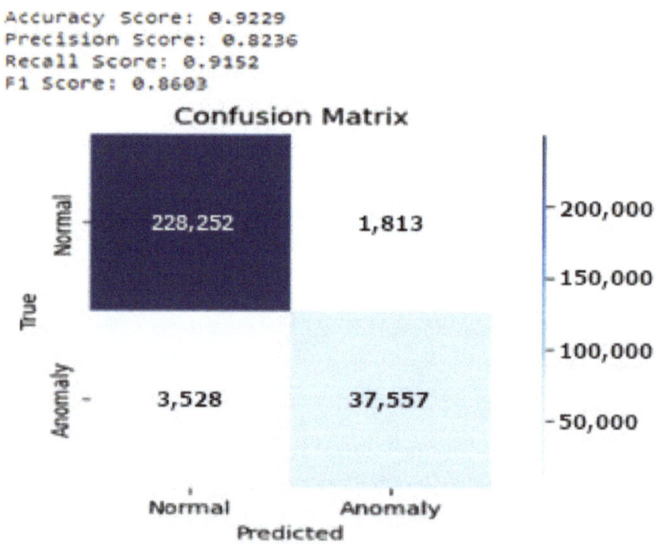

Figure 13. Confusion matrix and performance metrics.

Further, in Figure 14, which presents the precision–recall curve, we can observe that the curve begins with a high precision close to 1, indicating that the model was highly accurate in predicting anomalies, especially at lower levels of recall. As recall increases, precision declines sharply, suggesting that the model started to compromise accuracy for a broader coverage of potential anomalies.

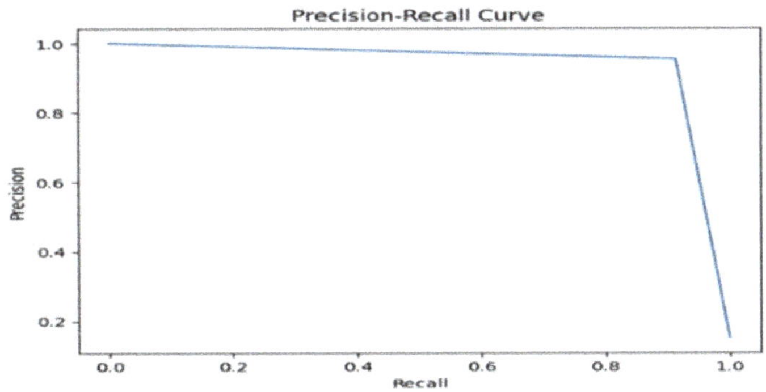

Figure 14. Precision–recall curve.

Figure 15 presents the ROC curve for the LSTM model used in anomaly detection. The curve plots the true positive rate (TPR) against the false positive rate (FPR) at various classification thresholds. The Area Under the Curve (AUC) is 0.95, indicating that the model effectively distinguished between normal and anomalous behavior. The steep rise of the ROC curve in the early stages reflects a high TPR with a minimal increase in the FPR, emphasizing the ability of the model to accurately detect anomalies with minimal false alarms, which is crucial in real-time monitoring systems for elderly care.

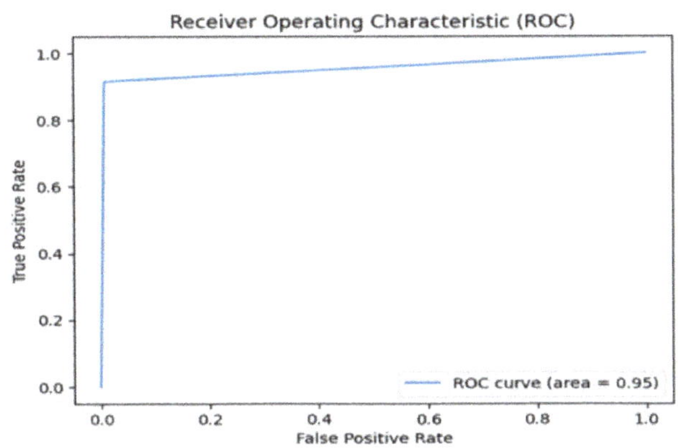

Figure 15. Receiver Operating Characteristic (ROC) curve.

Table 2 comprehensively compares the different anomaly detection methods used in related works. It highlights the accuracy, precision, recall, and F1 score of various models and methods employed across different studies, along with the datasets used. Each study's real-time context, key technologies, and targeted anomalies are detailed, offering a clear perspective on how our approach stands in comparison. This table underlines our study's unique approach in employing a combination of isolation forest and LSTM models to handle multi-activity anomaly detection, leveraging the CASAS TM029 [12] dataset to enhance real-time monitoring capabilities in smart homes.

Table 2. Comparison of different anomaly detection methods used in related works.

Study/ Paper	Model/ Method	Dataset	Accuracy (%)	Precision (%)	Recall (%)	F1 Score (%)	Real-Time	Context	Technologies	Key Anomalies
Our Paper	Isolation Forest + LSTM	CASAS TM029	92.29	82.36	91.52	86.03	Yes	Smart homes	Python [31], Node.js [34], Express.js [38]	Multi-activity (walking duration, sitting duration, sleep duration, wake-up events)
[14]	LSTM + SMOTE	Anomaly detection for fallen people	84.00	97.00 (normal), 19.00 (anomalous)	60.00 (anomalous)	91.00 (normal), 28.00 (anomalous)	Yes	Elderly care	Python [31], TensorFlow [48]	Fall detection
[13]	Parametric statistical method	Real-world IoT data; 9 participants	Not reported	Not reported	Not reported	Not reported	No	Smart homes	Python [31], custom stats software	Daily activity monitoring
[11]	Graph-based approach	Kyoto (CASAS)	Not reported	Not reported	Not reported	Not reported	No	Smart homes	Python [31], graph tools	Cognitive decline indicators
[15]	SOM + Markov models	MavHome	>75% (Anomaly detection)	Not reported	Not reported	Not reported	Yes	Smart homes	Python [31], SOM, Markov models	Inactivity; unusual presence
[19]	Binary sensors; activity maps	CASAS HH126 and Kasteren	Not reported	Not reported	Not reported	Not reported	No	Smart homes	Python [31]; custom algorithms	Behavioral changes

6.3. *Dashboard: Integration and Utility Monitoring*

6.3.1. Heatmap of Activities (Normalized)

The heatmap presented in Figure 16 illustrates the normalized intensity of four key activities, i.e., sitting duration, walking duration, wake-up count, and sleep duration over 10 days. The activities were normalized using min-max scaling, where the color scale ranges from blue (low intensity) to red (high intensity). This visualization allows for a comparative analysis of the intensity and frequency of each activity, making it easier to identify trends, shifts, or potential anomalies in daily behavior. This heatmap is integrated into the monitoring dashboard, enhancing its utility by providing a clear, color-coded overview of activity patterns that help caregivers make informed decisions. The insights gained from the heatmap analysis directly inform care strategies, enabling dynamic adjustments to care plans to better accommodate the day-to-day variability in the individual's activities and conditions.

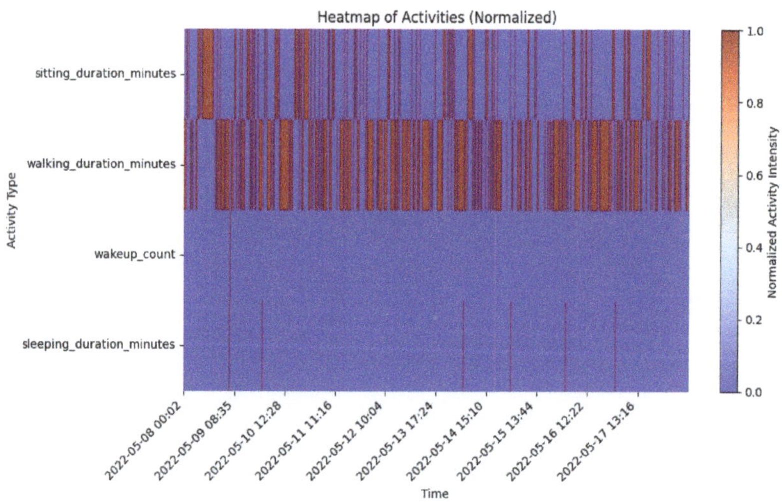

Figure 16. Heatmap of activities.

The x-axis represents the timeline, while the y-axis displays the four monitored activities. This graphic representation helps quickly identify periods of higher or lower activity levels and offers insight into the daily routines of elderly adults. The walking duration activity shows the highest overall intensity, as evidenced by the prevalence of red bands throughout the 10 days, suggesting that the individual frequently moved throughout the day, with regular fluctuations in activity levels. However, the heatmap also reveals sharp drops in walking activity, represented by the blue sections, which could indicate potential health concerns, such as mobility issues or fatigue. These deviations may be flagged by the anomaly detection models, particularly the LSTM model, as potential areas of concern. The sitting activity displays a more sporadic pattern, alternating high- and low-intensity periods. Red bands that indicate prolonged sitting suggest periods of reduced movement, which may be normal during specific times of day but could also signal potential risks, such as prolonged inactivity. The isolation forest model may flag longer sitting durations as potential anomalies, particularly if they coincide with a reduction in walking. Wake-up events are less frequent and are represented by sporadic red spikes, and they are generally expected to occur once a day. However, multiple wake-up events or an increased frequency of wake-up events, such as nighttime disruptions, might indicate sleep disturbances. Anomalies in wake-up patterns may indicate underlying health issues such as insomnia or restlessness, and the isolation forest model may highlight these. Sleeping patterns appeared relatively stable across the 10-day period, with consistent blue bands representing the times when the individual slept. Any significant reductions or fragmentations in sleeping patterns, visible as breaks in the blue sections, could signal potential sleep disorders or disturbances. These sleep irregularities could be captured by the LSTM model, providing early warnings of deteriorating sleep quality.

6.3.2. Time-Series Plots of Activity Patterns

The time-series plots presented in Figure 17 illustrate the daily activity patterns of four key activities: sitting duration, walking duration, wake-up count, and sleep duration over a 10-day observation period. These visualizations provide critical insights into temporal fluctuations in activity levels and allow for the identification of potential deviations from typical behavior. The monitoring dashboard, as shown in Figure 3, complements these plots by providing real-time visual feedback on these activities, allowing caregivers to promptly respond to detected anomalies directly from these interfaces. The x-axis represents the timeline across 10 days, while the y-axis depicts the normalized durations for sitting, walking, and sleeping or the count of wake-up events. Each activity is represented by a distinct color: sitting duration is shown in orange, walking duration in blue, wake-up count in green, and sleep duration in red, enabling a clear distinction between activities and facilitating the detection of trends or anomalies.

The time-series plot of the sitting duration reveals the frequency and length of time spent sitting over the 10-day period. The data show frequent short intervals of sitting interspersed with occasional extended periods. High variability was observed in the duration of sitting periods, suggesting normal day-to-day fluctuations in the daily routine. However, prolonged sitting periods may indicate potential health concerns, such as fatigue or reduced mobility, and can be classified as anomalies. The isolation forest model successfully flagged several instances where extended sitting durations deviated from the norm. These instances were flagged on the monitoring dashboard (as shown in Figure 3), highlighting them as anomalies for caregiver intervention. The walking duration plot shows regular movement throughout the day, indicating periods of mobility. The relatively consistent walking patterns suggest stable mobility, with only a few instances of reduced walking duration. However, significant drops in walking could signal mobility issues or health concerns. For example, on certain days, noticeable reductions in walking were flagged by the LSTM model as anomalies. These deviations could indicate potential fatigue or difficulties in movement, which are crucial to elderly care monitoring. These deviations were prominently displayed on the monitoring dashboard, allowing caregivers to quickly identify and address these mobility issues. The plot shows wake-up events as distinct spikes, representing transitions from sleep to wakefulness. Typically, one or two wake-up events were expected each day, corresponding to the normal sleep cycle of the individual. However, an increased number of wake-up events within a day may indicate disturbed sleep patterns, potentially signaling sleep disorders or other health-related concerns. The isolation forest model flagged days with increased wake-up counts as potential anomalies, drawing attention to possible sleep disturbances. The sleep duration plot shows regular sleep periods during the 10-day observation window, with most sleep occurring at night. Consistency in sleep patterns is a positive indicator of stable health. However, deviations from typical sleep patterns were observed on certain days, including shorter sleep durations or frequent interruptions, which could indicate sleep-related health concerns such as insomnia. The LSTM model successfully identified these deviations, flagging them as anomalies for further investigation. Additionally, these visualizations provided a comprehensive overview of the elderly individual's daily routines, enhancing the capability of caregivers to effectively monitor and respond to anomalies. Enhanced data analysis from the time-series plots has enabled the implementation of targeted interventions, such as mobility exercises and sleep quality improvements, to address the specific needs identified through anomaly detection.

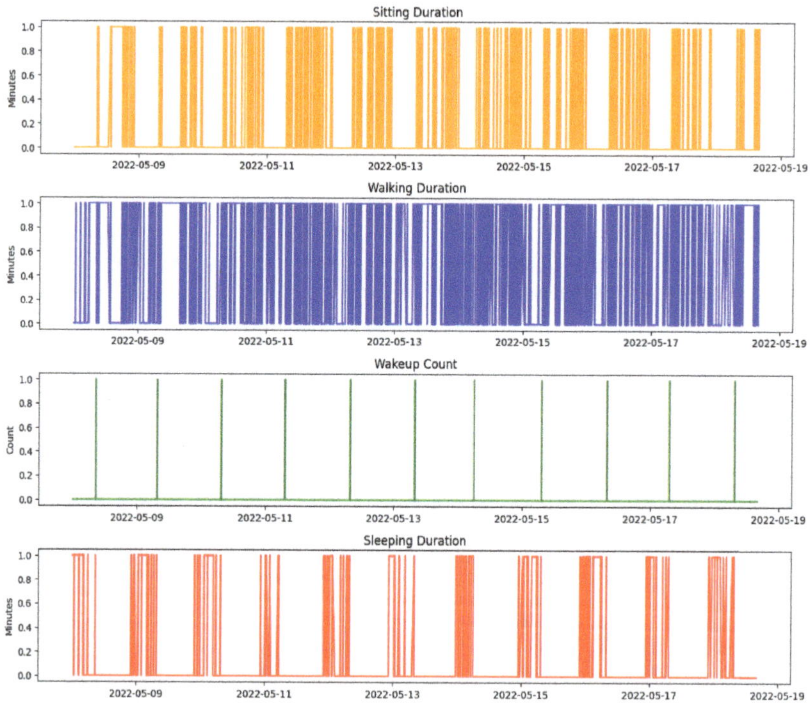

Figure 17. Time-series plots displaying the daily activity patterns over the 10-day period.

6.4. Lessons Learned

Through the development and implementation of our system, several key insights emerged, shaping our understanding of real-world deployment challenges and opportunities for future improvements.

Initially, the integration of ML models, particularly the isolation forest and LSTM models, demonstrated strong potential for identifying behavioral anomalies. By leveraging time-series analysis and anomaly scoring, we were able to flag potential cognitive health concerns early, enabling caregivers to take proactive measures. However, thanks to the modular nature of our system, experimenting with new ML techniques or optimizing the already used ones could improve efficiency in large-scale applications. In addition, we noticed that the variability of home environments, sensor placements, and user behaviors influenced the system's performance. Future improvements should focus on increasing the diversity of training data to enhance model adaptability and robustness across different real-world settings. Furthermore, despite the stated advantages of edge computing, increasing sensor deployment may increase computational demands. Therefore, any system modification or integration must always consider scalability. Moreover, we demonstrated that using non-wearable sensors can improve compliance, but privacy concerns remain a key focus area that needs to be managed and safeguarded in any future development. Finally, although the experiments demonstrated the system's efficacy in providing valuable insights to doctors and caregivers, further real-world experiments could expand the dataset to validate the system's effectiveness in cases of data variability. Additionally, they could motivate the addition of new real-time services to further assist all stakeholders, including patients.

These lessons provide a roadmap for future studies, ensuring that the system continues to evolve as a reliable and scalable solution for elderly care.

7. Conclusions

The global population is aging rapidly, with the number of people aged 65 and over expected to double by 2050. This demographic shift presents significant challenges to healthcare systems, emphasizing the need for innovative solutions to ensure the safety and well-being of elderly individuals living independently. Monitoring systems that detect health issues early are essential to enable timely intervention and reduce the strain on caregivers and healthcare resources.

Traditional monitoring approaches rely heavily on cloud computing and wearable devices, which face notable challenges, including high latency, privacy risks, and low compliance among elderly people due to discomfort or forgetfulness. These factors lead to unreliable data collection and hinder real-time decision-making.

In this paper, we present an innovative IoT edge-based framework designed for the remote monitoring of elderly adults that integrates non-wearable sensors, edge computing, and machine learning-based anomaly detection. Our approach leverages anomaly detection techniques to accurately identify behavioral deviations in elderly individuals. Unlike conventional methods, our system processes data locally within the smart home environment, enhancing privacy, reducing latency, and ensuring seamless real-time monitoring. Furthermore, the system offers innovative services, such as a doctor-focused dashboard and a caregiver application, which provide timely alerts and long-term behavioral insights, supporting healthcare professionals and caregivers in proactive decision-making. This design enables seamless, non-intrusive monitoring, eliminating the need for active participation from elderly individuals and effectively addressing the compliance challenges posed by wearable devices.

A prototype of the proposed framework was implemented to evaluate its feasibility and showcase its scalability, efficiency, and reliability in elderly care settings. In our experiments, two of the most frequently used ML anomaly detection models, i.e., isolation forest for spatial anomaly detection and LSTM for temporal analysis, were utilized to identify behavior anomalies in sleep quality, motor function, and social interactions, which are key indicators of cognitive decline.

The findings revealed that the system successfully detected anomalies in daily activities, delivering valuable insights into cognitive and physical health. The experimental evaluation demonstrated that the system achieved 92.2% accuracy, highlighting its effectiveness in early health risk detection and anomaly detection in elderly care.

Future work will include developing the system's unaddressed components, particularly focusing on enhancing predictive capabilities by integrating advanced machine learning models to improve the early detection of cognitive decline. We plan to utilize more recent relevant studies like the one conducted by Rahman et al. [49], in which a comprehensive review of ML approaches for anomaly detection in smart homes was presented. Additionally, we will expand the dataset to further study the best configuration of ML models to provide the most valuable outcomes to doctors and caregivers. The dashboard and application will also be further extended and adapted to provide suggestions and feedback to elderly individuals, enhancing engagement and usability. Finally, the edge-based nature of the designed system will be leveraged to further test its effectiveness in real scenarios, which will be conducted in collaboration with hospitals and doctors.

Our system has created a basis for developing comprehensive edge-based systems that aim to improve the quality of life of the elderly while also ensuring their safety and well-being through creative, innovative technological solutions.

Author Contributions: Conceptualization: A.A., T.M., I.S., S.C., D.G., C.D. and L.P.; Investigation, A.A., T.M., I.S., S.C., D.G., C.D. and L.P.; Methodology, A.A., T.M., I.S., C.D. and L.P.; Resources: A.A.; Software, A.A.; Supervision: T.M., I.S., C.D. and L.P.; Visualization: A.A.; Writing—original draft: A.A. and T.M.; Writing—review and editing: A.A., T.M., I.S., S.C., D.G. and L.P. All authors prepared the manuscript. They all critically edited the manuscript and approved the final draft. All authors have read and agreed to the published version of the manuscript.

Funding: Amir Ali utilized a Ph.D. research grant partially funded under Italian Ministerial Decree No. 352/2022—PNRR, Mission 4, Component 2, titled "From Research to Enterprise"—and Investment 3.3, titled "Introduction of Innovative Doctoral Programs that Respond to the Needs of Enterprises and Promote the Recruitment of Researchers from Enterprises"—an innovative doctoral scholarship with an industrial focus.

Informed Consent Statement: Not applicable.

Data Availability Statement: The original data presented in the study are openly available in [TM029 dataset from CASAS] at [https://casas.wsu.edu/datasets/].

Acknowledgments: We wish to thank Waleed Raza from the Department of Electrical Engineering and Computer Science, Embry-Riddle Aeronautical University, Daytona Beach, FL 32114, USA, for his valuable assistance.

Conflicts of Interest: Authors Simone Carrisi and Daniele Galli were employed by the company Gematica s.r.l. The remaining authors declare that the research was conducted in the absence of any commercial or financial relationships that could be construed as a potential conflict of interest.

References

1. United Nations, Department of Economic and Social Affairs, Population Division. World Population Prospects 2019: Highlights (ST/ESA/SER.A/423). 2019. Available online: https://population.un.org/wpp/ (accessed on 15 January 2025).
2. Beard, J.R.; Officer, A.; de Carvalho, I.A.; Sadana, R.; Pot, A.M.; Michel, J.P.; Lloyd-Sherlock, P.; Epping-Jordan, J.E.; Peeters, G.M.E.E.G.; Mahanani, W.R.; et al. The World report on ageing and health: A policy framework for healthy ageing. *Lancet* **2016**, *387*, 2145–2154.

3. Maudsley-Barton, S.; Yap, M.H. Objective Falls Risk Assessment Using Markerless Motion Capture and Representational Machine Learning. *Sensors* **2024**, *24*, 4593. [CrossRef] [PubMed]
4. Ruiu, P.; Caragnano, G.; Masala, G.L.; Grosso, E. Accessing Cloud Services through Biometrics Authentication. In Proceedings of the 2016 10th International Conference on Complex, Intelligent, and Software Intensive Systems (CISIS), Fukuoka, Japan, 6–8 July 2016; pp. 38–43. [CrossRef]
5. Fadda, E.; Plebani, P.; Vitali, M. Monitoring-Aware Optimal Deployment for Applications Based on Microservices. *IEEE Trans. Serv. Comput.* **2021**, *14*, 1849–1863. [CrossRef]
6. Fernández-Cardeñosa, G.; de la Torre-Díez, I.; López-Coronado, M.; Rodrigues, J.J. Analysis of cloud-based solutions on EHRs systems in different scenarios. *J. Med. Syst.* **2012**, *36*, 3777–3782. [CrossRef]
7. Regano, L.; Canavese, D.; Mannella, L. A Privacy-Preserving Approach for Vulnerability Scanning Detection. *Ceur Workshop Proc.* **2024**, *3731*.
8. Chan, M.; Estève, D.; Fourniols, J.Y.; Escriba, C.; Campo, E. Smart wearable systems: Current status and future challenges. *Artif. Intell. Med.* **2012**, *56*, 137–156. [CrossRef] [PubMed]
9. Chua, W.; Pajas, A.L.D.; Castro, C.S.; Panganiban, S.P.; Pasuquin, A.J.; Purganan, M.J.; Malupeng, R.; Pingad, D.J.; Orolfo, J.P.; Lua, H.H.; et al. Web Traffic Anomaly Detection Using Isolation Forest. *Informatics* **2024**, *11*, 83. [CrossRef]
10. Xu, W.; He, J.; Li, W.; He, Y.; Wan, H.; Qin, W.; Chen, Z. Long-Short-Term-Memory-Based Deep Stacked Sequence-to-Sequence Autoencoder for Health Prediction of Industrial Workers in Closed Environments Based on Wearable Devices. *Sensors* **2023**, *23*, 7874. [CrossRef]
11. Paudel, R.; Eberle, W.; Holder, L.B. Anomaly Detection of Elderly Patient Activities in Smart Homes using a Graph-Based Approach. In Proceedings of the 2018 International Conference on Data Science (ICDATA), Las Vegas, NV, USA, 30 July–2 August 2018; pp. 163–169.
12. CASAS Datasets. Available online: https://casas.wsu.edu/datasets/ (accessed on 15 January 2025).
13. Shahid, Z.; Saguna, S.; Åhlund, C. Detecting Anomalies in Daily Activity Routines of Older Persons in Single Resident Smart Homes: Proof-of-Concept Study. *JMIR Aging* **2022**, *5*, e28260. [CrossRef]
14. Gupta, B.B.; Gaurav, A.; Attar, R.W.; Arya, V.; Alhomoud, A.; Chui, K.T. LSTM Based Neural Network Model for Anomaly Event Detection in Care Independent Smart Homes. *CMES—Comput. Model. Eng. Sci.* **2024**, *140*, 2689–2706. [CrossRef]
15. Novák, M.; Bináš, M.; Jakab, F. Unobtrusive anomaly detection in presence of elderly in a smart-home environment. In Proceedings of the 2012 ELEKTRO, Rajeck Teplice, Slovakia, 21–22 May 2012; pp. 341–344. [CrossRef]
16. Kohonen, T. Essentials of the self-organizing map. *Neural Netw.* **2013**, *37*, 52–65. Twenty-fifth Anniversay Commemorative Issue. [CrossRef] [PubMed]
17. Rabiner, L.R. A tutorial on hidden Markov models and selected applications in speech recognition. *Proc. IEEE* **1989**, *77*, 257–286. [CrossRef]
18. Pérez-Toro, P.A.; Rodríguez-Salas, D.; Arias-Vergara, T.; Klumpp, P.; Schuster, M.; Nöth, E.; Orozco-Arroyave, J.R.; Maier, A.K. Interpreting acoustic features for the assessment of Alzheimer's disease using ForestNet. *Smart Health* **2022**, *26*, 100347. [CrossRef]
19. Susnea, I.; Pecheanu, E.; Sandu, C.; Cocu, A. A Scalable Solution to Detect Behavior Changes of Elderly People Living Alone. *Appl. Sci.* **2022**, *12*, 235. [CrossRef]
20. Sadhu, S.; Solanki, D.; Constant, N.; Ravichandran, V.; Cay, G.; Saikia, M.J.; Akbar, U.; Mankodiya, K. Towards a telehealth infrastructure supported by machine learning on edge/fog for Parkinson's movement screening. *Smart Health* **2022**, *26*, 100351. [CrossRef]
21. Mokhtari, G.; Zhang, Q.; Hargrave, C.; Ralston, J.C. Non-Wearable UWB Sensor for Human Identification in Smart Home. *IEEE Sensors J.* **2017**, *17*, 3332–3340. [CrossRef]
22. Singh, A.; Margaria, T.; Demrozi, F. CNN-based Human Activity Recognition on Edge Computing Devices. In Proceedings of the 2023 IEEE International Conference on Omni-layer Intelligent Systems (COINS), Berlin, Germany, 23–25 July 2023; pp. 1–4. [CrossRef]
23. Wang, K.; Kong, S.; Chen, X.; Zhao, M. Edge Computing Empowered Smart Healthcare: Monitoring and Diagnosis with Deep Learning Methods. *J. Grid Comput.* **2024**, *22*, 30. [CrossRef]
24. Rancea, A.; Anghel, I.; Cioara, T. Edge Computing in Healthcare: Innovations, Opportunities, and Challenges. *Future Internet* **2024**, *16*, 329. [CrossRef]
25. Liu, X.; Chen, A.; Zheng, K.; Chi, K.; Yang, B.; Taleb, T. Distributed Computation Offloading for Energy Provision Minimization in WP-MEC Networks with Multiple HAPs. *IEEE Trans. Mob. Comput.* **2024**, *24*, 2673–2689. [CrossRef]
26. Punugoti, R.; Dutt, V.; Anand, A.; Bhati, N. Exploring the Impact of Edge Intelligence and IoT on Healthcare: A Comprehensive Survey. In Proceedings of the 2023 International Conference on Sustainable Computing and Smart Systems (ICSCSS), Coimbatore, India, 14–16 June 2023; pp. 1108–1114. [CrossRef]

27. Bonino, D.; Castellina, E.; Corno, F. DOG: An Ontology-Powered OSGi Domotic Gateway. In Proceedings of the 2008 20th IEEE International Conference on Tools with Artificial Intelligence, Dayton, OH, USA, 3–5 November 2008; Volume 1, pp. 157–160. [CrossRef]
28. Details About the Protocol Used for Collecting the CASAS Datasets. Available online: https://casas.wsu.edu/smarthomestats/bthomas_dissertation.pdf (accessed on 15 January 2025).
29. Nguyen, M.; He, T.; An, L.; Alexander, D.C.; Feng, J.; Yeo, B.T. Predicting Alzheimer's disease progression using deep recurrent neural networks. *NeuroImage* **2020**, *222*, 117203. [CrossRef]
30. Obaid, H.S.; Dheyab, S.A.; Sabry, S.S. The Impact of Data Pre-Processing Techniques and Dimensionality Reduction on the Accuracy of Machine Learning. In Proceedings of the 2019 9th Annual Information Technology, Electromechanical Engineering and Microelectronics Conference (IEMECON), Jaipur, India, 13–15 March 2019; pp. 279–283. [CrossRef]
31. Python Programming Language (Version 3.13.1), Developed by the Python Software Foundation. The Software Was Sourced from Wilmington, Delaware, USA. Available online: https://www.python.org (accessed on 15 January 2025).
32. The Pandas Library. Available online: https://zenodo.org/records/13819579 (accessed on 15 January 2025).
33. Harris, C.R.; Millman, K.J.; van der Walt, S.J.; Gommers, R.; Virtanen, P.; Cournapeau, D.; Wieser, E.; Taylor, J.; Berg, S.; Smith, N.J.; et al. Array programming with NumPy. *Nature* **2020**, *585*, 357–362. [CrossRef]
34. Node.js (Version 22.14.0), Developed by OpenJS Foundation. The Software Was Sourced from San Francisco, California, USA. Available online: https://nodejs.org (accessed on 15 January 2025).
35. Famá, F.; Faria, J.N.; Portugal, D. An IoT-based interoperable architecture for wireless biomonitoring of patients with sensor patches. *Internet Things* **2022**, *19*, 100547. [CrossRef]
36. Wu, T.; Wu, F.; Qiu, C.; Redouté, J.M.; Yuce, M.R. A Rigid-Flex Wearable Health Monitoring Sensor Patch for IoT-Connected Healthcare Applications. *IEEE Internet Things J.* **2020**, *7*, 6932–6945. [CrossRef]
37. MySQL (Version 8.0), Developed by Oracle Corporation. The Software Was Sourced from Austin, Texas, USA. Available online: https://www.mysql.com (accessed on 15 January 2025).
38. Express.js (Version 4.21.2), Node.js Web Application Framework. Developed by the OpenJS Foundation. The Software Was Sourced from San Francisco, California, USA. Available online: https://expressjs.com (accessed on 15 January 2025).
39. Shilpa, V.;Vidya, A.; Pattar, S. MQTT based Secure Transport Layer Communication for Mutual Authentication in IoT Network. *Glob. Transit. Proc.* **2022**, *3*, 60–66.
40. Imane, S.; Tomader, M.; Nabil, H. Comparison Between CoAP and MQTT in Smart Healthcare and Some Threats. In Proceedings of the 2018 International Symposium on Advanced Electrical and Communication Technologies (ISAECT), Rabat, Morocco, 21–23 November 2018; pp. 1–4. [CrossRef]
41. Ding, Z.; Fei, M. An Anomaly Detection Approach Based on Isolation Forest Algorithm for Streaming Data using Sliding Window. *IFAC Proc. Vol.* **2013**, *46*, 12–17.
42. Lin, S.; Clark, R.; Birke, R.; Schönborn, S.; Trigoni, N.; Roberts, S. Anomaly Detection for Time Series Using VAE-LSTM Hybrid Model. In Proceedings of the ICASSP 2020—2020 IEEE International Conference on Acoustics, Speech and Signal Processing (ICASSP), Barcelona, Spain, 4–8 May 2020; pp. 4322–4326. [CrossRef]
43. Maurya, P. Web Development Using ReactJS. In Proceedings of the IEEE Conference on Web Development, Buffalo, NY, USA, 3–6 June 2020.
44. Raspberry Pi 4 Model B—Prototypal Board. Available online: https://www.raspberrypi.com/products/raspberry-pi-4-model-b/ (accessed on 15 January 2025).
45. Radogna, A.V.; Latino, M.E.; Menegoli, M.; Prontera, C.T.; Morgante, G.; Mongelli, D.; Giampetruzzi, L.; Corallo, A.; Bondavalli, A.; Francioso, L. A Monitoring Framework with Integrated Sensing Technologies for Enhanced Food Safety and Traceability. *Sensors* **2022**, *22*, 6509. [CrossRef] [PubMed]
46. Fawcett, T. An introduction to ROC analysis. *Pattern Recognit. Lett.* **2006**, *27*, 861–874.
47. Powers, D.M.W. Evaluation: From Precision, Recall and F-Measure to ROC, Informedness, Markedness & Correlation. *J. Mach. Learn. Technol.* **2011**, *2*, 37–63.
48. TensorFlow (Version 2.16.1), Developed by Google Brain. The Software Was Sourced from Mountain View, California, USA. Available online: https://www.tensorflow.org (accessed on 15 January 2025).
49. Rahman, M.M.; Gupta, D.; Bhatt, S.; Shokouhmand, S.; Faezipour, M. A Comprehensive Review of Machine Learning Approaches for Anomaly Detection in Smart Homes: Experimental Analysis and Future Directions. *Future Internet* **2024**, *16*, 139. [CrossRef]

sensors

MDPI

Article

Robust Access Control for Secure IoT Outsourcing with Leakage Resilience

Khaled Riad

Computer Science Department, College of Computer Sciences & Information Technology, King Faisal University, Al-Ahsa 31982, Saudi Arabia; kriad@kfu.edu.sa

Abstract: The Internet of Things (IoT) has revolutionized various industries by enabling seamless connectivity and data exchange among devices. However, the security and privacy of outsourced IoT data remain critical challenges, especially given the resource constraints of IoT devices. This paper proposes a robust and leakage-resilient access control scheme based on Attribute-Based Encryption (ABE) with partial decryption outsourcing. The proposed scheme minimizes computational overhead on IoT devices by offloading intensive decryption tasks to the cloud, while ensuring resilience against master secret key leakage, side-channel attacks, and other common security threats. Comprehensive security analysis demonstrates the scheme's robustness under standard cryptographic assumptions, and performance evaluations show significant improvements in decryption efficiency, scalability, and computational performance compared to existing solutions. The proposed scheme offers a scalable, efficient, and secure access control framework, making it highly suitable for real-world IoT deployments across domains such as smart healthcare, industrial IoT, and smart cities.

Keywords: Internet of Things (IoT); access control; leakage resilience; attribute-based encryption (ABE); outsourced decryption; master secret key protection; side-channel attacks; secure data outsourcing

1. Introduction

The Internet of Things (IoT) is a network consisting of various intelligent devices and gadgets, including sensors, smartphones, and RFID tags [1]. The IoT has the potential to have a positive impact on our daily lives through applications such as smart city development, smart grid technologies, and advancements in healthcare services using intelligent medical devices [2]. In addition to providing convenience, the IoT has also stimulated its own advancement by enabling smart devices to monitor, collect, transmit, store, and share data within the IoT network. However, the increased connectivity and data sharing in IoT environments has raised significant security and privacy concerns [3]. IoT devices suffer from limited resources, which are represented in limited computational power and battery life. Consequently, traditional security mechanisms may not be directly applicable or efficient in IoT settings. Researchers have proposed various access control schemes for managing and securing IoT data [4,5]. Existing access control methods, such as lightweight encryption and cryptography algorithms, perform data evaluation on IoT devices, leading to security threats and efficiency issues.

As a result of the relevance and level of sensitivity of the IoT, information that is to be outsourced on untrusted cloud web servers requires effective access control in order to secure the information before it is outsourced [6]. In an open network environment, it is necessary to prevent unauthorized entities from accessing the information [7,8]. One of the primary access control methods for IoT is to leverage cryptographic techniques such as encryption and access control policies to protect the data. However, the limited computing and storage capabilities of IoT devices pose challenges in implementing robust and efficient access management protocols [9]. For IoT data, the sheer amount of data received from the IoT devices poses some challenges. Outsourcing data storage and processing to cloud or fog computing platforms has become an interesting solution to address the resource constraints of IoT devices [6]. However, the outsourcing of data and access control mechanisms introduces new security and privacy concerns, as data and access control policies are now managed by third-party service providers [10]. To address this issue, researchers have proposed the concept of secure outsourcing access control, where computationally intensive tasks are offloaded to more powerful cloud or fog computing entities, while IoT devices retain control over the data access [11].

When it comes to controlling data access, attribute encryption is a highly suitable choice. The innovation of attribute-based encryption determines whether the access should be granted or denied based on whether the users' attributes fulfill the access control framework [12]. Consequently, there is now less reliance on cloud providers for handling data access control. Thus, this paper introduces a new form of secure IoT data outsourcing with resilient protection against leakage [13]. Attribute-Based Encryption (ABE) techniques have been extensively studied for secure data sharing in IoT. However, these schemes do not adequately address the issue of potential leakage of sensitive information when access control mechanisms are outsourced [14].

When initially considering cryptographic access control, the well-known Ciphertext Policy Attribute-Based Encryption (CP-ABE) method is often mentioned. However, ABE itself involves significant computational costs, particularly during decryption. To improve security performance in this phase, offline/online technology may be utilized [15]. Moreover, ABE schemes producing ciphertexts of constant size can help alleviate communication overhead, but they are not suitable for IoT terminals due to resource constraints [16]. Additionally, in an IoT environment where numerous terminals are deployed across distributed settings, several security issues such as side-channel attacks may arise when attackers gain access to the system's master private keys [17]. If these keys are compromised, the entire system will lose its security integrity, and data privacy and safety within the system would face serious threats [18,19].

1.1. Motivation

The rapid proliferation of IoT devices has revolutionized multiple sectors, including healthcare, smart cities, and industrial automation. However, this growth has also introduced substantial challenges in managing the security and privacy of outsourced data, particularly when leveraging cloud and fog computing platforms for storage and processing. Traditional access control schemes, such as CP-ABE, offer fine-grained access control but suffer from computational overhead, scalability issues, and susceptibility to master key leakage. Furthermore, existing approaches often neglect the potential information leakage during outsourced computations, especially when sensitive access policies and cryptographic keys are managed by third-party servers. This oversight creates significant vulnerabilities in real-world IoT deployments, where resource-constrained devices cannot perform intensive cryptographic computations locally.

Recent advancements have focused on improving secure communication and data handling frameworks for distributed systems. For example, techniques explored in [20,21] provide innovative solutions for secure and efficient communication in dynamic environments. However, these works primarily address communication efficiency and sensor integration rather than fine-grained access control and resilience to key leakage. My proposed scheme builds on these foundations by introducing a leakage-resilient access control mechanism that offloads intensive computations to cloud servers while preserving the confidentiality of access policies and data. Unlike existing schemes, my approach integrates mechanisms for protecting against master secret key leakage and optimizes performance through partial decryption outsourcing. This ensures that even in resource-constrained environments, IoT devices can securely manage outsourced computations without sacrificing efficiency or scalability.

By addressing these critical gaps, my proposed access control model not only enhances the security of outsourced IoT data but also provides a scalable and computationally efficient framework that aligns with modern IoT security requirements.

1.2. Contributions

The main contributions in this paper are summarized as follows:

- This paper introduces a secure outsourcing access control model with leakage resilience to be used in the IoT environment, which takes advantage of attribute-based encryption in the access control.
- This paper designs efficient mechanisms to protect the access control scheme against potential information leakage when data and access policies are outsourced to third-party cloud or fog computing providers.
- This paper optimizes the computational and communication overheads of the access control scheme to accommodate the resource-constrained IoT devices by leveraging techniques such as offline/online encryption and decryption outsourcing.

Finally, comprehensive pilot evaluation of a variety of configurations confirmed the effectiveness and efficiency of the proposed access control system.

1.3. Acronyms and Notations

To facilitate understanding of the technical content, this paper provides a summary of the acronyms and notations used throughout this paper. Nomenclature and Abbreviations lists the key acronyms and notations, along with their respective descriptions. These definitions will help readers navigate the proposed scheme and its associated terminology with ease.

1.4. Organization

The remaining part of this paper is structured as follows: Section 3 introduces the system model and security requirements. Section 2 presents an overview of related studies and provides a comprehensive comparison between the CP-ABE schemes currently proposed and my access control solution. Section 4 outlines the fundamental preliminaries and notations that were taken into account during the development of my access control method. Section 5 explains the general description of my proposed model and the security model. In Section 6, this paper details the construction process for the proposed multidimensional access control scheme, including its five compatible algorithms. The security analysis and proofs are introduced in Section 7. An examination of the performance of the proposed scheme has been presented in Section 8. Finally, the paper summary is provided in Section 9.

2. Related Work

The problem of secure outsourcing access control with leakage resilience for the IoT has been studied in the literature [22–24]. Several existing works have proposed ABE schemes to achieve fine-grained access control in the IoT and cloud computing environments [24,25]. For example, ref. [22] presents an efficient decentralized attribute-based access control scheme for cloud storage, which supports user revocation. It incorporates a homomorphic encryption scheme to enable secure data storage and automatic data retrieval in the cloud [26]. However, these schemes do not consider the potential information leakage that may occur when the access control data and policies are outsourced to the cloud.

To address the information leakage issue, several research investigations have been introduced in this direction. Ref. [27] permits defining flexible access policies while efficiently revoking user attributes. However, the computational and communication overhead of these schemes may still be too high for resource-constrained IoT devices. More recently, ref. [28] proposed an efficient and secure attribute-based access control scheme for mobile cloud storage, which supports outsourcing of key generation and decryption to the cloud. This scheme reduces the computational overhead for IoT devices, but it does not consider the expected leakage of access control resources. Ref. [29] presented a leakage-resilient functional scheme. However, this scheme may not be suitable for IoT scenarios due to its high computational complexity.

The authors in [30] present a scheme called LR-ORCLS, which is relevant to the paper's focus on secure outsourcing access control with leakage resilience. This scheme addresses the issue of leakage resilience in outsourced signatures, which is crucial for ensuring the integrity and authenticity of outsourced data. Ref. [31] contribute a detailed security analysis of for IoT environments. Their formal security verification using the AVISPA tool demonstrates the effectiveness of the proposed protocol in protecting against known attacks, aligning with the paper's emphasis on secure access control mechanisms in the IoT. Ref. [32] categorize proposals related to access control in IoT into two main categories, focusing on ensuring the authenticity, confidentiality, and integrity of data streams during transmission. This categorization provides a broader perspective on access control mechanisms in the IoT, which is valuable for understanding the different approaches in the literature. Ref. [33] investigate leakage tracing enabled access control over outsourced data, emphasizing the importance of revoking leaked credentials and preparing judicial evidence. This aligns with the paper's focus on leakage resilience and the need for robust access control mechanisms to address potential data breaches.

The authors in [34] propose a privacy-preserving smart IoT-based healthcare big data storage system with self-adaptive access control. This aligns with the paper's focus on access control and security in IoT environments, particularly in sensitive domains such as healthcare, highlighting the importance of privacy-preserving access control mechanisms. Ref. [35] study the role of IoT security in managing enterprise human resource information leakage, emphasizing the need for the efficient and safe management of enterprise information. This reference provides insights into the broader implications of IoT security, including access control for sensitive enterprise data. Ref. [36] propose a scheme for a power IoT, which focuses on secure outsourcing decryption to resist selective ciphertext attacks. This contributes to the understanding of advanced encryption and access control mechanisms in IoT environments.

The authors in [37] present PROUD, a verifiable privacy-preserving outsourced attribute-based signcryption scheme supporting access policy update for cloud-assisted IoT applications. This scheme addresses the challenges of secure data designcryption process outsourcing, aligning with the paper's focus on secure outsourcing access control. In summary, the related work section provides a comprehensive review of the literature on access control and security in IoT environments, encompassing various aspects such as leakage resilience, formal security analysis, privacy preservation, and encryption mechanisms. The synthesis of these references contributes to a deeper understanding of the current state of research in secure outsourcing access control for IoT. The work by [33] provides valuable insights into the security concerns related to outsourcing and cloud computing. It emphasizes the need for tracing leakage and access control over outsourced data, aligning with my focus on secure outsourcing access control with leakage resilience for IoT. The research serves as a foundational reference for understanding the importance of addressing potential data leakage and unauthorized access in outsourced environments, which is crucial for my study's context.

The authors in [38] introduce a scheme, which addresses the challenge of ensuring confidentiality and access control for outsourced IoT data. This work is relevant, as it aligns with my focus on secure outsourcing access control for the IoT and provides insights into practical solutions for addressing confidentiality. The reference provided speaks to the task by offering a relevant and recent study that addresses IoT data outsourcing challenges. This work provides valuable insights and potential methodologies that can be compared and contrasted with the proposed scheme in the paper. The work by [39] presents a fine-grained access control mechanism for the Energy Internet, addressing the security challenges arising from data leakage in the collection, transmission, and storage processes.

The work by [40] is relevant to the current study. The paper discusses the delegation of computation tasks to untrusted cloud servers without compromising the confidentiality of the underlying data. This aligns with the current research's objective of secure outsourcing access control with leakage resilience for the Internet of Things, as both works emphasize the importance of secure and scalable access control in cloud computing environments. The proposed methodology aligns with the work on access control for outsourcing risks [41]. This work discusses an access control model to mitigate the risks associated with outsourcing, which is relevant to the proposed methodology for securing access control in the context of outsourcing. The combination of cryptography and access control techniques in the proposed methodology resonates with the need for robust access control measures highlighted in this work.

The work by [42] addresses the security challenges introduced by edge computing in the context of e-health. The proposed modular exponential outsourcing algorithm and Mask Algorithm with blind pairs aim to preserve the privacy of sensitive data during outsourced computing. This reference is relevant to the task, as it discusses the use of outsourcing calculation verifiability, aligning with the theme of secure outsourcing access control for IoT devices. The proposed access control scheme in [24] is also related to my work. It addresses the need for secure data sharing and privacy protection in IoT by proposing a scheme that utilizes attribute authentication and threshold policy. Both papers emphasize the importance of access control, specifically focusing on the use of ciphertext attribute authentication and threshold policy to achieve secure data access.

The reference [43] discusses the use of CP-ABE for achieving fine-grained data access control in cloud-based IoT environments. This is relevant to the task, as it provides insights into the access control mechanisms for outsourced data, which is a key aspect of the paper's focus on secure outsourcing access control for IoT. The use of attribute-hiding policy in CP-ABE aligns with the paper's emphasis on leakage resilience, making it a valuable source for the Related Work section and detailed discussions. The reference provided by [44] discusses an efficient scheme. This is relevant to my work, as it addresses the issue of information leakage during data transmission, which is crucial in the context of secure outsourcing access control for IoT devices. By leveraging this approach, the proposed system can enhance the overall efficiency and security of data sharing, aligning with the objectives of the paper.

The work in [45] is relevant, as it discusses the secure outsourcing of Ciphertext Policy Attribute-Based Encryption, which aligns with my focus on secure outsourcing access control for the Internet of Things and provides insights into secure outsourcing mechanisms, which can inform my approach to ensuring leakage resilience in IoT access control systems. The work by [46] provides valuable insights into policy-based access control for constrained healthcare resources in the context of the IoT. The paper discusses the use of XACML in managing permissions within computer networks, which is highly relevant to the topic of secure outsourcing access control in IoT systems. Various other methods for carrying out gain access to administration in the cloud with security are produced collectively along two research lines: ABE and selective encryption processes, such as ([47,48]). Also, another set of contributions [49,50] applies access control in the cloud by introducing trust notation for granting or denying access.

As technological advancements in quantum computing pose significant threats to traditional cryptographic systems, it is crucial to assess whether the proposed access control scheme can withstand post-quantum adversaries [51]. Classical security primitives, such as AES-128, SHA-256, and ECDSA, rely on mathematical problems that can be efficiently solved by quantum algorithms, like Shor's algorithm [52]. In contrast, post-quantum cryptographic primitives, including AES-256, SHA-384, and algorithms from the CRYSTALS suite, are specifically designed to resist quantum attacks [53]. The proposed scheme primarily relies on attribute-based encryption and bilinear maps, both of which are vulnerable to quantum adversaries. However, the modular structure of the proposed scheme allows for the underlying cryptographic primitives to be replaced or upgraded without requiring substantial changes to the access control logic. By substituting bilinear pairings and encryption components with post-quantum secure primitives, such as lattice-based encryption schemes, the proposed scheme can achieve quantum resilience while maintaining its functionality and efficiency.

In summary, existing access control schemes for IoT and cloud computing either do not consider the leakage resilience or do not provide efficient solutions for resource-constrained IoT devices.

3. System Model and Security Requirements

In this section, the detailed system model is presented, clearly defining the roles and responsibilities of each entity in the architecture to ensure clarity and coherence.

3.1. System Architecture

The proposed access control scheme operates within a cloud-assisted IoT system that consists of four primary entities, each with distinct responsibilities, as shown in Figure 1:

1. Attribute Authority (AA)
    - The AA is a trusted entity responsible for managing and distributing cryptographic keys and user attributes.
    - It generates the Master Secret Key (MSK), Public Parameter (PPs), user-specific Secret Keys (SKs), and Transformation Keys (TKs).
    - The AA ensures that the attributes assigned to each user align with their access privileges.
    - Security Role: Ensures attribute integrity and prevents unauthorized attribute assignment.
2. Cloud Service Provider (CSP)
    - The CSP provides storage and computational resources for encrypted IoT data.
    - It handles ciphertext storage and performs partial decryption operations on behalf of IoT devices using TK.

- Security Role: Operates as an honest-but-curious entity, meaning it follows protocol specifications but may attempt to infer information from stored or processed data.

3. IoT Devices
   - IoT devices (e.g., sensors, actuators, and smart cameras) collect and generate sensitive data.
   - Due to limited computational power and energy constraints, IoT devices outsource intensive cryptographic computations to the CSP.
   - Security Role: Encrypt data before transmission, ensuring that sensitive information remains secure during storage and processing.
4. IoT Users
   - IoT users are authorized individuals or systems that need access to specific IoT data.
   - They possess user-specific SKs issued by the Attribute Authority.
   - Users download partially decrypted ciphertext from the CSP and perform final decryption locally to retrieve plaintext data.
   - Security Role: Ensure private key confidentiality and verify the correctness of decryption results.

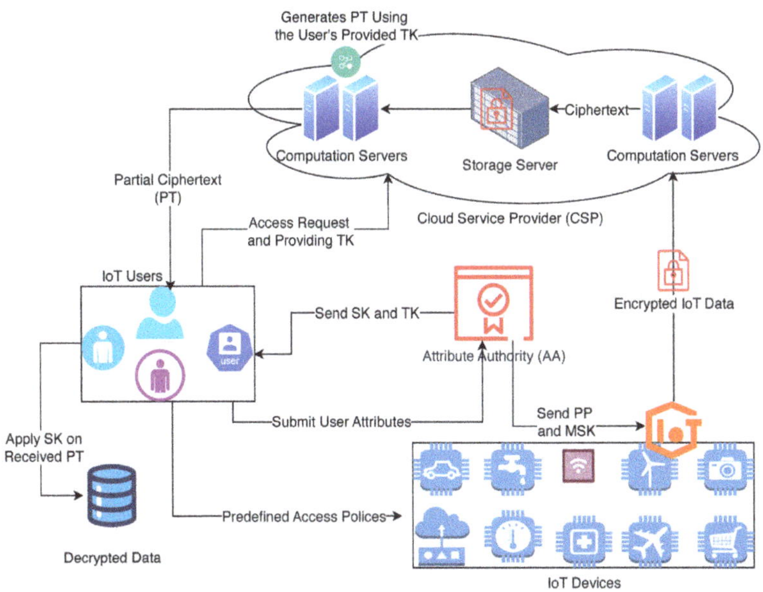

Figure 1. The system model showing the detailed communication steps among the four primary entities (Attribute Authority, Cloud Service Provider, IoT Devices, and IoT Users).

3.2. Workflow Overview

The overall workflow of the proposed system can be summarized in the following key steps:

1. System Initialization:
   - The AA generates (PPs) and the (MSK), which are distributed appropriately.
2. Key Generation:
   - The AA generates user-specific SKs and TKs based on their attribute sets.

3. Data Encryption:
    - IoT devices encrypt collected data using *PPs* and predefined access policies.
    - Encrypted data (ciphertext) are then uploaded to the *CSP* for storage.
4. Partial Decryption by the *CSP*:
    - Upon receiving a decryption request, the *CSP* uses the *TKs* associated with the user to perform partial decryption and generates a Partial Ciphertext (*PT*).
5. Final Decryption by IoT User:
    - The IoT user downloads the *PT* and applies their *SKs* to retrieve the plaintext message.

3.3. *Security Assumptions*

- Trusted Attribute Authority: The *AA* is considered fully trusted and will not deviate from protocol specifications.
- Honest-but-Curious *CSP*: The *CSP* follows the protocol honestly but may attempt to infer sensitive information from stored ciphertexts and access policies.
- Secure Communication Channels: All communication between entities (e.g., IoT devices, *CSP*, *AA*, and IoT users) occurs over secure communication channels protected by standard encryption protocols (e.g., TLS).
- Device Integrity: IoT devices are assumed to be secure against unauthorized access, though side-channel attack resilience is implemented as an added layer of security.

3.4. *Security Goals*

The proposed access control scheme should satisfy the following security requirements:

- Fine-grained Access Control: The scheme should support flexible and fine-grained access control, where access policies can be defined based on multiple attributes of IoT users.
- Leakage Resilience: The scheme should be resilient against potential information leakage when the access control data and policies are outsourced to the cloud service provider.
- Efficiency: The scheme should be computationally efficient for resource-constrained IoT devices by offloading the heavyweight cryptographic operations to the cloud service provider.
- Verifiability: The IoT users should be able to verify the correctness of the decryption results returned by the cloud service provider.

3.5. *Technical Challenges*

To achieve the above security requirements, the proposed access control scheme needs to address the following technical challenges:

- Designing a fine-grained attribute-based access control mechanism that can be efficiently outsourced to the cloud.
- Incorporating leakage-resilient techniques to secure the access control data confidentiality and policies when they are outsourced.
- Reducing the computational and communication overheads of the access control scheme to accommodate the resource-constrained IoT devices.
- Ensuring the verifiability of the decryption results returned by the cloud service provider.

4. Scheme Preliminaries

In this section, the basic knowledge and assumptions are introduced on which this paper relies on to construct the proposed new access control:

- Composite order Bilinear Groups
  The first composite order bilinear groups were introduced by Dan et al. [54]. This scheme randomly selects groups of $N = p_1p_2p_3$ order with different prime numbers p_1, p_2, and p_3. Let G express the composite order group. Let G_1, G_2, and G_3 express the three subgroups with order p_1, p_2, and p_3, respectively. The bilinear map with computable non-degenerate $G \times G \rightarrow G_T$ is characterized by the following properties:
  - Non-degenetate: $\forall g \in G\ e(g,g) \neq 1_{G_T}$
  - Bilinear: $\forall x, y \in Z_N \quad e(g^x, g^y) = e(g,g)^{xy}$

  To see this, if g generates G, $g^{p_1p_2}$ belongs to G_3, $g^{p_1p_3}$ belongs to G_2, and $g^{p_2p_3}$ belongs to G_1. $\forall \quad a \in G_1$, and $\forall \quad b \in G_2, e(a,b) = 1_{G_T}$.
- Vector Notation
  Choose G. Then, $\overrightarrow{u_1}, \overrightarrow{u_2} \in G^n$ is defined as follows:

$$e_n(\overrightarrow{u_1}, \overrightarrow{u_2}) = \prod_{i=1}^{n} e(u_{1i}, u_{2i}) \in G_T. \tag{1}$$

  where $\overrightarrow{u_1} =< u_{11}, u_{12}, \ldots, u_{1n} >$ and $\overrightarrow{u_2} =< u_{21}, u_{22}, \ldots, u_{2n} >$

Definition 1 (Discrete Logarithm Problem (DLP))**.** *Here is the finite field Z_p generated by the generator g. Select a group number $h \in Z_p$, and find an group element $a \in Z_p$ such that $g^a = h$ whenever such integer exist.*

5. Model Definition and Security Model

5.1. Model Definition

The proposed scheme is the extension of the Lewko scheme [55]. In the scheme, the decryption stage needs a lot of computation. The proposed scheme outsources the calculation in the decryption phase:

- $Ini(1^\lambda, U) \rightarrow (PP, MSK)$: This algorithm accepts λ (security parameter) as input and generates the PPs (public parameters) and the MSK (master secret key) as output.
- $KeyGen(MSK, S) \rightarrow (SK, TK)$: It accepts $S \in U$ and the MSK as inputs. It then generates the SK_S (secret key) of the user and the transformation key TK.
- $Enc(PK, (A, \rho), M) \rightarrow CT$: This algorithm accepts the PK, the access structure (A, ρ) and the message M as inputs and then generates the ciphertext CT.
- $Data \quad decryption$: The decryption algorithm contains the partial decryption algorithm on the cloud server and the decryption algorithm on the client side.
  - $Dec_{out}(TK, CT) \rightarrow PT$: The algorithm accepts CT, which is associated with the (A, ρ) and the transformation key TK as inputs, and then outputs the partial ciphertext PT.
  - $Dec(PT, SK) \rightarrow m$: The algorithm inputs the partial ciphertext PT and the user's private key SK and outputs the message M.

5.2. Security Model

In this section, the scheme's security has been defined through the game between A and B:

- Setup: The B process the $Ini(1^\lambda, U) \rightarrow (PP, MSK)$ to obtain the PK and the MSK. Then, the PK will be sent to the adversary A. The challenger also initializes an empty set D, where $D \subseteq 2^{U}$, and an empty table T, where $T \subseteq I \times 2^{U} \times SK \times TK$ (I represents a handle counter and TK represents the transformation key).

- Phase 1: A can repeat the upcoming inquiries to B.
  - $Create(S)$: A submits an attribute S to B to obtain the transformation key TK. B calls the $KeyGen(MSK, S) \rightarrow (SK, TK)$ to obtain the SK and the TK. B sets $i := i + 1$ and stores the entry (I, S, SK, TK) in T. Finally, TK is returned to the adversary.
  - $Corrupt(I)$: When the ith exists in the table T, B can obtain the entry (I, S, SK, TK), and return SK to A. When no such entry exists, the algorithm will be terminated. Finally, B sets $D := D \bigcup S$.
- Challenge: A sends a challenge access structure A^*, where the attribute set S in T is not satisfied with it, and two messages M_0, M_1 with the equal length to the challenger. The challenger randomly selects the two messages M_0, M_1 under A^* to encrypt; it then sends CT^* to A.
- Phase 2: It is the same phase 1 with the following restriction:
  - The attribute set whose private key will be asked is not satisfied with A^*.
  - A unable to request for decrypting of M_0 and M_1.
- Guess: The adversary outputs his guess b'.

  In this game, the advantage of the adversary is defined as

$$Adv_A := Pr|b' = b| - \frac{1}{2}. \tag{2}$$

6. Scheme Construction

The concrete scheme with outsourcing decryption and anti-master secret key leakage is as follows:

6.1. System Setup

$Ini(1^{\lambda}, U) \rightarrow (PP, MSK)$: This algorithm first selects a group of composite order $N = p_1p_2p_3$. Then, it randomly chooses the $\alpha, a \in Z_N$ and g_1 which belong to G_1. Let U represent attributes. The algorithm selects random $s_i \leftarrow Z_N, \forall i \in U$ and n, where $x_1, x_2, \ldots, x_n \in Z_N$. The algorithm also chooses $t^*, y_1, y_2, \ldots, y_n \in Z_N$ and $\overrightarrow{\rho} \in Z_N^{n+1}, \rho_{n+2} \in Z_N, \forall i \in U \rho_i{}' \in Z_N$ from the subgroup G_3.

Finally, The algorithm outputs PP: $PK = (N, g_1, g_3, g_1^{a}, e(g_1, g_1)^{\alpha}, g_1^{x_1}, g_2^{x_2}, \ldots, g_1^{x_n}, \forall i \in U \quad T_i = g_1^{s_i})$.

The master secret key MSK is as follows:

$$\begin{aligned} MSK &= (U, \overrightarrow{K_1^*}, L^*, \forall i \in U \quad K_i^*) \\ &= (U, < g_1^{y_1}, g_1^{y_2}, \ldots, g_1^{y_n}, g_1^{\alpha} g_1^{at^*} \prod_{i=1}^{n} g_1^{-x_i y_i} > \cdot g_3^{\overrightarrow{\rho}}, g_1^{t^*} g_3^{\rho_{n+2}}, \forall i \in S \quad T_i^{t^*} g_3^{\rho_i{}'}). \end{aligned} \tag{3}$$

6.2. Key Generation

$KeyGen(MSK, S) \rightarrow (SK, TK)$: It accepts an attribute set $S \in U$ and MSK. Then, it generates the SK and the TK. The algorithm randomly selects values $t, z, z_1, z_2, \ldots, z_n \in Z_N$. The TK is then represented as follows:

$$TK = (S, \overrightarrow{K_1^{z}}, L^{z}, \forall i \in S \quad K_i^{z}) \tag{4}$$

where

$$\overrightarrow{K_1} = \overrightarrow{K_1^*} \cdot < g_1^{z_1}, g_1^{z_2}, \ldots, g_1^{z_n}, g_1^{at} \prod_{i=1}^{n} g_1^{-x_i z_i} > \cdot g_3^{\overrightarrow{\rho}}, \tag{5}$$

$$L = L^* g_1^t g_3^{\rho_{n+2}}, \tag{6}$$

$$K_i = K_i^* T_i^t g_3^{\rho'} \tag{7}$$

The SK is represented as follows:

$$SK_s = z^{-1} mod \quad p \tag{8}$$

6.3. Data Encryption

$Enc(PK, (A, \rho), M) \rightarrow CT$: This algorithm inputs the public key PK, an access structure (A, ρ) ,where A is a $l \times n$ matrix, and ρ is a mapping $\rho(x) \rightarrow U$ from rows of matrix to the attributes and a message M. It outputs the final ciphertext. It randomly chooses $\overrightarrow{v} =< s, v_2, \ldots, v_n >\in Z_N^n$ and generates $\lambda_i = v \cdot A_i$. It also selects $r_x \in Z_N$ for A_x. Then, it randomly chooses r_i for $i \in [1, \ldots, l]$ and calculates C_i. The algorithm generates the partial CT.

$$CT = (C_0, \overrightarrow{C_1}, \{C_i, D_i\}_{i \in S}) \tag{9}$$

as:

$$C_0 = M(e(g_1, g_1)^{\alpha})^s, \tag{10}$$

$$\overrightarrow{C_1} =< (g_1^{x_1})^s, (g_1^{x_2})^s, \ldots, (g_1^{x_n})^s, g^s >, \tag{11}$$

$$\forall i \quad C_i = g^{a\lambda_i} T_i^{r_i}, \tag{12}$$

$$\forall i \quad D_i = g_1^{r_i} \tag{13}$$

Finally, the CT is represented as

$$CT = (C_0, \overrightarrow{C_1}, \{C_i, D_i\}_{i \in S}) \tag{14}$$

6.4. Data Decryption

The decryption algorithm contains the partial decryption algorithm on the cloud server and the decryption algorithm on the client side.

$Dec_{out}(TK, CT) \rightarrow PT$: The algorithm inputs the ciphertext CT associated with the access structure (A, ρ) and the transformation key TK and then outputs the partial ciphertext PT. It is a polynomial time to find a set $\sum_{i \in [1, \ldots, l]} \omega_i \lambda_i = s$. This algorithm calculates the outsourced decryption ciphertext PT as

$$PT = \frac{e_{n+1}(\overrightarrow{C_1}, \overrightarrow{K_1^z})}{\prod_{\rho(x) \in S}(e(C_x, L^z) e(D_x, K_x^z))^{\omega_i}} = e(g_1, g_1)^{\alpha s z} \tag{15}$$

The partial operations of the above equation are as follows:

$$\begin{aligned} e_{n+1}(\overrightarrow{C_1}, \overrightarrow{K_1^z}) &= e(\overrightarrow{C_1}, \overrightarrow{K_1^{*z}}) \cdot e(\overrightarrow{C_1}, (g_1^{z_1}, g_1^{z_2}, \ldots, g_1^{z_n}, g_1^{at} \prod_{i=1}^{n} g_1^{-x_i z_i})) \\ &= e(g_1, g_1)^{\alpha s z} e(g_1, g_1)^{aszt^*} e(g_1, g_1)^{aszt} \cdot e(C_x, L^z) e(D_x, K_x^z) \\ &= e(C_1, L^*)^z e(C_1, g_1^t g_3^{\rho_{n+2}})^z e(D_x, k_i^*)^z e(D_x, T_i g_3^{\rho'})^z \\ &= e(g_1, g_1)^{az\lambda_i t^*} e(g_1, g_1)^{azt\lambda_i} \end{aligned} \tag{16}$$

If the attribute sets satisfy the access control policy, it will be able to decrypt correctly. Otherwise, it will not obtain the correct plaintext in the next step.

$Dec(PT, SK) \rightarrow m$: The algorithm inputs the partial ciphertext PT and the user's private key SK_S and outputs the message M.

$$m = \frac{C_0}{PT^{SK_S}} \tag{17}$$

7. Security Analysis

In this section, the in-depth security analysis of the proposed access control scheme is presented, focusing on its resilience against various potential attack vectors, including unbreakable DLP, side-channel attacks, master secret key leakage, and other known cryptographic vulnerabilities.

7.1. Unbreakable DLP

Theorem 1. *Attackers cannot break the proposed system in polynomial time under the DLP.*

Proof. The proposed scheme is an extension of the Lewko scheme [55], which is denoted by $\prod_L = (Setup_L, Keygen_L, Encrypt_L, Decrypt_L)$. The scheme $\prod_L$ is (l_{MSK}, l_{SK}) master–leakage secure. The adversary cannot break it in polynomial time. If the adversary can break the DLP, the adversary can break the proposed scheme $\prod = (Ini, Keygen, Encrypt, Decrypt_{out}, decrypt)$ in polynomial time. The stimulator B will interact with the challenger C of the Lewko scheme and the adversary A. □

Setup: It is easy to find that the system parameters of the proposed scheme in initialization stage are the same as those of the Lewko's scheme. So, the stimulator B asks the challenger C of the scheme [55] about the system parameters. C returns the PK to B. Then, B sends it to A.

Phase 1: At this stage, when A sends S to B to obtain SK_S, B will send S to C and obtain the SK_S. The stimulator B selects a value $z \in Z_N$ randomly and computers $\vec{K_1^z}, L^z, \forall i \in S \quad K_i^z$. Let z^{-1} be the private key. The TK is represented as $TK = (S, \vec{K_1^z}, L^z, \forall i \in S \quad K_i^z)$. After that, it stores the entry (I, S, SK, TK) in table T and returns the transformation key TK and SK to the adversary A.

Challenge Phase: The adversary A sends to the B the messages M_0, M_1 and the A^*. B selects $b \in \{0, 1\}$. B sends M_b and A^* to the challenger C. The challenger C returns the ciphertext $CT = (C_0, C_1, \{C_i, D_i\}_{i \in S})$. B sends CT to A as the challenge ciphertext CT^*.

Phase 2: This phase is the same as Phase 1 with restriction that the attribute set of which private key will be asked is not satisfied with A^*. The adversary is unable to request to decrypt M_0 and M_1

Guess: The adversary outputs his guess b'.

If the adversary A has a probability advantage that cannot be ignored in this game, adversaries will solve the DLP. In other words, attackers cannot break the proposed system in polynomial time under the DLP.

7.2. Resilience Against Side-Channel Attacks

Side-channel attacks exploit physical information leakage during cryptographic operations, such as power consumption, electromagnetic emissions, or timing analysis, to deduce sensitive information like encryption keys or intermediate computation states. These attacks pose a significant threat to resource-constrained IoT devices, where operations are often conducted on limited hardware with fewer defenses against side-channel analysis. The proposed scheme incorporates several strategies to mitigate the risk of side-channel attacks:

- Outsourced Decryption: The most computationally expensive cryptographic operations, including bilinear pairing and modular exponentiation, are offloaded to cloud servers. This significantly reduces the number of resource-intensive computations performed on IoT devices, thereby minimizing the attack surface for side-channel adversaries.
- Constant-Time Algorithms: The cryptographic algorithms used in the proposed scheme have been implemented in a constant-time manner, ensuring that execution time does not vary with input values. This eliminates timing-based side-channel vulnerabilities.
- Transformation Key Isolation: The transformation key, used during the outsourced decryption phase, ensures that sensitive intermediate results are not exposed directly to the IoT device. Even if partial information is leaked from the IoT device, the master secret key remains secure, and the plaintext cannot be reconstructed without the transformation key and private key.
- Randomization of Intermediate States: During the key generation and decryption phases, random noise is added to the intermediate computation states, making it infeasible for adversaries to deduce meaningful patterns from observed emissions or power traces.

Proof of Security Against Side-Channel Attacks: Assume an adversary A attempts to extract secret information through side-channel analysis during decryption. Since the most vulnerable cryptographic operations are performed on the cloud server rather than the IoT device, the adversary's view is limited to observing lightweight operations such as final partial decryption. Furthermore, randomized intermediate states and constant-time execution ensure that no meaningful side-channel information can be extracted during local computations. As such, the adversary cannot derive the master secret key or plaintext solely through side-channel information, even if they have physical access to the IoT device.

7.3. Resilience Against Master Secret Key Leakage

The proposed scheme integrates anti-leakage techniques by leveraging composite order groups and key splitting mechanisms. Even if partial information about the master secret key is exposed, the attacker cannot reconstruct the full key without access to additional secure components distributed across different domains. This multilayered key protection mechanism ensures the confidentiality and integrity of the access control system even under master key exposure.

7.4. Resistance to Collusion Attacks

Collusion attacks occur when multiple unauthorized users pool their attribute keys to gain illegitimate access to encrypted data. The proposed scheme prevents such attacks by assigning unique transformation and private keys to each user based on their specific attribute set. Even if multiple users collude and combine their attribute keys, they cannot bypass the access policy or reconstruct the plaintext without satisfying the defined access structure.

7.5. Resistance to Replay Attacks

Replay attacks involve an adversary that reuses previously intercepted messages to gain unauthorized access. In the proposed scheme, ciphertexts are tied to unique access policies and transformation keys, rendering intercepted ciphertexts useless in future sessions. The scheme also uses randomization during encryption to ensure that each ciphertext is unique, even for identical plaintext.

Finally, the proposed access control scheme demonstrates robust security properties against a wide range of attack vectors, including side-channel attacks, master secret key leakage, collusion attacks, and replay attacks. By offloading resource-intensive cryptographic operations to cloud servers, incorporating randomization techniques, and employing constant-time algorithms, the proposed scheme significantly reduces vulnerabilities in resource-constrained IoT devices. Furthermore, reliance on established cryptographic hardness assumptions ensures resilience against traditional and quantum cryptographic attacks. This comprehensive analysis confirms the effectiveness of the proposed scheme in maintaining data confidentiality, access control integrity, and operational security in IoT environments.

8. Performance Analysis

In this section, the comprehensive performance evaluation of the proposed access control scheme has been presented, comparing it with multiple existing state-of-the-art schemes, including the Lewko et al. [55] scheme, Chase et al. [56]'s multiauthority ABE scheme, and Zhang et al. [57] constant-size ciphertext ABE scheme. The evaluation focuses on key performance metrics, including encryption time, decryption time, ciphertext size, and computational overhead on IoT devices.

8.1. Evaluation Metrics

1. Encryption Time: Measures the time required to encrypt a plaintext message under an attribute-based access policy.
2. Decryption Time: Measures the time taken to decrypt ciphertext and retrieve plaintext, including both partial decryption (outsourced to CSP) and final decryption (on IoT users' side).
3. Ciphertext Size: Represents the size of the encrypted data, directly impacting communication overhead.
4. Computational Overhead: Evaluates the computational burden on IoT devices, which is critical given their resource constraints.
5. Scalability: Assesses the performance impact when the number of attributes and users increases.

8.2. Comparative Analysis

The comparative results shown in Table 1 indicate the following:

- Decryption Time: The proposed scheme significantly reduced decryption time on IoT devices by outsourcing the most computationally intensive operations to the cloud server. In contrast, Lewko et al.'s scheme imposed a high computational burden on IoT devices during decryption.
- Ciphertext Size: Zhang et al.'s scheme achieved constant ciphertext size, which minimized communication overhead. However, it came at the cost of higher complexity in encryption. The proposed scheme maintained a good balance between compact ciphertexts and efficient encryption.

- Computational Overhead: Both the proposed scheme and Zhang et al.'s scheme exhibited low computational overhead on IoT devices, making them suitable for resource-constrained environments.
- Scalability: The proposed scheme is highly scalable due to its efficient attribute management and outsourced computation design, which ensures minimal performance degradation as the number of users and attributes increases.

Table 1. The detailed comparative analysis for the proposed scheme with three others schemes.

Scheme	Encryption Time	Decryption Time	Ciphertext Size	Computational Overhead (IoT)	Scalability
Proposed Scheme	Moderate	Low	Compact	Low	High
Lewko et al. [55]	Moderate	High	Large	High	Moderate
Chase et al. [56]	High	High	Moderate	High	Moderate
Zhang et al. [57]	Low	Moderate	Constant	Low	Moderate

8.3. Experimental Results

This paper conducted simulations to compare the proposed scheme against Lewko et al.'s scheme, Chase et al.'s scheme, and Zhang et al.'s scheme on an experimental setup emulating resource-constrained IoT devices (e.g., Raspberry Pi 4 with 4GB RAM and ARM Cortex-A72 CPU). The cloud environment was simulated using a high-performance server with multi-core architecture.

Figure 2—Decryption time vs. number of attributes:

- The proposed scheme demonstrated a significantly lower decryption time as the number of attributes increased thanks to the partial decryption outsourced to CSP.
- In Lewko et al.'s scheme, the decryption time scaled linearly with the number of attributes, causing performance degradation on the IoT devices.

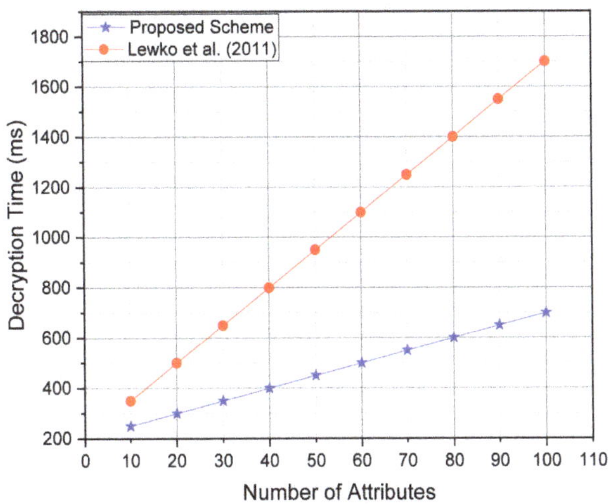

Figure 2. Decryption time against the number of attributes, ref. [55].

Figure 3—Encryption time comparison:

- The encryption time in the proposed scheme was moderate, and it remained competitive with Lewko et al.'s and Zhang et al.'s schemes.
- Chase et al.'s scheme exhibited higher encryption times, limiting its suitability for real-time IoT applications.

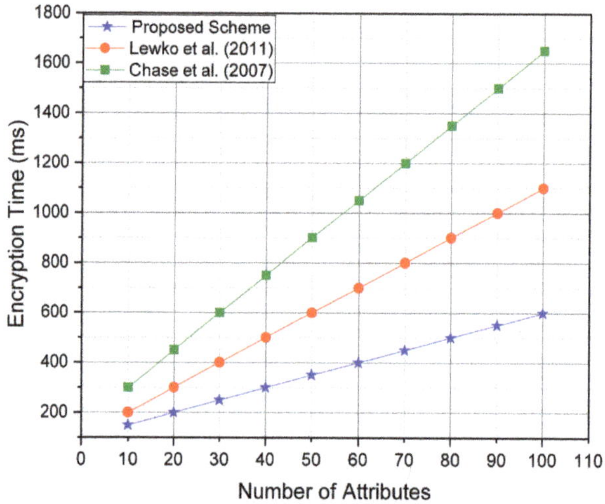

Figure 3. Encryption time comparison, refs. [55,56].

Figure 4—Ciphertext size comparison:

- Zhang et al.'s scheme achieved a constant ciphertext size, while the proposed scheme maintained compact ciphertexts with a slight increase as attributes grew.
- Lewko et al.'s ciphertext size grew significantly with more attributes, increasing storage and communication overhead.

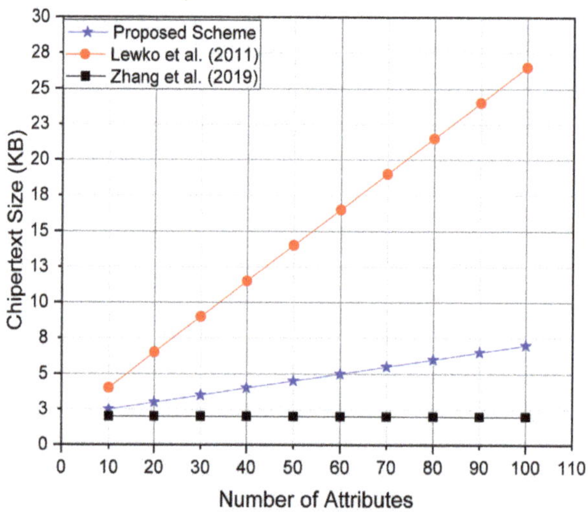

Figure 4. Ciphertext size comparison, refs. [55,57].

Finally, the experimental evaluation and comparative analysis demonstrate that the proposed scheme achieves superior performance in terms of decryption efficiency, computational overhead, and scalability compared to existing access control schemes. By leveraging outsourced decryption and optimizing attribute management, the proposed scheme addresses the limitations of traditional attribute-based encryption models while meeting the stringent requirements of IoT environments.

9. Conclusions

This paper presents a leakage-resilient access control scheme designed for resource-constrained IoT environments. By leveraging ABE with partial decryption outsourcing, the scheme significantly reduces the computational burden on IoT devices while ensuring fine-grained access control and resilience against master key leakage and side-channel attacks. The introduced security analysis confirms the scheme's robustness against common attack vectors, while performance evaluations demonstrate superior efficiency, scalability, and reduced computational overhead compared to existing solutions. The scheme is particularly well suited for real-world IoT applications, including smart healthcare, industrial IoT, and smart cities.

Future work will focus on integrating post-quantum cryptographic primitives, enhancing user privacy protections, and enabling cross-domain interoperability to further strengthen the scheme's adaptability and resilience in evolving IoT ecosystems.

Funding: This work was supported by the Deanship of Scientific Research, Vice Presidency for Graduate Studies and Scientific Research, King Faisal University, Saudi Arabia [Grant No. KFU250164].

Data Availability Statement: Data are contained within the article.

Acknowledgments: The author extends his appreciation to the Deanship of Scientific Research, Vice Presidency for Graduate Studies and Scientific Research, King Faisal University, Saudi Arabia [Grant No. KFU250164].

Conflicts of Interest: The author declares no conflicts of interest.

Nomenclature and Abbreviations

Acronyms and notations:

Symbol/Acronym	Description
Acronyms	
ABE	Attribute-Based Encryption
AA	Attribute Authority
CSP	Cloud Service Provider
IoT	Internet of Things
DLP	Discrete Logarithm Problem
CP-ABE	Ciphertext Policy Attribute-Based Encryption
Notations	
λ	Security Parameter
U	Universal Attribute Set
MSK	Master Secret Key
PP	Public Parameters
SK	Secret Key
TK	Transformation Key
A	Access Structure (matrix)
ρ	Mapping from Attributes to Rows of A
CT	Ciphertext
PT	Partial Ciphertext
g, g_1, g_3	Generators of Composite-Order Groups
$e(\cdot,\cdot)$	Bilinear Pairing
s	Random Secret Shared in the Sccess Structure
$v, \vec{v}$	Random vector used in Encryption
z	Random Value used in Key Generation
M	Plaintext Message

References

1. Zolfaghari, B.; Yazdinejad, A.; Dehghantanha, A.; Krzciok, J.; Bibak, K. The Dichotomy of Cloud and IoT: Cloud-Assisted IoT from a Security Perspective. *arXiv* **2022**, arXiv:2207.01590.
2. Cook, J.; ur Rehman, S.; Khan, M.A. Security and Privacy for Low Power IoT Devices on 5G and Beyond Networks: Challenges and Future Directions. *IEEE Access* **2023**, *11*, 39295–39317. [CrossRef]
3. Farman, H.; Jan, B.; Javed, H.; Ahmad, N.; Iqbal, J.; Arshad, M.; Ali, S. Multi-criteria based zone head selection in Internet of Things based wireless sensor networks. *Future Gener. Comput. Syst.* **2018**, *87*, 364–371. [CrossRef]
4. Huang, X.; Craig, P.; Lin, H.; Yan, Z. SecIoT: A security framework for the Internet of Things. *Secur. Commun. Netw.* **2015**, *9*, 3083–3094. [CrossRef]
5. Bate, K.O.; Kumar, N.; Khatri, S.K. Framework for authentication and access control in IoT. In Proceedings of the 2017 2nd International Conference on Telecommunication and Networks (TEL-NET), Noida, India, 10–11 August 2017. [CrossRef]
6. Alshehri, A.H.; Sandhu, R. Access Control Models for Cloud-Enabled Internet of Things: A Proposed Architecture and Research Agenda. In Proceedings of the 2016 IEEE 2nd International Conference on Collaboration and Internet Computing (CIC), Pittsburgh, PA, USA, 1–3 November 2016. [CrossRef]
7. Carlin, A.; Hammoudeh, M.; Aldabbas, O. Defence for distributed denial of service attacks in cloud computing. *Procedia Comput. Sci.* **2015**, *73*, 490–497. [CrossRef]
8. Belguith, S. Enhancing data security in cloud computing using a lightweight cryptographic algorithm. In Proceedings of the 11th International Conference on Autonomic and Autonomous Systems, Rome, Italy, 24–29 May 2015.
9. Li, F.; Rahulamathavan, Y.; Conti, M.; Rajarajan, M. Robust access control framework for mobile cloud computing network. *Comput. Commun.* **2015**, *68*, 61–72. [CrossRef]
10. Raptis, T.P.; Passarella, A.; Conti, M. Data Management in Industry 4.0: State of the Art and Open Challenges. *IEEE Access* **2019**, *7*, 97052–97093. [CrossRef]
11. Bertin, E.; Hussein, D.; Sengul, C.; Frey, V. Access control in the Internet of Things: A survey of existing approaches and open research questions. *Ann. Telecommun.* **2019**, *74*, 375–388. [CrossRef]
12. Bethencourt, J.; Sahai, A.; Waters, B. Ciphertext-Policy Attribute-Based Encryption. In Proceedings of the 2007 IEEE Symposium on Security and Privacy, Berkeley, CA, USA, 20–23 May 2007. [CrossRef]
13. Goswami, P.; Faujdar, N.; Debnath, S.; Khan, A.K.; Singh, G. Investigation on storage level data integrity strategies in cloud computing: Classification, security obstructions, challenges and vulnerability. *J. Cloud Comput.* **2024**, *13*, 45. [CrossRef]
14. Luqman, A.; Mahesh, R.; Chattopadhyay, A. Privacy and Security Implications of Cloud-Based AI Services: A Survey. *arXiv* **2024**, arXiv:2402.00896. [CrossRef]
15. Gorasia, N.; Srikanth, R.; Doshi, N.; Rupareliya, J. Improving Security in Multi Authority Attribute Based Encryption with Fast Decryption. *Procedia Comput. Sci.* **2016**, *79*, 632–639. [CrossRef]
16. Zhang, X.; Wu, F.; Wang, Y.; Zhao, W.; Wang, W. Multi-authority attribute-based encryption scheme with constant-size ciphertexts and user revocation. *Concurr. Comput. Pract. Exp.* **2018**, *31*, e4678. [CrossRef]
17. Bursztein, E.; Invernizzi, L.; Kral, K.; Moghimi, D.; Picod, J.M.; Zhang, M.Y. Generic Attacks against Cryptographic Hardware through Long-Range Deep Learning. *arXiv* **2023**, arXiv:2306.07249. [CrossRef]
18. Green, M.; Hohenberger, S.; Waters, B. Outsourcing the decryption of ABE ciphertexts. In Proceedings of the 20th USENIX Security Symposium (USENIX Security 11), San Francisco, CA, USA, 8–12 August 2011.
19. Roman, R.; Zhou, J.; Lopez, J. On the features and challenges of security and privacy in distributed internet of things. *Comput. Netw.* **2013**, *57*, 2266–2279. [CrossRef]
20. Ma, Y.; Ma, R.; Lin, Z.; Zhang, R.; Cai, Y.; Wu, W.; Wang, J. Improving Age of Information for Covert Communication With Time-Modulated Arrays. *IEEE Internet Things J.* **2024**, *12*, 1718–1731. [CrossRef]
21. Zhang, R.; Cheng, L.; Wang, S.; Lou, Y.; Gao, Y.; Wu, W.; Ng, D.W.K. Integrated Sensing and Communication with Massive MIMO: A Unified Tensor Approach for Channel and Target Parameter Estimation. *IEEE Trans. Wirel. Commun.* **2024**, *23*, 8571–8587. [CrossRef]
22. Chen, J.; Ma, H. Efficient decentralized attribute-based access control for cloud storage with user revocation. In Proceedings of the 2014 IEEE International Conference on Communications (ICC), Sydney, Australia, 10–14 June 2014. [CrossRef]
23. Yang, K.; Jia, X.; Ren, K. Attribute-based fine-grained access control with efficient revocation in cloud storage systems. In Proceedings of the 8th ACM SIGSAC Symposium on Information, Computer and Communications Security, Berlin, Germany, 4–8 November 2013. [CrossRef]
24. Zhang, Q.; Li, Y.; Yuan, J.; Gan, Y.; Luo, X. Access Control Based on Ciphertext Attribute Authentication and Threshold Policy for the Internet of Things. *Sensors* **2019**, *19*, 5237. [CrossRef]
25. Xu, S.; Yang, G.; Mu, Y.; Liu, X. A secure IoT cloud storage system with fine-grained access control and decryption key exposure resistance. *Future Gener. Comput. Syst.* **2019**, *97*, 284–294. [CrossRef]

26. Sowmiya, M.; Adimoolam, M. Secure cloud storage model with hidden policy attribute based access control. In Proceedings of the 2014 International Conference on Recent Trends in Information Technology, Chennai, India, 10–12 April 2014. [CrossRef]
27. Myint, P.W.W.; Hlaing, S.Z.; Htoon, E.C. A Policy Revocation Scheme for Attributes-based Encryption. In Proceedings of the 10th International Conference on Advances in Information Technology, Bangkok, Thailand, 10–13 December 2018. [CrossRef]
28. Deng, H.; Qin, Z.; Wu, Q.; Guan, Z.; Yin, H. Revocable Attribute-Based Data Storage in Mobile Clouds. *IEEE Trans. Serv. Comput.* **2022**, *15*, 1130–1142. [CrossRef]
29. Yu, Z.; Au, M.H.; Xu, Q.; Yang, R.; Han, J. Towards leakage-resilient fine-grained access control in fog computing. *Future Gener. Comput. Syst.* **2018**, *78*, 763–777. [CrossRef]
30. Tseng, Y.; Wu, J.; Huang, S.; Tsai, T. Leakage-resilient outsourced revocable certificateless signature with a cloud revocation server. *Inf. Technol. Control* **2020**, *49*, 464–481. [CrossRef]
31. Das, A.K.; Wazid, M.; Yannam, A.R.; Rodrigues, J.J.P.C.; Park, Y.H. Provably Secure ECC-Based Device Access Control and Key Agreement Protocol for IoT Environment. *IEEE Access* **2019**, *7*, 55382–55397. [CrossRef]
32. Miorandi, D.; Sicari, S.; Pellegrini, F.D.; Chlamtac, I. Internet of Things: Vision, Applications and Research Challenges. *Ad Hoc Netw.* **2012**, *10*, 1497–1516. [CrossRef]
33. Deng, H.; Wu, Q.; Qin, B.; Mao, J.; Liu, X.; Zhang, L.; Shi, W. Who Is Touching My Cloud. In *Computer Security, Proceedings of the 19th European Symposium on Research in Computer Security, Wroclaw, Poland, 7–11 September 2014*; Springer: Cham, Switzerland, 2014. [CrossRef]
34. Yang, Y.; Zheng, X.; Guo, W.; Liu, X.; Chang, V. Privacy-Preserving Smart IoT-based Healthcare Big Data Storage and Self-Adaptive Access Control System. *Inf. Sci.* **2019**, *479*, 567–592. [CrossRef]
35. Zeng, Z.; Zhang, J. Based on the role of internet of things security in the management of enterprise human resource information leakage. *Wirel. Commun. Mob. Comput.* **2021**, *2021*, 5936390. [CrossRef]
36. Cheng, R.; Wu, K.; Su, Y.; Li, W.; Cui, W.; Tong, J. An Efficient ECC-Based CP-ABE Scheme for Power IoT. *Processes* **2021**, *9*, 1176. [CrossRef]
37. Belguith, S.; Kaaniche, N.; Hammoudeh, M.; Dargahi, T. PROUD: Verifiable Privacy-Preserving Outsourced Attribute Based SignCryption Supporting Access Policy Update for Cloud Assisted IoT Applications. *Future Gener. Comput. Syst.* **2020**, *111*, 899–918. [CrossRef]
38. Liu, L.; Wang, H.; Zhang, Y. Secure iot data outsourcing with aggregate statistics and fine-grained access control. *IEEE Access* **2020**, *8*, 95057–95067. [CrossRef]
39. Miao, S.; Zhang, X.; Liu, Z. Fine-grained access control mechanism of energy internet. *Wuhan Univ. J. Nat. Sci.* **2022**, *27*, 231–239. [CrossRef]
40. Yu, S.; Wang, C.; Ren, K.; Lou, W. Achieving Secure, Scalable, and Fine-Grained Data Access Control in Cloud Computing. In Proceedings of the IEEE Conference on Computer Communications, San Diego, CA, USA, 14–19 March 2010. [CrossRef]
41. Meligy, A.; Diab, H.; Torky, M. An Access Control Model for Avoiding Outsourcing Risks. *Int. J. Comput. Appl.* **2012**. [CrossRef]
42. Zhang, L.; Gao, X.; Mu, Y. Secure data sharing with lightweight computation in e-health. *IEEE Access* **2020**, *8*, 209630–209643. [CrossRef]
43. Hao, J.; Huang, C.; Ni, J.; Rao, H.; Xian, M.; Shen, X. Fine-grained data access control with attribute-hiding policy for cloud-based iot. *Comput. Netw.* **2019**, *153*, 1–10. [CrossRef]
44. Xu, Z.; Zhang, S.; Han, H.; Dong, X.; Zhang, Z.; Wang, H.; Tian, W. Blockchain-aided searchable encryption-based two-way attribute access control research. *Secur. Commun. Netw.* **2022**, *2022*, 2410455. [CrossRef]
45. Nguyen, K.T.; Oualha, N.; Laurent, M. Securely Outsourcing the Ciphertext-Policy Attribute-Based Encryption. *World Wide Web* **2017**, *21*, 169–183. [CrossRef]
46. Pal, S.; Hitchens, M.; Varadharajan, V.; Rabehaja, T. Policy-based access control for constrained healthcare resources in the context of the internet of things. *J. Netw. Comput. Appl.* **2019**, *139*, 57–74. [CrossRef]
47. Riad, K. Multi-Authority Trust Access Control for Cloud Storage. In Proceedings of the 4th IEEE International Conference on Cloud Computing and Intelligence Systems (CCIS), Beijing, China, 17–19 August 2016; pp. 429–433. [CrossRef]
48. Riad, K. Revocation Basis and Proofs Access Control for Cloud Storage Multi-Authority Systems. In Proceedings of the 3rd IEEE International Conference on Artificial Intelligence and Pattern Recognition (AIPR), Lodz, Poland, 19–21 September 2016; pp. 118–127. [CrossRef]
49. Riad, K. Blacklisting and Forgiving Coarse-grained Access Control for Cloud Computing. *Int. J. Secur. Appl.* **2016**, *10*, 187–200. [CrossRef]
50. Riad, K.; Yan, Z. Multi-Factor Synthesis Decision-Making for Trust-Based Access Control on Cloud. *Int. J. Coop. Inf. Syst.* **2017**, *26*, 1–33. [CrossRef]
51. Nannipieri, P.; Crocetti, L.; Di Matteo, S.; Fanucci, L.; Saponara, S. Hardware Design of an Advanced-Feature Cryptographic Tile within the European Processor Initiative. *IEEE Trans. Comput.* **2023**, 1–14. [CrossRef]
52. Bernstein, D.J.; Lange, T. Post-quantum cryptography. *Nature* **2017**, *549*, 188–194. [CrossRef]

53. Campagna, M.; LaMacchia, B.; Ott, D. Post Quantum Cryptography: Readiness Challenges and the Approaching Storm. *arXiv* **2021**, arXiv:2101.01269.
54. Dan, B.; Goh, E.J.; Nissim, K. Evaluating 2-DNF formulas on ciphertexts. In *Theory of Cryptography, Proceedings of the Second Theory of Cryptography Conference, TCC 2005, Cambridge, MA, USA, 10–12 February 2005*; Springer: Berlin/Heidelberg, Germany, 2005; pp. 325–341.
55. Lewko, A.; Rouselakis, Y.; Waters, B. Achieving leakage resilience through dual system encryption. In *Theory of Cryptography, Proceedings of the 8th Theory of Cryptography Conference, TCC 2011, Providence, RI, USA, 28–30 March 2011*; Springer: Berlin/Heidelberg, Germany, 2011; pp. 70–88.
56. Chase, M. Multi-authority Attribute Based Encryption. In *Theory of Cryptography, Proceedings of the 4th Theory of Cryptography Conference, TCC 2007, Amsterdam, The Netherlands, 21–24 February 2007*; Vadhan, S.P., Ed.; Springer: Berlin/Heidelberg, Germany, 2007; pp. 515–534.
57. Zhang, Y.; Li, J.; Yan, H. Constant Size Ciphertext Distributed CP-ABE Scheme with Privacy Protection and Fully Hiding Access Structure. *IEEE Access* **2019**, *7*, 47982–47990. [CrossRef]

sensors

MDPI

Review

Flexible Phased Antenna Arrays: A Review

Waleef Ullah Usmani *, Francesco Paolo Chietera and Luciano Mescia

Department of Electrical and Information Engineering, Politecnico di Bari, Via E. Orabona 4, 70125 Bari, Italy; francescopaolo.chietera@poliba.it (F.P.C.); luciano.mescia@poliba.it (L.M.)
* Correspondence: w.usmani@phd.poliba.it

Abstract: Flexibility in phased antenna arrays open the world of new applications. Such arrays can conform to different shapes while ensuring performance in harsh conditions. The purpose of this study is to perform a detailed comparative analysis of numerous studies of flexible phased antenna arrays. This work summarizes the main manufacturing techniques and the commonly used materials with their properties. It also outlines the key challenges and future trends in the development of flexible phased antenna arrays. The paper concludes with research recommendations to address the identified technical issues.

Keywords: flexible antennas; phased antenna arrays; additively manufactured antennas

1. Introduction

During the past few decades, Phased Arrays (PAs) have been prioritized over traditional arrays due to specific features such as fast and reliable beamforming, the ability to generate multiple beams with minimum interference, the ability to rapidly steer beams electronically, coverage of large areas without mechanically moving parts, and the ability to dynamically adapt to different operational needs [1].

1.1. Historical Background

PAs faced challenges of cost, size, and component limitations in the implementation of Electronically Scanned Antenna Arrays (ESAAs) before 1980s [2–4]. These systems implemented Passive Electronically Scanned Arrays (PESAs) instead of Active Electronically Scanned Arrays (AESAs). The implementation of AESAs was made possible after the inclusion of Gallium Arsenide (GaAs)-based circuits at a monolothic scale during 1980s. Monolithic phase shifters utilizing GaAs technology have been actively incorporated in applications such as electronic warfare and Satellite Communications (SATCOM) ever since [5,6]. The advancement to incorporate multiple functionalities on a single chip was utilized for phase and gain control across different frequency bands [7–9], consequently reducing the size of the arrays. But the lower yield and higher costs of the solution posed a hurdle to realize complex architectural choices for array design. The limitations were addressed through the implementation of silicon-based Radio-Frequency Integrated Circuits (RFICs), which offer enhanced integration, reduced size, and improved performance in PA, leading to an increase in the number of their applications. Silicon-based RFIC allowed the realization of fully integrated multi-element PA systems for the first time. The idea of a fully integrated PA system is to have a Transmitter (TX) and Receiver (RX) capable of beamforming, gain, and phase control, with integrated on-chip power amplifiers, and Low-Noise Amplifiers (LNAs) [10–13].

1.2. Silicon-Based RFIC Technology and Diverse Applications

The advancement of silicon-based RFIC technology led to fully integrated PA systems operating across multiple frequency bands since the late 2000s. Notable examples include Ka-band TX/RX on-chip systems [14], Q-band PA TX based on RFIC [15], and system-level implementations in X/Ku-band using single silicon chips [16], demonstrating complex PA architectures for various applications [17]. This evolution enabled compact and power-efficient solutions for 5G devices [18], where machine learning techniques have also been explored for PA synthesis [19]. PA have also been incorporated in Wireless Power Transfer (WPT) systems, such as near-field neuro-modulation [20] and channel-based intelligent WPT with phase control for beam steering [21]. Further applications include medical imaging and meteorological remote sensing [22,23]. Beam steering and amplitude tapering previously required element-wise calibration, but this was addressed in 2017 by a fully functional RFIC-based PA with orthogonal phase and amplitude control for 5G communications [24]. These developments were later adopted in SATCOM applications [25,26]. Silicon-based RFICs enabled efficient and compact PA systems with wide-band and wide-scan performance, scalability, high Effective Isotropic Radiated Power (EIRP), and improved Cross-Polarization Discrimination (XPD) [27]. The use of RFIC beamformer chips with Wilkinson combiner networks on Printed Circuit Board (PCB) allowed polarization-agile PA systems with high polarization purity and simultaneous multi-beam control [28–30]. Integration of multiple functions on a single chip reduced discrete components, improving form factor and lowering cost [31]. In 2014, a wafer-scale approach combining all components on the same silicon wafer further enhanced array compactness [32]. By addresing size, cost, efficiency, and scalability, silicon-based RFIC technology has enabled the transition from bulky, mechanically steered systems to compact AESAs, opening opportunities in consumer electronics, automotive, and wearable technologies.

1.3. Flexibility in Phased Arrays

After the utilization of fully integrated silicon-based RFIC technology in PA, the next step was to introduce flexibility in PA systems. Flexible Phased Arrays (FPAs) are significantly important in terms of applications like avionics, automotive, warfare, aerospace or even wearable devices. Practical utilization of FPAs started in early 2000s with the development of active PA antenna using membrane technology [33] and Micro-Electro-Mechanical System (MEMS) micro-strip phase shifters on a flexible Liquid Crystal Polymer (LCP) substrate [34]. They led to the development of fully integrated PA system on a single chip with RF-MEMS switches on a flexible LCP substrate for the first time in 2008 [35]. When flexibility is introduced, the resulting deformation changes the geometry of the antenna array. For the preservation of the radiation pattern, a phase compensation mechanism was developed for a self-adapting FPA [36]. In a similar work, more recently, another self-adapting array was developed which was able to dynamically compensate for performance degradation due to real-time deformations [37]. Manufacturing of FPAs has evolved over time with the advancement in Additive Manufacturing (AM) technology. For example, Inkjet Printing (IP) was used for the fabrication of an FPA system using a flexible Kapton substrate in 2013 [38]. AM precisely deposits materials on flexible substrates, paving way for lightweight, and adaptable systems. Despite the revolutionary potential, the evolution of technology still faces some significant challenges such as material limitations, mechanical deformation effects, thermal management, and design complexity.

1.4. Problem Statement

Although the FPA concept encompasses a wide spectrum of shape-adaptive array architectures, much of the existing literature has focused on individual implementations or specific technologies, rather than providing a comprehensive comparison across different solutions. Given the rapid evolution of enabling materials, manufacturing processes (both subtractive and additive), and design methodologies, a comparative analysis is needed to identify the key differentiators and limitations of each approach. Such analysis can be useful to guide future developments by clarifying performance metrics, integration challenges, and areas where further innovation is required. Accordingly, the following Section 2 presents different manufacturing techniques exploited for producing FPA; Section 3 highlights the materials used for their realization; Section 4 focuses on the major challenges identified in the field of FPA; while Section 5 presents a comparative analysis of the various technologies based on key performance parameters. Finally, Section 6 concludes the review, trying to guide the reader through the future trends of this promising field.

2. Manufacturing Techniques

Advancements in FPA systems are largely driven by the development of different manufacturing techniques with the aim of achieving lightweight, flexible, and high performance in the designs. In this section, a structured analysis of the different technologies used in the manufacturing of FPA is presented.

2.1. *Additive Manufacturing*

One of the most widely utilized AM techniques in PAs is IP. IP is a contactless deposition method where drops of functional ink are deposited onto a substrate, enabling the creation of conductive traces and antenna elements directly onto flexible substrates. The process is composed of several steps including ink formulation, substrate preparation, drop-on-demand printing, and post-processing. Conductive inks, usually based on Silver Nanoparticles (SNPs), are utilized with flexible substrates such as PET and polyimide (Kapton). To achieve the conductive films, nanoparticles must be interconnected through as sintering process (post-processing). Ink design aims to minimize sintering temperature for compatibility with flexible substrates. IP is a low-cost fabrication process which enables rapid design and customization while minimizing the material waste, and it is compatible with bendable substrates, but post-processing may impose limitations on substrate choices. IP is extensively utilized to fabricate conductive traces, antenna elements, and interconnections. This technology can also be combined with other AM methods, such as Fused Filament Fabrication (FFF), which is widely used for prototyping conformal or geometrically complex, and potentially flexible, antenna structures [39,40]. For example, a wideband circularly polarized (CP) patch antenna array was fabricated on Polyethylene terephthalate (PET) film using SNP-based inks for the first time [41]. Similarly, a two-dimensional PA employed IP carbon nanotube thin-film transistor (CNT-TFT)-based phase shifters on Kapton substrates [38]. IP was combined with FFF and Stereolithography (SLA) to design a wearable FPA with integrated microfluidic cooling channel for effective thermal management [42]. In Figure 1a, the overall system layout is shown, while Figure 1b illustrates the difference in uniformity of the conductive inkjet-printed ink between the sample exposed to Ultra-Violet Ozone (UVO) treatment and the one not exposed. This work showed robust performance under bending and thermal stress.

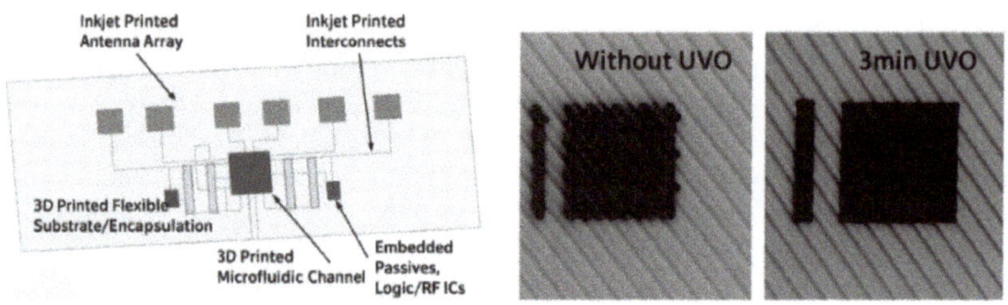

(a) (b)

Figure 1. (**a**) Proof-of-concept demonstration of flexible and wearable IP antenna array showing antennas, interconnects, flexible substrate, and microfluidic cooling channel. (**b**) Comparison of IP SNP ink on printed substrates with and without UVO surface treatment. Surface wetting can be enhanced by employing UVO treatment to increase the hydrophilicity of plastic surfaces [42].

In another solution for SATCOM and 5G millimeter wave applications, a tile-based solution of FPA was presented, exploiting IP for the fabrication of both the individual tiles and a flexible tiling layer [43]. Utilization of IP is not limited to only SATCOM and 5G millimeter wave applications, but its usefulness is also evident in case of Radio Frequency Identification (RFID) and sensing applications [44]. IP together with other AM techniques is useful for realizing low-profile and lightweight antenna elements. For example, a low-profile single-substrate multi-metal layer Ultra Wide Band (UWB) antenna was developed using Additive Manufacturing Electronics (AME) which is a particular kind of IP technology able to deal with both conductive SNP ink and dielectric (acrylate) materials, printing them in a multilayer configurations using an automated system [45]. The multilayer structure to achieve UWB is shown in Figure 2. The figure shows a three-layer structure of patch antenna. It consists of a driven patch and two stacked patches with truncated corners. The aim for the truncated corners is to induce CP.

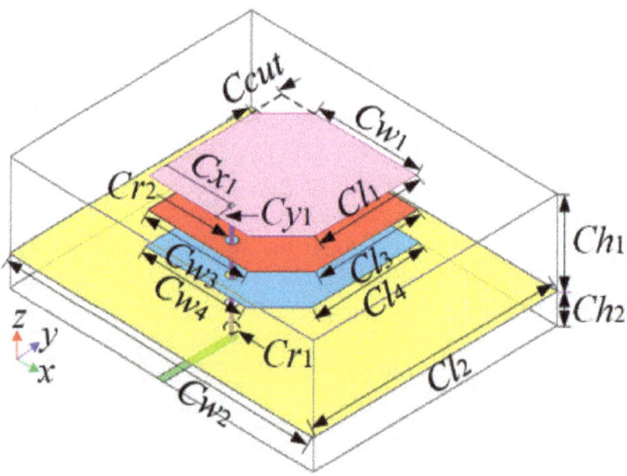

Figure 2. Three-layered structure of patch antenna (one driven patch and two stacked patches with truncated corners) used to ahieve UWB. Technique of truncated corners is used to induce CP [45].

This was the first demonstration of UWB, low-profile antennas using AME and integrating conductive and dielectric layers. In another work, fully IP PA was fabricated on a Kapton flexible substrate. The array was integrated with CNT-TFTs as switching elements for beam steering [46]. This was one of the earliest demonstrations of lightweight and conformal PA, but IP was already being utilized in fabrication of ultra-thin flexible substrates [47]. Another implementation of AM-based UWB antenna for FPAs was presented in [48]. A stacked patch antenna configuration was used to achieve wide bandwidth while the top and bottom patches were printed with IP onto a PET substrate. UWB was also achieved in the development of an ultrathin and flexible array using a three-layer laminate design and Frequency Selective Surface (FSS) [49]. IP has been widely utilized in numerous works due to its low cost and scalability. In addition to IP, additive fabrication facilitates the creation of complex, multifunctional PAs. Additive fabrication of multilayer flexible PCBs using Pyralux© All Polymide (AP) laminates was utilized in the fabrication of a 256-element FPA system with the aim of mechanical flexibility [50]. The front and back view of the final prototype of FPA system with 256-elements is presented in Figure 3. A similar design was presented for a lightweight FPA system integrating advanced electromagnetic components with self-deployable and self-contained power system. The PA used Pyralux®, AP8545R flexible polyimide substrate for mechanical flexibility [51]. In another work, an AM tile-based conformal array was designed by utilizing a computer vision based calibration of the array which was experimentally validated under symmetric and asymmetric bending configurations [52].

Figure 3. Photograph of the FPA antenna (front and back view). The FPA is equiped with deformation correction and is able to transfer RF- direct current (DC) wireless power transfer of about 80 mW at 1 m away from the transmitter [50].

2.2. Hybrid Rigid–Flexible Designs

The combination of rigid components with flexible substrates offers a practical balance between structural stability and mechanical adaptability. This concept is known as rigid–flex antenna array (RFAA), which is a hybrid architecture. It employs rigid radiators fabricated on substrates like Rogers®RO4350, integrated with flexible polyimide circuitry. These designs facilitate rollable and deployable configurations, such as for small satellite (SmallSat) applications. Rigid areas provide mechanical support for antenna elements and soldered active components, while flexible circuits route RF signals. In some circumstances, capacitive feed mechanisms eliminate the need for conductive vias, simplifying assembly while maintaining performance during repeated mechanical deformation [53]. In a similar attempt, a combination of flexible DuPont Pyralux®AP PCBs (flexible substrate) and rigid Rogers Duroid®5880 PCBs was recently utilized in a novel checkerboard configuration. The flexible material and rigid one were arranged in an alternating pattern to enable flexibility in both the x and y axes without compromising overall mechanical integrity. Novel checkerboard design provided the system with the ability to conform to a cylindrical geometry with a radius of 50 mm, expanding the range of potential applications [54].

Another work focusing on hybrid design for SATCOM demonstrated a deployable PA system where a 32-element Ka-band active PA was implemented on a flexible four-layer LCP substrate [55]. This trend of utilizing deployable PAs continued with more novelties, and in 2023, a heterosegmented LCP substrate was used combining six-layer and two-layer LCP substrates to achieve deployability and flexibility [56]. Other novel strategies have also been used to counter different challenges in FPAs. For example, to achieve a scalable production, a novel strategy of thermal drawing process was utilized for the design of a photonics based FPA for the first time. It integrated photonics and RF systems embedded in Polyetherimide (PEI) fiber carriers. This novel design achieved broadband matching and stable performance with flexibility across the ultra-high-frequency (UHF) band [57].

The diverse range of manufacturing techniques for FPAs reflected the adaptability of these methods to a large variety of application requirements, but along with the manufacturing techniques, the choice of materials used in the design of FPAs is also critically important.

3. Mostly Used Materials

An analysis of various studies on material characterization for FPAs reveals consistent trends in the selection of materials for both conductive and dielectric layers. Copper evidently emerges as the most widely used conductive material, particularly in traditional PCB-based implementations, where its high electrical conductivity and excellent RF performance make it the preferred choice. Its effectiveness also extends to rigid–flex approaches, ensuring low-loss transmission and reliable integration. Although chromium is also reported as a conductive material in some cases, its use is significantly less common [50,51,54,58–61]. In contrast, SNP ink is gaining attention for its compatibility with AM techniques, especially in the advanced manufacturing of flexible electronics. Despite its lower conductivity compared to copper, SNP remains a viable alternative for RF applications due to its excellent processability and suitability for printed electronics [43,44,62]. Its compatibility with IP processes makes it particularly attractive for complex geometries and conformal surfaces [41,42,63].

In terms of dielectric materials, the analysis highlights two main categories: semi-rigid dielectric substrates and flexible polymers compatible with flexible applications. Among semi-rigid substrates, Rogers® holds a leading position due to a series of patents covering PTFE-based laminates composed of resin mixtures with various fillers, yet free of internal glass fabric. This composition allows controlled mechanical flexibility under specific bending conditions, making them suitable for semi-conformal implementations. In particular, Rogers® RT5880 is extensively used in high-frequency applications due to its low dielectric constant and minimal loss tangent, which are critical for preserving signal integrity and reducing RF losses [50,51,54,59,61]. Comparable performance is also offered by other PTFE-based laminates (with or without woven glass reinforcement) such as Rogers® 4003C, Rogers® 3003, and Taconic® FastFilm™, which combine suitable dielectric characteristics with moderate flexibility [44,62,64]. However, fully flexible substrates are required for highly flexible or deployable applications such as space-deployable arrays. Several polymers have been widely adopted in the reviewed studies. Kapton® HN, a polyimide, combines good dielectric properties with excellent mechanical flexibility and thermal stability. This allows it to conform to non-planar surfaces and endure high-temperature processes without significant degradation [62]. LCP is another commonly used substrate due to its low dielectric constant ($\approx$2.9–3.2) and low loss tangent ($\approx$0.002–0.005), together with its natural flexibility and high dimensional stability under bending [65–67]. An example of such a single antenna element fabricated using the LCP flexible substrate and the antenna array fabricated using the LCP substrate demonstrating the bending capability of such a design is presented in Figure 4. Polydimethylsiloxane (PDMS) and PEI fibers have also been explored for their conformability and thermal resistance, making them suitable for integration on cylindrical or deformable surfaces [57,68].

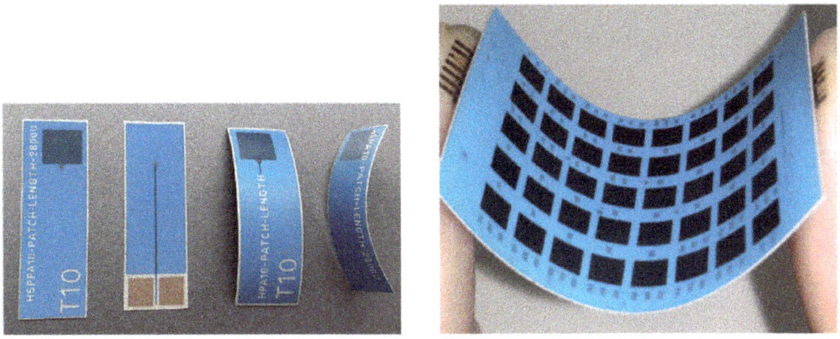

(**a**) (**b**)

Figure 4. (**a**) Photograph of a single antenna element using flexible LCP substrate with bending demonstration of the single element. (**b**) Bending demonstration of a square lattice planar antenna array made of LCP substrate [65].

An example of a flexible radiating element and an array configuration is shown in Figure 4. Other notable flexible solutions include DuPont Pyralux®-based laminates, which combine thin polyimide films with copper cladding, enabling multilayer flexible designs with good electrical and mechanical performance. The comparison of these materials, including their mechanical, electrical, and application-driven motivations, is provided in Table 1.

Table 1. Material properties.

Materials Used	Mechanical Properties	Electromagnetic Properties	Reason for Use
Chromium	Durable, Thin	Highly Conductive	Conductivity
Rogers® RT5880	Semi-flexible, Lightweight	Dielectric constant = 2.2, Loss tangent = 0.0009	Low loss, flexibility
DuPont Pyralux®	Flexible, Lightweight	Dielectric constant = 3.4, Loss tangent = 0.003	Flexible, thin, low profile
Copper	Durable	Conductive, Low resistivity	Conductivity, reliability, compatible with PCB standard manufacturing
LCP	Foldable, Lightweight, Deployable	Dielectric constant = 3.0–3.3, Loss tangent = 0.0049	Low loss, flexible
Rogers® 4003C	Rigid, Lightweight	Dielectric constant = 3.38–3.55, Loss tangent = 0.0027	Low loss
Rogers® 3003	Semi-Flexible, Durable, Lightweight, Thermally-conductive	Dielectric Constant = 3.00, Loss tangent = 0.0013	Low dielectric loss and flexible
SNP ink	Flexible, Printable	Conductive, low resistivity	Printability, thermal stability
Taconic® FastFilm™	Durable, Lightweight	Dielectric constant = 3.55, Loss tangent = 0.0027	Low loss, flexibility
Rogers® 3003	Semi-Flexible, Durable, Lightweight, Thermally conductive	Dielectric Constant = 3.00, Loss tangent = 0.0013	Low dielectric loss and flexibility
Kapton® HN Polyimide	Lightweight, Flexible, Durable, Thermostable	Dielectric constant = 3.4, Loss tangent = 0.002–0.005	Thermal stability

Table 1. *Cont.*

Materials Used	Mechanical Properties	Electromagnetic Properties	Reason for Use
SNP ink	Flexible, Printable	Conductive, low resistivity	Printability, thermal stability
Taconic® FastFilm™	Durable, Lightweight	Dielectric constant = 3.55, Loss tangent = 0.0027	Low loss, flexibility
PP	Lightweight, Robust	Dielectric constant = 2.2–2.34, Loss tangent = 0.0002–0.001	Low loss, robustness
PET	Flexible for small thickness, Durable	Dielectric constant = 3.0–3.4, Loss tangent = 0.02	Cost-effectiveness
PDMS	Flexible, Conformable	Dielectric constant = 2.65, Loss tangent = 0.031	Flexibility and thermal stability
PEI fiber	Flexible, Durable	Dielectric constant = 3.0, Loss tangent = 0.0012	Thermal stability

4. Major Challenges

Although FPAs offer significant potential for a wide range of advanced applications, their benefits do not come without critical challenges that must be addressed to ensure the performance of the final device. This section reviews the main issues identified across relevant studies.

4.1. Mechanical Deformation and Its Impact on Performance

FPAs are often deployed on either curved surfaces or dynamically deforming surfaces, consequently, introducing phase distortions, impedance mismatches, and degradation of beamforming accuracy. Arrays mounted on deformable surfaces suffer from phase variation as a result of geometric distortions. The array factor describes how the geometrical combination of multiple antenna elements affects the overall radiation pattern. It represents the impact of the geometry of the array, amplitude and phase of the excitation signals applied to each radiating element. Generally, in an FPA, the radiating elements are properly arranged on a curved surface. So, the array factor is given by

$$AF(\theta,\phi) = \sum_{n=1}^{N} a_n e^{jk[x_n(u-u_s)+y_n(v-v_s)+z_n(w-w_s)]} \tag{1}$$

with

$$\begin{aligned} u &= \sin\theta\cos\phi \\ u_s &= \sin\theta_s\cos\phi_s \\ v &= \sin\theta\sin\phi \\ v_s &= \sin\theta_s\sin\phi_s \\ w &= \cos\theta \\ w_s &= \cos\theta_s \end{aligned} \tag{2}$$

In (1)–(2), a_n is the amplitude of the nth radiating element, $k = 2\pi/\lambda$ is the wave number at the frequency f, x_n, y_n, and z_n are the position coordinates of the nth element relative to the center oh the global coordinate system, θ and ϕ are the elevation and azimuth angles pertaining an arbitrary observation point, θ_s and ϕ_s are the scanning angles, u_s, v_s, and w_s are the scanning directions in the u-v-w space.

Unlike the planar array, in the FPA, the z-axis coordinate of each radiating element changes according to its position over a 3D surface. As a result, the radiation pattern of FPA differs from that of the planar one even with the same number, shape, and spacing of the radiating elements. Phase of the signal is determined by its position relative to a reference point, so the variation due to the mechanical deformation affects the array factor. This variation can alter the conditions for constructive and destructive interference leading to the distortion in main lobe and side lobe level (SLL). This issue can be dealt with compensation mechanisms and calibration techniques for the impact of mechanical deformation on electromagnetic performance. For example, two studies utilized mutual coupling measurements to sense shape distortions and compensate for phase and one of them implements a self-calibration mechanism [69,70]. Similarly, another study addressed this challenge by integrating flexible resistive sensor and analog circuitry to monitor surface deformations and adjust phase shifters for real-time radiation pattern correction in dynamically changing conditions [36]. The problem of shape reconstruction was dealt with, in a different way with the introduction of Frequency Shift Key (FSK) radar approach to reconstruct the geometry of FPAs by estimating inter-element spacing [62]. For the preservation of the radiation pattern, one approach was implemented using phase in 2022 [71]. Moreover, frequent bending and rolling of FPAs can degrade electrical and mechanical integrity. Repetitive mechanical stress can induce small cracks in printed metallic traces leading to the disruption of the current flow, reducing the signal integrity which affects the gain and beam steering capability. Substrates such as PET or Rogers laminates experience fatigue over repeated flexing/bending, which can result in loss of elasticity. A solution to this challenge can be the use of a rigid–flex architecture which was briefly discussed in the previous sections. It combines rigid substrate tiles mounted on flexible dielectric laminates, so defining the precise region in which the device can bend and obtaining flexible and lightweight antenna arrays. For example, an RFAA maintained performance after 50 roll–unroll cycles in the application in small satellite lightweight apertures [53]. A similar study showed an array conformed to a cylindrical geometry, maintaining performance under deformation by alternating rigid–flex configuration [54]. With the aim of completely eliminating the rigid phase shifter circuitry to improve flexibility and durability under bending, a novel technique was implemented which utilized silver-coated ferrite particles to achieve tunable phase shifts by aligning the particles using a static magnetic field. The use of static magnetic fields to activate phase shifters eliminated the need for external biasing, significantly reducing complexity [59]. In another approach, a bending radius as small as 5 mm was achieved in 2021 by utilizing Taconic®fastFilm™ to develop a low-loss flexible PCB-based TX/RX module that maintains high RF performance under bending [64]. In general, RFAA is an efficient solution for FPA. Beamformer IC can be attached to the rigid parts of the device without the problem of bending degradation. While this is an efficient solution, maintaining electrical continuity between the flexible and rigid sections without degrading performance is still a key challenge. Furthermore, flexibility in RFAA can only be achieved as a global property with tiles that locally remain rigid but can conform globally due to the presence of the rigid tiles. Moreover, bending among the two axes (around a sphere) still remains challenging. The discussed design strategies as well as pros and cons of their phase compensation methods are summarized in Table 2. Future trends of research to face the mechanical deformation challenges lie in the development of advanced algorithms for phase compensation based on Field Programmable Gate Array (FPGA) technology.

Table 2. Comparison among different designs and phase compensation strategies.

Paper	Design	Phase Compensation	Advantages	Disadvantages
[69]	Folded dipoles on copper ground	Spiral Match algorithm, Semi-Definite Relaxation	No additional sensors, Modular	1D bending, Inaccuracy in non-uniform bends
[70]	Flexible substrate with copper PCB	Mutual coupling used to infer relative phase center (differential phase measurement)	No need for external sensors, Operates under bending, High mechanical flexibility	Variations in antenna/feed network not accounted for
[36]	Microstrip patch array printed on a flexible substrate	Phase compensation based on the data of sensor deformation detection	Low complexity, Pattern recovery under deformation	Limited to predefined geometrical shapes

Table 2. *Cont.*

Paper	Design	Phase Compensation	Advantages	Disadvantages
[62]	Probe-fed patch antennas mounted on an AM acrylonitrile butadiene styrene (ABS) frame	A phase-based distance measurement, FSK-based measurement using phase differences at two frequencies	Sensor-free approach, Applicable to arbitrary deformations	Reduced accuracy at close spacing
[71]	Patch array built on thin flexible composite laminate made of E-glass fiber and a highly conductive mesh-style fabric	Phase shifters are implemented as extended microstrip line lengths, introducing phase delays	High mechanical flexibility, Low ohmic and dielectric losses	Only suitable for fixed deformations
[53]	Rigid–flex flexible membrane for RF and DC routing, with rigid tiles as substrates for antenna and components	No phase compensation but element-level electronic beam steering	Ultra-lightweight, Highly flexible, Active component integration	No active beam steering demonstrated
[54]	Rigid–flex checkerboard structure combining rigid PCBs and flexible substrate	No dynamic phase compensation	Flexible in 2D, Dual polarization	Lacks real-time phase control
[59]	Patch array with integrated magnetic particles based Composite Right/Left-Handed (CRLH) metamaterial phase shifters	Phase control is achieved by tuning the permeability (per-element phase tuning)	Real-time tunability, Conformal flexibility	Complex fabrication, Magnetic field source required
[64]	Double-layer flexible PCB	Integrated digital phase shifters	Low loss under bending, Integrated active electronics	No dynamic phase calibration, Assembly complexity

4.2. Electrical Performance Limitations

Achieving consistent electrical performance, including low insertion loss, wide bandwidth, and stable impedance matching, is a core challenge in FPA designs. Flexible substrates have different dielectric properties compared to traditional rigid materials, which lead to variations in signal propagation speed, impedance, and power loss. Flexible materials such as polyimide and Kapton® exhibit higher dielectric losses compared to rigid substrates. This was evident in the design of PAs printed by IP which encountered a significantly performance-degrading insertion loss of 8.17 dB [38]. The utilization of CNT-TFTs as switching elements for phase shifters introduced a high insertion loss reducing overall system performance. Similarly, in the design of self-adapting FPA, gain degradation (loss of 0.6–1.8 dB) due to phase shifter insertion loss was found [36]. The impact of insertion loss was minimized in different studies by optimizing the design parameters. For example, curved microstrip lines exhibited only a 0.3 dB increase in insertion loss compared to planar configurations, indicating negligible performance degradation [64]. Moreover, a highly efficient PA system with insertion loss below 0.1 dB was demonstrated through direct on-package fabrication, eliminating traditional connectors and thus shortening the signal path [42].

Bandwidth is also an important electrical performance parameter. Some applications require wide-band operation which is often constrained by coupling effects and impedance mismatches, especially on flexible platforms. In an attempt to enhance the bandwidth, stacked-patch design and inverted-F feed structures were utilized to ensure impedance matching and bandwidth enhancement. This solution delivered the widest bandwidth (17.5%) among active bendable PA antennas while maintaining scanning up to 60° [72]. Among other methodologies to achieve a wider bandwidth, an important technique is the use of proper feeding mechanism. In the design of a flexible Ka-band active PA on LCP substrate to support SATCOM applications, a novel feeding mechanism of proximity-fed patch antennas leveraged the proximity effect to achieve a wider bandwidth [67]. With the integration of LCP layers, bandwidth limitations emerge as a major challenge in such applications that is mitigated by using proximity feed design [55]. An interesting field of research regards the use of AME platforms to achieve precise multilayered structures while maintaining full custom design.

Impedance matching in FPA is another critical parameter. In [61], authors designed periodic rectangular microstrip patches with an inset feed. The design reduced mutual coupling between elements by modifying the patch dimensions and utilized a two-level multiple-section feeding network for improved impedance matching and spectral performance.

Wide angle scanning requires the achievement of a stable beam steering over wide angles, but this often results in grating lobes or gain reduction. This hurdle was mitigated by using a simplified phase-shifting network [73]. The suppression of gratin lobes is also possible with another technique of utilization of a triangular lattice configuration, particularly the offset rows in the triangular configuration mitigated the grating lobes, but this resulted in the increased design complexity for modular integration [72].

4.3. Fabrication and Manufacturing Challenges

Fabricating FPAs involves significant challenges related to substrate handling, multilayer integration, and precision assembly, all of which are critical for achieving flexibility. In particular, multilayer PCB integration introduces issues such as alignment accuracy and interlayer crosstalk. For example, a deployable phased-array transmitter using a four-layer LCP substrate integrated RF, ground plane, Serial Peripheral Interface (SPI) control, and patch antennas to achieve compactness. However, this configuration revealed the inherent difficulties in maintaining uniformity during the multilayer lamination process [65]. These critical challenges include layer-to-layer alignment sensitivity due to LCP dimensional instability under thermal cycling. This can be mitigated by utilizing low-temperature lamination techniques to reduce the expansion/contraction of LCP or by using pre-conditioned LCP sheets to stabilize thermal expansion prior to assembly.

Scalability was identified as another challenge, and fabrication methods such as IP, while cost-effective, are limited in scalability for large arrays. This can be dealt with by the division of multilayer architecture into separate modules connected through flexible interposers to reduce complexity. Another way is to utilize the modular rigid–flex architectures to isolate sensitive components from deformation zones.

Furthermore, the design of wideband stacked patch antenna array using IP emphasized the challenges in achieving uniformity and consistency in multilayer designs [41]. The fabrication process involved IP on PET. It posed the challenge of Silver conductivity limitation, photonic curing, stack alignment (the ground and radiating layers were printed on separate PET films, necessitating accurate alignment during assembly). IP is an efficient fabrication technique, but it also introduces alignment and attachment challenges [63]. A possible solution for future works can be the utilization of conformal pick-and-place systems with 3D vision feedback to map surface topology and guide placement.

In addition to aligning and attaching ICs on 3D-printed surfaces, addressing the surface roughness of 3D-printed materials to improve the metallization quality posed another challenge [42]. The surface roughness of printed materials (that depends on the printing technology as well as on the working frequency of the final device) has a significant impact on metallization quality. It can cause non-uniform copper deposition, leading to higher surface resistance and degradation of RF performance. Sanding and surface finishing techniques are utilized to counter this to smooth the substrate before electroplating.

4.4. Thermal Management

The inclusion of active components, such as RFICs and amplifiers, introduces significant thermal challenges such as heat dissipation in FPAs. Power amplifiers in particular convert a significant amount of power into heat, leading to localized hot spots. Furthermore, thin and flexible substrates (LCP and PET film) used in FPA have limited thermal conductivity, making heat management critical, impeding heat dissipation compared to rigid PCBs or metal-backed substrates. Such temperature variations can cause mechanical stress on the components due to mismatched coefficients of thermal expansion between components. This phenomenon can lead to the risks of delamination or cracking, especially in laminated or 3D-printed structures. Moreover, electrical performance of RFIC can be significantly affected, which can lead to shift in frequency responses, increase in noise figure, and reduction in linearity. Consequently, without efficient thermal management, long-term exposure to high temperatures reduces component reliability. This issue was addressed with various methodologies. For example, microfluidic cooling channels were embedded to reduce the temperature of the IC in short time interval by using water flow [42,63]. Thermal management is also critical in high-power configuration of FPAs. For example, in a 256-element FPA system for WPT, with high-power configuration delivering power wirelessly at a 1 m distance, thermal management was achieved through aluminum heat spreaders in high-power system and flat ceramic spreaders in the flexible system [50]. Similarly, for an FPA system with 16 radiators, Power Synthesis and Control Unit (PSCU) was particularly used to ensure stable voltage and current distribution during operation with sufficient temperature control [51]. Thermal insensitivity is a desirable feature. In another work, a thermally and mechanically insensitive antenna array was developed using a PDMS substrate that overcomes the challenge of thermal interference due to the low value of the thermal coefficient of PDMS substrate by reducing the effects of anisotropic thermal expansion [68]. Different thermally resilient materials and integrated cooling channels integrated with dynamic cooling systems can be viable solutions for efficient thermal management.

4.5. Environmental and Operational Challenges

The real-world deployment of FPAs requires not only high electromagnetic performance but also robust operation under various environmental and mechanical stress conditions, including extreme temperatures, humidity, moisture exposure, and mechanical deformation from stretching and bending. So, a crucial area of research focuses on maintenance of consistent performance in face of such challenges. This has led to the exploration of environmentally resilient materials and mechanically adaptive design strategies. As an example, low-loss substrates have been utilized to reduce the dielectric losses under variation of environmental conditions such as moisture absorption and thermal fluctuations [61]. Real-time phase calibration techniques have been introduced to tackle the problem of environmental variability affecting array performance, specifically in beamforming and signal alignment. Such systems have continuous element-wise monitoring and phase readjustment in response to the changes in environmental conditions, leading to dynamic beam correction [50]. In recent years, FPA have been widely utilized in disaster response and field surveillance equipment. In such applications, self-sufficient power solutions are of high significance. One such innovative approach integrated Photovoltaic (PV) concentrators directly with power amplifiers. This enabled a standalone operation. This is considered as a significant advancement for long-duration deployments without access to conventional power infrastructure dedicated to remote deployment [51]. Although not for remote environments, another study was conducted for a smart and wearable textile-based antenna array. The solution incorporated textile-based microstrip antennas with direction of arrival estimation for dynamic beam correction, achieving a low-cost and scalable structure [74].

5. Performance Comparison

Following the analysis of the main challenges in FPA design, a comparative study was carried out based on key performance parameters. The results are summarized in Table 3, which provides a compact overview of the surveyed works on FPAs. The table includes multiple parameters essential for evaluating FPA performance: operating frequency (freq), array size, bandwidth (BW), gain (in dB), maximum beam steering angle, insertion loss, beam width, flexibility characteristics, bending radius or curvature (flex curve), bending performance, polarization type (pol), and thermal performance (thermal perf). A multi-dimensional evaluation is necessary due to the variability introduced by different application scenarios. Among the most critical parameters, bandwidth reflects frequency agility, while the maximum steering angle quantifies beam reconfigurability. Insertion loss is analyzed to assess signal power loss and system efficiency. Beam width is relevant for resolution and directionality. Flexibility-related information is divided into how flexibility is achieved (e.g., flexible substrates, additive manufacturing, rigid–flex integration), the achieved bending radius, and the impact of bending on performance. Lastly, the table reports the polarization (e.g., linear or circular) and thermal behavior, including maximum operating temperature, cooling strategies, and any observed performance degradation under thermal stress.

It is possible to extract some information regarding the performance, design choice, and practical implications of FPA based on the comparative analysis across multiple works. FPAs are being designed over a range of frequencies highlighting the diversity in applications such as UHF communications, SATCOM, 5G/6G communications, and radar systems. The size of the arrays also varies from small and compact 1×4 arrays to large 16×16 configurations depending on the type of application. This reflects the trade-off between size of the array, gain of the array, and beam control. Moreover, the gain of the array varies from small values of 1 dB or 6 dB for smaller arrays to higher values as high as 60.5 dB (EIRP) for higher frequencies and large scale designs. Moreover, beam scanning is a major factor in PA, with a substantial number of works with a steering angle range up to $\pm 70°$. This indicates strong reconfigurability. The wide scanning range comes with an increase in the value of insertion loss and a degradation in the gain, especially at extreme angles. Based on the type of applications, bandwidth also varies, with some designs supporting UWB operation while others are optimized for narrowband applications. This suggests a trade-off between bandwidth and parameters such as structural complexity and fabrication methodologies. Furthermore, the major focus of this comparative analysis was flexibility. Flexibility in FPA is primarily achieved with the utilization of flexible substrates such as polymide and LCP, sometimes using AM like IP to define the conductive traces on the flexible substrates. Another approach for obtaining flexibility was introduced in the form of hybrid rigid–flex design. Performance under bending is also analyzed for the analysis with bending radii ranging from 3 cm to 50 cm, with most studies having stability in performance under mechanical deformation. Moreover, in some circumstances, phase compensation techniques are implemented to mitigate performance degradation. Ultimately, thermal performance is discussed, particularly for high-power and space applications. Several arrays demonstrate good thermal stability up to 400 °C or are labeled as compatible for space applications, highlighting robustness under extreme conditions. The key trade-offs are between the flexibility and performance parameters such as bandwidth, gain, and beam steering. In cases of high flexibility, gain or bandwidth can be compromised to a certain extent. The field of FPAs has evolved to include high-performance, thermally stable, and reconfigurable antenna arrays with considerable mechanical flexibility. Emerging trends highlight the adoption of AM, microfluidic cooling systems for thermal management, advanced phase compensation techniques, and hybrid rigid–flex designs. The comparative analysis emphasizes the need to balance flexibility, electromagnetic performance, and environmental resilience to meet the requirement for real-world applications.

Table 3. Performance comparison.

NO.	Freq. (GHz)	Size	BW (GHz)	Gain (dB)	Steering	Insertion Loss (dB)	Beam Width	Flexibility	Flex Curve	Bending	Pol.	Thermal Perf.
[33]	1.26	2×4	0.08	Rx 17, Tx 30	0°–15°	–	–	Flexible sub.	–	Good	–	–
[34]	14	–	4	–	–	0.59 (1–bit)	–	Flexible Sub. LCP	–	Curling addressed	–	LCP Properties
[35]	14	2×2	≈0.7	7.75	±12°	4.57	–	Flexible Sub	–	Minimal impact	–	Max 150 °C
[36]	2.45	1×4	≈0.02	5.9 (flat)	≈30°–45°	2–3	Broad, Narrow	Flexible Sub	30°–45° wedge, 10 cm cylinder	Autonomous correction	–	–
[37]	2.25	1×8	100 MHz	14.79 dB	±40°	–	≈77°	Flex Sub	13.2 cm	–	CP	–
[38]	5	4×4	–	14.6	±34°	8.17	–	Flexible Sub AM	–	Demonstrated	–	–
[43]	19	2×2 tile	–	16	±50°	–	15°–25°	Flexible Sub AM	–	Retain perf.	–	Stable ≈ 400 °C
[61]	10	1×4	1.3	12.4	±30°	–	Varies	Semi Flex	20 cm	Phase compensation	–	–
[41]	7.7–8.3	4×4	0.6	16.9	–	Varies	–	Flex Sub AM	≥1.2 cm	Gain reduction	RHCP	–
[42]	24–40	–	Wide Band	Max ≈ 10.09	±37°	–	Broad	Flex Sub AM	3 cm	Minimal shift	Circular	Microfluidic Cooling
[63]	24–40	–	Wide Band	Max ≈ 10.09	±37°	–	Broad	Flex Sub AM	3 cm	Stable for 10,000 bending cycles	Circular	Microfluidic Cooling
[44]	30	5×1 patches	0.6	12.5	±70°	1.17	–	Flex Sub AM	3 cm	Minimal RCS variation	LP (Dual)	Robust
[45]	Sub-6	2×2 array	0.47	7.7	–	–	–	Flex Sub 3D	–	–	CP	–
[46]	4.99	1×4	–	–	±27°, ±45°	–	–	Flex Sub	6.5 cm	24 cm	–	–
[47]	4.79–5.04	1×4	0.25	4.5	±20°	–	5°, HPBW	Flex Sub AM	5 cm, 25 cm	Stable	CP	Moderate Heating
[48]	22.3–42.5	2×4	20.2 GHz	13 dBi	–	–	–	Flex Sub	–	Effective	–	Temp. stable
[49]	2–18	6×6	15 GHz	–	–	–	–	Flex Sub	20 cm	–	–	–
[50]	10	30×30 cm^2 sheet, 256 elements	UWB	≈60.5 EIRP	±10° azimuth, ±3° elevation	–	2D Scan	Flex Sub	23 cm	Effective	LP	Temp. stable
[51]	9.4–10.4	4×4 tile	1.0	≈37.1 EIRP	±56.4° without GLs	–	14°–22.5°	Flex Sub	3 cm	Stable	LP (Dual)	Space Compatible
[52]	19.5	2×1 and 4×1	–	≥7 dB	−45° to +45°	–	–	Flex Sub	up to 45°	–	–	–
[53]	9.4–10.4	16×16	0.8	15.2	±30°	–	Consistent	Rigid Flex	3.8 cm	50 roll–unroll cycles	LP	Space Compatible
[54]	10.7–12.75	8×4 (checkerboard)	2.05	16.3	±60°	–	Consistent	Rigid Flex	5 cm	Measured on cylinder	LP (Dual)	–
[55]	28	32 elements	≈ 0.8	EIRP ≈ 41	±50°	–	+10° at edges	Flex Sub	Deployable	–	LP	–

Table 3. *Cont.*

NO.	Freq. (GHz)	Size	BW (GHz)	Gain (dB)	Steering	Insertion Loss (dB)	Beam Width	Flexibility	Flex Curve	Bending	Pol.	Thermal Perf.
[56]	26.25	8×8	≈ 0.5	46.7	±10° to +20°	–	7°	Flex Sub	Foldable	±10° bent	–	–
[57]	0.42–0.45	Linear Array	Broad UHF	≈5	±15°	< 3	19.5°	Flex Sub	–	–	Dual polarized	Minimal Deformation
[58]	24	10×10	0.67	29.4	149.8° (θ), 120° (ϕ)	–	Broadside, ±30°	Flex Sub AM	4.5 cm	Stable	LP	Consistent
[59]	2.45	1×4	0.5	9.52	±30°	≈1	Broadside, ±30°	MP CRLH Phase Shifter	30 cm, 10 cm	Minimum variation	LP	Minimum Deformation
[60]	5.8	1×4	5–6	7.38 (0°), 6.72 (30°), 6.2 (45°)	±20°	−2 to −6	Adjusted	MP CRLH Phase Shifter	30°, 45°	Minimum variation	LP	–
[62]	2.5	–	0.03	–	–	–	–	Flex Sub AM	–	12% wavelength deviation	LP	–
[64]	8–12	–	–	TX 15.32–21.92, RX 14.87–20.57	–	2.5 to −5.6	–	Flex Sub PCB	PCB 4 cm, TX/RX 3.5 cm	Stable	LP	–
[65]	28	8×4	1.0	19.5 (0°)	±50°	–	–	Origami, Flex Sub	–	–	LP	–
[66]	10	16×16	–	–	–	–	–	Flex PCB	–	–	LP	Space Compatible
[67]	28	16×16	1.0	≈3 (element), 58.16 EIRP	–	–	–	Flex Sub	–	–	LP	–
[68]	2.44	Single Antenna	0.33	1.6	–	–	Omni directional	Flex Sub	2 cm to 7 cm	Shift 2.8%	LP	Insensitive (20–100 °C)
[69]	2.5	8 elements	–	–	–	–	–	Flex Sub	–20.6 cm to 20.6 cm	error ≈ 0.04λ	LP	–
[70]	10	8 elements	0.48	–	±60°	–	–	Flex Sub	±12 cm	Stable	LP	–
[71]	5.8	4×1	5 MHz	11.9 dB (flat), 8.7 dB (bent)	–	–	–	Flex Sub	5 cm	–	–	–
[72]	10.7–12.75	8×4	2.05	30.7–34.8	±60°	≈ 8.6	–	Flex Sub	5 cm	Stable	Dual Pol.	–
[73]	2.4	1×4 subarrays	0.074	≈15	360° azimuth	≈ 7.2	≈ 17–23° (xz–yz)	Balloon Shaped	50 cm	loss ≈ 0.2 dB	LP	–
[74]	2.45	–	240 MHz to 270 MHz	4.8 dB, 4.3 dB	80°	–	–	Flex Sub	–	–	LHCP	–

6. Conclusions and Future Trends

This paper analyzed several critical aspects related to the design and development of FPAs. An overview was provided of the most commonly used materials and technologies, as well as the major limitations currently encountered in their implementation.

A first key direction explored involves the use of flexible substrates and AM techniques, such as IP, to realize conductive elements. While this approach is widespread, it typically enables only passive array configurations due to the inherent difficulty in implementing multilayer structures (integrating both conductive and dielectric layers). These multilayers are essential for embedding the control and RF electronics required in active array systems.

Currently, the AME platform Dragonfly by Nanodimensions is the only one that supports 3D-printed multilayer architectures of both conductive and dielectric materials. However, they are limited by the narrow range of compatible dielectric materials and the lack of robust evidence supporting their integration with standard soldering processes which remain fundamental for high-performance RF systems.

From a manufacturing point of view, a promising alternative lies in the use of rigid–flex PCB technologies. These offer a trade-off between structural flexibility and the reliability of conventional fabrication processes. This area is expected to witness significant developments in the near future, especially as the demand for conformal and adaptive antennas increases.

Another key consideration is the impact of mechanical deformation on electromagnetic performance. The non-planar positioning of radiating elements in flexible arrays can degrade radiation characteristics. Addressing this issue likely requires the development of compensation techniques, such as phase correction and real-time electronic reconfiguration, to maintain optimal performance under deformation.

Furthermore, the design of arrays based on unconventional geometries, such as concentric or concentric–hexagonal lattices, opens new opportunities for performance improvement, particularly in terms of SLL mitigation. However, these configurations pose new challenges for the design and implementation of complex feeding networks, which requires dedicated research efforts.

In conclusion, the data presented in this review can serve as a starting point for identifying the target specifications for various application scenarios involving FPAs. They also provide a useful reference for guiding future research activities in this promising technology field.

Author Contributions: Conceptualization, F.P.C. and L.M.; methodology, W.U.U., F.P.C. and L.M.; formal analysis, W.U.U.; investigation, W.U.U.; data curation, W.U.U., F.P.C. and L.M.; writing—original draft preparation, W.U.U.; writing—review and editing, F.P.C. and L.M.; visualization, W.U.U., F.P.C. and L.M.; supervision, F.P.C. and L.M.; All authors have read and agreed to the published version of the manuscript.

Funding: This research received no external funding

Institutional Review Board Statement: Not applicable.

Informed Consent Statement: Not applicable.

Data Availability Statement: Not applicable.

Conflicts of Interest: The authors declare no conflicts of interest.

Abbreviations

The following abbreviations are used in this manuscript:

AESA	Active Electronically Scanned Array
AM	Additive Manufacturing: Layer-by-layer fabrication of 3D objects using digital designs
AME	Additive Manufacturing Electronics
AP	All Polymide: High-temperature, flexible substrate made entirely of polyimide
CP	Circular Polarization
CNT-TFT	Carbon Nanotube Thin-Film Transistors: Transistors made of carbon nanotubes for electronics
DC	Direct Current
EIRP	Effective Isotropic Radiated Power
ESA	Electronically Steerable Antenna
ESAA	Electronically Steerable Antenna Array
FFF	Fused Filament Fabrication: 3D printing method using melted thermoplastic filaments
Flex Sub	Flexible Substrate
FSK	Frequency Shift Key: Modulation scheme where data are encoded by varying frequency
FSS	Frequency Selective Surface: Surface that selectively transmits or reflects specific frequencies
FPA	Flexible Phased Array
FPGA	Field Programmable Gate Array: reconfigurable integrated circuit enabling fast operations
GaAs	Gallium Arsenide: High-speed semiconductor used in RF and microwave electronics
IC	Integrated Circuit
IMU	Internal Measurement Unit: Sensor module measuring orientation and motion
IP	Inkjet-printing: Non-contact printing technique depositing functional inks for electronics
LCP	Liquid Crystal Polymer: Thermoplastic with excellent RF and mechanical properties
LHCP	Left-Hand Circular Polarization
LNA	Low-Noise Amplifier
LP	Linear Polarization
MEMS	Micro-Electromechanical System
PA	Phased Array
PCB	Printed Circuit Board
PDMS	Polydimethylsiloxane: Flexible, biocompatible silicone used in soft electronics
PEI	Polyetherimide: High-strength, heat-resistant polymer used in aerospace and electronics
PESA	Passive Electronically Scanned Array
PET	Polyethylene terephthalate: Flexible and transparent polymer used in substrates
PP	Polypropylene: Thermoplastic used for flexible and chemical-resistant packaging
PSCU	Power Synthesis and Control Unit: Module managing power distribution in arrays
PTFE	Polytetrafluoroethylene: Non-stick, low-loss dielectric material used in RF applications
PV	Photovoltaic
RFAA	Rigid–flex antenna array: Hybrid antenna combining rigid and flexible sections
RHCP	Right-Hand Circular Polarization
RFIC	Radio Frequency Integrated Circuit
RFID	Radio Frequency Identification
RX	Receiver
SATCOM	Satellite Communication
SLA	Stereolithography: 3D printing method using photopolymerization with a laser
SLL	Side Lobe Level
SmallSat	Small Satellite
SNP	Silver Nanoparticle: Nanoscale silver particles used in conductive inks
SPI	Serial Peripheral Interface
TX	Transmitter
UAV	Unmanned Aerial Vehicle
UHF	Ultra-High Frequency
UVO	Ultra-Violet Ozone: Surface treatment method to modify materials

UWB	Ultra Wide Band
VSWR	Voltage Standing Wave Ratio
WPT	Wireless Power Transfer
XPD	Cross-Polar Discrimination: Ratio of co-polarized to cross-polarized signal components

References

1. Chaloun, T.; Boccia, L.; Arnieri, E.; Fischer, M.; Valenta, V.; Fonseca, N.J.; Waldschmidt, C. Electronically Steerable Antennas for Future Heterogeneous Communication Networks: Review and Perspectives. *IEEE J. Microwaves* **2022**, *2*, 545–581. [CrossRef]
2. Epstein, B.R.; Olsson, R.H., III; Rotman, R. Phased Array Technologies [Scanning the Issue]. *Proc. IEEE* **2016**, *104*, 482–486. [CrossRef]
3. Mohr, M.C.; Monaghan, S. Circularly Polarized Phase Shifter for Use in Phased Array Antennas. *IEEE Trans. Microw. Theory Tech.* **1966**, *14*, 672–683. [CrossRef]
4. Mohr, M.C.; Monaghan, S. Polarization Insensitive Phase Shifter for Use in Phased Array Antennas. In Proceedings of the G-MTT International Microwave Symposium Digest, Boston, MA, USA, 8–11 May 1967.
5. Levy, D.; Noblet, A.; Bender, Y.W. A 2–18 GHz monolithic phase shifter for electronic warfare phase array applications. In Proceedings of the 10th Annual IEEE (GaAs IC) Symposium, Gallium Arsenide Integrated Circuit. Technical Digest, Nashville, TN, USA, 6–9 November 1988.
6. Hodges, N.E.; Yam, M.H. A precise analog phase shifter for SHF SATCOM phased arrays. In Proceedings of the GaAs IC Symposium Technical Digest, Miami Beach, FL, USA, 4–7 October 1992.
7. Smith, J.K. N-Bit L-Band Phase Shifters for Phased Arrays. In Proceedings of the IEEE MTT-S International Microwave Symposium Digest, Boston, MA, USA, 31 May–3 June 1983.
8. Pao, C.K.; Chen, J.C.; Lan, G.L.; Wang, D.C.; Wong, W.S.; Herman, M.I. V-band monolithic phase shifters. In Proceedings of the 10th Annual IEEE (GaAs IC) Symposium, Gallium Arsenide Integrated Circuit. Technical Digest, Nashville, TN, USA, 6–9 November 1988.
9. Jacomb-Hood, A.W.; Windyka, J.A.; Booth, M.S.; Snow, K.H.; Adams, S.B. C-band phase and gain control MMIC. In Proceedings of the 10th Annual IEEE (GaAs IC) Symposium, Gallium Arsenide Integrated Circuit. Technical Digest, Nashville, TN, USA, 6–9 November 1988.
10. Guan, X.; Hashemi, H.; Hajimiri, A. A fully integrated 24-GHz eight-element phased-array receiver in silicon. *IEEE J. Solid-State Circuits* **2004**, *39*, 2311–2320. [CrossRef]
11. Hajimiri, A.; Komijani, A.; Natarajan, A.; Chunara, R.; Guan, X.; Hashemi, H. Phased array systems in silicon. *IEEE Commun. Mag.* **2004**, *42*, 122–130. [CrossRef]
12. Hajimiri, A.; Hashemi, H.; Natarajan, A.; Guan, X.; Komijani, A. Integrated Phased Array Systems in Silicon. *Proc. IEEE* **2005**, *93*, 1637–1655. [CrossRef]
13. Komijani, A.; Hajimiri, A. A 24 GHz, +14.5 dBm fully-integrated power amplifier in 0.18/spl mu/m CMOS. In Proceedings of the IEEE 2004 Custom Integrated Circuits Conference (IEEE Cat. No.04CH37571), Orlando, FL, USA, 6 October 2004.
14. Kang, D.W.; Kim, J.G.; Min, B.W.; Rebeiz, G.M. Single and Four-Element Ka-Band Transmit/Receive Phased-Array Silicon RFICs With 5-bit Amplitude and Phase Control. *IEEE Trans. Microw. Theory Tech.* **2009**, *57*, 3534–3543. [CrossRef]
15. Koh, K.J.; May, J.W.; Rebeiz, G.M. A Q-band (40–45 GHz) 16-element phased-array transmitter in 0.18-micro-m SiGe BiCMOS technology. In Proceedings of the 2008 IEEE Radio Frequency Integrated Circuits Symposium, Atlanta, GA, USA, 15–17 June 2008.
16. Atesal, Y.A.; Cetinoneri, B.; Koh, K.; Rebeiz, G.M. X/Ku-band 8-element phased arrays based on single silicon chips. In Proceedings of the 2010 IEEE MTT-S International Microwave Symposium, Anaheim, CA, USA, 23–28 May 2010.
17. Kim, S.; Gudem, P.S.; Larson, L.E. A 44-GHz 8-element phased-array SiGe HBT transmitter RFIC with an injection-locked quadrature frequency multiplier. In Proceedings of the 2010 IEEE Radio Frequency Integrated Circuits Symposium, Anaheim, CA, USA, 23–28 May 2010.
18. Choi, P.; Antoniadis, D.A.; Fitzgerald, E.A. Towards Millimeter-Wave Phased Array Circuits and Systems For Small Form Factor and Power Efficient 5G Mobile Devices. In Proceedings of the 2019 IEEE International Symposium on Phased Array System and Technology (PAST), Waltham, MA, USA, 15–18 October 2019.
19. Danesh, S.; Araghi, A.; Khalily, M.; Xiao, P.; Tafazolli, R. Millimeter Wave Phased Array Antenna Synthesis Using a Machine Learning Technique for Different 5G Applications. In Proceedings of the 2020 International Symposium on Networks, Computers and Communications (ISNCC), Montreal, QC, Canada, 20–22 October 2020.
20. Saha, N.; Mahbub, I. Design, Modeling, and Simulation of a 2.4 GHz Near-field Phased-Array based Wireless Power Transfer System for Brain Neuromodulation Applications. In Proceedings of the 2022 IEEE Texas Symposium on Wireless and Microwave Circuits and Systems (WMCS), Waco, TX, USA, 19–20 April 2022.

21. Lee, K.; Kim, J.; Seo, J.; Yu, H.; Cha, C. Channel-based Phase and Power Controllable Intelligent Wireless Power Transfer Architecture Using 4 by 4 Planar Array Antennas. In Proceedings of the 2018 IEEE International Conference on Industrial Engineering and Engineering Management (IEEM), Bangkok, Thailand, 16–19 December 2018.
22. Sadeghpour, S.; Zilonova, E.; D'Hooge, J.; Kraft, M. A Novel 6 MHz Phased Array Piezoelectric Micromachined Ultrasound Transducer (pMUT) with 128 Elements for Medical Imaging. In Proceedings of the 2021 IEEE International Ultrasonics Symposium (IUS), Xi'an, China, 11–16 September 2021.
23. Kelly, P.K.; Kumar, M.; McCaskey, R.; Maddocks, E.; Rhodes, J.; Chandrasekar, V. A Novel Portable Phased Array Radar for Meteorological Remote Sensing. In Proceedings of the 2022 IEEE International Symposium on Phased Array Systems and Technology (PAST), Waltham, MA, USA, 11–14 October 2022.
24. Sadhu, B.; Tousi, Y.; Hallin, J.; Sahl, S.; Reynolds, S.K.; Renström, Ö.; Sjögren, K.; Haapalahti, O.; Mazor, N.; Bokinge, B.; et al. A 28-GHz 32-Element TRX Phased-Array IC With Concurrent Dual-Polarized Operation and Orthogonal Phase and Gain Control for 5G Communications. *IEEE J. Solid-State Circuits* **2017**, *52*, 3373–3391. [CrossRef]
25. Jacomb-Hood, A.; Lier, E. Multibeam active phased arrays for communications satellites. *IEEE Microw. Mag.* **2000**, *1*, 40–47. [CrossRef]
26. Silverstein, S.D. Application of orthogonal codes to the calibration of active phased array antennas for communication satellites. *IEEE Trans. Signal Process.* **1997**, *45*, 206–218. [CrossRef]
27. Low, K.K.W.; Zihir, S.; Kanar, T.; Rebeiz, G.M. A 27-31-GHz 1024-Element Ka-Band SATCOM Phased-Array Transmitter with 49.5-dBW Peak EIRP, 1-dB AR, and ±70° Beam Scanning. *IEEE Trans. Microw. Theory Tech.* **2022**, *70*, 1757–1768. [CrossRef]
28. Aljuhani, A.H.; Kanar, T.; Zihir, S.; Rebeiz, G.M. A 256-Element Ku-Band Polarization Agile SATCOM Receive Phased Array with Wide-Angle Scanning and High Polarization Purity. *IEEE Trans. Microw. Theory Tech.* **2021**, *69*, 2609–2628. [CrossRef]
29. Jin, S.; Liu, D.; Cui, Z.; Mei, C.; Yang, Y.; Wu, B. A Scalable Ka-Band 256-Element Transmit Dual-Circularly-Polarized Planar Phased Array for SATCOM Application. In Proceedings of the 2022 International Symposium on Networks, Computers and Communications (ISNCC), Shenzhen, China, 19–22 July 2022.
30. Gultepe, G.; Low, K.; Ma, Q.; Rebeiz, G.M. A 1024-Element K-band Commercial SATCOM User Terminal with a G/T of 10.2 dB/K and Two Simultaneous Beams. In Proceedings of the 2022 IEEE International Symposium on Phased Array Systems and Technology (PAST), Waltham, MA, USA, 11–14 October 2022.
31. Rebeiz, G.M.; Koh, K.J. Silicon RFICs for phased arrays. *IEEE Microw. Mag.* **2009**, *10*, 96–103. [CrossRef]
32. Rebeiz, G.M.; Shin, W.; Golcuk, F.; Inac, O.; Zihir, S.; Gurbuz, O.; Edwards, J.; Kanar, T. Wafer-Scale Millimeter-Wave Phased-Array RFICs. In Proceedings of the 2014 IEEE Compound Semiconductor Integrated Circuit Symposium (CSICS), La Jolla, CA, USA, 19–22 October 2014.
33. Moussessian, A.; Del Castillo, L.; Huang, J.; Sadowy, G.; Hoffman, J.; Smith, P.; Hatake, T.; Derksen, C.; Lopez, B.; Caro, E. An active membrane phased array radar. In Proceedings of the IEEE MTT-S International Microwave Symposium Digest, Long Beach, CA, USA, 17 June 2005.
34. Kingsley, N.; Wang, G.; Papapolymerou, J. 14 GHz microstrip MEMS phase shifters on flexible, organic substrate. In Proceedings of the 2005 European Microwave Conference, Paris, France, 4–6 October 2005.
35. Kingsley, N.; Ponchak, G.E.; Papapolymerou, J. Reconfigurable RF MEMS Phased Array Antenna Integrated Within a Liquid Crystal Polymer (LCP) System-on-Package. *IEEE Trans. Antennas Propag.* **2008**, *56*, 108–118. [CrossRef]
36. Braaten, B.D.; Roy, S.; Nariyal, S.; Al Aziz, M.; Chamberlain, N.F.; Irfanullah, I.; Reich, M.T.; Anagnostou, D.E. A Self-Adapting Flexible (SELFLEX) Antenna Array for Changing Conformal Surface Applications. *IEEE Trans. Antennas Propag.* **2013**, *61*, 655–665. [CrossRef]
37. Zhi; Huang, X.; Zhang, K.; Zhang, Q.; Zhu, B.; Zhu, Z.; Ran; L. A Self-Adaptive Phased Array Conformed on a Highly Flexible Aperture. *IEEE Trans. Microw. Theory Tech.* **2025**, *73*, 1821–1836. [CrossRef]
38. Subbaraman, H.; Pham, D.T.; Xu, X.; Chen, M.Y.; Hosseini, A.; Lu, X.; Chen, R.T. Inkjet-Printed Two-Dimensional Phased-Array Antenna on a Flexible Substrate. *IEEE Antennas Wirel. Propag. Lett.* **2013**, *12*, 170–173. [CrossRef]
39. Montisci, G.; Mura, G.; Muntoni, G.; Casula, G.A.; Chietera, F.P.; Aburish-Hmidat, M. A Curved Microstrip Patch Antenna Designed From Transparent Conductive Films. *IEEE Access* **2023**, *11*, 839–848. [CrossRef]
40. Muntoni, G.; Mura, G.; Casula, G.A.; Chietera, F.P.; Aburish-Hmidat, M.; Montisci, G. A Transparent Curved Microstrip Patch Antenna. In Proceedings of the 2023 IEEE Conference on Antenna Measurements and Applications (CAMA), Genoa, Italy, 15–17 November 2023; pp. 698–700. [CrossRef]
41. Castro, A.T.; Sharma, S.K. Inkjet-Printed Wideband Circularly Polarized Microstrip Patch Array Antenna on a PET Film Flexible Substrate Material. *IEEE Antennas Wirel. Propag. Lett.* **2018**, *17*, 176–179. [CrossRef]
42. Hu, K.; Zhou, Y.; Sitaraman, S.K.; Tentzeris, M.M. Additively manufactured flexible on-package phased array antennas for 5G/mmWave wearable and conformal digital twin and massive MIMO applications. *Sci. Rep.* **2023**, *13*, 12515. [CrossRef]

43. Hu, K.; Soto–Valle, G.; Cui, Y.; Tentzeris, M.M. Flexible and Scalable Additively Manufactured Tile-Based Phased Arrays for Satellite Communication and 50 mm Wave Applications. In Proceedings of the 2022 IEEE/MTT-S International Microwave Symposium—IMS 2022, Denver, CO, USA, 19–24 June 2022.
44. Hester, J.G.; Tentzeris, M.M. Inkjet-Printed Flexible mm-Wave Van-Atta Reflectarrays: A Solution for Ultralong-Range Dense Multitag and Multisensing Chipless RFID Implementations for IoT Smart Skins. *IEEE Trans. Microw. Theory Tech.* **2016**, *64*, 4763–4773. [CrossRef]
45. Li, M.; Yang, Y.; Iacopi, F.; Nulman, J.; Chappel-Ram, S. 3D-Printed Low-Profile Single-Substrate Multi-Metal Layer Antennas and Array with Bandwidth Enhancement. *IEEE Access* **2020**, *8*, 217370–217379. [CrossRef]
46. Pham, D.T.; Subbaraman, H.; Chen, M.Y.; Xu, X.; Chen, R.T. Light Weight and Conformal 2-Bit, 1 × 4 Phased-Array Antenna with CNT-TFT-Based Phase Shifter on a Flexible Substrate. *IEEE Trans. Antennas Propag.* **2011**, *59*, 4553–4558. [CrossRef]
47. Li, W.T.; Hei, Y.Q.; Grubb, P.M.; Shi, X.W.; Chen, R.T. Inkjet Printing of Wideband Stacked Microstrip Patch Array Antenna on Ultrathin Flexible Substrates. *IEEE Trans. Components Packag. Manuf. Technol.* **2018**, *8*, 1695–1701. [CrossRef]
48. Hu, K.; Tentzeris, M.M. Additively Manufactured Ultra-Wideband Cavity-Backed Stacked Patch Antenna for 5G mmWave Conformal Phased Array Systems. In Proceedings of the 2024 IEEE International Symposium on Antennas and Propagation and INC/USNC-URSI Radio Science Meeting (AP-S/INC-USNC-URSI), Firenze, Italy, 14–19 July 2024.
49. Chen, X.; Li, K. Ultrathin and Flexible Ultrawideband Antenna Array Based on Integrated Impedance Matching Line. *IEEE Antennas Wirel. Propag. Lett.* **2023**, *22*, 960–964. [CrossRef]
50. Gal-Katziri, M.; Fikes, A.; Hajimiri, A. Flexible active antenna arrays. *npj Flex. Electron.* **2022**, *6*, 85. [CrossRef]
51. Hashemi, M.R.M.; Fikes, A.C.; Gal-Katziri, M.; Abiri, B.; Bohn, F.; Safaripour, A.; Kelzenberg, M.D.; Warmann, E.L.; Espinet, P.; Vaidya, N.; et al. A flexible phased array system with low areal mass density. *Nat. Electron.* **2019**, *2*, 195–205. [CrossRef]
52. Joshi, M.; Hu, K.; Lynch, C.A.; Cui, Y.; Soto-Valle, G.; Tentzeris, M.M. Computer Vision Enabled Calibration of Additively Manufactured Conformal Phased Arrays Utilizing 3-D Depth Sensing Camera. *IEEE Trans. Microw. Theory Tech.* **2024**, *72*, 6728–6738. [CrossRef]
53. Moulder, W.F.; Das, R.N.; Maccabe, A.C.; Bowen, L.A.; Thompson, E.M.; Bell, P.J. Rigid-Flexible Antenna Array (RFAA) for Lightweight Deployable Apertures. In Proceedings of the 2020 14th European Conference on Antennas and Propagation (EuCAP), Copenhagen, Denmark, 15–20 March 2020.
54. Gan, T.H.; Liu, A.; Tan, P.K.; Lu, J. A Flexible Ku-Band Checkerboard Phased Array Antenna. In Proceedings of the 2024 International Applied Computational Electromagnetics Society Symposium (ACES), Orlando, FL, USA, 19–22 May 2024.
55. Wang, X.; You, D.; Fu, X.; Lee, H.; Li, Z.; Awaji, D.; Pang, J.; Shirane, A.; Sakamoto, H.; Okada, K. A Flexible Implementation of Ka-Band Active Phased Array for Satellite Communication. In Proceedings of the 2022 IEEE/MTT-S International Microwave Symposium—IMS 2022, Denver, CO, USA, 19–24 June 2022.
56. You, D.; Xi, F.; Herdian, H.; Wang, X.; Narukiyo, Y.; Fadila, A.A.; Lee, H.; Ide, M.; Kato, S.; Li, Z.; et al. A Ka-Band 64-Element Deployable Active Phased-Array TX on a Flexible Hetero Segmented Liquid Crystal Polymer for Small Satellites. *IEEE Microw. Wirel. Technol. Lett.* **2023**, *33*, 903–906. [CrossRef]
57. Kataria, C.Y.; Cantley, L.E.; Fenn, A.J.; Terranova, D.F.; Yegnanarayanan, S.S.; Zhang, B. Design of a Flexible Thermally-Drawn Photonics Receive Linear Dipole Phased Array Antenna at UHF. In Proceedings of the 2022 IEEE International Symposium on Phased Array Systems and Technology (PAST), Waltham, MA, USA, 11–14 October 2022.
58. Kakaraparty, K.; Mahbub, I. A 24 GHz Flexible 10 × 10 Phased Array Antenna for 3D Beam Steering Based V2V Applications. In Proceedings of the 2022 IEEE International Symposium on Phased Array Systems and Technology (PAST), Waltham, MA, USA, 11–14 October 2022.
59. Ayaz, M.; Ullah, I.; Iftikhar, A.; Abbas, S.M.; Braaten, B.D.; Khattak, S.; Alathbah, M. A Flexible Conformal Phased Array With Embedded Magnetic Particles Based Composite Right/Left Handed Metamaterial Phase Shifters. *IEEE Access* **2024**, *12*, 129147–129159. [CrossRef]
60. Ayaz, M.; Ullah, I. A Singly Curved Conformal Phased Array with Integrated Particles-Based CRLH Phase Shifters. *Electron. J.* **2023**, *27*, 43–49. [CrossRef]
61. Baudh, R.K.; Sahu, S.; Parihar, M.S.; Kumar, V.D. A Semi-flexible Conformal Beam Scanning Broadband Wearable Antenna Array for Radar Application. In Proceedings of the 2023 IEEE 7th Conference on Information and Communication Technology (CICT), Jabalpur, India, 15–17 December 2023.
62. Rutan-Bedard, S.I.; Winkler, C.; Kugler, K.; Webb, K.M.; Roy, S. A Frequency Shift Key Radar Approach for the Reconstruction of Flexible Phased Array Geometry. In Proceedings of the 2022 IEEE International Symposium on Antennas and Propagation and USNC-URSI Radio Science Meeting (AP-S/URSI), Denver, CO, USA, 10–15 July 2022.
63. Hu, K.; Callis, T.W.; Tentzeris, M.M. Additively Manufactured Flexible On-Package Phased Antenna Arrays With Integrated Microfluidic Cooling Channels for 5G/mmWave System-on-Package Designs. *IEEE Microw. Wirel. Technol. Lett.* **2023**, *33*, 899–902. [CrossRef]

64. Liu, M.; Zhou, Y.; Tu, Z.; Zhang, C.; Chen, K.; Luo, J. A novel Conformal T/R Module Based on Low Loss Flexible Printed Circuit Board. In Proceedings of the 2021 CIE International Conference on Radar (Radar), Haikou, Hainan, China, 15–19 December 2021.
65. You, D.; Wang, X.; Herdian, H.; Fu, X.; Lee, H.; Ide, M.; Da Gomez, C.; Li, Z.; Mayeda, J.; Awaji, D.; et al. A Ka-Band Deployable Active Phased Array Transmitter Fabricated on 4-Layer Liquid Crystal Polymer Substrate for Small-Satellite Mount. *IEEE Access* **2023**, *11*, 69522–69535. [CrossRef]
66. Mizrahi, O.S.; Fikes, A.; Hajimiri, A. Flexible Phased Array Sheets: A Techno-Economic Analysis. *arXiv* **2023**, arXiv:2302.03562. [CrossRef]
67. You, D.; Awaji, D.; Shirane, A.; Sakamoto, H.; Okada, K. A Flexible Element Antenna for Ka-Band Active Phased Array SATCOM Transceiver. In Proceedings of the 2020 IEEE Asia-Pacific Microwave Conference (APMC), Hong Kong, China, 8–11 December 2020.
68. Nan, X.; Kang, T.; Zhang, Z.; Wang, X.; Zhang, J.; Lei, Y.; Gao, L.; Cui, J.; Xu, H. Flexible Symmetric-Defection Antenna with Bending and Thermal Insensitivity for Miniaturized UAV. *Micromachines* **2024**, *15*, 159. [CrossRef] [PubMed]
69. Mizrahi, O.S.; Fikes, A.; Hajimiri, A. Flexible Phased Array Shape Reconstruction. In Proceedings of the 2021 IEEE MTT-S International Microwave Symposium (IMS), Atlanta, GA, USA, 7–25 June 2021.
70. Fikes, A.C.; Safaripour, A.; Bohn, F.; Abiri, B.; Hajimiri, A. Flexible, Conformal Phased Arrays with Dynamic Array Shape Self-Calibration. In Proceedings of the 2019 IEEE MTT-S International Microwave Symposium (IMS), Boston, MA, USA, 2–7 June 2019.
71. Elmobarak, H.A.; Himdi, M.; Castel, X.; Rahim, S.K.A.; Geok, T.K. Flexible Patch Antenna Array Operating at Microwaves Based on Thin Composite Material. *IEEE Access* **2022**, *10*, 115663–115672. [CrossRef]
72. Gan, T.H.; Liu, A.; Tan, P.K.; Lu, J. A Bendable Wideband Dual-Polarization Conformal Phased-Array Antenna. *IEEE Antennas Wirel. Propag. Lett.* **2023**, *22*, 1952–1956. [CrossRef]
73. Ding, C.; Gao, X.; Tang, W.; Liu, H.; Zhu, C.; Bu, X.; An J. A Balloon Conformal Array Antenna of Broadened Bandwidth, High Gain, and Steerable Directional Beam Fed by a Flexible Phase Shifting Network. *IEEE Trans. Antennas Propag.* **2023**, *71*, 8525–8536. [CrossRef]
74. Soh, P.J.; Van den Bergh, B.; Xu, H.; Aliakbarian, H.; Farsi, S.; Samal, P.; Vandenbosch, G.A.; Schreurs, D.M.P.; Nauwelaers, B.K. A smart wearable textile array system for biomedical telemetry applications. *IEEE Trans. Microw. Theory Tech.* **2013**, *61*, 2253–2261. [CrossRef]

sensors

MDPI

Review

From IOTA Tangle 2.0 to Rebased: A Comparative Analysis of Decentralization, Scalability, and Suitability for IoT Applications

Pierre Sedi Nzakuna [1,*], Vincenzo Paciello [1,*], Aimé Lay-Ekuakille [2,3], Angelo Kuti Lusala [4], Salvatore Dello Iacono [5] and Antonio Pietrosanto [1]

1 Department of Industrial Engineering, University of Salerno, 84084 Fisciano, Italy
2 Department of Innovation Engineering, University of Salento, 73100 Lecce, Italy
3 Department of Electronics, Institut Supérieur Pédagogique Technique de Kinshasa, Kinshasa, Democratic Republic of the Congo
4 Department of Electrical and Computer Engineering, Polytechnic Faculty, University of Kinshasa, Kinshasa XI BP 127, Democratic Republic of the Congo
5 Department of Information Engineering, University of Brescia, 25121 Brescia, Italy
* Correspondence: psedinzakuna@unisa.it (P.S.N.); vpaciello@unisa.it (V.P.)

Abstract: The Internet of Things (IoT) demands scalable, secure, and feeless distributed ledger technologies (DLTs) to enable seamless machine-to-machine transactions. The IOTA DLT was developed to fulfill this vision through its feeless Directed Acyclic Graph (DAG) named the Tangle, whose announced upgrade to IOTA 2.0 promised feeless microtransactions and coordinator-free (Coordicide) decentralization via a Nakamoto Consensus mechanism and a Mana anti-spam system. However, its delayed decentralization and scalability limitations hindered ecosystem growth and practical IoT adoption, leading to a new ledger architecture named IOTA Rebased. This paper critically analyzes this architectural pivot and its implications for IoT applications, contrasting the abandoned IOTA 2.0 protocol—a leaderless, feeless DAG designed for the IoT—with the adoption of a Move Virtual Machine-based, object-oriented ledger secured by a Delegated Proof-of-Stake consensus via the Mysticeti protocol in IOTA Rebased. We evaluate IOTA Rebased trade-offs: enhanced programmability and speed versus compromised IoT suitability due to fees, and explore mitigation strategies such as sponsored transactions, lightweight clients, and hierarchical tiered transaction architecture to align IOTA Rebased with IoT environments where microtransactions are prevalent. A use case analysis is provided for the integration of IOTA Rebased in IoT scenarios. This study underscores the tension between technological innovation and decentralization, offering insights for balancing scalability with the unique demands of the IoT.

Keywords: Internet of Things; IOTA; distributed ledger technologies; blockchain; directed acyclic graph

1. Introduction

The Internet of Things (IoT) refers to a vast network of interconnected physical objects—ranging from everyday household items to sophisticated industrial sensors—each equipped with unique identifiers that enable autonomous data exchange [1]. As these devices proliferate, ensuring secure, scalable, and efficient communication becomes increasingly challenging, particularly due to centralized vulnerabilities and resource constraints inherent in traditional architectures [2]. Distributed Ledger Technologies (DLTs) emerge as a transformative solution in this context by facilitating decentralized, tamper-proof, and transparent data management. DLTs provide robust mechanisms for ensuring data integrity and trust through immutable records and consensus protocols that eliminate single points of failure in the network [1]. Adaptive consensus techniques within DLTs can achieve fairness among heterogeneous devices, making them particularly suited for the high-frequency, low-value transactions typical of IoT applications [3].

IOTA emerged in this context as an open-source, feeless DLT specifically designed to meet the demands of IoT applications, offering high scalability through its innovative Directed Acyclic Graph (DAG) structure named the Tangle [4]. The DAG structure of the Tangle, contrasting with the traditional blockchain structure, allowed IOTA to process a higher volume of microtransactions with zero fees, thus achieving high scalability since all nodes in the network acted as validators by validating existing transactions in the network before inserting their own. Additionally, IOTA implemented a lightweight Proof-of-Work (PoW) mechanism well suited for the low-power requirements of IoT hardware, thereby eliminating the cost barrier for IoT devices with limited computational resources. This approach made IOTA a promising solution to address the decentralized, scalable, and secure communication needs of the IoT and machine-to-machine transactions [5,6].

However, the IOTA Tangle faced the blockchains trilemma that they cannot be decentralized, scalable, and secure at the same time. The Coordinator was then introduced in the IOTA Tangle as a temporary bootstrapping mechanism to enhance security and finality by issuing milestones to reach consensus. This solution came at the cost of introducing a centralized element into what was inherently meant to be a decentralized system, thereby creating a potential single point of failure in the network [2]. To address this contradiction, IOTA 2.0 was designed with the goal of executing "Coordicide", an update aimed at removing the coordinator from the network, to allow full decentralization while preserving the scalability and security that are critical for the IoT environment [7,8]. In addition, IOTA 2.0 aimed at introducing a novel Nakamoto Consensus mechanism with a reputation system known as Mana and a sharding-inspired approach for scalability to facilitate seamless, feeless microtransactions across billions of connected devices.

Despite the groundbreaking vision of IOTA 2.0, practical issues and changing network demands prevented its deployment. In response, the IOTA Foundation has shifted its focus toward a reimagined protocol known as IOTA Rebased [9], which departs significantly from its predecessors by incorporating transaction fees to sustain network operations and leverages the Mysticeti DAG structure [10] in replacement of the Tangle to bolster throughput and security. While these changes enhance certain aspects of network performance and resilience, they also introduce complexities, particularly the reintroduction of fees, which may affect the feasibility of microtransactions that are critical to IoT applications.

This paper provides a systematic comparative analysis of the architectural evolution of IOTA, grounded in a structured framework that contrasts the decentralized, feeless promises of IOTA 2.0 with the fee-driven realities of IOTA Rebased. We derive evaluation criteria from IoT requirements—scalability, microtransaction feasibility, decentralization, and hardware efficiency—to assess trade-offs between these protocols and reveal how the shift from a leaderless DAG to a staking-driven object-oriented ledger impacts IoT adoption. Our analysis synthesizes insights from peer-reviewed case studies and protocol whitepapers while aligning findings with established DLT literature. In addition, this work proposes actionable mitigation strategies, such as sponsored transactions and tiered architectures, to reconcile enhanced programmability in Rebased with IoT microtransaction demands. Furthermore, we critically evaluate the consequences of abandoning feeless transactions, a cornerstone of the original vision of IOTA, on resource-constrained IoT ecosystems. By dissecting the strengths and limitations of both protocols through this rigorous comparative analysis, this study aims to equip researchers and practitioners with insights to bridge the gap between IoT-friendly decentralization and modern ledger governance.

The remainder of the paper is structured as follows: Section 2 summarizes the evolution of the IOTA Tangle. Section 3 reviews the promises and technical foundations of IOTA 2.0. Section 4 explains the new IOTA Rebased architecture, including its consensus mechanism and tokenomics. Section 5 presents a comparative analysis of IOTA 2.0 with IOTA Rebased, especially in the context of IoT applications. Section 6 proposes recommendations for the integration of IOTA Rebased in the IoT—according to its original mission of enabling efficient and cost-effective machine-to-machine transactions—with a use case analysis. Section 7 proposes potential future research directions, and finally, the last Section 8 concludes this work.

2. Evolution of the IOTA Tangle

2.1. The Tangle

The DAG structure of the Tangle overcomes the scalability limitations of traditional blockchains by allowing each transaction to reference multiple predecessors simultaneously, instead of only one as in blockchains. This design in the Tangle enabled several transactions to be attached and validated concurrently, thus significantly enhancing the network throughput and efficiency [11]. Figure 1 shows the difference between the DAG structure of the Tangle and the single-chain structure of blockchains, with the Tangle solving the bottleneck and throughput issues in blockchains.

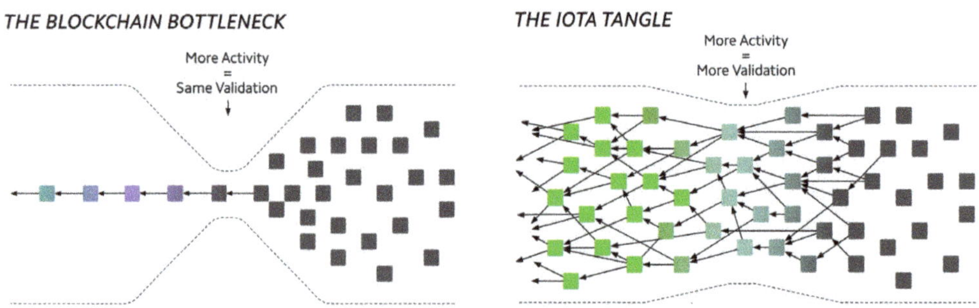

(a) (b)

Figure 1. The DAG structure of the Tangle solves the scalability issue in blockchains expressed by the bottleneck. Black squares represents transactions waiting to be added in the ledger. (**a**) The single-chain structure of blockchains, where only one new transaction can be added to the ledger at a time. (**b**) The DAG structure of the IOTA Tangle, allowing multiple transactions to be simultaneously added to the ledger. Green squares represents finally-confirmed transactions.

Vertices in the Tangle DAG are called blocks, and edges are called *references*. A new block in the Tangle is known as a tip (dark gray blocks on Figure 1b). When first issued, a tip is unconfirmed and must reference at least two earlier blocks to validate them. As additional blocks are issued and reference the tip, peer nodes in the network gradually increase its confirmation confidence. The more a tip is referenced by subsequent blocks, the more reliable it becomes, thereby raising its likelihood of reaching final confirmation [12]. The computational effort required to validate a block determines its weight. The cumulative weight of a block is calculated by adding its own weight to those of all transactions that directly or indirectly approve it, which increases its significance within the ledger.

Consequently, a tip remains unconfirmed until its cumulative weight reaches a specified threshold, signifying that it has garnered sufficient network validation [2].

When issuing a new tip in the network, a node selects the two blocks to reference using a weighted Markov Chain Monte Carlo random walk algorithm. This algorithm employs a regulating parameter α, which prevents the selection process from always favoring the tip with the highest cumulative weight, thereby ensuring a more balanced and secure tip selection [12]. Consensus within the Tangle is achieved when nodes collectively agree on which blocks should be included and validated in the ledger, with the cumulative weight of a block serving as a crucial indicator of its legitimacy [2]. The Tangle relies on a Proof-of-Authority (PoA) consensus mechanism—leveraging the verifiable identity of each node as a form of stake—along with a lightweight PoW that must be performed for every transaction. This combination not only helps establish trust in the network but also protects it from flooding attacks by malicious entities attempting to overwhelm the system with spurious blocks [11].

The Coordinator—a node plugin controlled by the IOTA Foundation—functions as a centralized finality mechanism by periodically issuing signed blocks called milestones. These milestones are trusted by other nodes in the network and serve as confirmation checkpoints, so that any block in the Tangle that is either directly or indirectly referenced by a milestone is deemed confirmed [13]. However, this centralized approach is controversial within the cryptocurrency community because it grants the IOTA Foundation the authority to effectively decide which transactions receive priority, potentially freeze funds, or even ignore certain transactions. Moreover, the reliance on the coordinator creates a single point of failure because if it were to stop functioning, the entire confirmation process in the network would come to a halt [2].

Network roles are divided among full nodes, perma nodes, and clients. Nodes (both full and perma) use a consensus algorithm to validate transactions. Perma nodes hold the entire ledger history, while full nodes focus on the current ledger state. To send transactions, clients must connect to the ledger via either a full or perma node. The Tangle network ecosystem includes the Mainnet (production network), Shimmer (staging network for pre-Mainnet protocol testing), and the public testnet (developer test bed) [13].

Unlike classical modeling tools such as finite-state automata and Petri nets, which routinely use cycles to represent recurring control flows and resource replenishment, the Tangle is built on an enforced acyclicity guaranteeing that once a transaction is confirmed, it cannot be revisited, thereby eliminating infinite-loop scenarios and delivering deterministic, irreversible settlement. While Petri nets model concurrency through tokens and synchronizing transitions inside potentially cyclic structures [14], the Tangle captures concurrency by recording the causal order between independent transactions, allowing many tips to be approved in parallel. By trading the expressiveness of cycles for a growing, one-way DAG, the Tangle achieves the high throughput and rapid finality demanded by large-scale ledger applications.

2.2. Network Protocol Updates

The IOTA Tangle has evolved through four major protocol iterations since its initial release in 2016. The initial version, later on known as IOTA 1.0, employed a trinary data representation coupled with a probabilistic consensus mechanism based on random walkers that navigated toward the tips to select candidate transactions for approval [4]. This early design, however, was particularly susceptible to conflict spamming attacks. In response, the first Chrysalis protocol update was introduced, upgrading the ledger to the IOTA 1.5 version, and mitigating these vulnerabilities by ignoring conflicting transactions during tip selection. By imposing a deterministic order on the Tangle, the Chrysalis update enabled the use of milestones to confirm transactions by accepting only the first non-conflicting message that maintained the integrity of the ledger.

The second Chrysalis upgrade marked a pivotal transformation for IOTA by shifting the data representation of the ledger from trinary to binary, thereby streamlining data processing and interoperability with modern systems. The operating model shifted from the traditional account model to an Unspent Transaction Output (UTXO) framework, allowing for more precise and conflict-resistant tracking of funds. Additionally, the upgrade replaced the Winternitz One-Time Signature scheme with the more robust Edwards-curve Digital Signature Algorithm (EdDSA), thereby enhancing both security and computational efficiency considering IoT devices with limited resources. Protective measures were also introduced for microtransactions by ensuring recipient addresses could resist dust attacks (wherein negligible amounts of funds are sent to clutter and potentially exploit the network) [13].

The Stardust upgrade introduced the support of multiple assets beyond the base IOTA token. This update refined the dust protection in the ledger by introducing a more granular protection mechanism based on precise byte cost calculations through storage deposits. Furthermore, Stardust rebranded the basic unit of the ledger from messages to blocks, reflecting a maturation of the underlying architecture. It also laid the foundation for IOTA Smart Contracts (ISCs) by incorporating Alias Outputs for efficient anchoring of smart contracts by guaranteeing there is always exactly one such output for each chain of the UTXO ledger [15], and adopted the Bench32 format for address encoding, which provides strong error correction. The single coordinator node was replaced with a distributed validator committee using an off-chain Byzantine Fault Tolerant (BFT) consensus mechanism, a leap towards the full decentralization goal of IOTA. Table 1 summarizes the four IOTA Tangle protocol versions and their main characteristics.

Table 1. Protocol versions of the IOTA Tangle [13].

Protocol	Release Date	Description	Main Features
IOTA 1.0	July 2016	The legacy version of the IOTA Tangle, using a probabilistic consensus algorithm based random walkers to select most suitable tips for approval.	Trinary data representation.
Chrysalis 1 (IOTA 1.5)	August 2020	The White-Flag approach replaces the heavy probabilistic consensus algorithm, addressing the conflict spamming attack by enforcing a deterministic ordering of the Tangle.	White-Flag confirmation.
Chrysalis 2 (IOTA 1.5)	April 2021	Transitioned from trinary to binary data representation, replaced the account model with the UTXO model, migrated from Winternitz One-Time Signature to EdDSA signature scheme, and implemented measures to guard against dust transactions.	Binary data representation, UTXO, EdDSA.
Stardust	October 2023	Introduced support for multiple assets support, implemented byte cost-based granular dust protection through storage deposits, evolved vertices from messages to blocks, and prepared for ISC anchoring via Alias Outputs, replacement of the single Coordinator node with a validation committee.	Multiple assets, smart contracts, byte cost-based dust protection, blocks as vertices, Bench32 addresses, validation committee.
IOTA Rebased	May 2025	Reimagined ledger architecture replacing the Tangle with the Mysticeti DAG, introducing transaction fees, MoveVM-based smart contracts, and a Delegated Proof-of-Stake (DPoS) consensus model via the Mysticeti protocol.	Object-oriented ledger, Mysticeti protocol, transaction fees, MoveVM integration, DPoS consensus.

The Shimmer Ethereum Virtual Machine (EVM) was released along with the Stardust protocol to support ISCs, allowing developers to leverage existing Ethereum smart contracts and tools on the Shimmer EVM chain while demonstrating noticeable performance, such as high Transactions Per Second (TPS) and low latency [16]. The Shimmer EVM integrated the support of Layer 1 native assets and Non-Fungible Tokens (NFTs), with the aim of allowing the creation and operation of multiple individual Layer 2 networks running in parallel in future upgrades [17]. Following the release of the Shimmer EVM, ISCs underwent extensive improvements, leading to the release of IOTA EVM by June 2024, a Layer 2 solution designed to bring smart contract capabilities to the IOTA network. IOTA EVM ensures full EVM compatibility and adds innovative features leveraging the unique Layer 1 Native Asset Framework of IOTA. IOTA EVM leverages its unique Layer 1 for parallel processing, enabling horizontal scalability and boosting multi-chain performance. It offers seamless interoperability via easy Solidity contract deployment across EVM and non-EVM chains. Furthermore, native randomness and resistance against Maximal Extractable Value (MEV) enhance fairness and security by preventing front-running through unpredictable transaction ordering, promoting market integrity [18].

3. IOTA 2.0: Promises and Challenges

3.1. Core Concepts

IOTA 2.0 was conceived as a paradigm shift from the coordinator-dependent architecture of its predecessors, aiming to achieve full decentralization while preserving scalability and security for IoT applications. This subsection delineates the foundational pillars of the IOTA 2.0 design, which sought to eliminate centralized oversight through leaderless consensus, dynamic resource allocation, and cryptoeconomic incentives.

3.1.1. Leaderless Full Decentralization

The flagship objective of IOTA 2.0 was to remove the centralized coordinator from the network. The removal of this single point of control—and potentially of failure—, referred to as Coordicide, was intended to pave the way for a trustless and fully decentralized network. In the absence of the coordinator, every network participant would contribute to the validation process, thus performing autonomous transaction validation in the ledger. This approach was designed to enhance network resilience by distributing trust and security across nodes [19].

Various challenges needed to be tackled to achieve effective removal of the coordinator, summarized as building blocks in Figure 2. Firstly, node accountability was implemented through unique global node IDs, combined with an innovative Sybil protection mechanism that avoided requiring operators to risk or reveal funds. By reliably identifying which node sent a message, the network could automatically organize itself through auto peering and effectively discourage harmful actions by implementing measures like rate control against malicious nodes. Secondly, a node discovery and auto-peering mechanism was proposed to facilitate easy network entry for new nodes while also securing the peering process against targeted attacks, thereby protecting the network from Eclipse attacks. Thirdly, a rate control mechanism was designed to prevent congestion by selectively allowing transactions to propagate based on statistics related to the originating node. To issue transactions, the latter had to solve a cryptographic puzzle to issue a transaction whose difficulty dynamically scaled, increasing with recent transaction frequency and decreasing with the Mana owned by that node, thus regulating throughput. Finally, to achieve consensus, two voting mechanisms were proposed to allow nodes to determine the current network status by querying the opinions of other nodes [7].

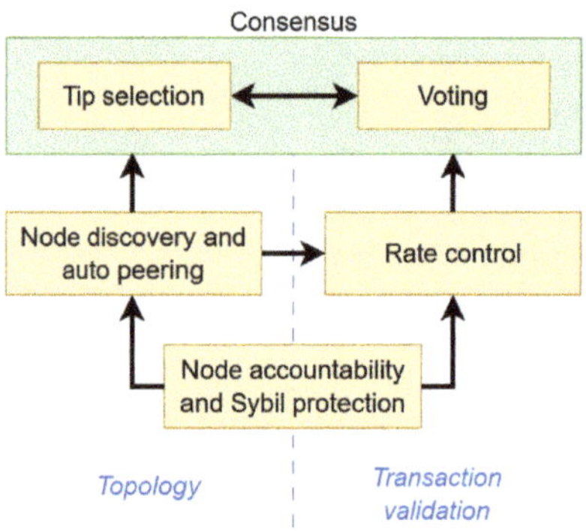

Figure 2. Building blocks in the Coordicide architecture [7].

3.1.2. Consensus Mechanism

While previous network protocol versions relied on the centralized coordinator and the weighted Markov Chain Monte Carlo to achieve consensus to finally confirm objects in the ledger, IOTA 2.0 initially relied on a combination of Fast Confirmation of On-the-Heavy Branch (FCOB) and Fast Probabilistic Consensus (FPC) as a probabilistic leaderless binary voting protocol, lightweight and scalable to provide rapid and energy-efficient consensus [19]. Instead of relying on resource-intensive mining, FPC leverages random sampling and iterative voting to quickly achieve agreement among network participants, with this procedure being terminated locally if a node has not changed its opinion over a specified period of time or if some maximal amount of rounds is reached [20]. The probabilistic nature of the consensus not only boosted efficiency but also added a layer of unpredictability that helps protect the network from coordinated attacks, thereby implementing security through randomness by ensuring that no single entity could easily manipulate the validation process. The voting power in FPC is proportional to the node reputations, and several improvements were implemented to decrease the failure rates significantly and allow IOTA to withstand adversaries with higher weight [21].

However, while FPC offers advantages in terms of simplicity and resistance to certain attacks, it was primarily effective for conflict resolution and did not provide a comprehensive solution for achieving global consensus across the network. In addition, the voting process performed via direct queries between the peers required an additional communication layer. Recognizing the limitations of FPC, the IOTA Foundation transitioned to a more robust consensus model inspired by the Nakamoto Consensus, adapting the traditional longest-chain rule from blockchain systems to the Tangle DAG structure [22,23]. In this model, validators, selected through a staking mechanism, issue specialized validation blocks at regular intervals. These validation blocks carry the validator stake weight and contribute to the approval weight of referenced blocks. The approval weight determines the consensus status of transactions and blocks, guiding the network toward a unified ledger state. This approach enables an asynchronous, leaderless, and scalable consensus process, allowing for parallel transaction validations without the need for a global total order [23].

To manage conflicting transactions and ensure network coherence in this adaptation of the Nakamoto Consensus, the protocol employs consensus flags and slot commitment chains. Consensus flags indicate the confidence level in the validity of a block, ranging from pre-acceptance to finalization. Slot commitment chains allow nodes to track and compare their views of the ledger, facilitating agreement on the most valid chain (with the heaviest cumulative weight). In cases of divergence, nodes adopt the chain with the highest cumulative weight, ensuring convergence to a single, consistent ledger state [23].

The Distributed Random Number Generation (dRNG) mechanism—inspired by voting-based Byzantine consensus protocols [24]—was added to address synchronization and prevent adversarial exploits by resolving scenarios where nodes might otherwise remain indecisive due to conflicting valid opinions. By generating synchronized external randomness at predetermined epochs, the protocol periodically aligns the state across nodes, breaking potential stalemates and facilitating rapid convergence towards consensus. This reduces vulnerabilities to adversarial tactics aimed at prolonging indecision within the network, ensuring more robust and predictable consensus outcomes [22].

3.1.3. Reputation and Resource Management

To incentivize honest participation, IOTA 2.0 proposed a reputation system based on Mana. This mechanism assigns a reputation score to nodes based on their historical behavior and the volume of legitimate transactions they process, thereby acting as a critical resource governing access to network resources, securing the ledger, and ensuring fairness without the need for traditional transaction fees [20]. Mana was intended to ensure that nodes with higher reputation and investment in the network would have a more significant influence in the consensus process. This design was critical in mitigating Sybil attacks and maintaining overall network integrity.

In IOTA 2.0, Mana is accrued by holding IOTA tokens, with its amount dependent on both token quantity and holding duration. Mana serves the two following main functions:

- Stored Mana: Linked to token outputs and transferable, this type of Mana gradually decays over time to encourage continuous participation.
- Block Issuance Credits (BIC): Non-transferable credits enabling users to issue blocks, with usage dynamically adjusted based on network congestion, deterring spam and misuse.

Additionally, users can stake tokens to act as validators, directly contributing to consensus and network stability, or delegate tokens to trusted validators. Validators and delegators earn rewards based on their network contributions and performance, incentivizing reliable participation. To further avoid centralization, Mana and delegation rewards diminish as more tokens concentrate with a single validator, promoting equitable distribution and active engagement across the network [25].

3.1.4. Smart Contracts

IOTA 2.0 featured ISCs, a scalable multi-chain smart contract framework anchored to the main ledger (Layer 1). ISC was proposed to support multiple parallel programmable ledgers (Layer 2 chains), each leveraging the UTXO model and quasi-Turing complete programmability to enhance scalability and efficiency. Cross-chain interactions occurred trustlessly via Layer 1 anchors, preserving decentralization and security. Additionally, ISC employed advanced cryptographic techniques such as threshold signatures and Verkle trees for secure state validation and efficient transaction processing, enabling practical, high-throughput smart contract applications [15].

3.1.5. Security Enhancements

IOTA 2.0 incorporated advanced cryptographic techniques to secure transactions and protect against fraudulent activities. The decentralized validation process, combined with the FPC consensus and Mana-based reputation, was expected to offer a high level of security without sacrificing speed. By distributing both the validation responsibility and the control of network resources, IOTA 2.0 aimed to build a system that was inherently resistant to many conventional attack vectors found in centralized or less DLTs, such as the consequences of denial-of-service (DoS) attacks [26].

Beyond ledger-specific cryptography, the security posture of any IoT-centric DLT must be viewed in the wider cyber-physical-systems (CPS) threat landscape. Comprehensive surveys [27,28] catalogue attack types (sensor spoofing, control-loop hijacking, eavesdropping, etc.) and defense measures (defense-in-depth, data cleaning, elastic model predictive control, etc.) that span both cyber and physical domains. The layered security model proposed for IOTA 2.0—combining lightweight PoW, reputation-weighted rate control, and slot-anchored finality—aligned with the multi-tier mitigation strategies highlighted in these surveys.

3.2. *Protocol Layers*

IOTA 2.0 adopted a modular, layered architecture comprising three main layers—network, communication, and application—each building upon the functionalities of the one beneath it. At the foundational network layer, a peer-to-peer (P2P) overlay facilitates secure and efficient peer discovery and neighbor selection through mechanisms like autopeering, allowing nodes to connect automatically. This layer also manages message propagation via a gossip protocol to maintain network connectivity. Built upon this is the communication layer, responsible for the dissemination and processing of blocks that form the Tangle DAG structure. It employs rate control and congestion control modules to regulate data flow and prevent network congestion. Finally, at the highest level, the decentralized application layer encompasses core ledger functionalities, including the ledger engine responsible for maintaining the state of the ledger and resource management through Mana. Additionally, this layer supports decentralized applications (dApps) and smart contracts, enabling developers to build and deploy various services directly onto the IOTA platform [29]. Figure 3 presents the layered architecture of the IOTA 2.0 protocol.

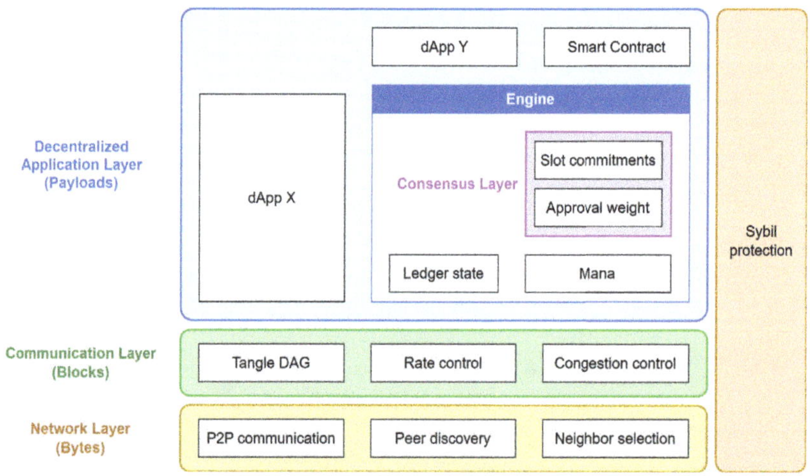

Figure 3. Layered architecture of the IOTA 2.0 protocol [29].

3.3. *Unmet Goals and Challenges of IOTA 2.0*

Despite its pioneering vision, IOTA 2.0 encountered unresolved challenges that impeded its deployment. This subsection examines critical hurdles, including consensus and finalization complexities and prolonged development cycles.

3.3.1. Consensus Mechanism Complexities

The FPC mechanism and Nakamoto Consensus implemented in IOTA 2.0 represented a substantial improvement over the coordinator-based system. However, it introduced complexity that posed challenges to seamless network operation. Under challenging network conditions—such as network partitions or when validators go offline—nodes may perceive different blocks being accepted, creating divergent slot commitment chains. This can lead to network forks where different nodes maintain inconsistent views of the ledger state. While IOTA 2.0 implements a Chain Switching Rule to resolve these inconsistencies, the very existence of temporary forks introduces uncertainty and operational complexity. The protocol requires nodes to track and update their slot commitment chains while checking commitments from other nodes, adding significant computational overhead [30].

3.3.2. Finalization Process Complexity

The finalization process in IOTA 2.0 required a supermajority (more than two-thirds of voting power) of the validators committee to reference slot commitments in their validation blocks. All validation blocks had to be issued within the same slot and be confirmed to achieve finalization. Each node computed finalization flags based on its own locally maintained copy of the Tangle [30]. This rigorous finalization process enhanced security but introduced latency that might have impacted time-sensitive applications. The complexity of this mechanism creates potential barriers to adoption for developers unfamiliar with the unique consensus approach of IOTA 2.0.

3.3.3. Protracted Development Cycle

One of the most significant challenges limiting the adoption of IOTA 2.0 has been its extended development timeline. The journey toward a fully decentralized IOTA began years ago, with initial testing network deployments occurring as early as 2020. The development progressed through multiple stages, beginning with the Pollen testing network (the first testing phase of IOTA 2.0), then advancing to the Nectar test network (a full implementation of IOTA 2.0 on an incentivized testing network), with the final aim of implementing the Honey main network (including all modules and final specifications of the Coordicide) [31]. However, as of early 2024, the full implementation of IOTA 2.0 appeared to remain a work in progress, with the IOTA Foundation still making progress in the research and development of IOTA 2.0. By the end of 2024, the IOTA 2.0 project had finally been abandoned after 7 years of continuous development, and the ledger switched to IOTA Rebased following a vote by IOTA token holders, with the IOTA Foundation acknowledging that this switch would make time-to-market for full decentralization shorter than IOTA 2.0 plans [32,33]. This extended development cycle of the IOTA 2.0 protocol has dampened enthusiasm among potential adopters and developers, potentially leading to exploring other blockchain solutions with more stable and established protocols during that period.

3.3.4. Architectural Refinements and Optimizations

The IOTA Foundation has acknowledged the need for substantial protocol refinements, indicating that the initial design may have been suboptimal. Continuous modifications suggested that the original architecture may have contained inefficiencies or unnecessary complexity that required remediation. In September 2021, the foundation reported working on optimizing the consensus layer along with the communication layer to enhance protocol efficiency and reduce its complexity, with the majority of changes involving optimization of the data flow and code simplification [34]. Such iterative refinements, while necessary for protocol improvement, have extended the timeline for a production-ready implementation and created uncertainty around the final form and capabilities of IOTA 2.0.

3.3.5. Smart Contracts Functionality Implementation

IOTA 2.0 was designed to support smart contracts, addressing a significant limitation of the original protocol. However, the implementation of this functionality faced delays due to the unique architecture of the IOTA 2.0. The introduction of the Shimmer EVM followed by the IOTA EVM launch suggested that the original smart contract design may not have met market requirements, raising concerns regarding the scalability, security and user-adaptability of the protocol design. While the integration of the Sui MoveVM—a novel smart contract language for programming blockchains using objects abstraction [35]—was considered feasible by the IOTA Foundation, its implementation in IOTA 2.0 architecture would have been a multi-year effort that could have caused additional undesirable delays due to breaking protocol changes [33]. Commendable progress was made in ledger performance, but the necessity of implementing fundamental revisions for Layer 1 programmability in IOTA 2.0 raised pertinent questions about the overall stability and developmental maturity of the protocol, with the IOTA Foundation recognizing that it could not afford additional years of research and development with uncertain outcomes [32].

3.3.6. Competition from Other Blockchains

IOTA 2.0 entered a highly competitive landscape dominated by established smart contract platforms like Ethereum and rapidly growing alternatives such as Solana and Sui [36,37]. These platforms offer mature developer ecosystems, extensive tooling, and proven scalability that have attracted significant developer communities and user bases. Continuous design revisions in IOTA 2.0 made it challenging to compete with other well-established Layer 1 blockchains that offered consistent performance metrics.

3.3.7. Hardware and Resource Constraints

The IOTA design philosophy of maintaining low hardware requirements to accommodate unpaid node operators created a fundamental tension with performance objectives. While this approach aligns with the initial IOTA vision of enabling IoT applications on resource-constrained devices, it simultaneously limits the ability of the protocol to scale to meet competitive performance benchmarks. This trade-off may be particularly problematic for attracting institutional users and developers building high-throughput applications who prioritize performance over hardware efficiency.

4. IOTA Rebased: A Technical Overview

4.1. Technical Foundation

IOTA Rebased reimagines the architecture of the protocol to prioritize scalability and programmability. This subsection details its technical pillars: an object-oriented DAG ledger for parallel transaction processing, Mysticeti consensus for Byzantine fault tolerance with sub-second finality, and MoveVM-based smart contracts for enhanced programmability.

4.1.1. Object-Based DAG Ledger

IOTA Rebased fundamentally restructures the previous Tangle implementation by transitioning from the UTXO model to an object-based DAG ledger inspired by the Sui Network. This enables enhanced programmability and parallel transaction processing and significantly enhances throughput capabilities. The object model in IOTA Rebased differentiates between shared and owned objects. Shared objects represent resources that multiple entities can access and modify according to predefined rules, while owned objects are exclusively controlled by their designated owners. This distinction facilitates fine-grained access control and enables more complex transaction patterns within the network [9,33].

4.1.2. Mysticeti Consensus Protocol

IOTA Rebased integrates the Mysticeti protocol [10], a BFT consensus protocol optimized for low latency and high throughput using an uncertified DAG structure. Delegated Proof of Stake (DPoS) is used as a consensus mechanism in which token holders delegate their voting power to validators who participate in block production and finalize transactions by executing the Mysticeti protocol [38]. Mysticeti solves the high latency issue in certified DAGs by eliminating the need for explicit certificates, achieving block commitment within the known lower bound of 3 message rounds. Furthermore, it enhances efficiency by committing each block independently without waiting for the entire wave to conclude and minimizes CPU overhead by requiring only a single signature generation and verification per block.

The Mysticeti protocol is proposed in two classes: Mysticeti-C, which is the main protocol version, and Mysticeti-FPC, which adopts a fast commit path that lowers the latency in asset transfer. Both Mysticeti protocol classes reach decisions by interpreting the structure of the DAG, utilizing only block messages. Mysticeti protocol classes operate in a sequence of logical *rounds*. Honest validators propose a unique signed block each round, incorporating transactions and references to recent blocks from prior rounds. A block is signed and sent once it references at least $2f+1$ blocks from the previous round, with f being the number of validators. Clients submit transactions to validators; if a transaction is not finalized within a specific time frame, the client resubmits it to a different validator. Byzantine validators might attempt to equivocate by sending multiple or no blocks to different parties. Such validators are neutralized through DAG pattern analysis, akin to fault predictability in labeled Petri nets [39]. Whereas Petri nets use transition invariants to anticipate faults, Mysticeti dynamically identifies adversarial blocks by requiring validators to reference prior rounds' certificates. This structural approach ensures Byzantine resilience without relying on static fault models, aligning with the unpredictable nature of decentralized networks.

For a block to be correct, it must include the author A and signature on the block contents, a round number r, a list of transactions, and at least $2f+1$ hashes h of blocks from the previous round (plus potentially other previous rounds), with the first hash linking to the previous block of A. A block $B \equiv (A,r,h)$ is valid if the signature is valid, A is a validator, all hashes point to distinct valid blocks from previous rounds, the first block links to another block from A, and the sequence of past blocks contains $2f+1$ blocks from the immediate past round $r-1$. A block B' supports a past block $B \equiv (A,r,h)$ if, when tracing back from B' through hashed blocks, B is the first block found for validator A at round r. While a block might reference others (from different validators) that support equivocating blocks at the same round, at most one of these equivocating blocks can receive support from $2f+1$ validators.

Mysticeti protocol classes operate primarily by involving the identification of two main patterns by validators:

- The skip pattern (Figure 4a) is identified at round $r+1$ when at least $2f+1$ blocks do not support a specific block (A,r,h). Regardless of the number of proposals for the slot (including none), the skip pattern is identified if, across all proposals, we observe $2f+1$ subsequent blocks that either do not support it or support no proposal at all.
- The certificate pattern (Figure 4b) is identified at round $r+1$ when at least $2f+1$ blocks support a block $B \equiv (A,r,h)$. In this case, block B is considered certified, guaranteeing its availability and that no other certified block may exist for the same

spot (A,r). Any subsequent block (like one at round $r+2$) that includes this pattern within its history is called a certificate for block B.

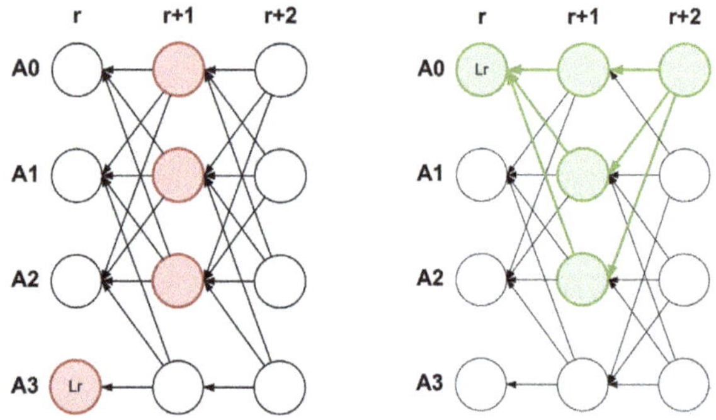

Figure 4. The main DAG patterns identified by validators in the Mysticeti protocol [10]. (**a**) In this skip pattern (formed by red blocks), blocks $(A_0, r+1, \cdot), (A_1, r+1, \cdot), (A_2, r+1, \cdot)$ do not support (A_3, r, L_r). (**b**) In this certificate pattern (formed by green blocks), block $(A_0, r+2, \cdot)$ serves as a certificate for (A_0, r, L).

In IOTA Rebased, Mysticeti is used with a validator committee of initially 150 permissionless slots, selected through staking by token holders. Validators collectively determine transaction validity and finality in 3 message rounds, ensuring economic alignment between network security and stakeholder interests [9,33]. This new consensus scheme in IOTA is designed for fast finality to reduce the risk of rollbacks, energy efficiency compared with the PoW, and parallel transaction processing while maintaining strong security guarantees [38]. The introduction of a leader-based DAG-BFT mechanism allows validators to propose blocks asynchronously, making IOTA Rebased achieve 0.5-s finality with more than 50,000 TPS with sub-second finality [33].

The IOTA Foundation reported significant progress on the development of an evolution of Mysticeti, named the *Starfish* protocol, a partially synchronous DAG-based BFT protocol that incorporates the security properties of certified DAGs, the efficiency of uncertified DAG approaches, and linear amortized communication complexity. The key innovation in Starfish is the Encoded Cordial Dissemination (ECD), a push-based dissemination strategy that combines Reed-Solomon erasure coding with Data Availability Certificates (DACs). This lowers communication costs per byte, ensuring suitability for large-scale DLT deployments. Starfish experimental results demonstrated improved robustness and stable latency over Mysticeti, especially under steady-state and Byzantine scenarios, along with potentially higher throughput reportedly up to 150,000 TPS. The Starfish upgrade will include a fairer top-stakers committee selection process that allows higher-stake validators to replace underperforming ones [40].

4.1.3. Transactions Life Cycle

The transaction life cycle in IOTA Rebased is designed as follows. Firstly, the transaction is created by a client, whose wallet signs the transaction with the private key and submits the signed transaction to an IOTA full node for processing. The full node, not having the full view of all transactions within the network, broadcasts the transaction to all validators, who check its validity, sign it in the case it is valid, and return the signature to the full node after having locked the involved client-owned objects to prevent double-spending. The full node collects signatures in parallel from a supermajority (quorum) of validators to certify the transaction and establish its certificate. Once certified, transactions are sent to validators for execution, who first verify the certificate signatures to confirm the validity of the transaction and prevent double-spending. If no shared objects are involved, validators will instantly execute the transaction (fast path consensus); otherwise, they will submit it to the consensus layer to order the transaction with others sharing the same object and then execute it. After a transaction, validators sign its *effects* (details about objects involved, amount spent, and transaction status) and send them back to the full node. Once the full node receives signed effects from a majority of validators, it creates an *effects certificate*, guaranteeing that the transaction is final and will be included in a future checkpoint—the final stage in the life of that transaction. Validators submit executed transactions to the consensus layer for universal ordering, and checkpoints are then created using this ordered list of transactions and their corresponding effects [41]. Figure 5 summarizes transaction life cycle in IOTA Rebased.

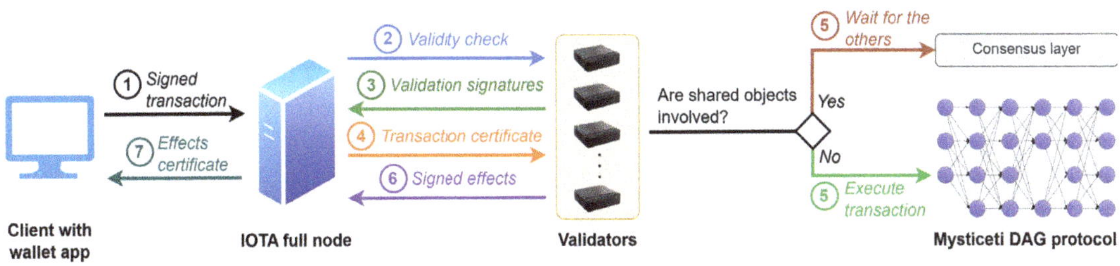

Figure 5. Transactions life cycle in IOTA Rebased.

4.1.4. Smart Contracts Virtualization

In this object-based design, each transaction creates, modifies, or deletes objects, which are stateful entities representing assets, smart contracts, or metadata. This enables Layer 1 programmability and simplifies complex use cases like supply chain tokenization. Resource-oriented programming is enabled as the MoveVM resource model ensures type safety and eliminates reentrancy vulnerabilities, critical for financial applications. Developers define custom assets (e.g., tokens, NFTs) as immutable resources, enforced at the protocol level.

The Rebased architecture supports dual virtualization: Layer 1 smart contracts powered by the MoveVM and Layer 2 IOTA EVM compatibility. The MoveVM provides static verification, formal validation of contract logic, and robust ownership guarantees through its linear type system, thereby enhancing security and simplifying development for financial instruments and other complex applications. The integration of IOTA EVM as a parallel execution environment allows developers to port Solidity-based dApps while leveraging the scalability of IOTA. This EVM compatibility at Layer 2 ensures seamless migration for existing Ethereum-based dApps, broadening ecosystem appeal while supporting diverse development needs. A future *MultiVM* upgrade aims to unify MoveVM and IOTA EVM under a single Layer 1 security model [9].

4.2. Tokenomics

IOTA Rebased introduces a revised economic model to sustain network operations and incentivize participation. This subsection examines its fee-driven structure with adaptive gas pricing, staking rewards for validators and delegators, and controlled inflation to balance token supply.

4.2.1. Fee Model

IOTA Rebased introduces an initial *gas fee* of about 0.005 IOTA per average transaction, with an adaptive fee burn mechanism to balance inflationary and deflationary dynamics based on network activity. Storage deposits (refundable) are also required for on-chain data, discouraging spam. To mitigate IoT friction, developers and businesses can subsidize fees for users via the IOTA Gas Station, a tool allowing dApps to cover user costs. This preserves accessibility for microtransactions through sponsored transactions [42].

4.2.2. Stacking and Inflation

The tokenomics model in IOTA Rebased is designed to incentivize network participation and ensure long-term sustainability. Validators, who secure the network, will be rewarded with newly minted tokens in the form of staking rewards. Delegators share rewards proportionally to incentivize participation. The initial annual inflation rate is estimated to be 6% in the first year and then gradually decline to a constant rate, with burnt fees (0.005 IOTA per transaction) offsetting inflation [9].

Overall, the revised tokenomics are designed to provide a self-sustaining economic model that supports validator operations and network maintenance while offering a clear incentive structure for participants. The fee-based model, combined with staking and dynamic validator committees, introduces a robust economic deterrent against attacks. The cost associated with misbehavior or malicious activities is increased, thereby enhancing overall security.

5. Comparative Analysis of IOTA 2.0 with Rebased

5.1. Consensus Mechanism

Nakamoto Consensus on a DAG in IOTA 2.0 is designed to maximize decentralization by involving all network nodes in validation. Conversely, the Mysticeti DPoS-based model in Rebased sacrifices some of this decentralization in favor of higher processing speed and scalability. While the first consensus mechanism prioritizes decentralization, the latter puts network efficiency at the forefront.

In IOTA 2.0, consensus is gradually achieved as more transactions add weight to valid blocks, which may result in longer confirmation times. The Mysticeti approach in Rebased, however, offers near-instant finality through its validator committee, making it more suitable for applications that require rapid transaction processing. While the first focuses on finality, the latter prioritizes network throughput.

While both consensus mechanisms strive to secure the network, Nakamoto Consensus on a DAG depends on the organic growth of cumulative weight to prevent double-spending. In contrast, Mysticeti utilizes a validator committee that can efficiently manage Byzantine faults, with potential vulnerabilities if the committee becomes unbalanced or subject to collusion.

Overall, Nakamoto Consensus on a DAG in IOTA 2.0 emphasized community-driven, fully decentralized validation at the cost of complexity, whereas Mysticeti consensus in IOTA Rebased focuses on achieving high throughput and low latency in a more secure consensus, albeit with a somewhat more centralized validator model.

5.2. Transaction Economics

One of the cornerstone promises of IOTA 2.0 was the ability to conduct feeless microtransactions, with the Mana system preventing network spamming without monetary costs. This design was particularly attractive for IoT devices, where high-frequency, low-value transactions are common. IOTA Rebased, on the other hand, introduces minimal transaction fees based on a gas pricing model. These fees serve as a mechanism to discourage spam, incentivize validators and delegators, and maintain the economic integrity of the network. This transition from feeless to fee-based transactions represents a significant departure from the original IoT-friendly vision, undermining IoT usability. Although fees help create a sustainable economic model and improve network security, they also risk limiting the utility of IOTA for high-volume, low-cost IoT applications in the Big Data era.

5.3. Scalability

IOTA 2.0 was designed to allow parallel processing of transactions and enable the network to scale naturally with increased activity, as more transactions contributed to the validation process. However, while this approach promoted decentralization and resilience, it may have experienced variable confirmation times, especially under high load, as the system relied on the accumulation of approval weight to finalize transactions.

On the other hand, the Mysticeti protocol with DPoS in IOTA Rebased, along with a committee of initially 150 permissionless validators processing transactions, achieves high throughput—reportedly over 50,000 TPS—with sub-second finality. This structure allows for rapid transaction processing and is well-suited for applications requiring high-speed confirmations. However, the reliance on a fixed validator set introduces a degree of centralization, which could impact network adaptability and resilience.

Overall, IOTA Rebased offers superior performance metrics suitable for applications demanding high-speed transactions, while IOTA 2.0 provides a more decentralized and adaptable framework that scales with network activity.

5.4. Security Considerations

IOTA 2.0 sought to enhance security by distributing the validation process across all network participants and using Mana as a reputation system to mitigate Sybil attacks. The removal of the centralized coordinator was intended to eliminate single points of failure in the network. With IOTA Rebased, security is enforced through economic incentives. Validators are required to stake tokens, and misbehavior carries financial penalties. The periodic reconfiguration of validator committees further secures the network against long-term attacks.

The two models represent different approaches to achieving security. IOTA 2.0 focused on distributed trust and reputation, whereas IOTA Rebased relies on economic disincentives and structured validator selection. Although the staking mechanism in IOTA Rebased enhances economic security, it also introduces new challenges in ensuring equitable participation across diverse IoT devices.

5.5. Implications for IoT Applications

The feeless model of IOTA 2.0 was ideally suited for the high-frequency, low-value transactions typical in IoT environments. The reintroduction of fees in IOTA Rebased may complicate this scenario, as even minimal fees can accumulate and affect the cost-efficiency of microtransactions in the case of large amounts of data. While IOTA 2.0 encouraged broad participation through autonomous validation, IOTA Rebased requires IoT devices or their proxies to engage in staking and validation processes. This may impose additional resource requirements on devices not originally designed for such tasks. Although the optimized consensus provided by the Mysticeti protocol in Rebased offers improved transaction finality and network adaptability, these benefits must be weighed against potential increases in operational complexity and the economic barriers introduced by transaction fees in specific cases.

Table 2 summarizes the advantages and drawbacks of the IOTA Rebased protocol with respect to the previously proposed IOTA 2.0 protocol.

Table 2. Comparison of IOTA Rebased with the proposed IOTA 2.0.

Feature	IOTA 2.0	IOTA Rebased
Ledger model	UTXO-based Tangle DAG with simpler microtransaction logic	Object-oriented ledger enabling Layer 1 smart contracts at the cost of microtransaction logic
Hardware requirements	Light nodes could operate on low-power IoT devices	MoveVM and Mysticeti demand higher computational resources, limiting edge devices participation
Consensus	Nakamoto on DAG, prioritizes decentralization	DPoS with Mysticeti, high throughput and low latency, 0.5-s finality
Transaction fees	Feeless, ideal for microtransactions	Minimal fees, supports economic sustainability
Scalability	Theoretically high TPS, scaling with network activity, potentially variable confirmation times	Over 50,000 TPS, fixed set of validators, suited for high-speed transactions
Security	Decentralized validation with Mana for reputation	Economic security via staking and periodic validator refresh
Smart contracts	Layer 2 with limited Layer 1 support	Native Layer 1 with Layer 2 solutions, MoveVM and IOTA EVM, supports complex dApps
Governance	Permissionless, decentralized, fully inclusive	On-chain voting, DPoS validators committee, faster but less inclusive
IoT suitability	Optimized for low-cost, high-frequency transactions	Economic model may hinder low-value transactions in IoT setups

6. Challenges and Recommendations for Adopting IOTA Rebased in IoT

6.1. Key Challenges

Adopting IOTA Rebased in IoT ecosystems introduces critical challenges that conflict with its original mission of enabling frictionless machine-to-machine transactions. This subsection identifies barriers such as transaction fees undermining microtransactions, staking requirements incompatible with resource-constrained devices, and technical complexity in bridging the object-oriented model of Rebased with legacy IoT protocols.

6.1.1. Reintroduction of Transaction Fees

The feeless design of IOTA 2.0 was specifically tailored for the high-frequency, low-value transactions common in IoT ecosystems. With IOTA Rebased introducing even minimal fees, the economic viability of microtransactions may be compromised, especially in the context of Big Data with millions of IoT devices making billions of transactions. Even low fees could accumulate when millions of IoT devices are transacting, potentially deterring their use in real-time applications and breaking the business case for millions of tiny telemetry messages. Additionally, IoT devices often operate with very tight energy and computational budgets. Adding fee calculations and fee-burning mechanisms could increase overhead, complicating the seamless integration of devices with the network.

6.1.2. Staking and Validator Participation

Many IoT devices are resource-constrained and may lack the capacity to participate directly in staking or validation processes. This forces reliance on proxy nodes or gateway solutions, which can introduce additional points of failure and potential centralization risks. In addition, the DPoS model and staking requirements may favor participants with substantial token holdings and robust infrastructure. This could potentially lead to a concentration of network control among a few entities, counteracting the original decentralization promise of IOTA.

Hardware requirements for validators in IOTA Rebased have been initially set to 24-core CPUs, 128 GB RAM, 4 TB NVMe, and 1 Gbps uplink, plus a 2M IOTA stake per validator [43]. Such hardware is orders of magnitude beyond what typical IoT gateways can host, forcing edge operators to rely on remote data center nodes. This limitation hinders the possibility of IoT devices to participate in network throughput and scaling by acting as validators in a fully decentralized leaderless DLT.

6.1.3. Technical Complexity and Integration

Integrating Move-based smart contracts directly on Layer 1 adds a level of complexity in the context of IoT. While Move offers enhanced security and programmability, its adoption requires a steep learning curve and robust developer support. In an IoT context, where seamless interoperability is key, ensuring that legacy IoT devices and software can communicate with the new protocol poses a noticeable challenge.

6.1.4. Security and Network Scalability

While staking and fee-burning mechanisms bolster economic security, they can also deter smaller participants around the world from joining the network. Ensuring robust security without sacrificing the global inclusive goal of DLTs for IoT networks is a delicate balance.

The use of Mysticeti consensus in replacement of Nakamoto on DAG must be rigorously tested under IoT-scale high transaction loads, since the network throughput no longer scales with network activity. While IOTA Rebased promises high scalability (over 50,000 TPS), large-scale IoT deployments may involve millions of devices generating transactions. The transition from a DAG-based architecture to an object-based ledger with a DPoS consensus mechanism raises questions about scalability under extreme conditions. Any latency or stability issues in high-load scenarios could impede the real-time communication essential for large-scale, high-load IoT deployments.

6.2. *Strategic Recommendations*

To mitigate the challenges of adopting IOTA Rebased in IoT environments, this subsection proposes actionable strategies. While the proposed approaches aim to align Rebased with IoT demands, they necessitate trade-offs between decentralization, efficiency, and economic viability.

6.2.1. Subsidization and Tiered Fee Structures

Sponsored transactions can help preserve accessibility for microtransactions, as well as other models where fees for IoT microtransactions are partially or fully subsidized by network incentives or partnerships [44]. This approach could help preserve the low-cost transaction benefits that are critical for IoT use cases. In addition, designing a fee structure that adjusts based on transaction size or frequency can help ensure that low-value transactions incur negligible costs while maintaining the economic balance of the network. Still, while the new IOTA Gas Station lets integrators sponsor fees, it shifts operational expenses to application owners instead of the protocol.

6.2.2. Lightweight Client Protocols and Proxies

To address resource constraints in IoT devices, specialized lightweight client implementations can allow resource-constrained IoT devices to interact with the IOTA Rebased network indirectly, through proxies [45]. In such hierarchical tiered transaction architecture, these proxies would handle staking on behalf of multiple resource-constrained IoT devices, reducing the burden on individual IoT nodes, but at the cost of more centralization. However, attention should be paid to implementing redundancy in order to avoid these proxies becoming single points of failure. Efficient caching mechanisms should be considered to minimize redundant computations for repetitive transactions common in IoT applications.

6.2.3. Batch Transactions

The batch transactions paradigm [46–48] can be implemented on a case-by-case basis to minimize fee overhead for frequent low-value transactions in IoT deployments where real-time data transfers are not required. This could be combined with fee estimation frameworks, creating predictive models that help IoT systems estimate the necessary budget for transaction fees in various network conditions. This implies reevaluating how value is captured in IoT applications to shift from many small transactions to fewer higher-value aggregated transactions.

6.2.4. Adaptive Transaction Scheduling

In a hierarchical transaction architecture, intelligent adaptive transaction scheduling can be implemented to prioritize time-sensitive operations while delaying non-critical transactions in the case of network congestion [45,49]. This strategy can be combined with data processing and aggregation at the tier edge to reduce the volume of transactions that need to be recorded on the main ledger. Value-based transaction filtering algorithms on the tier edge can help determine which data are valuable enough to warrant on-chain storage versus what can be processed off-chain.

6.3. *Use Cases Analysis*

This subsection evaluates the applicability of IOTA Rebased through real-world IoT scenarios, including smart energy grids, supply chain tracking, and healthcare monitoring. These use cases underscore the potential of Rebased in enterprise-scale applications, albeit with adaptations to preserve its original vision for decentralized machine economies.

6.3.1. Smart Energy Grids and Peer-to-Peer Energy Trading

Energy microgrids typically involve numerous small energy transactions between prosumers (producer-consumers). IOTA Tangle has been explored for secure and scalable P2P energy trading [50–52]. The introduction of fees in IOTA Rebased could make very small energy trades uneconomical. To overcome this constraint, one could aggregate energy transactions over time periods (e.g., hourly or daily) rather than recording each kilowatt-hour exchange. Sponsored transactions can be implemented, where energy service providers cover transaction fees as part of their service offering. In addition, staking rewards from a community energy pool can be used to offset transaction costs.

6.3.2. Supply Chain Tracking and Provenance

Supply chains involve multiple stakeholders and numerous tracking events. Several works in the literature explored the use of IOTA Tangle in supply chain tracking and provenance certification [53,54]. Adding fees to each tracking update in IOTA Rebased could increase operational costs. An efficient implementation strategy could include selective on-chain recording, where only critical handoffs and quality measurements are recorded on-chain. Batch updates can be used for non-critical tracking events, further reducing operational expenses. In addition, a cost-sharing model can be implemented where each supply chain participant stakes IOTA tokens and shares in both staking rewards and transaction fees.

6.3.3. Vehicle Communication Systems

Vehicle-to-everything communications often require real-time, high-frequency data exchanges that may become prohibitively expensive if each message incurs a fee. The IOTA Tangle was proposed for securing V2X communications in scalable decentralized networks [55,56]. With the fee-based model of IOTA Rebased, a tiered data model can be implemented, where safety-critical messages use direct on-chain transactions with sponsored fees. Off-chain channels can be utilized for routine telemetry data with periodic checkpoints. Additionally, to encourage participation, a token economics model can be developed, where vehicles earn tokens by providing valuable road condition data that offsets the costs of their transactions.

6.3.4. Healthcare Remote Monitoring

Continuous patient monitoring generates high volumes of data, most of which is routine but occasionally includes critical alerts. IOTA Tangle demonstrated suitability in the context of secure healthcare data management and sharing [57,58]. In IOTA rebased, with the introduction of fees, one could implement smart filtering algorithms that only record clinically significant changes on-chain. Healthcare providers or insurance companies can be involved to cover transaction sponsoring on IOTA Rebased. Further design can implement value-based data models where the importance of the data determines its recording strategy.

7. Potential Research Directions

The evolution of IOTA from a feeless, IoT-centric DAG to a fee-driven, object-oriented ledger necessitates targeted research to reconcile its original vision with modern scalability and security demands. Below, we outline key directions to advance the applicability of the IOTA in IoT ecosystems.

7.1. Adaptive Indexing and QUERY Optimization for DAGs

The branching structure of the IOTA DAG makes efficient hierarchical data retrieval more complex than in linear blockchains, particularly for resource-constrained IoT devices. Future work should explore machine learning-driven indexing techniques, such as graph neural networks trained on transaction metadata, to predict transaction locations dynamically. However, concurrent branches in the IOTA DAG necessitate dynamic index updates, posing challenges for machine learning model accuracy. Hybrid approaches combining learned models with heuristic DAG traversal could mitigate the computational overhead of real-time updates. Adapting blockchain-optimized frameworks like Telex—a two-level learned index [59]—to the IOTA DAG could further enhance query efficiency, though challenges remain in balancing model accuracy with the concurrent validation processes in IOTA.

7.2. Lightweight Privacy-Preserving Mechanisms

Enabling encrypted rich queries on IOTA without compromising IoT device efficiency requires novel privacy-preserving techniques. Oblivious RAM (ORAM) protocols, paired with Trusted Execution Environments like Intel SGX or ARM TrustZone, could obfuscate access patterns while ensuring forward and backward privacy in the DLT [60,61]. Simplified ORAM variants optimized for edge devices, coupled with lattice-based cryptography, may reduce computational overhead. Such mechanisms would allow IoT devices to securely query ledger data (e.g., supply chain events) without exposing sensitive access patterns.

7.3. Scalability Under Extreme IoT Loads

Validating the performance of IOTA Rebased at IoT scale with millions of devices generating thousands of transactions per second requires stress-testing frameworks. Sharded validator committees and adaptive consensus protocols (e.g., dynamically adjusting block sizes in Mysticeti) could enhance throughput. Simulation tools modeling high-density IoT deployments such as smart cities would identify bottlenecks in transaction finality and network resilience. Recently, reinforcement learning-based approaches have proven to improve decentralization and efficiency of DAG-based DLTs, ensuring stable scalability and high security resilience [62].

7.4. Energy-Efficient Participation

Resource-constrained IoT devices struggle with staking and validation tasks. Delegated validation mechanisms, where edge gateways act on behalf of low-power devices, could democratize participation. Energy-efficient consensus variants, such as proof-of-uptime [63], might further optimize computational demands while maintaining security.

7.5. Convergence with Other Emerging Technologies

Synergies with 5G, 6G, edge AI, and federated learning present opportunities to optimize IOTA for next-generation IoT. Federated learning frameworks could leverage IOTA for secure model aggregation across edge devices [64,65], while the ultra-low latency in 5G and 6G networks might enhance sub-second finality of Mysticeti in vehicular networks.

8. Conclusions

The transition from IOTA Tangle 2.0 to IOTA Rebased marks a pivotal moment in DLTs, reflecting broader tensions between ideological purity and pragmatic evolution. While IOTA 2.0 pioneered a feeless, leaderless DAG architecture tailored for the IoT, its complexity and delayed smart contract functionality hindered its adoption. IOTA Rebased, leveraging MoveVM and Mysticeti DAG protocol, prioritizes enterprise scalability, sub-second finality, and cross-chain interoperability, but at the cost of abandoning the IOTA foundational promise of feeless transactions and secure data exchange.

Our comparative analysis reveals that this evolution introduces both opportunities and challenges, particularly in reconciling the original IoT-friendly vision with the practical demands of maintaining a secure, scalable, and economically viable distributed ledger. Mitigating these issues will demand, among other recommendations of this paper, a tiered architecture where edge gateways speak for sensor swarms, along with fee-sponsorship models. Future research and development will need to address these trade-offs to ensure that IOTA continues to serve as a robust platform for machine-to-machine communication in an increasingly connected world.

Author Contributions: Conceptualization, P.S.N. and V.P.; methodology, P.S.N.; validation, A.L.-E. and A.K.L.; writing—original draft preparation, P.S.N.; writing—review and editing, S.D.I. and V.P.; supervision, A.P. All authors have read and agreed to the published version of the manuscript.

Funding: This research received no external funding.

Institutional Review Board Statement: Not applicable.

Informed Consent Statement: Not applicable.

Data Availability Statement: Not applicable.

Conflicts of Interest: The authors declare no conflicts of interest.

Abbreviations

The following abbreviations are used in this manuscript:

IoT	Internet of Things
DLTs	Distributed Ledger Technologies
DAG	Directed Acyclic Graph
PoW	Proof-of-Work
PoA	Proof-of-Authority
UTXO	Unspent Transaction Output
EdDSA	Edwards-curve Digital Signature Algorithm
ISC	IOTA Smart Contracts
BFT	Byzantine Fault Tolerant
EVM	Ethereum Virtual Machine
TPS	Transactions Per Second
NFTs	Non-Fungible Tokens
MEV	Maximal Extractable Value
FCOB	Fast Confirmation of On-the-Heavy Branch
FPC	Fast Probabilistic Consensus
dRNG	distributed Random Number Generation
BIC	Block Issuance Credits
DoS	Denial of Service
CPS	Cyber-Physical Systems
P2P	Peer-to-Peer
dApps	Decentralized Applications
DPoS	Delegated Proof-of-Stake
ECD	Encoded Cordial Dissemination
DAC	Data Availability Certificate
ORAM	Oblivious RAM

References

1. Zhu, Q.; Loke, S.W.; Trujillo-Rasua, R.; Jiang, F.; Xiang, Y. Applications of Distributed Ledger Technologies to the Internet of Things: A Survey. *ACM Comput. Surv.* **2019**, *52*, 120. [CrossRef]
2. Silvano, W.F.; Marcelino, R. Iota Tangle: A cryptocurrency to communicate Internet-of-Things data. *Future Gener. Comput. Syst.* **2020**, *112*, 307–319. [CrossRef]
3. Vigneri, L.; Welz, W. On the Fairness of Distributed Ledger Technologies for the Internet of Things. In Proceedings of the 2020 IEEE International Conference on Blockchain and Cryptocurrency (ICBC), Toronto, ON, Canada, 2–6 May 2020; pp. 1–3. [CrossRef]
4. Popov, S. The Tangle. In *White Paper*; 2018. Available online: https://cryptoverze.s3.us-east-2.amazonaws.com/wp-content/uploads/2018/11/10012054/IOTA-MIOTA-Whitepaper.pdf (accessed on 9 April 2025).
5. Soltani, R.; Zaman, M.; Joshi, R.; Sampalli, S. Distributed Ledger Technologies and Their Applications: A Review. *Appl. Sci.* **2022**, *12*, 7898. [CrossRef]
6. Akhtar, M.M.; Rizvi, D.R.; Ahad, M.A.; Kanhere, S.S.; Amjad, M.; Coviello, G. Efficient Data Communication Using Distributed Ledger Technology and IOTA-Enabled Internet of Things for a Future Machine-to-Machine Economy. *Sensors* **2021**, *21*, 4354. [CrossRef] [PubMed]
7. Popov, S.; Moog, H.; Camargo, D.; Capossele, A.; Dimitrov, V.; Gal, A.; Greve, A.; Kusmierz, B.; Mueller, S.; Penzkofer, A.; et al. The Coordicide. In *White Paper*; 2020. Available online: https://files.iota.org/papers/20200120_Coordicide_WP.pdf (accessed on 9 April 2025).
8. Fartitchou, M.; Boussouf, J.; Makkaoui, K.E.; Maleh, Y.; Allali, Z.E. IOTA Tangle 2.0: An Overview. *Edpacs* **2023**, *68*, 15–26. [CrossRef]
9. IOTA Foundation. IOTA Rebased. Technical View. Available online: https://blog.iota.org/iota-rebased-technical-view (accessed on 9 April 2025).
10. Babel, K.; Chursin, A.; Danezis, G.; Kichidis, A.; Kokoris-Kogias, L.; Koshy, A.; Sonnino, A.; Tian, M. Mysticeti: Reaching the limits of latency with uncertified dags. *arXiv* **2023**, arXiv:2310.14821. [CrossRef]
11. Kahmann, F.; Honecker, F.; Dreyer, J.; Fischer, M.; Tönjes, R. Performance Comparison of Directed Acyclic Graph-Based Distributed Ledgers and Blockchain Platforms. *Computers* **2023**, *12*, 257. [CrossRef]

12. Fan, C.; Ghaemi, S.; Khazaei, H.; Chen, Y.; Musilek, P. Performance Analysis of the IOTA DAG-Based Distributed Ledger. *ACM Trans. Model. Perform. Eval. Comput. Syst.* **2021**, *6*, 10. [CrossRef]
13. Sedi Nzakuna, P.; Paciello, V.; Lay-Ekuakille, A.; Kuti Lusala, A.; Iacono, S.D. Empirical Analysis of the IOTA Tangle Ledger in the Stardust Stage. In Proceedings of the 2024 IEEE International Symposium on Measurements & Networking (M&N), Rome, Italy, 2–5 July 2024; pp. 1–6. [CrossRef]
14. Staines, A.S. Concurrency and Petri Net Models. *Science* **2022**, *18*, 19. [CrossRef]
15. Drąsutis, E. IOTA Smart Contracts. In *White Paper*; 2021. Available online: https://files.iota.org/papers/ISC_WP_Nov_10_2021.pdf (accessed on 9 April 2025).
16. Muntaha, S.T.; Ahmed, Q.Z.; Khan, F.A.; Zaharis, Z.D.; Lazaridis, P.I. Hybrid Blockchain-Based Multi-Operator Resource Sharing and SLA Management. *IEEE Open J. Commun. Soc.* **2025**, *6*, 362–377. [CrossRef]
17. IOTA Foundation. Shimmer's Roadmap and Evolution. Available online: https://blog.shimmer.network/shimmers-roadmap-and-evolution (accessed on 14 April 2025).
18. IOTA Foundation. IOTA's EVM Mainnet Unleashed. Available online: https://blog.iota.org/iotas-evm-mainnet-launch (accessed on 14 April 2025).
19. Sealey, N.; Aijaz, A.; Holden, B. IOTA Tangle 2.0: Toward a Scalable, Decentralized, Smart, and Autonomous IoT Ecosystem. In Proceedings of the 2022 International Conference on Smart Applications, Communications and Networking (SmartNets), Palapye, Botswana, 29 November–1 December 2022; pp. 1–8. [CrossRef]
20. IOTA Foundation. IOTA 2.0 Research Specifications. Available online: https://github.com/iotaledger/IOTA-2.0-Research-Specifications (accessed on 13 April 2025).
21. Müller, S.; Penzkofer, A.; Kuśmierz, B.; Camargo, D.; Buchanan, W.J. Fast Probabilistic Consensus with Weighted Votes. In *Proceedings of the Future Technologies Conference (FTC) 2020, Volume 2*; Arai, K., Kapoor, S., Bhatia, R., Eds.; Springer: Cham, Switzerland, 2021; pp. 360–378. [CrossRef]
22. Müller, S.; Penzkofer, A.; Polyanskii, N.; Theis, J.; Sanders, W.; Moog, H. Tangle 2.0 Leaderless Nakamoto Consensus on the Heaviest DAG. *IEEE Access* **2022**, *10*, 105807–105842. [CrossRef]
23. IOTA Foundation. A New Consensus Model: Nakamoto Consensus on a DAG. Available online: https://blog.iota.org/a-new-consensus-model-iota20 (accessed on 13 April 2025).
24. Popov, S.; Buchanan, W.J. FPC-BI: Fast Probabilistic Consensus within Byzantine Infrastructures. *J. Parallel Distrib. Comput.* **2021**, *147*, 77–86. [CrossRef]
25. IOTA Foundation. IOTA 2.0 Tokenomics: Mana, Accounts, Staking and Delegation. Available online: https://wiki.iota.org/learn/protocols/iota2.0/core-concepts/mana (accessed on 13 April 2025).
26. Fartitchou, M.; Lamaakal, I.; Maleh, Y.; El Makkaoui, K.; El Allali, Z.; Pławiak, P.; Alblehai, F.; A. Abd El-Latif, A. IOTASDN: IOTA 2.0 Smart Contracts for Securing Software-Defined Networking Ecosystem. *Sensors* **2024**, *24*, 5716. [CrossRef]
27. Yu, Z.; Gao, H.; Cong, X.; Wu, N.; Song, H.H. A Survey on Cyber–Physical Systems Security. *IEEE Internet Things J.* **2023**, *10*, 21670–21686. [CrossRef]
28. Humayed, A.; Lin, J.; Li, F.; Luo, B. Cyber-Physical Systems Security—A Survey. *IEEE Internet Things J.* **2017**, *4*, 1802–1831. [CrossRef]
29. IOTA Foundation. IOTA 2.0 Data Flow. Available online: https://wiki.iota.org/learn/protocols/iota2.0/core-concepts/data-flow (accessed on 13 April 2025).
30. IOTA Foundation. Finality Explained: How Nodes Sync the Ledger. Available online: https://blog.iota.org/finality-explained-iota20 (accessed on 13 April 2025).
31. IOTA Foundation. IOTA 2.0: Introducing Pollen, Nectar and Honey. Available online: https://blog.iota.org/iota-2-0-introducing-pollen-nectar-and-honey-de7b9c4c8199 (accessed on 15 April 2025).
32. IOTA Foundation. IOTA Rebased Governance Proposal. Available online: https://govern.iota.org/t/iota-rebased-governance-proposal/1768 (accessed on 15 April 2025).
33. IOTA Foundation. IOTA Rebased Fast Forward. Available online: https://blog.iota.org/iota-rebased-fast-forward (accessed on 15 April 2025).
34. IOTA Foundation. IOTA 2.0 Details on Current Status and Next Steps. Available online: https://iota-news.com/iota-2-0-details-on-current-status-and-next-steps (accessed on 15 April 2025).
35. Welc, A.; Blackshear, S. Sui Move: Modern Blockchain Programming with Objects. In Proceedings of the SPLASH 2023: Companion Proceedings of the 2023 ACM SIGPLAN International Conference on Systems, Programming, Languages, and Applications: Software for Humanity, New York, NY, USA, 22–27 October 2023; pp. 53–55. [CrossRef]
36. Yakovenko, A. Solana: A New Architecture for a High Performance Blockchain v0. 8.13. In *White Paper*; 2018. Available online: https://solana.com/solana-whitepaper.pdf (accessed on 9 April 2025).
37. The MystenLabs Team. The Sui Smart Contracts Platform. In *White Paper*; 2022. Available online: https://docs.sui.io/paper/sui.pdf (accessed on 9 April 2025).

38. IOTA Foundation. Consensus on IOTA. Available online: https://docs.iota.org/about-iota/iota-architecture/consensus (accessed on 15 April 2025).
39. Cong, X.; Yu, Z.; Fanti, M.P.; Mangini, A.M.; Li, Z. Predictability Verification of Fault Patterns in Labeled Petri Nets. *IEEE Trans. Autom. Control* **2025**, *70*, 1973–1980. [CrossRef]
40. Polyanskii, N.; Mueller, S.; Vorobyev, I. Starfish: A high throughput BFT protocol on uncertified DAG with linear amortized communication complexity. *Cryptol. ePrint Arch.* 2025, *preprint*.
41. IOTA Foundation. Transaction Life Cycle. Available online: https://docs.iota.org/about-iota/iota-architecture/transaction-lifecycle (accessed on 16 April 2025).
42. IOTA Foundation. Q1 2025 Progress Report. Available online: https://blog.iota.org/q12025-progress-report (accessed on 15 April 2025).
43. IOTA Foundation. FAQ - IOTA Documentation. Available online: https://docs.iota.org/about-iota/FAQ (accessed on 17 April 2025).
44. Chen, E.; Xiao, Z.; Liang, J.; Chen, D.; Hung, P.; Chalkias, K.K. Buck You: Designing Easy-to-Onboard Blockchain Applications with Zero-Knowledge Login and Sponsored Transactions on Sui. *arXiv* **2024**, arXiv:2404.03845. [CrossRef]
45. Oktian, Y.E.; Lee, S.G.; Lee, H.J. Hierarchical Multi-Blockchain Architecture for Scalable Internet of Things Environment. *Electronics* **2020**, *9*, 1050. [CrossRef]
46. Bagga, P.; Sutrala, A.K.; Das, A.K.; Vijayakumar, P. Blockchain-based batch authentication protocol for Internet of Vehicles. *J. Syst. Archit.* **2021**, *113*, 101877. [CrossRef]
47. Lee, H.; Yoon, C.; Bae, S.; Lee, S.; Lee, K.; Kang, S.; Sung, K.; Min, S. Multi-Batch Scheduling for Improving Performance of Hyperledger Fabric Based IoT Applications. In Proceedings of the 2019 IEEE Global Communications Conference (GLOBECOM), Waikoloa, HI, USA, 9–13 December 2019; pp. 1–6. [CrossRef]
48. Mišić, J.; Mišić, V.B.; Chang, X.; Qushtom, H. Adapting PBFT for Use with Blockchain-Enabled IoT Systems. *IEEE Trans. Veh. Technol.* **2021**, *70*, 33–48. [CrossRef]
49. Qiu, X.; Chen, W.; Tang, B.; Liang, J.; Dai, H.N.; Zheng, Z. A Distributed and Privacy-Aware High-Throughput Transaction Scheduling Approach for Scaling Blockchain. *IEEE Trans. Dependable Secur. Comput.* **2023**, *20*, 4372–4386. [CrossRef]
50. Zhang, X.; Zhu, X.; Ali, I. Performance Analysis of IOTA Tangle and a New Consensus Algorithm for Smart Grids. *IEEE Internet Things J.* **2024**, *11*, 6396–6411. [CrossRef]
51. Mullaney, C.; Aijaz, A.; Sealey, N.; Holden, B. Peer-to-Peer Energy Trading meets IOTA: Toward a Scalable, Low-Cost, and Efficient Trading System. In Proceedings of the 2022 IEEE/ACM 15th International Conference on Utility and Cloud Computing (UCC), Vancouver, WA, USA, 6–9 December 2022; pp. 399–406. [CrossRef]
52. García-Hernández, J.; Marín-Collazos, L.G.; Jiménez-Estévez, G.; Mendoza-Araya, P. Distributed Ledger Technologies Based Microgrid Energy Management Using IOTA Tangle. In Proceedings of the 2021 IEEE CHILEAN Conference on Electrical, Electronics Engineering, Information and Communication Technologies (CHILECON), Valparaíso, Chile, 6–9 December 2021; pp. 1–5. [CrossRef]
53. Raman, S.E.; Venkatramaraju, D. Secured and Transparent Transactions Using IOTA Tangle Distributed Ledger Technology in Dairy Supply Chains. In Proceedings of the 2024 International Conference on E-mobility, Power Control and Smart Systems (ICEMPS), Thiruvananthapuram, India, 18–20 April 2024; pp. 1–6. [CrossRef]
54. Leivadaros, S.; Kornaros, G.; Coppola, M. Secure Asset Tracking in Manufacturing through Employing IOTA Distributed Ledger Technology. In Proceedings of the 2021 IEEE/ACM 21st International Symposium on Cluster, Cloud and Internet Computing (CCGrid), Melbourne, Australia, 10–13 May 2021; pp. 754–761. [CrossRef]
55. Halgamuge, M.N. Malicious Data Traceability and Throughput Optimization Using Blockchain (IOTA Smart Contracts) for Next Generation Internet of Vehicles. *IEEE Trans. Veh. Technol.* **2024**, *73*, 18102–18119. [CrossRef]
56. Lozano Domínguez, J.M.; Mateo Sanguino, T.J. Review on V2X, I2X, and P2X Communications and Their Applications: A Comprehensive Analysis over Time. *Sensors* **2019**, *19*, 2756. [CrossRef] [PubMed]
57. Abdullah, S.; Arshad, J.; Khan, M.M.; Alazab, M.; Salah, K. PRISED tangle: A privacy-aware framework for smart healthcare data sharing using IOTA tangle. *Complex Intell. Syst.* **2023**, *9*, 3023–3041. [CrossRef]
58. Akbulut, S.; Semantha, F.H.; Azam, S.; Pilares, I.C.A.; Jonkman, M.; Yeo, K.C.; Shanmugam, B. Designing a Private and Secure Personal Health Records Access Management System: A Solution Based on IOTA Distributed Ledger Technology. *Sensors* **2023**, *23*, 5174. [CrossRef]
59. Wu, H.; Tang, Y.; Shen, Z.; Tao, J.; Lin, C.; Peng, Z. TELEX: Two-Level Learned Index for Rich Queries on Enclave-based Blockchain Systems. In *IEEE Transactions on Knowledge and Data Engineering*; IEEE Computer Society: Washington, DC, USA, 2025; pp. 1–16. [CrossRef]
60. Wei, W.; An, B.; Qiao, K.; Shen, J. A Blockchain-Based Multi-Users Oblivious Data Sharing Scheme for Digital Twin System in Industrial Internet of Things. *IEEE J. Sel. Areas Commun.* **2023**, *41*, 3318–3332. [CrossRef]

61. Cao, H.; Li, R.; Tian, W.; Xu, Z.; Xiao, W. Blockchain-based accountability for multi-party oblivious RAM. *J. Parallel Distrib. Comput.* **2020**, *137*, 224–237. [CrossRef]
62. Alshaikhli, M.; Al-Maadeed, S.; Saleh, M. Enhancing Scalability and Network Efficiency in IOTA Tangle Networks: A POMDP-Based Tip Selection Algorithm. *Computers* **2025**, *14*, 117. [CrossRef]
63. Avalanche Foundation. How to Stake. Available online: https://github.com/ava-labs/builders-hub/blob/master/content/docs/nodes/validate/how-to-stake.mdx (accessed on 26 May 2025).
64. Mazzocca, C.; Romandini, N.; Montanari, R.; Bellavista, P. Enabling Federated Learning at the Edge through the IOTA Tangle. *Future Gener. Comput. Syst.* **2024**, *152*, 17–29. [CrossRef]
65. Javed, F.; Zeydan, E.; Mangues-Bafalluy, J.; Dev, K.; Blanco, L. Blockchain for Federated Learning in the Internet of Things: Trustworthy Adaptation, Standards, and the Road Ahead. *arXiv* **2025**, arXiv:2503.23823. [CrossRef]

sensors

MDPI

Review

Innovative Driver Monitoring Systems and On-Board-Vehicle Devices in a Smart-Road Scenario Based on the Internet of Vehicle Paradigm: A Literature and Commercial Solutions Overview

Paolo Visconti [1,*], Giuseppe Rausa [1], Carolina Del-Valle-Soto [2], Ramiro Velázquez [3], Donato Cafagna [1] and Roberto De Fazio [1,3]

[1] Department of Innovation Engineering, University of Salento, 73100 Lecce, Italy; giuseppe.rausa@unisalento.it (G.R.); donato.cafagna@unisalento.it (D.C.); roberto.defazio@unisalento.it (R.D.F.)
[2] Facultad de Ingeniería, Universidad Panamericana, Zapopan 45010, Mexico; cvalle@up.edu.mx
[3] Facultad de Ingeniería, Universidad Panamericana, Aguascalientes 20296, Mexico; rvelazquez@up.edu.mx
* Correspondence: paolo.visconti@unisalento.it; Tel.: +39-0832-29-7334

Abstract: In recent years, the growing number of vehicles on the road have exacerbated issues related to safety and traffic congestion. However, the advent of the Internet of Vehicles (IoV) holds the potential to transform mobility, enhance traffic management and safety, and create smarter, more interconnected road networks. This paper addresses key road safety concerns, focusing on driver condition detection, vehicle monitoring, and traffic and road management. Specifically, various models proposed in the literature for monitoring the driver's health and detecting anomalies, drowsiness, and impairment due to alcohol consumption are illustrated. The paper describes vehicle condition monitoring architectures, including diagnostic solutions for identifying anomalies, malfunctions, and instability while driving on slippery or wet roads. It also covers systems for classifying driving style, as well as tire and emissions monitoring. Moreover, the paper provides a detailed overview of the proposed traffic monitoring and management solutions, along with systems for monitoring road and environmental conditions, including the sensors used and the Machine Learning (ML) algorithms implemented. Finally, this review also presents an overview of innovative commercial solutions, illustrating advanced devices for driver monitoring, vehicle condition assessment, and traffic and road management.

Keywords: internet of vehicles; intelligent transport system; wearable device; driver monitoring; traffic management; OBD-II; vehicle condition monitoring

1. Introduction

In recent years, growing urbanization and the increasing number of vehicles on the road have introduced new challenges related to transportation safety, efficiency, and sustainability, leading to growing interest in the Internet of Vehicles (IoV) [1]. Within this framework, the monitoring systems of driver, vehicle, and road conditions have become key components of Intelligent Transportation Systems (ITSs) and smart cities [2]. These systems leverage advanced technologies, including sensors, IoT devices, and Artificial Intelligence (AI) to collect, analyze, and utilize real-time data. Their objectives include optimizing traffic flows, enhancing road safety, safeguarding drivers, and reducing environmental impact (Figure 1) [3]. The automotive industry has recently undergone rapid advancements in driver assistance technologies, notably with the development of advanced driver assistance systems (ADASs) [4,5]. These systems represent a crucial step toward future mobility [6]. However, addressing road safety, transportation efficiency, sustainability, and increasingly complex driving conditions requires innovative solutions, particularly to ensure the safety of all road users [7]. In this regard, Driver Monitoring Systems (DMSs) are gaining prominence for their ability to detect health anomalies, drowsiness, intoxication, and unsafe driving behaviors [8]. Recent studies indicate that human errors account for approximately 90% of road accidents, many of which could be prevented with effective monitoring systems [9,10]. These systems rely on both intrusive and non-intrusive sensors to monitor drivers' biophysical parameters (such as Electrocardiograms (ECGs), photoplethysmography (PPG), Electrodermal Activity (EDA), heart rate (HR), Blood Alcohol Concentration (BAC), respiratory rate (RR), and oxygen saturation (SpO_2)) and cameras to detect facial cues like yawning, blink rate, and head movements. Using Machine Learning (ML) algorithms and real-time data analysis, these systems enable timely interventions, preventing accidents and saving lives [11–15]. Research has also addressed the recurring tragedy of infants being accidentally left in vehicles. Innovative systems now exist that can detect an infant's cries and send alert signals through mobile applications, prompting immediate action [16,17]. Investing in such technologies enhances road safety and fosters a more responsible and conscientious driving environment.

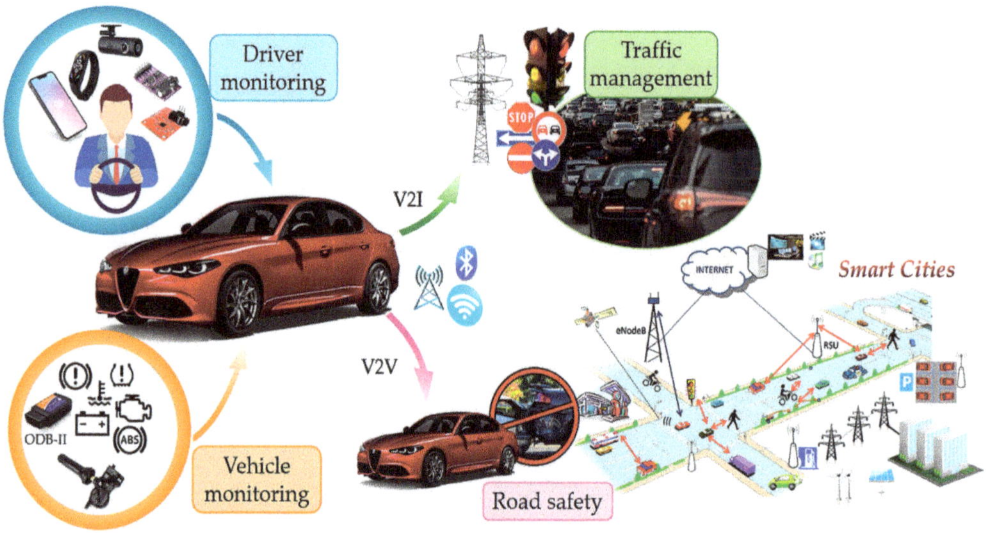

Figure 1. Summary picture of the IoV paradigm in the smart city scenario for road safety and transportation efficiency purposes: driver, vehicle, road, and traffic monitoring systems.

In this context, advanced on-board diagnostic (OBD-II) technologies play a pivotal role, not only enhancing vehicle reliability and durability but also directly improving road user safety [18,19]. Many innovative systems utilize data collected through the vehicle's OBD-II system to monitor its status. Another significant aspect of these technologies is emissions monitoring, which is increasingly important given the tightening of environmental regulations worldwide. Real-time detection of polluting emissions helps minimize environmental impact and optimize resource use [20]. Tire pressure and wear monitoring extend beyond routine maintenance. Proper tire management improves fuel efficiency and reduces the risk of accidents caused by blowouts or loss of control [21]. Similarly, vehicle stability is a fundamental safety factor. Real-time monitoring systems can detect anomalies in driving parameters, preventing potentially hazardous situations [22].

Traffic and road condition monitoring are essential for ensuring urban mobility's safety, efficiency, and sustainability. Advanced systems incorporating sensors, cameras, and real-time communication technologies collect and analyze data on traffic patterns and infrastructure conditions [23]. This information exchange has given rise to interactive Vehicle-to-Infrastructure (V2I) communication networks, leveraging protocols like Wi-Fi, LoRa (Long Range), 4G/5G, and other technologies to support seamless connectivity and enhanced mobility [24]. Another critical aspect is traffic management, which focuses on optimizing vehicle flow, reducing congestion, and enhancing the driving experience through intelligent planning and innovative solutions such as smart traffic lights and dynamic signaling systems [25]. Additionally, vehicular traffic has been harnessed as a source of energy, enabling the development of systems that harvest, store, and supply energy to power road infrastructure [26]. Monitoring road conditions is equally vital, allowing for the timely detection and resolution of issues like potholes and asphalt deterioration, thereby improving both safety and comfort for road users [27]. Implemented cooperatively, all previous practices enhance urban living standards and support environmental sustainability by fostering smarter, more responsible mobility solutions [28].

Vehicle health monitoring systems are an essential aid to the driver since they provide safety support, such as DMSs used to detect driver anomalies, like anxiety, drowsiness, fatigue, or drunkenness, which are all conditions that significantly reduce attention while driving. DMSs combined with a traffic monitoring and management system create smart frameworks for improving road safety, as by monitoring the vehicles' flow on the roads they can provide alternative routes in case of emergency, reducing the risk of accidents. Also, the driver's condition information can be jointly processed and evaluated along with the vehicle parameters to implement smarter DMSs, determining a cumulative driver risk that also takes into account the vehicle conditions (brake wear, tire pressure and wear, etc.). For these reasons, the proposed manuscript explores recent advancements in intelligent driver, vehicle, and traffic monitoring systems to bring out insights and trends for the development of road safety systems in the near future.

This manuscript aims to provide the reader with an overview of IoV solutions in the ITS and smart cities paradigm, analyzing three different macro categories:

- Driver monitoring: various on-board applications for driver monitoring are presented, focusing on detecting health anomalies, drowsiness, stress, and impairment due to alcohol consumption. Additionally, the intrusiveness of these applications and their impact on user ergonomics are analyzed (Section 2).
- Vehicle condition monitoring: on-board technical solutions for vehicle condition monitoring (i.e., tire pressure and wear, exhaust emissions, and vehicle stability) are examined alongside innovative diagnostic systems and devices for detecting road conditions. Additionally, systems for monitoring driving style and driver behavior are analyzed (Section 3).
- Road and environmental monitoring, traffic management: an overview is presented of innovative off-board systems for traffic management, environmental monitoring, and road monitoring, including predictive maintenance solutions. These systems also offer capabilities for detecting road congestion and providing drivers with alternative routes (Section 4).

To enhance the manuscript's scientific contribution, each section includes a comparative analysis of the reviewed articles, along with summary tables highlighting their key features. Additionally, Section 5 presents innovative commercial solutions developed by international companies for monitoring drivers, vehicles, and traffic conditions.

The structure of this review paper is as follows. Section 1 introduces the topics covered and explains the methodology used for selecting the scientific articles. Section 2 provides an overview of on-board solutions for driver monitoring, organized into three subsections focusing on applications for detecting driver status. Section 3 examines on-board solutions for vehicle condition monitoring, divided into five subsections covering vehicle diagnostics, asphalt monitoring systems, driving style and behavior analysis, tire monitoring, and emissions monitoring. Section 4 explores solutions for traffic monitoring and management, as well as road and environmental condition monitoring. Section 5 reviews innovative commercial applications related to the discussed topics. Finally, Section 6 presents comments and conclusions, emphasizing potential future applications.

Selection Method of Analyzed Articles Based on PRISMA Methodologies

This section outlines the criteria used for selecting scientific articles, focusing on key aspects such as relevance to the covered topics, publication year, and redundancy with other works. The objective of this review is to provide readers with a comprehensive, state-of-the-art overview of IoV and ITS paradigm applications. The article selection process followed the PRISMA methodology, ensuring the reliability and applicability of the chosen approach [29,30]. The selection process began with evaluating the title of each candidate article to identify keywords related to the topic. Next, the abstract was reviewed to assess its alignment with the issues addressed in this review. If the article appeared relevant, it was thoroughly read and analyzed. Additional sources were consulted to resolve ambiguities for articles with unclear topics. If the content remained unclear after further investigation, the article was excluded.

Figure 2a illustrates the selection process, which included assessing the title's relevance, the abstract's affinity, and the manuscript's scientific value. Figure 2b depicts the main keywords used to filter the literature. This methodology was consistently applied across all topics discussed in this review.

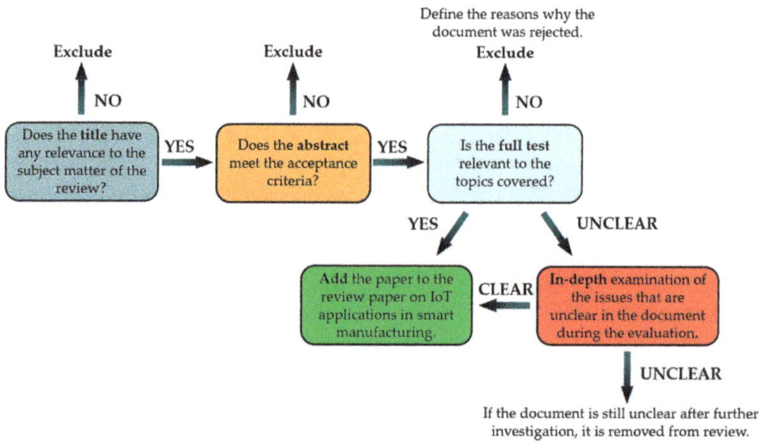

(a)

Figure 2. *Cont.*

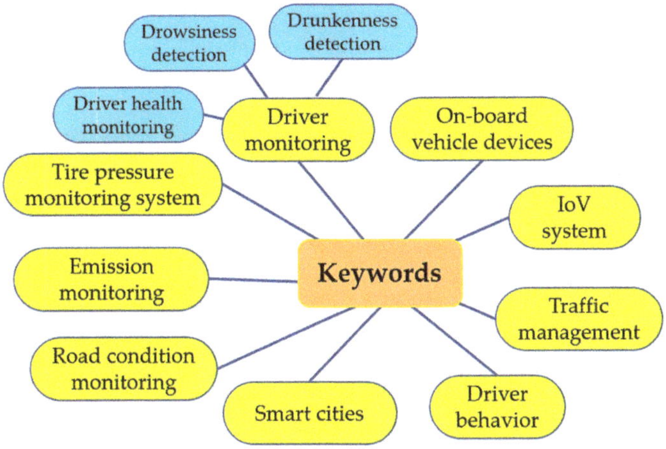

(**b**)

Figure 2. Document selection method: (**a**) description of articles' selecting method with topics related to the presented review paper, (**b**) main keywords to filter the documents found in the literature.

To provide an accurate analysis of the topics covered in this manuscript (i.e., on-board systems for driver and vehicle monitoring, and off-board systems for traffic and road management and monitoring), the authors have reviewed 119 documents, including research articles, conference papers, review articles, and websites. The primary sources for these articles were Elsevier, MDPI, and IEEE. In Figure 3a, the selected articles are categorized by publisher, while in Figure 3b, they are sorted by type.

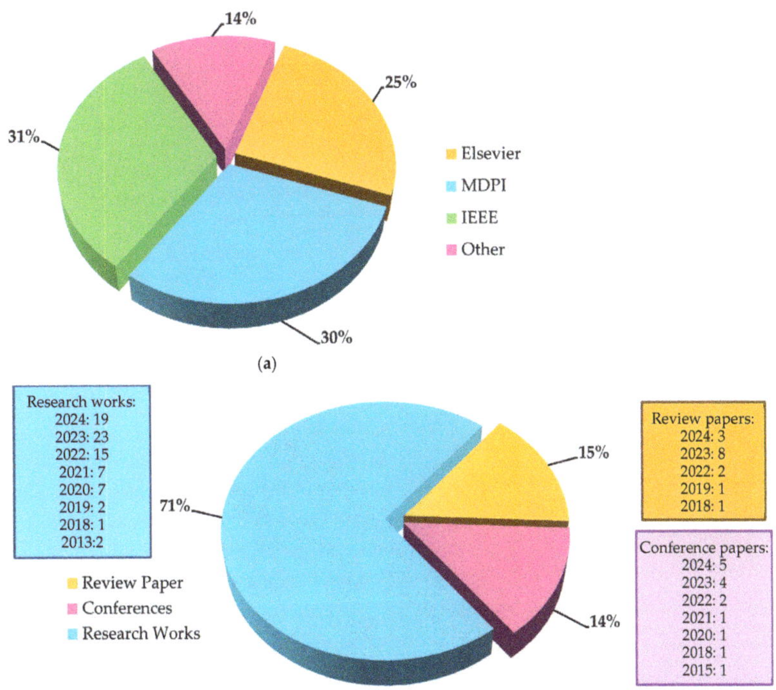

(**b**)

Figure 3. (**a**) Selected articles sorted by publishers, (**b**) selected articles sorted by typology.

2. On-Board Driver Monitoring Systems for Road Safety Applications

In recent years, road safety has become increasingly important, driven by the need to reduce the number of accidents. In this context, on-board DMSs are emerging as a promising solution to improve the safety of vehicles and occupants [31]. Different methods have been proposed in the literature to detect the driver's condition, starting from the acquisition of biophysical signals or video/images, allowing the detection of drowsiness, stress, tiredness, and distraction of the driver, i.e., all aspects that reduce the attention to driving and considerably increase the risk of accidents [32]. The primary goal of DMSs is to identify potentially hazardous driving behaviors and provide immediate alerts to the driver; they can also record and analyze the driving data to prevent accidents, promoting a culture of responsible road safety. Figure 4 illustrates key devices used in DMSs, including wearable and non-wearable sensors for collecting biophysical data, and cameras to capture facial features, head movements, and eye behavior, such as blink rates and eyelid drooping. ML algorithms process these data to determine stress, drowsiness, alcohol effects, and the driver's health condition. A further feature of DMSs is the capability to report hazardous conditions to the driver, authorities, and emergency services. Below, the methodologies proposed in the literature are analyzed: the intrusive methods involve wearable devices, while non-intrusive ones rely on image processing and biosignal acquisition by sensors integrated into the seatbelt, steering wheel, and seat.

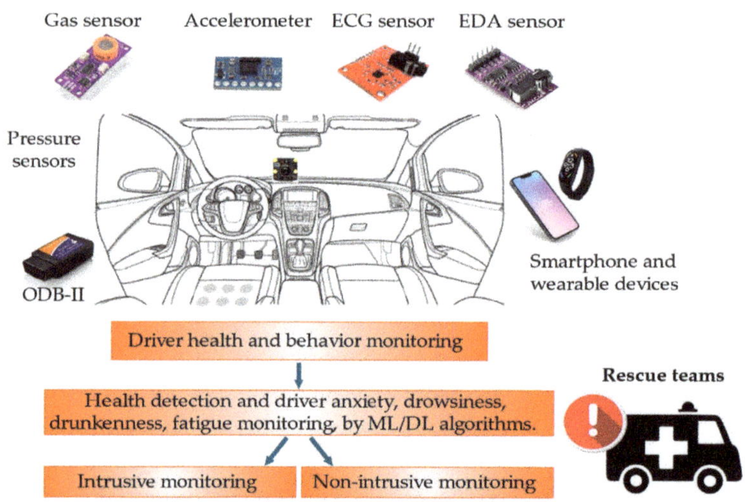

Figure 4. Example of an innovative DMS based on intrusive and non-intrusive methods for the acquisition and processing of biophysical and behavioral driver parameters by using sensors and cameras integrated into the cockpit and wearable devices.

Monitoring the driver's condition is crucial for road safety, enabling early detection of driver anomalies, like drowsiness or drunkenness. These systems, integrated into modern vehicles, not only improve personal safety but also complement the vehicle's advanced diagnostic systems, which detect malfunctions and analyze driving style to prevent risky situations. In addition, systems for monitoring vehicle stability play an essential role in ensuring road safety, guaranteeing constant control over the vehicle and driving conditions. These technologies, working in synergy, also contribute to a more efficient traffic management, optimizing urban mobility and reducing the risk of accidents thanks to the collection and analysis of data on driving behavior and vehicle conditions.

This section explores innovative DMSs highlighted in recent literature. It is organized into four subsections: driver health monitoring, driver drowsiness monitoring, driver drunkenness monitoring, and comparative analysis of the reported articles.

2.1. Driver Health Monitoring Systems to Detect Stress, Anxiety, and Other Biophysical Parameters

This subsection analyzes the applications developed by the authors for monitoring driver health during driving operations. In this regard, Kaur et al. proposed an IoV-Health system dedicated to monitoring the health conditions of the driver [11]. The monitoring occurs with six wearable sensors, such as an Electrocardiogram (ECG) and Blood Pressure (BP), Body Temperature (BT), heart rate (HR), Respiration Rate (RR), and alcohol sensors. Based on the data acquired by the sensors, the system can manage four emergency states: heart failure, stress, drowsiness, and drunkenness. Additionally, the system designed by the authors issues alerts to nearby vehicles, hospitals, and ambulances according to the detected alarm level, ensuring prompt emergency assistance for the driver. The authors used different ML algorithms for the classification of the recorded data, obtaining high percentages of accuracy through Random Forest (RF) (99.86%), ensemble classifier (EC) (99.83%), and decision tree (DT) (99.7%) algorithms. Finally, the authors observed that their system's efficiency surpassed similar systems by more than 50%.

Affanni et al. presented a smart necklace for monitoring the driver's well-being [12], including a sensor that monitors HR and blood oxygen saturation (SpO_2). The collected data are sent to a mobile application via Wi-Fi for continuous real-time viewing. The authors focused on the necklace's comfort, obtaining excellent results from the driving tests they carried out. The algorithm for the data processing is based on the "Mountaineer algorithm" for pick detection in ECG signals [33]. Gong et al. in [13] adopted remote photoplethysmography (rPPG) based on a monochrome camera to measure HR. They proposed an innovative approach called "Quality-guided Spectrum Peak Screening" (QSPS) to eliminate noise interference related to rapidly changing lighting conditions, head movements, or vibrations of the vehicle. The QSPS framework detects PPG signals on multiple regions of the face, in order to attenuate the residual noise through a spectral peak screening algorithm. With this innovative approach, the authors demonstrated that QSPS has greater robustness than infrared methods, achieving a Mean Absolute Error (MAE) of 4.32 bpm, a Root Mean Square Error (RMSE) of 6.15 bpm, and a Pearson correlation coefficient of 0.943 in the nighttime test and 0.906 in the daytime test for 10 min time monitoring.

Jiao et al. conducted an experimental study for real-time detection of driver fatigue status to improve driving and traffic safety [34]. The authors conducted an experimental study to gather data, including fatigue levels measured using the Karolinska Sleepiness Scale [35], as well as Heart Rate Variability (HRV) and Electrodermal Activity (EDA) parameters. Broad statistical analysis revealed significant variations in several HRV and EDA features across different fatigue levels. Using various ML techniques, the Light Gradient Boosting Machine (LGBM) classifier demonstrated the best performance for binary classification, achieving an accuracy of 88.7% with HRV and EDA features as inputs. For three-class classification, accuracy slightly dropped to 85.6% with the RF classifier. These findings highlight the effectiveness of combining HRV and EDA features to capture diverse physiological responses to fatigue, enhancing overall detection accuracy.

Leicht et al. in [36] evaluated non-intrusive methods for monitoring HR and RR, including a hybrid imaging approach, under both simulated and real driving conditions. The feasibility of these methods was assessed by comparing data from non-intrusive sensors with reference sensors. In laboratory settings, magnetic induction and photoplethysmography sensors, integrated into seatbelts, along with hybrid imaging combining visual and thermal data, were tested for RR detection. In real driving conditions, HR and RR were monitored using hybrid imaging and a seat-integrated capacitive ECG across both urban and rural scenarios. The laboratory tests demonstrated reliable RR detection with all three sensor technologies. In real-world rural driving, both HR and RR were reliably detected. However, only RR detection is feasible in urban driving due to motion artifacts affecting the capacitive ECG, which impairs HR detection.

The system proposed in [37] allows driver and vehicle monitoring, stating that the driver affects the vehicle's performance and vice versa. The proposed system integrates on-board and remote vehicle sensors to develop algorithms that estimate pollutant emissions, fuel consumption, driving behavior, and driver health. The main contribution lies in analyzing the interaction between these factors and between driver behavior and vehicle performance, conducting an experimental analysis with different drivers and routes, and the results are implemented in a mobile application. Unlike commercial systems, this approach utilizes standard sensors and established algorithms tested interactively. During the system design, principal component analysis was employed to reduce the number of variables for training and testing, minimizing Bluetooth data transfer between the biometric wristband, smartphone, and vehicle's central computer, as shown in Figure 5. Experimental results indicate the system accurately predicts fuel consumption 84% of the time, pollutant emissions 89%, and driving behavior 89%. Notably, the study revealed significant correlations between the driver's heart condition and traffic conditions.

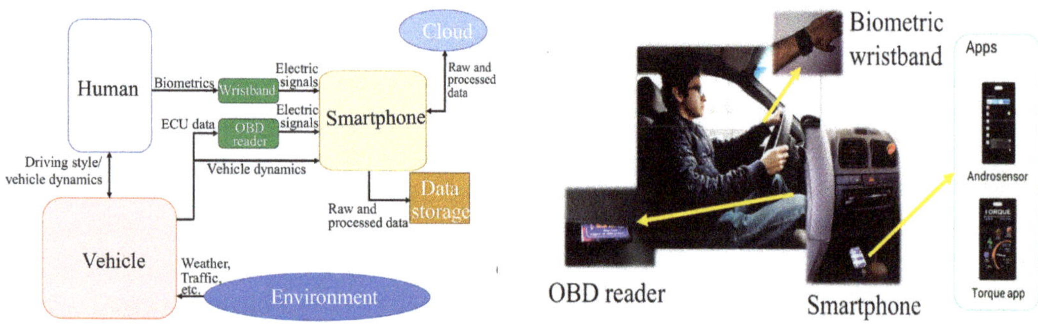

Figure 5. The system proposed in [37] integrates on-board and remote vehicle sensors to develop algorithms that estimate pollutant emissions, fuel consumption, driving behavior, and driver health.

The model proposed by Škorić in [38] explores the classification of stress levels using capacitive Electrocardiogram (cECG) signals recorded during driving with non-intrusive acquisition systems featuring various hardware configurations. The ML model developed for this purpose relies on four features, all derived from detecting the R peak (the local maximum of the ECG signal), which is recognized as the most reliably identified point even in lower-quality cECG recordings. These features were chosen for their low computational complexity, facilitating real-time application. The model was evaluated using three datasets of driving-related recordings: high-quality ECG captured with direct skin contact electrodes, medium-quality cECG obtained via a portable cushion with electrodes operating through clothing, and lower-quality cECG from car seat-embedded electrodes also functioning through clothing. The model demonstrated very high accuracy: 100% for high-quality ECG, 96.67% for medium-quality cECG, and 98.08% for lower-quality cECG.

2.2. *On-Board Systems for Driver Drowsiness Monitoring*

To mitigate the risk of accidents caused by driver drowsiness, Ebrahimian et al. introduced a novel methodology for multi-level drowsiness detection using Convolutional Neural Networks (CNN) and Long Short-Term Memory (LSTM) networks [39]. This approach relies on HR and Respiratory Rate Variations (RRV) data. The authors tested their model on data from 30 participants in a driving simulation, achieving effective results with the combined CNN and LSTM models. Specifically, they reported an accuracy of 91% for a three-level drowsiness classification and 67% for a five-level classification.

Similarly, Alguindigue et al. conducted research aimed at reducing drowsy-driving-related accidents by monitoring sleep-induced attention loss [40]. To assess driver sleepiness, they explored various indicators, including HRV, percentage of eyelid closure over time (PERCLOS), blink rate and percentage, and EDA signal. They employed three deep learning (DL) algorithms: a Sequential Neural Network (SNN) for HRV, a 1D-CNN for EDA, and a Recurrent Convolutional Neural Network (CRNN) for eye-tracking data. The models using HRV and EDA provided high accuracy, achieving 98.28% and 96.32%, respectively, promoting real-time neuro-adaptive systems to detect driver drowsiness. In [41], a different approach is used to detect driver drowsiness based on the Johns Drowsiness Scale (JDS). This method utilizes Optalert glasses equipped with an IR transmitter/receiver to measure eyelid velocity and blink duration through reflectance oculography. The JDS provides ratings on a scale from 1 to 10, derived by the mean and standard deviation of eyelid movement duration and relative velocity during blinks. According to this, drivers are classified as alert if their score falls between 0 and 4 and drowsy between 5 and 10.

Sukumar et al. proposed a model to determine driver stress and drowsiness by detecting ECG signals, head movements by accelerometers, and hand contact with the steering wheel by interdigitated capacitive sensors [14]. By integrating these data, the authors gained insights into driving behavior; to assess stress level, they analyzed ECG signals using the Synchronized Continuous Wavelet Transform (CWT-SST) to evaluate HRV. The same algorithm was applied to detect drowsiness level through head movement data. Their model provided an accuracy of 91.82% in stress detection and 97.30% in identifying drowsiness level. Amidei et al. developed a model to detect driver drowsiness that provides timely warnings [42]. Their approach uses the Empativa E4 wristband that monitors the skin conductance; moreover, the authors tested three ML algorithms for classifying drowsiness, finding that the RF one performed best, achieving an accuracy of 84.1%.

To enhance vehicle and driver safety, Díaz-Santos et al. proposed a model combining facial recognition and drowsiness detection [43], which operates in two phases: the first identifies the driver before the vehicle starts, and the second monitors eye activity for fatigue and drowsiness signs (Figure 6). For driver identification, the authors utilized OpenCV, a computer vision library that detects facial features such as eyes, nose, and mouth. Drowsiness detection relies on prolonged eye closure by a CNN model. The system was developed in Python, employing the Keras deep learning framework and OpenCV. The proposed model achieved 100% recognition accuracy under ideal conditions, which decreased slightly to 97.5% when the driver wore glasses. Khan et al. implemented a non-intrusive DMS using computer vision techniques [44]. The proposed framework operates on four levels: an embedded system, edge computing, cloud computing, and a user interface. In the edge computing layer, facial features such as eye, mouth, and jaw movements are analyzed to detect drowsiness signs. The system then sends notifications to relevant authorities, including general information and drowsiness score, to assess the driver's condition; after testing, the model provided high robustness with 96% overall accuracy.

Figure 6. Innovative model for facial recognition and driver drowsiness detection proposed in [43].

Safarov et al. in [15] applied a combination of deep learning (DL) and computer vision algorithms to detect driver drowsiness. They trained their model using custom datasets and tested it with multiple subjects. Eye-blink frequency and mouth region coordinates were captured using facial landmarks (Figure 7). These landmarks were analyzed in real time to monitor fluctuations in eye-blinking rates and mouth movements. Their real-time experiments revealed a correlation between yawning and prolonged eye closure, which were classified as signs of drowsiness. The model demonstrated strong performance, achieving 95.8% accuracy for detecting closed eyes, 97% for open eyes, 84% for yawning, 98% for detecting right-sided head falling, and 100% for left-sided head falling. Additionally, the system enabled real-time eye state analysis, effectively classifying eyes into "Open" or "Closed" categories based on threshold values.

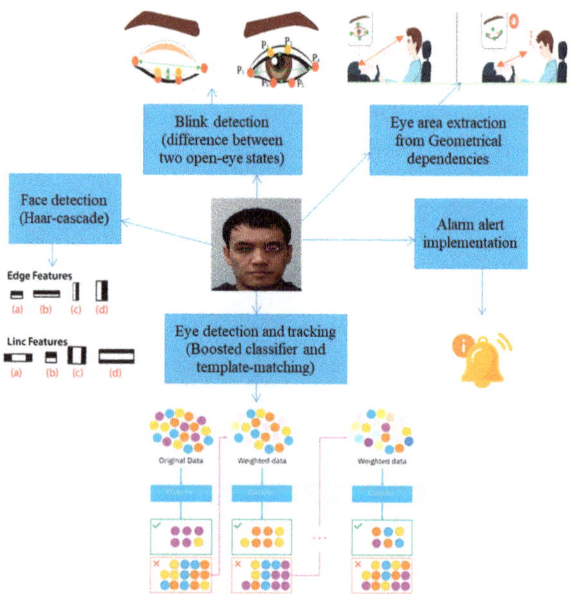

Figure 7. Main features of the method proposed by the authors in [15]; the Haar–Cascade classifier was trained to detect faces and extract the related features (a–d). After detection, the authors captured the coordinates of facial landmarks and exported them to a comma-separated value (csv) file, based on the proposed classes.

The proposed approach in [15] may not be suitable for drivers wearing sunglasses. Sunglasses can obstruct the model's ability to accurately detect eye landmarks, which are crucial for measuring eye blinks and assessing driver alertness. Building on the concept of non-intrusive monitoring, Kundinger et al. developed an ML approach for detecting driver drowsiness using physiological data from a sensor embedded in a wrist-worn bracelet (Figure 8) [45]. They compared the bracelet's performance against a medical-grade ECG device. Among the binary classification algorithms tested, the K-Nearest Neighbor (KNN) model achieved the highest accuracy at 92.13%.

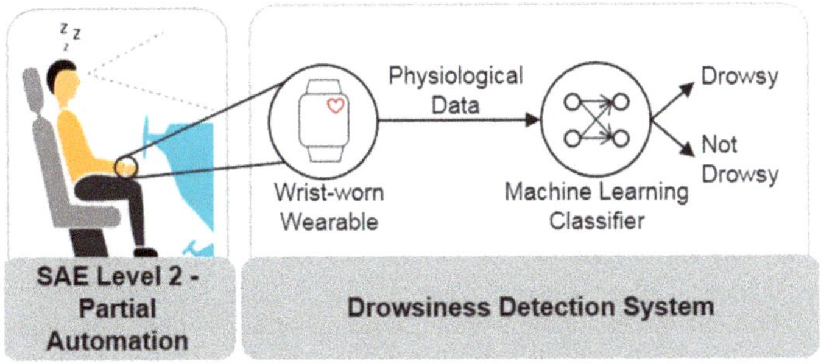

Figure 8. Architecture proposed in [45] for non-intrusive monitoring of the driver to detect the state of drowsiness by processing the ECG signals via ML algorithms.

Vyas et al. introduced the "DriverSense" system, designed to enhance the safety of drivers on two- or three-wheeled vehicles [46]. The framework relies on a wearable device equipped with mobility sensors (gyroscope, accelerometer, GPS) and biosensors able to collect real-time biosignals (EEG-electroencephalogram, PPG, ECG, SpO_2). By combining the detected signals, the designed system can assess the driver's stress level and driving behavior in real time, classifying the first as normal or stressed and the second one into three categories: slow, normal, and aggressive. The authors reported classification accuracies of 86.05% for driving behavior and 88.24% for stress detection.

2.3. On-Board System for Driver Drunkenness Monitoring

Drunkenness is a significant cause of traffic accidents. To address this issue, Ku-mar et al. proposed an IoT-based system designed to detect driver intoxication and prevent the car engine from starting [47]. Their model utilizes advanced sensors to detect alcohol on the driver's breath (using an MQ-3 sensor), a Raspberry Pi microcontroller, and a Global Positioning System (GPS) module for tracking the vehicle's location. When the driver attempts to start the car, the system checks for alcohol levels. If the detected level exceeds legal limits, the microcontroller activates a buzzer, warning the driver that the engine will shut off shortly. Simultaneously, a Global System Mobile (GSM) module sends the vehicle's position to a monitoring center or designated contacts. Similarly, Liu et al. developed an intoxication detection system integrated into the vehicle's steering wheel for improved data accuracy [48]. This system uses an array of sensors, as shown in Figure 9, and processes the data through a weighted fusion algorithm. If the sensors detect signs of intoxication, a signal is sent to the microcontroller, which then acts on the ignition relay to prevent the engine from starting.

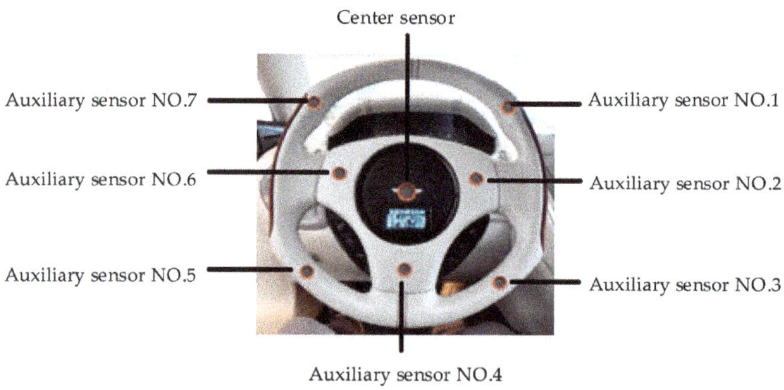

Figure 9. MQ3 gas sensor array integrated into the steering wheel used for the model proposed by the authors in [48]. They also integrated an Organic Light-Emitting Diode (OLED) display screen showing the alcohol concentration level, indicating the level of drunkenness.

Swarna et al. proposed an architecture known as the Improved IoT-based Driver Safety Model (IIoTDSM) to prevent driving under the influence of alcohol [49]. This system integrates alcohol detection sensors into the vehicle's steering wheel. Key components include a Wi-Fi module, an Arduino microcontroller, an MQ3 gas sensor, and a blink sensor. If the system detects that the driver's Blood Alcohol Concentration (BAC) exceeds legal limits, it prevents the vehicle from starting and alerts the appropriate authorities.

In response to the rise of shared car usage, Wang et al. developed a model to detect intoxicated individuals inside the vehicle or determine if the driver is drunk [50]. Their system features an electronic nose equipped with gas sensors to measure alcohol concentration in the vehicle's air. The detection process involves two steps: first, assessing the overall alcohol level in the car, and second, determining if the driver is intoxicated. These steps are illustrated in Figure 10. The authors reported high accuracy, achieving 99.44% in the first and 100% in the second steps, with a sampling time of just five seconds.

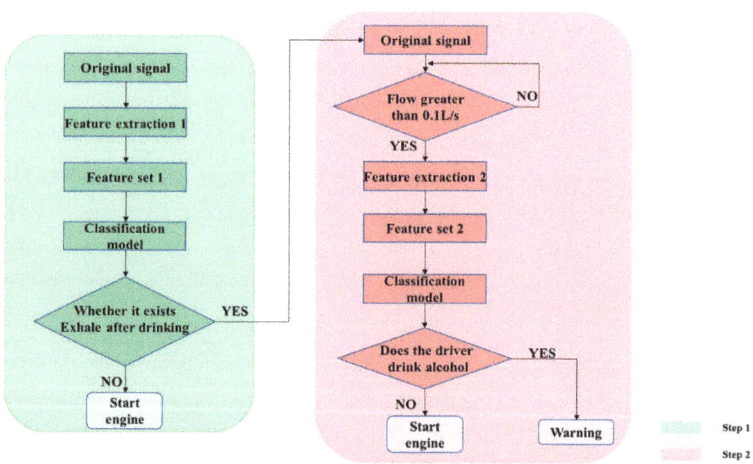

Figure 10. Flowchart of the model proposed in [50], which checks whether the driver has drunk alcohol, preventing the engine from being started if the outcome is positive.

In [51], the MQ-3 sensor is used to measure BAC through the driver's breath and surrounding air, comparing the readings with the legal limits in respective countries. The system includes a proximity sensor to prevent the MQ-3 sensor from being covered. If drunkenness is detected, a flashing red light alerts the driver and the vehicle will not start. Furthermore, Cho et al. conducted a study using chromatography to detect BAC and assess the drunkenness level [52]. Although this method is more complex, it offers an alternative to semiconductor sensors, overcoming the short duration limitations and lack of stability.

2.4. Comparative Analysis of On-Board Driver Monitoring Systems

The literature analysis has shown that the primary challenge lies in accurately detecting anomalies in driver health, drowsiness, and intoxication using intrusive and non-intrusive methods. Various approaches are discussed to assess the driver's condition, involving monitoring biophysical parameters such as heart rate (HR), Heart Rate Variability (HRV), Respiratory Rate (RR), and Electrodermal Activity (EDA), and facial expressions such as blinking frequency, eyelid closure duration, and excessive head movements.

The methods examined in Section 2.3 focus on measuring BAC from breath analysis by MQ-3 sensors, preventing the engine from starting if the limit is exceeded. Advanced systems also employ cameras to analyze head movements or facial expressions to enhance accuracy. Various ML algorithms, such as Random Forest (RF) and Convolutional Neural Networks (CNN), are commonly used to assess the model accuracy, with CNNs particularly suited for real-time image analysis. The reviewed methodologies demonstrated high accuracy, indicating their readiness for commercial implementation. A crucial feature of these systems is maintaining driver comfort, ensuring that monitoring does not interfere during driving. Non-intrusive approaches using cameras or sensors integrated into belts or seats are preferred, as they preserve comfort despite requiring higher computational power. Ongoing research aims to enhance the DMS accuracy in assessing health conditions, for instance by combining biophysical data, postural changes, and driving behavior. In this regard, inertial sensors on the steering wheel can measure the steering force, piezo-resistive sensors in the seatbelt can monitor the respiratory activity, and the driver's posture can be monitored by pressure sensors in the seat and the stress level by ECG electrodes placed on the steering wheel.

Section 5.1 discusses commercial DMSs that detect drowsiness and distractions, which are major contributors to accidents. Tables 1–3 summarize the key features of the reviewed studies, including applications, data acquisition devices, monitored parameters, intrusiveness levels, alarm systems, ML algorithms used, and accuracy rates.

Table 1. Main characteristics of driver health monitoring systems in the literature.

Reference	Application	Acquisition Device	Detected Parameters	Intrusiveness	Alarm Systems	AI Algorithm	Accuracy
K. Kaur et al. [11]	Health monitoring	Six unspecified wearable sensors	ECG, BP, BT, HR, RR, BAC	High	Yes	RF, EC, DT	99.86% (RF) 99.83% (EC) 99.70% (DT)
A. Affanni et al. [12]	Well-being monitoring	Smart necklace	HR, blood oxygen saturation (SpO_2)	Medium	No	"Mountaineer algorithm"	N.A.
Z. Gong et al. [13]	Heart rate monitoring (rPPG)	Camera	HR (rPPG)	Low	No	LAB algorithm by SenseTime	Pearson correlation coefficients: 0.943 (night) 0.906 (day)
Y. Jiao et al. [34]	Fatigue detection	Accelerometers	HVR, EDA	Low	No	LGBM, RF	88.70% (LGBM) 85.60% (RF)
L. Leicht et al. [36]	Health monitoring	MI coil, PPG sensor in seatbelt, camera, capacitive ECG sensor into seat	HR, RR	Medium	No	Pan–Tompkins algorithm, FaceLAB 5	N.A.
A.E. Campos-Ferreira et al. [37]	Health monitoring	Biometric wristband	HR, skin temperature	Medium	No	Principal Component Analysis (PCA)	N.A.
T. Škorić [38]	Health monitoring	Direct contact electrodes, capacitive electrodes in cushion and car seat	ECG, cECG	High (direct contact), Low (indirect contact)	No	KNN, SVM, ANN	100 % 96.67 % 98.08 % ([1])

([1]) The accuracy values refer to the authors' three ECG signal detection methods, direct contact (100%), and indirect contact (96.67% in the case of sensors embedded into cushion and 98.08% in the case of sensors embedded into car seat); N.A.: Not Available.

Table 2. Main characteristics of driver drowsiness detection systems in the literature.

Reference	Application	Acquisition Device	Detected Parameters	Intrusiveness	Alarm Systems	AI Algorithm	Accuracy
S. Ebrahimian et al. [39]	Drowsiness detection	eWave32D, thermal camera	HR, RRV	Medium	No	CNN-LSTM	91% ([1]) 67% ([2])
J. Alguindigue et al. [40]	Drowsiness detection	Smart band, smart glasses	HRV, EDA, PERCLOS	Medium	No	SNN, 1D-CNN, CRNN	98.28% ([3]) 96.32% ([4])
A.S. BaHammam et al. [41]	Drowsiness detection (JDS)	Optalert glasses	Eyelid velocity, blink duration.	Low	No	Optalert™ Proprietary pattern recognition algorithm	N.A. (*)
N. Sukumar et al. [14]	Detection of stress and drowsiness levels	Accelerometer, capacitive sensor	ECG, head movement, IDT	Medium	No	CWT-SST	91.82% ([5]) 97.30% ([6])
A. Amidei et al. [42]	Drowsiness detection	Smart band	Skin conductance	Medium	No	RF	84.1%
S. Díaz-Santos et al. [43]	Drowsiness detection	Camera	Facial recognition	Low	No	OpenCV CNN	97.5%
M. A. Khan et al. [44]	Drowsiness detection	Camera	Facial features	Low	No	Unspecified	96%
F. Safarov et al. [15]	Drowsiness detection	Camera	Blink rate	Low	Yes	Computer vision algorithm	95.8% ([7]) 97% ([8])
T. Kundinger et al. [45]	Drowsiness detection	Smart band	ECG	Medium	No	KNN	92.13%
D. A. Vyas et al. [46]	Detection of stress and aggression level of the driver	Mobility sensors, sensors for biophysical signals	EEG, PPG, ECG, SpO_2	High	Yes	KNN, RF	86.05% (KNN) 88.24% (RF)

([1]) Three-level drowsiness classification; ([2]) five-level drowsiness classification; ([3]) HRV detection; ([4]) EDA detection; ([5]) stress recognition; ([6]) drowsiness detection; ([7]) drowsy eye detection; ([8]) open eye detection; (*) N.A.: Not Available.

Table 3. Main characteristics of drunkenness detection methods in the literature.

Reference	Application	Acquisition Device	Detected Parameters	Intrusiveness	Alarm Systems	Algorithm	Performance
T.S. Kumar et al. [47]	Drunkenness detection	MQ-3 sensor, GPS module	BAC	Low	Yes	Threshold-based algorithm	85–94%
J. Liu et al. [48]	Drunkenness detection	MQ-3 sensor	BAC	Low	Yes	Data fusion and threshold-based algorithm	N.A. (*)
M. Swarna et al. [49]	Drunkenness detection	MQ-3 sensor, GPD module	BAC	Low	Yes	IIoTDSM	Efficiency: 98.70%
F. Wang et al. [50]	Driver and passenger drunkenness detection	Array of 21 gas sensors (1)	BAC	Low	Yes	KNN, SVM, RF	99.44% (I Step) 100% (II Step)
J. Akanni et al. [51]	Drunkenness detection	MQ-3 sensor	BAC	Low	Yes	Threshold-based algorithm	N.A. (*)
Y. Cho et al. [52]	Drunkenness detection	Spectroscopic module	BAC	Low	Yes	Absorbance vs alcohol characteristic	0.9908 (2) 0.9938 (3)

(1) Reported in Table 1 of Ref. [50]; (2) R-squared value for first-order equation; (3) R-squared value for second-order equation;.(*) N.A.: Not Available.

Future challenges in the field of driver health monitoring concern various technological, ethical, and practical aspects to achieve the following goals:

- High accuracy and reliability of parameter detection by using non-intrusive devices integrated into the vehicle and wearable devices;
- Integration with embedded ML/DL models to process large data amounts in real time;
- The development of systems that work in synergy with vehicle diagnostics, autonomous driving, and traffic management solutions;
- Privacy and high security of user data against cyber-attacks and unauthorized use.

Technological advancements are pushing toward systems implementing multi-modal analysis for detecting and simultaneously processing driver, vehicle, and traffic data. Recently several sensing technologies have been proposed in the literature for discretely detecting the driver's biophysical parameters, addressing critical aspects of safety and well-being in the driving context. Flexible piezoelectric sensors offer a promising approach for real-time posture analysis; their integration into car seats enables continuous monitoring without being intrusive to detect fatigue or discomfort early on [53]. Also, radiofrequency and optical signal reflection can be exploited to extract driver biophysical parameters; radar and LiDAR devices have demonstrated capability in measuring respiratory and cardiac parameters (e.g., HR, HRV), providing accurate and real-time physiological data while maintaining driver comfort [54]. Finally, cameras installed in the vehicle cockpit can be used to measure heart and respiratory parameters; for example, estimating HR using images acquired from an iPPG (imaging PPG) camera is a novel and non-invasive technique, particularly useful in dynamic environments like vehicles, as it avoids direct contact with the driver [13]. These approaches highlight the potential of combining different sensing modalities to create a robust and comprehensive DMS. Future work could focus on integrating these technologies into a unified platform, addressing challenges such as data fusion, reliability in varying conditions, and ensuring user privacy.

3. Smart Monitoring Systems Based on Vehicle Information

In recent years, vehicle safety has become a priority in the automotive industry, leading vehicle manufacturers to develop and implement advanced technologies for monitoring and enhancing road safety. These systems protect vehicle occupants while improving interaction with the surrounding environment. One key safety feature is the Tire Pressure Monitoring System (TPMS) [55], which alerts drivers to under- or over-inflated tires, improving vehicle safety. Additionally, sophisticated systems can assess road conditions in real time, detecting hazards such as potholes, ice, or other surface irregularities. Driver-assistance technologies, including stability and braking control, employ sensors and cameras to continuously monitor vehicle performance and the surrounding environment, and automatically intervene in critical situations, providing a more controlled and safer driving experience. These advancements are often integrated into broader smart mobility strategies, where connectivity and data collection are crucial for improving road safety and minimizing accident risks. Such innovations contribute to a safer, more informed, and more responsible driving experience. A comprehensive vehicle monitoring model relies on engine and battery health data, tire pressure and wear, and oil and coolant temperatures, all collected via the on-board diagnostics (OBD-II) system and then processed by ML algorithms to assess the vehicle safety level, as illustrated in Figure 11. Furthermore, this information allows predictive maintenance to guarantee optimal vehicle performance and efficiency.

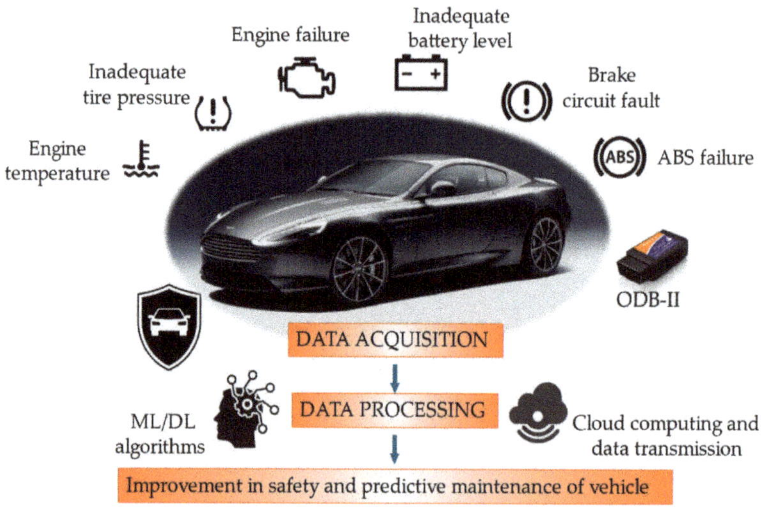

Figure 11. Vehicle monitoring based on ML algorithms for processing data acquired by sensors for detecting and predicting vehicle anomalies.

Integrating vehicle status information with driver and road conditions can significantly improve driving safety by creating a holistic and reactive system that can prevent accidents and optimize risk management. Through vehicle diagnostic tools and driver monitoring, it is possible to achieve adaptation to a driving style; furthermore, by detecting fatigue or drowsiness signs, the diagnostic system can reduce the vehicle's speed until it stops. Also, automatic adaptation of vehicle parameters can be performed based on real-time information on road conditions (e.g., slippery pavement, heavy traffic, or sharp curves) by means of the driving assistance systems, such as traction control and stability, and by warning the driver to slow down in case of dangerous road conditions or brake/suspension failures.

3.1. On-Board Smart Monitoring Systems for Vehicle Diagnosis

The first approach for monitoring vehicle conditions uses intelligent diagnostic systems that detect anomalies and malfunctions before the failure occurs. Yang proposes a model to enhance fault diagnosis in electric vehicles by a system based on the Controller Area Network (CAN) bus and diagnostic communication architecture compliant with the AUTOSAR standard [18]. This approach aims to improve software reusability and the portability of fault diagnosis systems for electric vehicles. By leveraging AUTOSAR, developers can focus more on functionality rather than system architecture. The system provides a fault detection response time of 0.0217 s and a detection accuracy of 98.70%.

Kim et al. developed a diagnostic algorithm to effectively identify and localize faults of a 4WIS4WID vehicle which, employing redundant hardware and fault-tolerant controls, allows continuous operation even in the presence of faults [56]. The limited degrees of freedom compared to the number of potential failures complicate the data analysis process. The algorithm uses residual analysis to compare vehicle behavior with sensor data; these residuals are converted into three-channel images using histogram analysis and logical functions to train a CNN algorithm to detect and locate faults. The Recursive Least Squares (RLS) method also classifies fault types and vehicle output in response to detected issues.

Kannan et al. proposed a fault diagnosis system using a PIC microcontroller within a Predictive Maintenance (PdM) framework, highlighting the crucial role of extending vehicle life [57]. The system collects and processes engine, transmission, and control unit data, utilizing IoT and ML technologies to monitor vehicle health and accurately identify faults. The system performs better than existing approaches by generating detailed fault information and enabling more effective and proactive vehicle management.

Min et al. developed a framework focused on sensors' self-diagnosis, as a sensor is reliable when it works correctly [58]. Their approach uses a residual consistency-checking algorithm to identify and isolate faulty sensors by leveraging their redundancy. The system employs a Denoising Sparse Autoencoder (DSAE) integrated with a soft threshold-shrinking block, enhancing feature representation and anomaly detection accuracy. Experimental results confirm that the developed solution effectively detects and isolates sensor faults, contributing to more robust and reliable monitoring.

To enhance accident prevention, Shermila et al. proposed an automotive black box that continuously monitors and records data from on-board sensors to uncover the actual causes of accidents [59]. This system provides information, such as speed and acceleration during and after an accident, to verify if safety devices functioned correctly in case of impact; it integrates different sensing devices, including collision sensors to detect the accident, temperature and gas sensors to identify anomalies derived from the impact, and GPS to determine the vehicle's position. In case of emergency, sensor data are transmitted to law enforcement and medical services via GSM and an internet network. Figure 12 illustrates the system architecture through a block diagram.

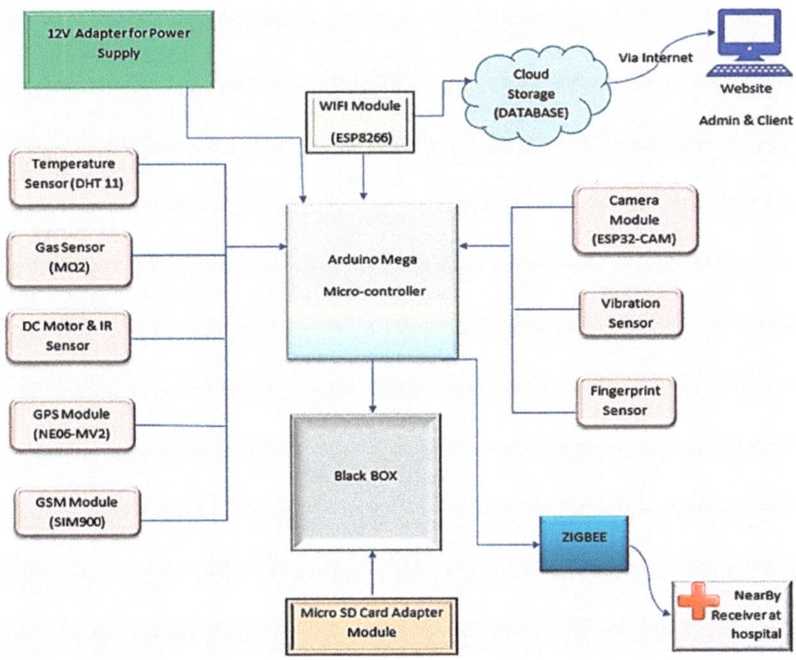

Figure 12. Architecture of the vehicle monitoring system proposed in [59].

3.2. Innovative Systems for Detecting Vehicle Instability Behavior While Driving in Adverse Asphalt Conditions (Icy Asphalt, Uneven Asphalt, and Aquaplaning)

Driving in difficult environmental conditions poses significant risks, especially at high speeds. Ensuring vehicle stability on icy or wet roads, where the lack of friction between asphalt and tires leads to the loss of the vehicle's control, is essential to increase driver safety. Fichtinger et al. [60] developed a method combining various aquaplaning detection models, including monitoring of the slip slope variations (similar to Gustafsson's work [61]), changes in rolling resistance, wheel spin-up/down detection, and analysis of wheel speed variance, acceleration, and rolling resistance using CUSUM (Cumulative Sum Control Chart). Their approach utilizes a minimal set of sensors, focusing on Electronic Stability Control (ESC) data and driving torque.

Hu et al. propose an advanced system for identifying slippery road conditions using Connected Vehicle (CV) technology [22]. Their model leverages data from connected vehicles and employs deep learning algorithms, particularly LSTM networks. These networks, trained with simulated data from VISSIM software and optimized using Bayesian algorithms, accurately classify road conditions into dry, snowy, and icy categories, achieving Success Rates (SRs) of 100%, 99.06%, and 98.02%, respectively. A key innovation is the system's ability to improve driving behavior: issuing timely warnings helps drivers adjust speed and maintain safe distances, reducing potential accidents by up to 90%. The system effectively detects black ice, a critical factor in preventing hazardous driving incidents.

Lee et al. [62] developed an advanced black ice detection system for autonomous vehicles, notoriously difficult to detect, by using CNN algorithm to identify it accurately. To enhance the system's performance, the authors applied data augmentation techniques, training the CNN with images representing various road conditions prone to black ice formation. The model achieved a detection accuracy of 96%, showing high efficacy to reduce accidents in icy conditions. Another critical issue affecting urban roads is the presence of potholes and cracks, which can make driving uncomfortable and damage vehicles. Bhadraray et al. addressed this problem by creating a model that adjusts vehicle speed and trajectory to avoid potholes while maintaining the travel route [63]. They simulated the system using a TurtleBot3 Burger robot equipped with an RGB camera and trained the YOLOv4-Tiny model on a dataset of 2160 images to detect potholes. Additionally, they used Canny edge detection to identify road boundaries within the region of interest.

Raja et al. introduced the Smart Pothole Avoidance Strategy (SPAS) that uses the Deep Deterministic Policy Gradient (DDPG) algorithm for dynamic pothole management [64]. A key innovation of SPAS is its hybrid recognition model, combining audio and visual feedback from passengers through a mechanism called Speech and Gesture (HRM-SG). This interactive feedback refines the DDPG system's capabilities, enhancing pothole avoidance accuracy and improving passenger comfort. SPAS optimizes lane changing and speed adjustments based on sensor data and passenger input, making travel safer and more comfortable. Experimental results show significant improvements over existing methods with a 10–15% increase in pothole avoidance accuracy, a 10–12% boost in comfort, and 8–10% faster convergence. This work highlights advancements in autonomous vehicle technology and the value of integrating human feedback to optimize navigation performance.

3.3. *Smart Systems for Driver Style and Behavior Monitoring*

Sensor data acquisition enables the identification of a driver's behavior, distinguishing aggressive and irrational styles [65]. Nouh et al. introduced *SafeDriver*, a Dynamic Driver Profile (DDP) system, which uses historical data on traffic violations and accidents to classify drivers into four risk levels [66]. The goal was to improve driver behavior and road safety by alerting other vehicles connected through the IoV. The authors derived driver profiles using data from the Saudi Traffic Points System Regulation (STPSR) and applied DL techniques to process them. Figure 13 shows the architecture for driver classification.

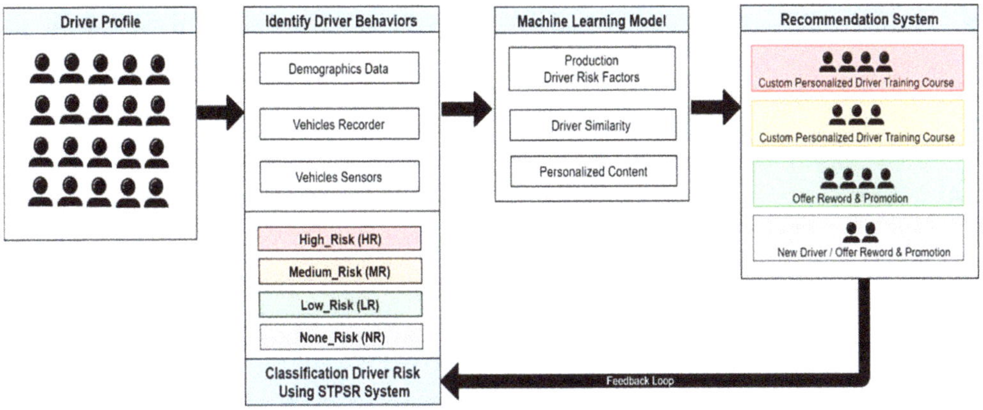

Figure 13. Architecture proposed in [66] based on different functional blocks.

Similarly, Mohammed et al. developed a real-time driving style recognition model using electronic monitoring devices to detect the vehicle speed, engine RPM, coolant temperature, and geolocation coordinates [9]. These data are transmitted to a server, visualized through a virtual dashboard, and stored as driving history and route records. Shichkina et al. proposed an algorithm that processes twenty descriptive features to monitor driving behavior [67]. By accessing data from the diagnostic system (OBD-II), the model gathers information on environmental conditions, driver actions, and vehicle performance, to be analyzed by a Kohonen network, which clusters drivers based on driving style and associated risk levels. Chhabra et al. developed a federated learning approach for monitoring driver behavior, where data from connected vehicles enhance the algorithm performance [68]. By using CNN-LSTM and CNN-Bi-LSTM deep learning models, they analyzed data collected from aboard-smartphone sensors and in-car devices to classify driver behavior accurately. In another study, Malik et al. focused on analyzing driving patterns using real-world data gathered via the vehicle OBD port [69], employing hierarchical clustering, k-means, ANN, and multilayer perceptrons to identify driving behavior patterns and categorize data into meaningful groups. A key innovation is the use of Inter-Class-ReLU activation function, which strengthened the model's robustness and classification accuracy.

Lattanzi et al. explored driver behavior detection using data from on-board sensors [70]. By applying Support Vector Machines (SVM) and a feed-forward Neural Network (NN), they extracted descriptive features to identify driving patterns, enhance road safety, and prevent accidents caused by risky behaviors. Dong et al. detected fatigue and distracted driving by vision-based techniques and ML models [71]. For fatigue detection, they used facial feature analysis by the RF algorithm to assess driving conditions; instead, to classify distracted driving behaviors, the authors implemented a CNN model, obtaining 91% and 97.5% accuracy values, respectively.

3.4. On-Board Innovative System for Tire Monitoring

Vasantharaj et al. proposed an innovative indirect TPMS that leverages existing vehicle sensors to measure tire pressure indirectly [55]. Their model also detects grip conditions and pressure loss based on speed, enhancing overall performance monitoring.

Màrton et al. introduced an advanced indirect TPMS (iTPMS) that uses modern signal processing techniques combined with a CNN for detecting pressure-related eigenfrequencies [21]. By integrating wavelet-Fourier transforms with CNN-based pattern recognition, their approach demonstrates superior accuracy and efficiency in experiments, marking a significant step forward in AI-driven TPMS technologies.

Shan et al. presented another indirect TPMS utilizing wheel speed sensors and on-board hardware [72]. They detail a method that processes wheel speed signals through denoising, rim error filtering, and resampling. By analyzing wheel speed spectra and tire vibration characteristics, the system accurately identifies pressure-related frequencies.

Huo et al. developed a tire pressure warning system using magnetoelectric wheel speed sensors to collect and analyze tire speed data [73]. This system, based on beam and frequency techniques, allows early detection of dangerous conditions. Experimental results show improved accuracy and stability compared to traditional methods. Focusing on tire–road interaction, M. Pomoni investigated factors influencing tire–asphalt contact and advocates for the use of intelligent tires [74]. These advanced tires enhance driver perception and responsiveness, improving safety in varying road conditions.

Yu et al. proposed a tire wear monitoring system that goes beyond the traditional methods [75]. Using a magnetic angular velocity sensor operating at 600 Hz and a mini host device for data transfer, they apply the Pacejka model to analyze tire behavior. Data features are extracted by a transformer-based encoder followed by classification through a fully connected layer to define the tires' wear degree, achieving a 97.77% accuracy, outperforming the Recurrent Neural Networks (RNNs) and SVMs by 5.13% and 4.75%, respectively, even at low sampling rates.

3.5. Innovative Emissions Monitoring Systems

Focusing on vehicle emissions control, Anusha et al. introduced a cloud-based model for monitoring and analyzing pollutant emissions using neural networks (NNs) [76]. Their proposed architecture enables real-time emissions tracking and alerts drivers to emergencies. The system gathers data from air quality monitoring stations, vehicle-installed sensors, and remote sensing devices. These data undergo preprocessing, including handling missing values and removing outliers, to create a robust dataset for neural network training. Once processed, the model extracts key features such as vehicle type, engine specifications, maximum speed, and road conditions.

Sao et al. present an innovative method to predict emissions of carbon dioxide (CO_2), nitrogen oxides (NO_x), and carbon monoxide (CO) from diesel vehicles using ANN [77]. They evaluate six operational parameters (vehicle speed, engine speed, engine torque, coolant temperature, air/fuel ratio, and intake airflow) collected via on-board diagnostics (OBD) as predictors. The study finds that predictive accuracy improves with additional parameters, though the extent varies depending on which parameters are included. Moreover, the accuracy differs based on vehicle type, with models featuring after-treatment devices showing lower predictive precision compared to those without.

Andrade et al. propose a continuous emissions monitoring system using the OBD-II interface to indirectly measure vehicle emissions [78]. To enhance data accuracy, they developed a soft-sensor approach that processes engine combustion metrics such as fuel injection and airflow to estimate CO_2 emissions. Two distinct algorithms, each with tailored mathematical formulations, handle input-specific data. Additionally, an unsupervised TinyML method removes outliers, improving sensor accuracy without relying on cloud connectivity. The results confirm the system's viability, providing emissions measurements in gCO_2/km. Figure 14 illustrates the emissions monitoring model developed by authors.

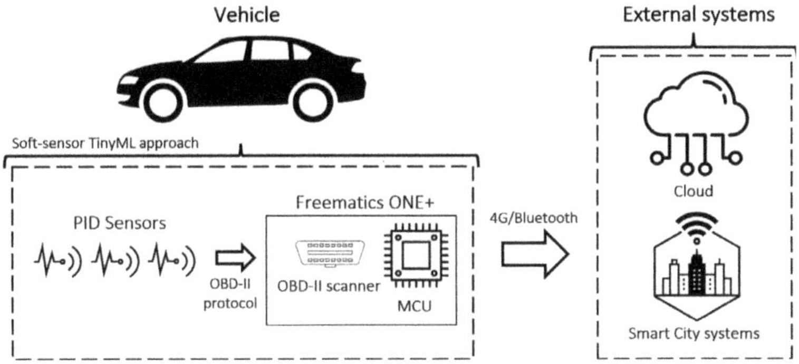

Figure 14. Emissions monitoring architecture proposed in [78].

3.6. Comparative Analysis of Smart Monitoring Systems Based on Vehicular Information

Monitoring vehicle conditions and driver behavior is essential in reducing road accidents, controlling emissions, and enhancing driver safety. The first step is to establish a diagnostic architecture connected to vehicle one, by integrating redundant hardware, ML algorithms capable of solving multiple simultaneous faults, and self-diagnostic systems for sensors in order to identify faults and enable predictive maintenance. Table 4 highlights the key features of the diagnostic systems analyzed in the literature.

Driving on slippery roads, e.g., covered with ice or water, significantly reduces driving safety, because the reduced grip can lead to losing vehicle control; thus, developing systems that can detect hazardous conditions is essential. The systems discussed in Section 3.2 can identify aquaplaning conditions by detecting wheel slippage or icy roads using cameras and DL algorithms trained on web images and real-time road photos. Additionally, cameras installed on the vehicle can detect potholes on the road, which, when combined with speed modulation and trajectory modification systems, contribute to innovative safety solutions. Table 5 summarizes the key features of the systems analyzed.

Monitoring and analyzing driving styles have become essential for enhancing road safety and optimizing vehicle management. The proposed solutions for monitoring driving behavior and improving safety are promising, incorporating advanced technologies such as deep learning (DL), federated learning, and clustering. However, each approach has limitations that could be addressed through innovations in sensors, data processing, and optimization of predictive models. Future improvements could focus on leveraging mobile technologies for data collection, integrating more advanced sensors for more precise detection of driving styles, and evolving ML algorithms to better accommodate various driver behaviors and environmental factors. Furthermore, the use of a monitoring box, which records the driving data, communicates the vehicle's position, and alerts emergency services in case of accident, as outlined in [59], is vital. Table 6 summarizes the key features of the driving style monitoring systems analyzed in the literature.

Recent research on tire pressure monitoring systems (TPMSs) has led to innovative approaches that enhance vehicle safety, reliability, and operational efficiency by integrating advanced technologies and intelligent methods. Some proposed models utilize existing vehicle sensors to detect tire grip and pressure conditions, which, combined with ML algorithms such as CNNs, offer a valid solution to reduce the need for additional hardware, also providing accurate results. However, challenges such as computational complexity, sensitivity to variable driving conditions, and reliability in complex scenarios remain. For this purpose, it will be necessary to ensure consistent performance in different operating conditions, reduce hardware requirements for broader accessibility, and enhance integration in new vehicles. This can be achieved by developing more robust, interference-resistant sensors and optimizing AI algorithms in software solutions. Table 7 summarizes the main characteristics of the tire monitoring models discussed in this section.

Recent research on vehicle emissions monitoring and forecasting has shown significant advancements. The solutions proposed by Anusha et al., Sao et al., and Andrade et al. represent crucial steps toward more intelligent and responsive emissions monitoring systems. These approaches stand out for their integration of advanced technologies such as cloud computing, neural networks, and Machine Learning (ML), which enable real-time monitoring and accurate forecasting. However, to enhance reliability, scalability, and accuracy, certain challenges remain, including input data quality, the complexity of forecasting models, and compatibility across various vehicle types. Future improvements could focus on optimizing data models, integrating advanced sensor fusion techniques, and adapting the system to accommodate a broader range of vehicles and operating conditions, ensuring more precise and efficient emissions monitoring. Table 8 summarizes the key features of the emissions monitoring models from the reviewed articles.

Section 5.2 examines commercially available devices for monitoring vehicle conditions, including innovative anti-collision devices and intelligent sensors.

As highlighted in Tables 4–8, most of the systems developed for monitoring vehicle conditions, emissions, tires, and driving style are connected to the vehicle's OBD-II that provides a lot of parameters such as speed, engine temperature, engine speed, transmission data, and more. When combined with ML methods, these data enhance the effectiveness of monitoring systems. Additionally, OBD-II enables the detection and diagnosis of mechanical issues by storing error codes that can identify specific faults, allowing for timely interventions and reducing the risk of significant damage. A monitoring system that integrates with OBD-II can also help ensure lower emissions and regulatory compliance. In terms of driving safety, OBD-II-based systems can provide valuable insights, such as detecting dangerous driving behaviors or responding to emergencies. In summary, connecting to the OBD-II system offers access to a rich, standardized data source that can greatly enhance the capabilities and effectiveness of vehicle monitoring systems.

Vehicle monitoring systems face several challenges, concerning technological, operational, economic, and regulatory aspects, including the following:

- Providing high accuracy and reliability by sensors used for a long time, comprising the development of self-diagnosis models to verify their correct functioning;
- Ensuring the interoperability and integration with other vehicle systems, given the heterogeneity of technologies and standards between the different manufacturers;
- The management and processing of large amounts of data to develop predictive models to identify possible failures, the control of emissions, or tire wear;
- To guarantee data protection from cyber-attacks;
- The containment of system costs.

In the future, it will be necessary to develop systems capable of performing global monitoring of vehicle conditions, including the prediction of the failure chain, such as suspension wear due to worn tires. Regarding systems to detect driving style, the future is complex, involving the development of software solutions to adapt the vehicle parameters in real time and automatically to driving style by accessing the vehicle diagnostic system (OBD-II), thus limiting the driver's control.

Table 4. Main features of vehicle diagnostic systems reported in Section 3.1.

Reference	Application	Sensors	Vehicular Parameters	Driving Parameters	Vehicle OBD-II Interaction	Detection Model	Performance
X. G. Yang et al. [18]	Fault diagnosis	Unspecified ([1])	Unspecified ([1])	Unspecified ([1])	Yes	AUTOSAR	Detection rate: 98.70%, Response time: 0.0217 s
J. Kim et al. [56]	Fault diagnosis	6-D IMU sensor	Tire longitudinal and lateral force	Speed, heading	No	CNN	Accuracy: 99.50 %
Kannan et al. [57]	Fault diagnosis in a PdM context	On-board sensors (OBDII)	Engine data, transmission data, control unit	N.A.	Yes	LSTM	Accuracy: 99.62% Sensitivity: 100% Specificity: 100%
Min et al. [58]	Framework on sensor self-diagnosis	Unspecified ([2])	Unspecified ([2])	Unspecified ([2])	No	DSAE	AUC_ROC: 0.8516 F1-score: 0.6701
J.P. Shermila et al. [59]	Automotive black box system	DHT 111, MQ-2, SW-18010p	Engine temperature	Speed, acceleration	Yes	Proprietary algorithm	Accuracy: 98% Recall: 90% Specificity: 85% Precision: 75%

([1]) Ref. [18] studies a diagnostic software without explicitly referring to specific types of sensors; ([2]) the model proposed in [58] is an algorithm for identifying and isolating faulty sensors, and as such, it does not refer to specific types of sensors.

Table 5. Main features of innovative systems for detecting vehicle instability conditions described in Section 3.2.

Reference	Application	Sensors	Detection Model	Performance
A. Fichtinger et al. [60]	Aquaplaning detection	MARWIS ([1])	Threshold-based algorithm	1.5 s detection time
J. Hu et al. [22]	Slippery road conditions detection	Ground speed, wheel speed, ground acceleration, and wheel acceleration	LSTM	SR: 100% (dry road), SR: 99.06% (snowy road), SR: 98.02% (icy road)
H. Lee et al. [62]	Black ice detection	Camera	CNN	Accuracy: 96%
S. Bhadraray et al. [63]	Detection of potholes on the road surface	Turtlebot3 Burger robot with RGB camera	YOLOv4-Tiny	N.A. (*)
G. Raja et al. [64]	Smart pothole avoidance system (SPAS)	Camera	DDPG and HRM-SG	Accuracy: 94.2%

([1]) Mobile Advanced Road Weather Information Sensor; (*) N.A.: Not Available.

Table 6. Main features of systems for driver style and behavior monitoring reported in Section 3.3.

Reference	Application	Sensors	Vehicular Parameters	Driving Parameters	Vehicle OBD-II Interaction	Detection Model	Performance
R. Nouh et al. [66]	Driver risk classification	Speed sensor and accelerometer	-	Speed, acceleration	No	Collaborative Filtering (CF)	MAE: 88.41% MAPE: 90.18% MSE: 95.12%
K. Mohammed et al. [9]	Driving style tracking	GPS, on-board sensors (OBD-II)	Engine speed, coolant temperature, geolocation coordinates	Speed	Yes	-	N.A. (*)
Y. Shichkina et al. [67]	Driving style recognition	Stereo sensor, camera, on-board sensors (OBD-II)	Mass airflow, engine load, intake manifold pressure	Speed, throttle position, acceleration	Yes	Detectron2	N.A. (*)
R. Chhabra et al. [68]	Driver behavior classification	Accelerometer, gyroscope	-	Acceleration and angular speed	No	CNN-LSTM, CNN-Bi-LSTM	CNN-Bi-LSTM: 87%
M. Malik et al. [69]	Driving behavior	On-board sensors (OBD-II)	Engine RPM, engine load, coolant temperature	Speed and throttle position	Yes	ANN	Accuracy: 98.68%
E. Lattanzi et al. [70]	Driving behavior	On-board sensors (OBD-II)	Engine speed, engine load	Speed, brake pedal pressure, throttle position, steering wheel angle	Yes	SVM FFNN	Accuracy: 95%
B.T. Dong et al. [71]	Fatigue detection, distracted behavior	S3FD	-	Facial features	No	RF CNN	Accuracy: 91% for fatigue detection, 97.5% for distracted behavior

(*) N.A.: Not Available.

Table 7. Main features of the tire monitoring models reported in Section 3.4.

Reference	Application	Sensors	Vehicular Parameters	Driving Parameters	Vehicle OBD-II Interaction	Detection Model	Performance
A. Vasantharaj et al. [55]	TPMS	WSS, pressure sensors	Wheel speed	-	Yes	DSAE-ANN	Accuracy: 98.69%
Z. Màrton et al. [21]	iTPMS	WSS	Gear shift, sharp turn detection	Acceleration, deceleration	No	CNN, HWFT-64	Accuracy: 97.32% (HWFT-64)
T. Shan et al. [72]	iTPMS	WSS, on-board hardware	Vertical and circumferential vibrations, wheel speed	-	No	Threshold-based model	N.A. (*)
B. Huo et al. [73]	TPMS	Magnetoelectric WSS	Wheel speed	-	No	Threshold-based model	Accuracy: 79.3% (Radius); 95.6% (Frequency); 94.5% (Composite)
H. Yu et al. [75]	Tire wear detection	Magnetic angular velocity sensor, CAN ([1])	Engine torque, engine speed, yaw rate, four-wheel angular velocity	Acceleration, brake	No	Transformer model	Accuracy: 97.77%

([1]) Controller Area Network (CAN); (*) N.A.: Not Available.

Table 8. Main features of the emissions monitoring system reported in Section 3.5.

Reference	Application	On-Board Sensors	Off-Board Sensors	Vehicular Parameters	Driving Parameters	Environmental and Other Parameters	Vehicle OBD-II Interaction	Detection Model	Performance
N. Anusha et al. [76]	Monitoring and analysis of polluting emissions	Emissions sensors (particulate, NO_x, CO_2), GPS, IoT devices, OBD-II	Lidar, cameras, air quality sensors, weather stations	Vehicle type, engine size, emissions level	Speed	Temperature, weather conditions, traffic density, road type	Yes	NN	N.A. (*)
J. Sao et al. [77]	Emissions prediction of diesel vehicles	PEMS ([1])	Not considered	Engine torque, speed, coolant temperature fuel/air ratio, intake air mass flow	Speed	Not considered	Yes	ANN	N.A.
P. Andrade et al. [78]	CO_2 emissions estimate	Engine Control Unit (EUC) sensors	Not considered	RPM, intake air temperature, mass air flow	Speed	Not considered	Yes	TEDA-based algorithm	Volumetric efficiency: 80%

([1]) Portable Emission Measurement System (PEMS); (*) N.A. Not Available.

4. Environmental, Road, and Traffic Conditions Monitoring

In recent years, Intelligent Transportation Systems (ITSs) have become increasingly significant in the context of urban mobility and traffic management. Central to this innovative approach is the real-time monitoring of traffic flow and environmental conditions. By tracking traffic, authorities can gather vital data on vehicle behavior, congestion, and accidents, enhancing road infrastructure management and ensuring greater safety for road users [79]. Similarly, monitoring environmental factors like air quality and weather conditions is crucial for driver safety and comfort and for promoting sustainable mobility. Together, these aspects support the development of intelligent solutions that streamline traffic, reduce emissions, and elevate the urban quality of life. ITSs, therefore, stand as a key element for future mobility. Figure 15 illustrates how smart cities can enhance traffic management by integrating vehicles, infrastructure, and pedestrians.

Figure 15. Architecture of a smart city highlighting the interconnectivity between vehicles, infrastructures, and pedestrians within an urban center.

Road and traffic conditions have a significant impact on driver and vehicle health, affecting overall safety and performance; complex roads, heavy traffic, delays, roadworks, and poor roads can significantly increase levels of stress and fatigue as well as distraction and nervousness, increasing the risk of accidents. Traffic conditions and poor roads cause faster vehicle wear (e.g., of tires, suspension, and shock absorbers), as well as continuous stop-and-go on congested roads causing thermal stress to the engine and increased brake wear. Difficult road and traffic conditions produce negative effects on the driver (stress, fatigue) and the vehicle (wear, malfunctions), creating high-risk situations for accidents.

4.1. Smart Systems for Traffic Monitoring

In the context of road monitoring, Lilhore et al. designed and implemented an Adaptive Traffic Management (ATM) system utilizing IoT technologies and ML algorithms to tackle issues such as congestion, pollution, and delays in urban transportation [25]. This system consists of three core components (vehicles, infrastructure, and events) and applies various scenarios to address diverse transportation challenges. It employs the DBSCAN clustering method to detect anomalies and dynamically adjusts traffic lights based on traffic volume. Experimental results indicate that ATM surpasses traditional traffic management methods by reducing waiting times, alleviating congestion, and enhancing road safety, thereby contributing to smarter transportation planning in urban environments.

Saleem et al. developed an intelligent traffic congestion control system, known as FITCCS-VN, specifically designed for smart city vehicular networks (VNs) [80]. This system leverages ML techniques to analyze and manage increasing traffic congestion and accidents. Collecting data on traffic flow and route availability facilitates more efficient navigation and reduces congestion. The results show that the model achieves 95% accuracy with a 5% error rate, outperforming previous approaches and offering innovative services to help drivers optimize traffic flow. Traditional ML algorithms require the analysis of sensitive user data, posing privacy risks; Federated Learning (FL) provides a solution by ensuring data security during training. In Ref. [81], the authors introduced a cooperative edge caching scheme for next-generation vehicular networks that combines an elastic FL algorithm for personalized content prediction using an adversarial autoencoder (AAE), and a multiagent deep reinforcement learning (MADRL)-based approach for collaborative caching among SBSs (Small-Cells Base Stations), optimizing the cost and efficiency. Experimental results show improved cache hit ratio, personalized predictions, and reduced operational costs compared to traditional schemes.

Mandal et al. proposed an automated real-time traffic surveillance system employing Deep Convolutional Neural Networks (DCNNs) and a graphical user interface [82]. Their system aims to simplify the tasks of human operators in traffic management centers, enabling quicker and more proactive responses to mitigate accidents and congestion. Using an extensive video database, the model detects traffic jams, tracks stationary vehicles, and counts vehicles by applying pixel-level segmentation and object detection techniques. The results demonstrate that the system performs effectively under various environmental conditions, maintaining reliability even with reduced visibility or adverse weather.

For traffic management in Riyadh, Humayun et al. propose an intelligent traffic management system that integrates IoT, cloud computing, 5G, and big data to enhance real-time traffic efficiency [83]. The system gathers data on road conditions and accidents and then communicates alert messages. This model is structured into three levels. In the first level, the Information Acquisition one, data are collected via vehicle-mounted sensors that capture images and roadside infrastructure that monitors vehicle counts. In the second Network level, information is transmitted over a 5G Wi-Fi network to the cloud for processing. In the third level, the Application Level, dashboards integrate Google Maps and smartphone applications to provide real-time traffic updates, aiding drivers in navigation.

Putra et al. introduce an Intelligent Traffic Monitoring System (ITMS) framework designed to assess drivers' punctuality in reaching their destination, even when unexpected events occur [84]. Their ITMS leverages IoT technology to measure speed using four types of sensors: motion, ultrasonic, passive infrared (PIR), and speed sensors. The system includes four monitoring tools: RFID tags for vehicles, roadside sensors, IP addresses for vehicle connectivity, and QR codes, which allow easy scanning by readers installed throughout the city. Sarrab et al. propose an IoT-based system for real-time traffic data collection and processing [85]. This system updates road congestion and accidents through roadside message units, enabling drivers to adjust their routes accordingly. The proposed model uses Wi-Fi to connect roadside sensors with a cloud server, ensuring timely and accurate information dissemination. Also, the authors proposed an algorithm for vehicle detection and physical length estimation based on magnetic sensors, which detect the disturbances in the Earth's magnetic field caused by moving vehicles. A threshold is applied to identify vehicle presence based on significant magnetic field fluctuations. Two magnetic sensors are placed at a known distance; the vehicle speed (v_i) is calculated by measuring the travel time between two sensors. The vehicle magnetic length (vml_i) is derived as follows:

$$vml_i = v_i \cdot ot_i \tag{1}$$

where v_i is the vehicle speed and ot_i the occupancy time. The vehicle's physical length (vpl_i) is estimated by subtracting the detection zone length (idz) from the magnetic length:

$$vpl_i = vml_i - idz \tag{2}$$

with:

$$idz = v_i \cdot \left(T^A_{dep} - T^A_{arr} \right) - d_{A-B} \tag{3}$$

where T^A_{arr} and T^A_{dep} are the arrival and departure times, whereas d_{A-B} is the distance between the sensor nodes A and B.

Barbosa et al. introduce an innovative system called LightSpaN, a CNN-based model designed for efficient and rapid training [86]. The system's performance is assessed using SUMO software and real-world IoT data, achieving a remarkable accuracy rate of 99.9% in identifying emergency vehicles. Their solution improves vehicle detection speed and significantly reduces waiting and travel times. LightSpaN is trained on images from sources such as the COCO API, VOC datasets, Google Images, and IMAGINET.

Kheder et al. [87] propose enhancements to traffic monitoring systems by integrating vehicular cloud computing (VCC) and IoT-assisted robotics (IoRT). Their architecture incorporates IoT sensor nodes and cameras to collect real-time traffic data. Two deep learning models are implemented: a modified version of LeNet-5 for traffic sign recognition and Inception-V3 for detecting traffic lights. Additionally, the authors optimize ultrasonic sensor positioning to prevent accidents. The processed data are transmitted to the cloud and made accessible via a mobile app for drivers and commuters. Results show the LeNet-5 model achieves 99.12% and 99.78% accuracy, while Inception-V3 reaches 98.6%, surpassing other traffic monitoring systems significantly.

Dhingra et al. present an integrated fog and cloud computing framework to address real-time IoT data analytics challenges in urban traffic monitoring and traffic light management [88]. Since traditional cloud computing can suffer from delays due to data overload and network congestion, the authors propose using fog computing nodes (miniature computers) to pre-process real-time data from distributed sensors. This method reduces latency and enhances system responsiveness, leading to more effective management of traffic congestion and accident detection. Additionally, the system connects to Twitter to send congestion alerts, blending technological advancements with practical tools to improve urban mobility. Despite implementing monitoring systems like SAHER in Saudi Arabia, drivers continue to circumvent penalties and violate traffic laws.

Khan et al. propose an advanced traffic surveillance system leveraging Un-manned Aerial Vehicles (UAVs) and 5G technology to tackle road accidents [89]. The key innovation lies in the combination of UAVs and 5G, offering a more dynamic and responsive solution than traditional traffic monitoring systems. Their results indicate that reducing common traffic violations significantly lowers accident rates, leading to fewer fatalities and injuries. Figure 16 illustrates the UAV system's monitoring capabilities.

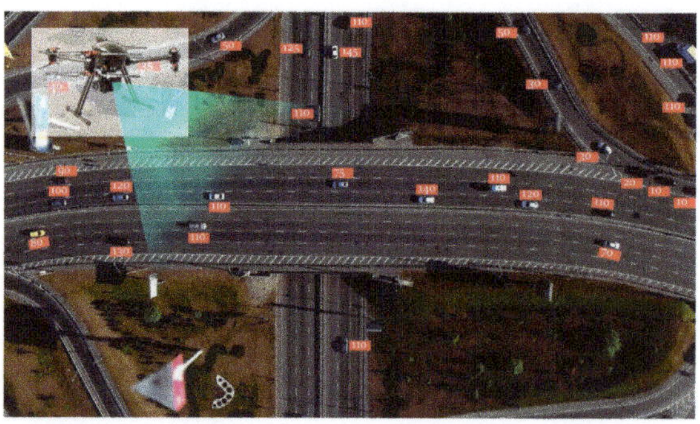

Figure 16. UAV system proposed in [89], able to monitor vehicles' movement by detecting and highlighting the speeds.

Singh et al. [90] introduce a framework that integrates traffic management with air quality prediction for smart cities using crowdsourced data. Their approach involves creating Navigation Reference Spatial Data (NRSD) derived from GPS and OpenStreetMap trajectories, ensuring spatial reliability. These data feed into a real-time traffic density analysis utilizing k-means clustering and Graham scanning techniques. For air quality assessment, the framework employs a Bayesian classifier based on eight standard parameters. An analysis of three major cities in North India demonstrates 98% accuracy, reflecting a 3.8% improvement over previous models. Additionally, the system suggests optimal routes that avoid traffic in 87% of cases. The framework's innovation lies in its integrated use of crowdsourced data and advanced spatial–temporal analysis to enhance real-time traffic management and air quality monitoring.

4.2. Off-Board Systems for Road and Environmental Monitoring

For road condition monitoring, Åstrand et al. presented a prototype system for road condition monitoring in underground mine tunnels in Sweden [91]. Their proposed system aims to improve production efficiency and reduce vehicle wear. The system consists of three main components. The first is a localizer that uses an extended Rao–Blackwellized particle filter that combines vehicle sensor data with Wi-Fi signal strength from access points. The second is road monitoring using two methods, one based on a Kalman filter in synergy with a vehicle suspension system model and another using road condition measurements based on power spectral density. Finally, re-routing is planned using a rescheduling algorithm to make automatic decisions about roads in poor condition, integrating maintenance activities into the short-term production schedule to minimize operational disruption.

Ye et al. in [92] developed an intelligent monitoring system for road pavements using IoT technology. The project addresses several challenges related to using sensors for infrastructure monitoring, such as the limited lifespan of sensors, damage caused by sensor insertion, and managing the large amount of data generated. The proposed system includes a sensor network, cloud platform, communication, and autonomous power supply (Figure 17). Specifically, it focuses on cement concrete pavement and integrates sensors with the material structure to collect data on energy, temperature, humidity, noise, wind, and vibration. These data support pavement design and maintenance and promote synergistic applications between vehicles and roads. The authors plan to continue exploring the use of IoT in road maintenance, safety, and material design.

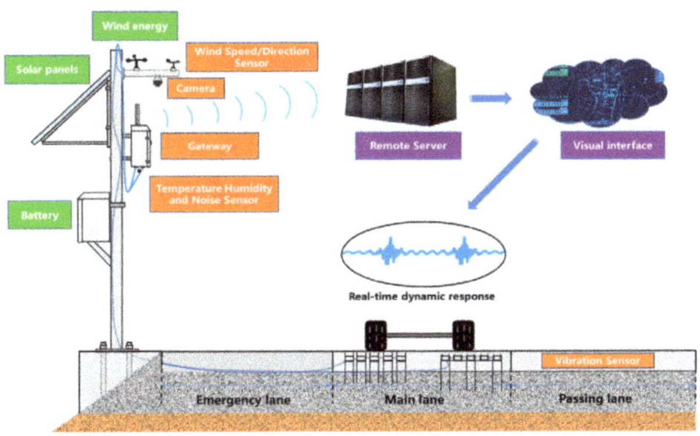

Figure 17. Monitoring system for road pavements using IoT technology proposed in [92].

Abdelmalak et al. [93] proposed an innovative method for adaptive and intelligent pothole detection on asphalt, highlighting the importance of integrated approaches for efficient repair of road inspection systems. The method, which integrates advanced software and hardware solutions, has been evaluated by the Pothole Detection Dataset (PDD) and tested by different computer vision models, including VGG19Net, ResNet-50, GoogLeNet and AlexNet. Among them, the most effective was AlexNet, with a 92.15% accuracy, sensitivity of 91.38%, F1 score of 96.52%, and processing time of 279.35 s.

Ye et al. [94] focused on road surface condition monitoring, presenting an IoT-based architecture to enhance service quality and road infrastructure maintenance. They developed a vibration monitoring prototype including wireless sensor nodes, a gateway, a remote server, and a browser interface, and based on multiple layers: data acquisition, pre-processing, processing, interaction, energy management, and networking. The study highlights critical aspects such as data pre-processing, wireless communication, and visualization. The prototype enables the use of IoT-based solutions in traffic and environmental monitoring systems, also improving road maintenance management.

Gaspar et al. [95] addressed the growing demand for safe forest roads, due to increased tourism, climate change, and forest management needs. They proposed enhancing road infrastructure management through systematic data collection and diagnostic activities by designing a temperature profile measurement probe and a simulation model to validate the probe's effectiveness. This device will be deployed in forest road environments to support ongoing infrastructure monitoring and maintenance.

Hajder et al. [96] developed an automated, maintenance-free system for dosing chemical reagents used in winter road maintenance to combat slipperiness caused by water crystallization, a problem that significantly reduces friction and increases the accident risk. The proposed system collects data during salt spreader operations to determine the optimal reagent density for specific road sections. These data are provided to the thermodynamic models to calculate the ideal reagent concentration and to NNs to detect anomalies. Unlike traditional methods, the goal of the proposed system is to enhance application efficiency, reduce chemical overuse, and minimize the environmental impact.

Chen et al. [97] addressed the challenge of road ice, a major traffic safety concern, by proposing a detection and forecasting system utilizing IoT technology. Their innovation includes a low-power ice sensor that monitors road conditions and transmits data to an IoT gateway via LoRa technology. The system features a distributed algorithm on the gateway that identifies ice formation trends over time, adapting to various road conditions. The algorithm uses road data provided from ice capacitive and temperature sensors, time stamps, and sensors IDs to monitor the icy road's risks. It switches between two monitoring modes: *Commfreq_Monitor_Model* (low-frequency sampling/transmission) for temperature range [+2 °C ÷ +5 °C] and *Highfreq_Monitor_Model* (high-frequency sampling/transmission) for temperatures below +2 °C. The sensors autonomously switch modes and evaluate icing risks by analyzing temperature trends and sensor data. Additionally, the authors developed a deep NN model called Trans-CGAN to provide accurate ice predictions, even with unbalanced datasets, by combining a time series encoder–decoder structure and generative adversarial networks (GAN)-based classification. Time-series data are processed with position encoding and multi-head attention to extract patterns, while GAN components balance the class distribution by generating positive samples. The proposed multi-head self-attention is a collection of general self-attention mechanisms:

$$AttentionOutput = Attention(Q, K, V) \tag{4}$$

where Q, K, and V are the matrices constituted by query, key, and value vectors. This approach improves classification accuracy by leveraging generated and real data to optimize the model performance, thus enhancing the road safety.

Kotus et al. presented a novel method for detecting wet road surfaces using an acoustic vector sensor (AVS) [98]. Their technique analyzes sound intensity in the frequency domain, identifying acoustic events related to vehicle movements. In detail, the following equation was employed to calculate the sound intensity:

$$I(\omega) = \frac{1}{2\rho r\omega} Im(P_1 \cdot P_2^*) \tag{5}$$

where ρ is the air density, r the spacing between pressure sensors, and P_i the Fourier transform of the pressure p_i, whereas Im indicates the imaginary parts of a complex spectrum and asterisks, the complex conjugation.

By calculating the sound direction for different spectral components, the system filters out irrelevant noise and estimates the road surface condition from the sound intensity spectrum. In particular, the algorithm identifies acoustic events, such as vehicle sounds, by detecting increases in sound intensity above background noise, determined by an exponential filter with long averaging time:

$$I_n[k] = \alpha \cdot I_n[k-1] + (1-\alpha) \cdot I_n[k-\delta] \tag{6}$$

where α is the averaging factor, k the sample index, and δ the update delay.

Afterward, the algorithm evaluates the frequency spectrum of these sounds, with higher intensity in specific high-frequency bands (above 2.5 kHz) serving as a clear indicator of wet road conditions. To ensure reliability, the results are smoothed using filters, reducing noise and accounting for variations in traffic and environmental conditions. In real tests, the algorithm demonstrated an accuracy of about 89% in precision, recall, and F1 score. The study concludes that this technology can be integrated into smart city systems for efficient road water detection, contributing to safer and more responsive urban environments.

4.3. Comparative Analysis of Traffic Management and Road Condition Monitoring

The traffic management systems proposed in the analyzed literature have demonstrated excellent performance, utilizing diverse acquisition technologies such as strategically positioned cameras, ultrasonic sensors, RFID tags, and magnetic sensors. A critical feature of any effective traffic management model is a seamless interconnection between vehicles, infrastructure, and users, which helps reduce congestion by providing drivers with real-time information and suggesting alternative routes to avoid traffic. Such real-time updates are particularly beneficial for urban and extra-urban road networks, enhancing overall mobility and driver awareness. An intriguing aspect of the model discussed in [25] is its dynamic traffic light management capability, which improves traffic flow and prevents queue formation. Effective control of traffic lights within urban centers is essential for smooth traffic operations and minimizing delays. Another significant advantage of these systems is their ability to monitor traffic violations (speed limit breaches, dangerous driving, and other infractions) through strategically placed cameras and sensors, fostering greater driver responsibility and road safety by reducing accidents. In this regard, Section 5.3 provides an overview of various commercial hardware and software solutions for traffic control and management. Table 9 summarizes the key common features of the systems proposed in the articles reviewed in Section 4.1.

Monitoring road surface conditions is equally important. Cameras and sensors properly placed can assess road wear and facilitate predictive maintenance, ensuring safer driving conditions and extending the road lifespan. Table 10 compares the proposed systems based on performance metrics, communication technologies, involved entities, and detection sensors, offering a comprehensive evaluation of their capabilities.

Table 9. Main features of traffic management systems reported in Section 4.1.

Reference	Application	IoT Sensors and Technologies	Communication Technology	Entities Involved in Communication	Monitoring Algorithm	Performance
U.K. Lilhore et al. [25]	Traffic management	Cameras, RFID tags, GPS module, WSNs ([1])	IP-based internet 3G/4G, LTE, Wi-Fi, NFC, Bluetooth, ZigBee	Vehicle, roadside infrastructure, and events	DBSCAN clustering method	N.A. (*)
M. Saleem et al. [80]	Traffic management	IoV sensors	Internet (IoV)	Vehicle, roadside infrastructure	FITCCS-VN	Accuracy: 95% Miss rate: 5%
V. Mandal et al. [82]	Traffic monitoring	Cameras	N.A. (*)	Traffic monitoring centers	YOLO	Accuracy: 93.7% F1-score: 0.8333 RMSE: 154.7741 S3: 0.4034
M. Humayun et al. [83]	Traffic management	Cameras, magnetic sensors	5G/Wi-Fi	Vehicle, roadside infrastructure, mobile user.	Threshold-based method	N.A. (*)
A.S. Putra et al. [84]	Traffic management	Motion, ultrasonic, PIR, and speed sensor	Internet network, cellular wireless technology	Vehicle, mobile user	-	N.A. (*)
M. Sarrab et al. [85]	Traffic management	Magnetic sensors, NodeMCU	Wi-Fi	Roadside infrastructure	Threshold-based method	Functionality evaluation: 80%
R. Barbosa et al. [86]	Traffic monitoring and vehicle identification	Image sensor	5G	Roadside infrastructure	Light-SpaN	Performance: 99.9%
M.Q. Kheder et al. [87]	Traffic monitoring systems with VCC	IR, GPS, ultrasonic sensor, cameras	Wi-Fi, Bluetooth, cellular networks	Vehicle, commuters, Google's firebase	LeNet-5 Inception-V3	Performance: 99.7% Performance: 98.6%
S. Dhingra et al. [88]	Urban traffic monitoring and light management	HC-SR04 Ultrasonic Sensor	Wi-Fi, BLE	Roadside infrastructure	Cloud computing	N.A. (*)
N.A. Khan et al. [89]	Traffic surveillance system based on UAVs	Cameras, GPS module	5G	UAV, driver, base station	Threshold-based method	N.A. (*)
S. Singh et al. [90]	Navigation reference spatial data	Open street map, GPS module	Unspecified	Roadside infrastructure	Bayesian classifier	Accuracy: 98% Performance: 87%

([1]) Wireless Sensor Nodes (WSNs); (*) N.A. Not Available.

Table 10. Main features of environmental and road condition monitoring systems reported in Section 4.2.

Reference	Application	IoT Sensors and Technologies	Communication Technology	Entities Involved in Communication	Monitoring Algorithm	Performance
M. Åstrand et al. [91]	Road condition monitoring	Inertial and wheel speed sensors	Wi-Fi	WiFi access points	Kalman filter-based model	N.A. (*)
Z. Ye et al. [92]	Vehicle type, speed, and weight; pavement health status and environment monitoring	Vibration, temperature, humidity, and wind sensors, cameras	5G	Roadside infrastructure (5G gateway), remote server	Threshold-based model	±1 cm localization accuracy, 100 % survival rate
M.E.S. Abdelmalak et al. [93]	Pothole detection on asphalted roads	Cameras, IMU, GPS module	LoRa	Unmanned ground vehicles (AGV) and ground station	AlexNet	Accuracy: 92.15% Sensitivity: 91.38% F1-score: 96.52%
Z. Ye et al. [94]	Road monitoring system	WSNs (acceleration sensing nodes and gateway)	LoraWand, UDP, HTTP communication protocol	Front-end devices and back-end platform	Time features processing	Vehicle speed and whee-lbase error < 2%
G. Gaspar et al. [95]	Road infrastructure management	Temperature sensors	Generic wireless transmission technology	DBAR logger, server, database	N.A.	N.A. (*)
M. Hajder et al. [96]	Road maintenance	Temperature sensors, humidity sensors	3G and 4G/LTE communication	Unspecified	LSTM	Accuracy: 89%
Z. Chen et al. [97]	Road icing detection and prediction	Ice detection sensor, DS18B20 temperature sensor	Wireless communication based on Semtech SX1278	Roadside infrastructure	Position encoding multi-head attention mechanism, GAN-based classification	Precision: 0.743 Recall: 0.811 F1-score: 0.767
J. Kotus et al. [98]	Wet road detection	Acoustic vector sensor	Unspecified	Roadside infrastructure	Threshold-based model, spectral analysis	Accuracy: 89%

(*) N.A. Not Available.

Tables 9 and 10 highlight that not all the articles analyzed consider the critical role of communication between vehicles and infrastructure (V2I), which is essential in the context of smart cities. First, V2I improves traffic management by enabling better coordination of traffic lights and faster adaptation to real-time road conditions; it also allows the reduction of waiting times, alleviation of congestion, and improvement of vehicle energy efficiency, leading to lower CO_2 emissions. Second, V2I enhances road safety by enabling vital information exchange, such as weather alerts or accident reports, allowing drivers to adjust their behavior and avoid potential accidents. Additionally, V2I is key to integrating autonomous vehicles with smart infrastructures into urban environments, enabling route optimization and better interoperability, making cities more accessible, and reducing parking needs. In conclusion, V2I is a technological advancement essential in transforming cities into smarter, more livable places that can address mobility and environmental challenges.

Current technologies for traffic monitoring and management, although advanced, have some limitations that hinder their effectiveness, concerning technological aspects related to data accuracy, limited coverage on the road network, and fragmented data caused by discontinuous communication between different monitoring platforms. The growing amount of data coming from sensors, connected vehicles, and mobile devices requires high computing power and sophisticated algorithms to support real-time analysis; furthermore, the poor quality of data (noisy or incomplete data) can lead to errors and ineffective decision management. Moreover, connected monitoring systems can be vulnerable to cyber-attacks, making data privacy an ongoing challenge. The enhancement of infrastructures with advanced sensors (better performance and resistance to environmental conditions), the big data techniques, predictive analytics, and the adoption of digital twin models are emerging technologies that offer new opportunities for traffic monitoring and management. A further challenge is that non-centralized and distributed systems may not be able to determine alternative routes based on information about the level of traffic.

5. Cutting-Edge Commercial Solutions for Monitoring Vehicle, Driver, and Traffic Conditions

This section describes the most recent commercial hardware and software solutions in the field of advanced driver and vehicle monitoring systems, as well as out-of-vehicle solutions for monitoring traffic, environmental, and road conditions.

5.1. Overview of Commercial Solutions for Driver Condition Monitoring

Many companies are focusing on designing a DMS based on cameras or sensors to monitor driver behavior and alertness. By processing in-cockpit data in real time through advanced algorithms, the DMS detects signs of distraction or drowsiness (main causes of accidents), preserving driver privacy by processing data locally. In May 2024, Smart Eye Co. (Göteborg, Sweden) introduced the Smart Eye Pro 12 eye-tracking system, featuring Profile ID and advanced drowsiness detection capabilities to enhance safety and the user experience (Figure 18) [99]. The Profile ID function assigns each user a unique identifier based on facial features, removing the need for manual identification. The drowsiness detection technology uses algorithms to assess it on a scale from 4 to 9, in line with upcoming 2026 regulations mandating such features for vehicle safety. The Smart Eye Pro 12 release also includes gaze accuracy and pupil tracking for heightened robustness and reliability. The system features a multi-camera setup, with two to eight cameras to be placed horizontally or vertically, so enabling 360-degree head and eye tracking. AI-based emotion software developed by Affectiva Co. (Boston, MA, USA) allows the Smart Eye's DMS to capture a variety of features in real time [99]:

1. Driver identification: it recognizes the driver to adjust the vehicle's settings accordingly and ensures that only authorized drivers can use the car.
2. Distraction, drowsiness, and attention: it monitors the driver's eye, head, and facial movements to detect distraction and catch early signs of fatigue, ensuring attention remains on the road.
3. Dangerous behavior: it identifies distracting behaviors such as eating, drinking, smoking, or using a mobile phone.
4. Object detection: it detects objects within the vehicle and monitors how the driver interacts with them.
5. Activity detection: it monitors the driver's activities to determine if they are focused on driving or engaged in other tasks.
6. Body posture: it monitors the driver's posture, movements, and interactions with objects or vehicle interfaces.
7. Facial expression analysis: Affectiva's Emotion AI interprets the driver's facial expressions to assess his/her mood, emotions, and behaviors.
8. Health status: it assesses the driver's health status by analyzing body posture along with eye, head, and facial movements.

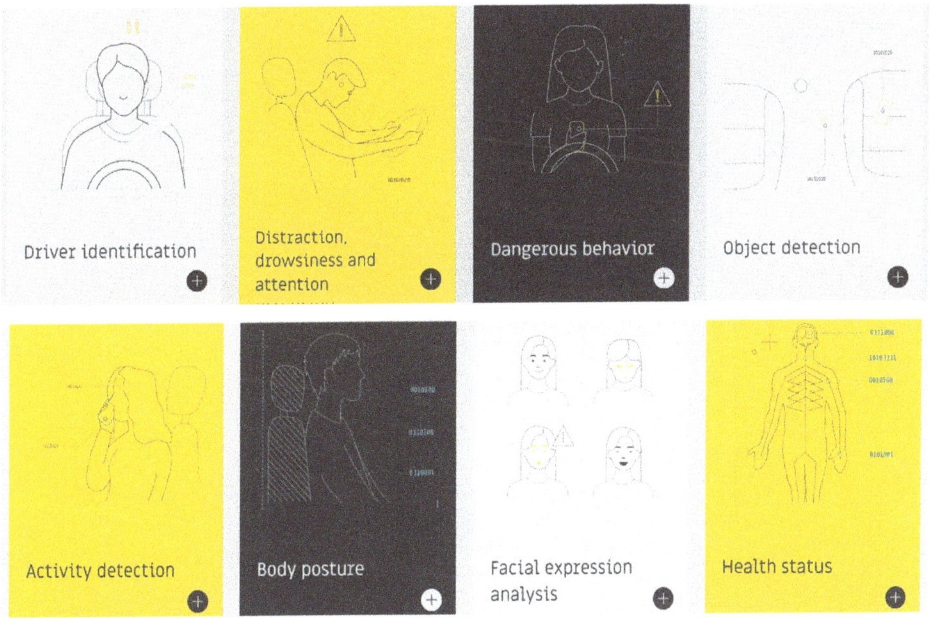

Figure 18. Key features of the "Smart Eye Pro 12" DMS released in 2024 by Smart Eye company [99].

Recently the Continental Engineering Co. (Frankfurt, Germany) proposed the "Cabin Sensing Solution" with technologies and detection capabilities, including a DMS, radar, cockpit monitoring system, and thermal camera (Figure 19a,b,d) [100]. The DMS detects the driver's drowsiness and distraction, and supports facial recognition, emotion detection, and advanced controls (Figure 19c). Its main component is the DMS software algorithm, which operates on dedicated hardware by a CAN-FD interface or is integrated into a third-party High-Performance Computer (HPC). A 1MP camera with LED illumination is available as an encapsulated unit or integrated within the instrument cluster [100].

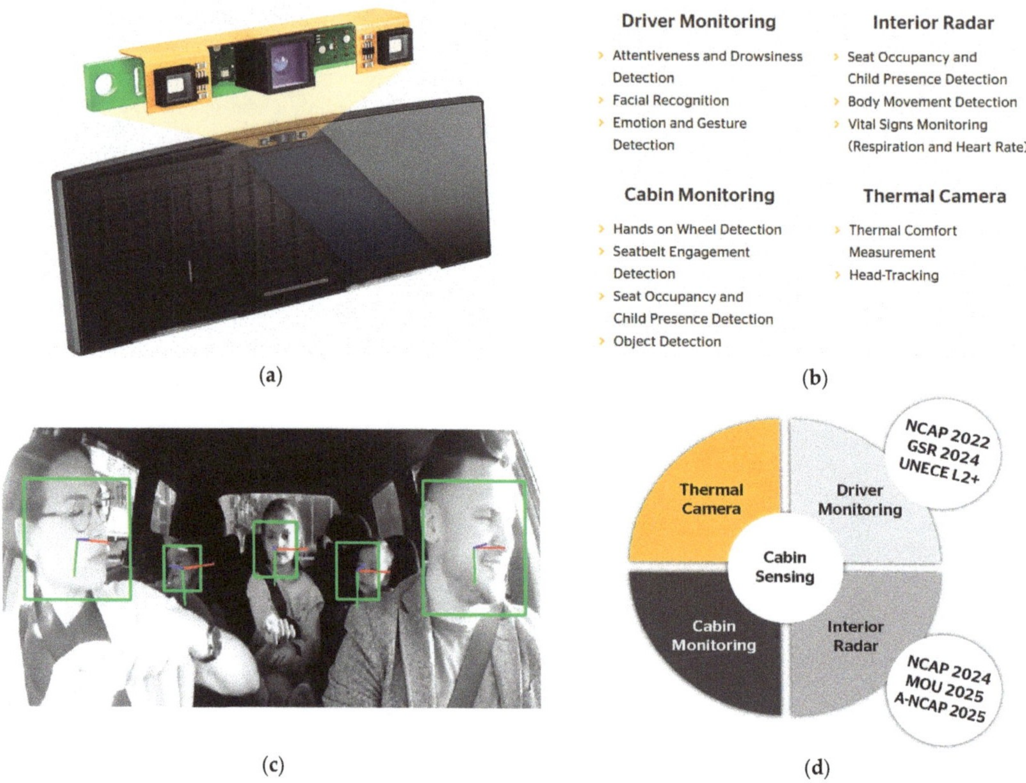

Figure 19. (**a**) Interior camera system enabling driver and vehicle to interact seamlessly; (**b**) developed technologies and detection capabilities from Continental Engineering; (**c**) hardware and software solutions for face gesture detection; (**d**) key features of "Cabin Sensing" solution and standards met [100].

The interior radar enables the child presence detection (CPD) in compliance with the EuroNCAP 2024 standard (more details at the website https://cdn.euroncap.com/en/ (accessed on 16 December 2024)). In addition, it tracks body movements and vital signs, including RR and HR, to detect seat occupancy and any children [100]. The monitoring system covers the driver, all passengers, and the full cabin, detecting the presence of passengers and objects, seat occupancy and seatbelt use, and assessing the occupants' posture and status for Level 4 and 5 vehicles. For example, it checks if the driver is focused and their hands are on the wheel. Also, the system includes driver identification (Driver ID) for personalization and support for video conferencing [100]. The thermal camera measures the temperature of the head, forehead, and body, calculating a Thermal Comfort Value (TCV) for both driver and passengers. This numeric scale reflects their thermal sensation, from very cold to very hot, and can be valuable input for optimizing climate control [100].

Magna International Inc. (Aurora, ON, Canada) recently launched a Driver and Occupant Monitoring System (DOMS) that uses ADAS cameras and interior mirrors to detect driver distraction, drowsiness, and fatigue, aiming to reduce driving distractions and thus crash risks (Figure 20). The Magna DOMS, integrated into the rearview mirror, analyses the driver's head, eye, and body movements to detect signs of distraction or fatigue. By using cameras and infrared sensors, it can detect the loss of concentration of the driver, warning him by visual or audible alerts [101]. In particular, the proposed technology is able to distinguish between routine driving actions, such as checking the side mirrors, and real distraction actions; once the system recognizes them, it alerts the driver with acoustic or visual signals. A camera in the interior rearview mirror allows monitoring of the driver but also other occupants. This system can include features like child presence detection, seatbelt use verification, and passenger identification, enabling personalized settings based on user preferences [101]. Infrared ethanol sensors positioned near the driver monitor the alcohol and CO_2 levels in the driver's diluted exhalations to determine any BAC level above the legal limit. Magna meets the Euro NCAP and General Safety Regulation requirements [101].

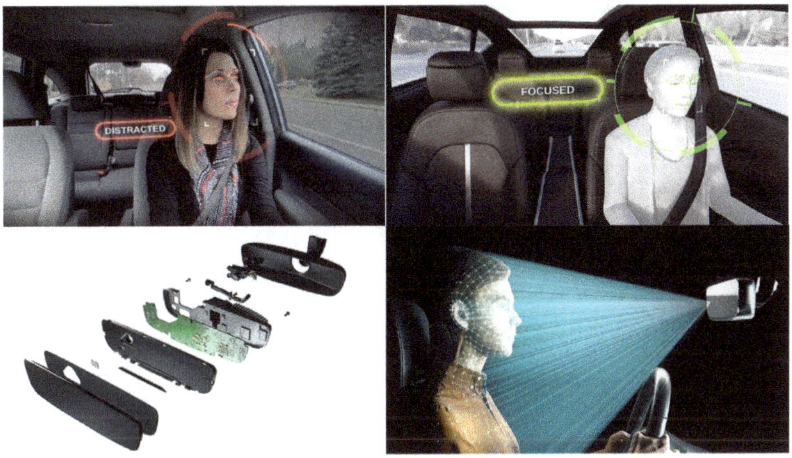

Figure 20. Key features of Magna DMS: distracted driver detection, drowsy detection, occupant detection, child presence/seat detection, occupant classification, properly worn seatbelt detection.

The last system, proposed by Speedir Inc. (Santee, CA, USA), is an AI dash cam with eye-tracking functionality to detect drowsy driving. It currently operates as a standalone tool and does not integrate with typical fleet management platforms. The Driver Fatigue Monitoring System (DFMS) features a solution for preventing drowsy driving [102]; equipped with a night vision camera and AI facial recognition algorithms, this system tracks signs of drowsiness and distraction using infrared sensors to monitor the eye positioning (blink rate, closed eyes, eye direction), head movements (e.g., nodding off), and yawning, even when the driver is in low light or wearing glasses. Speedir's DMS is powered by pre-trained AI software, requiring no Wi-Fi for operation or updates (Figure 21). This system enhances driver safety by alerting users to fatigue or distraction. The key features of the system are as follows:

- Utilizing AI algorithms, Speedir's DMS detects drowsy or distracted driving by tracking eye and head movements, issuing audible alerts for cell phone use, falling asleep, or other distractions.
- With infrared technology, the system performs day or night, unaffected by reflective light or any glasses the driver may wear.
- The DMS is designed for universal plug-and-play USB installation, fitting seamlessly in any vehicle.
- Privacy is protected as it operates without internet or cloud connectivity.

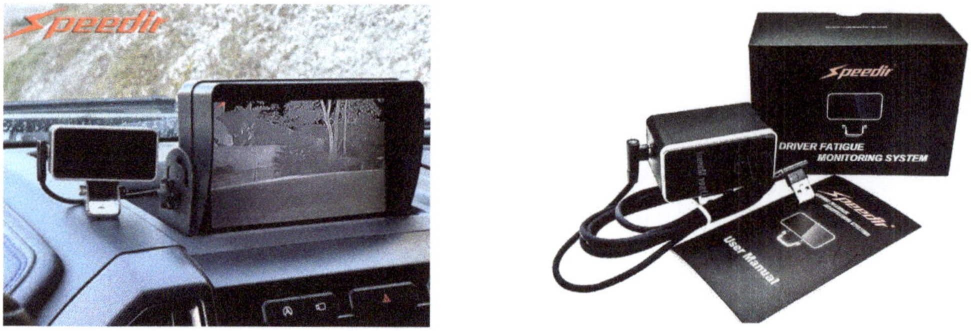

Figure 21. Driver Fatigue Monitoring System for driver's safety by Speedir [102].

Table 11 compares the commercial DMSs previously analyzed, considering their detection capabilities, integrated sensing devices, monitored human body and car parts, and developed detection algorithms, offering a comprehensive evaluation of potentialities.

Table 11. Comparison of commercial DMSs illustrated in Section 5.1.

Commercial DMS	Detection Capabilities	Sensing Devices	Body and Vehicle Parts Monitored	Installation Area	Detection Algorithm
Smart Eye Pro 12 by SmartEye [99]	Driver identification, distraction, drowsiness, dangerous behaviors, body posture, facial expressions	1 MP cameras	Body posture, head and eye position, eye gaze and blinking, pupil diameter	Front of the cockpit	AI-based emotion
Cabin Sensing Solution by Continental Engineering [100]	Facial expressions and recognition, driver distraction, drowsiness, head and forehead temperature, RR, HR	1 MP camera, radar, thermal camera	Head and eye position, seatbelt, seat occupancy	Front of the cockpit	Proprietary software
Driver and Occupant Monitoring System by Magna International Inc. [101]	Driver identification, distraction, drowsiness, fatigue, facial expressions, ethanol environmental concentration	1.7 MP camera, IR movement and ethanol sensors	Head and eye position, seatbelt, seat occupancy	Rear-view mirror and steering block	Proprietary software
Driver Fatigue Monitoring System by Speedir Inc. [102]	Driver distraction, drowsiness, facial expressions	Night vision camera, IR sensors	Head and eye position, eye gaze and blinking, pupil diameter	Front of the cockpit	Pre-trained AI software

5.2. Overview of Commercial Solutions for Monitoring Vehicle Conditions

An In-Vehicle Monitoring System (IVMS) consists of one or more electronic devices and purpose-designed software installed in a vehicle to capture and record information about vehicle performance and driver behavior. Often referred to as the "black box" of the automotive world, an IVMS typically logs live data such as speed, acceleration, braking patterns, seatbelt usage, and more. These data provide a picture of activities across a single vehicle or fleet of vehicles. Moreover, a robust IVMS evaluates driver and vehicle performance against pre-defined safety standards and criteria. IVMSs range in complexity: some connect via the vehicle's OBD-II port for easy plug-and-play functionality, while others require installation and enable real-time tracking and data downloads. IVMSs offer a wide range of advantages that enhance the safety, efficiency of vehicles, and cost management in the case of the fleet of vehicles:

- Enhanced driver safety: improve safety by providing feedback on unsafe driving behaviors.
- Improved driving behavior: discourage unsafe actions like speeding, harsh acceleration, or abrupt braking.
- Crash detection and response: this allows identification and a quick reaction to accidents, ensuring timely assistance to distressed drivers.
- Vehicle performance tracking: monitor fuel consumption, mileage, engine health, and other crucial factors to identify potential issues before they become costly problems, extending the lifespan of vehicles.
- Compliance with health and safety standards: help employers meet legal obligations by monitoring driver hours and addressing fatigue, demonstrating a commitment to safety and duty of care.
- Fewer incidents and crashes: reduce accidents by raising awareness of risk factors and improving driving habits.
- Lower insurance premium: offer detailed insights into incidents, enabling efficient claims handling and accurate allocation of responsibility (many insurers provide premium discounts for fleets using IVMS).
- Enhanced load and vehicle security: help protect cargo and vehicles from theft or tampering by tracking their locations in real time.
- Reduced fuel costs and environmental impact: lower fuel expenses and maintenance costs by providing insights into fuel consumption, promoting efficient routing, and reducing idling.
- Live GPS tracking and geo-fencing: GPS tracking provides an overview of vehicle locations, and, at the same time, geo-fencing enables the setting of boundaries for vehicle usage, such as speed limits in specific areas or restricting access to certain zones.

Many companies are focusing on developing new IVMSs based on radar, cameras, GPS, or multi-sensor setups. In the following section, some cutting-edge commercial solutions are reported. Trakm8 is a technology company (Birmingham, UK) focused on automotive telematics solutions and connected vehicles, including fleet cameras, multi-camera systems, dash cams, and related software. The 4G RH600 camera offers telematics capabilities, including driver behavior scoring, vehicle health alerts, driver ID, and upcoming ADAS features to enhance road safety and reduce risks (Figure 22a). Connectivity technical specifications are Bluetooth Low Energy (BLE 4.2), GPS, 4G technology, vehicle CAN 1, and Tacho CAN 2 interfaces [103]. The RH600 camera (manufactured by Trakm8 Holdings PLC, Birmingham, UK) can be used as a traditional dash cam in two different ways, with a single lens facing the road/driver or a dual lens facing both the road and driver. The RH600 features the Trakm8's ConnectedCare vehicle health software solution, designed to reduce breakdowns and non-start risks, manage service and repair costs, and enhance residual vehicle value. Trakm8 Insight is a software platform that processes data from the RH600 camera and displays it via a desktop portal or mobile app. The RH600 camera will implement ADAS features, including driver distraction and fatigue detection, forward collision warnings, and following distance warnings.

Trakm8's technology also integrates the RoadHawk DC-4 forward-facing dash cam (Figure 22b), equipped with Wi-Fi, GPS, and a 1440p Quad HD (2K) camera (30 fps frame rate), which can capture road signs and license plate numbers, so allowing drivers to save photos and videos via a gesture sensor. Moreover, it also features a built-in G-Sensor that automatically allows drivers to save road accident video for later review [104]. The systems designed by Trakm8 also include the RH800 4G Mobile Digital Video Recorder (MDVR), which supports up to four cameras and features telematics functionality, including GPS tracking, driver behavior, and vehicle health monitoring (Figure 22c) [105]. The Trakm8 Insight telematics system can measure speed, over-revving, and harsh acceleration and braking. Trakm8 Insight telematic utilizes GPS, fuel consumption, driving trends, and dash cam data to offer information supporting management processes.

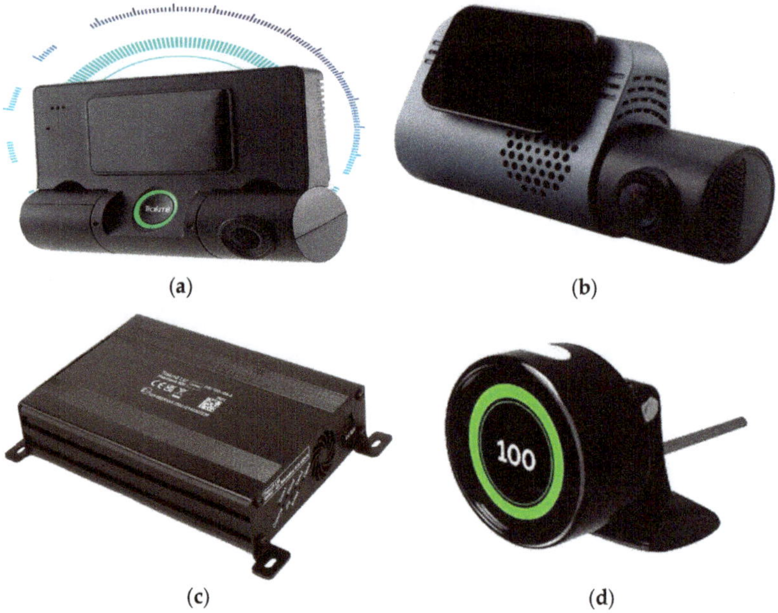

Figure 22. Solution for vehicle condition monitoring by Trakm8: (**a**) 4G Integrated Telematics Camera RH600 [103]; (**b**) RoadHawk DC-4 Dash Cam [104], (**c**) RH800 4G Mobile Digital Video Recorder [105]; (**d**) ACC750 Driver ID and Feedback device [106].

Trakm8's ConnectedCare vehicle health solution provides diagnostic data by accessing the CANbus network of cars and commercial vehicles. This system identifies Diagnostic Trouble Codes (DTCs) or engine fault codes characteristic of specific vehicle issues and displays them on the Insight software platform. The Insight platform allows remote monitoring of different vehicle warning lights, such as washer fluid level, ABS, traction control, Diesel Particulate Filter (DPF), AdBlue levels, and tire pressure, and it alerts vehicle owners and fleet managers to potential issues early for proactive maintenance.

Trakm8's Driver ID authentication software allows drivers to log into their assigned vehicles using standard door entry cards, employee ID cards, or NFC-enabled mobile devices. Drivers can utilize the ACC750 Driver ID, a combined driver identification and feedback solution connected to different Trakm8 telematics solutions via Bluetooth or a wired connection (Figure 22d) [106]. The ACC750 features a multi-function button that allows drivers to switch between business and private travel modes. Holding the button for three seconds triggers an alert sent to the paired telematics device, which then relays the message to the server for monitoring through a web portal or mobile app.

Another leading IVMS available on the market is the TrackoBit GPS tracking software proposed recently by InsightGeeks Solutions Pvt. Ltd (Noida, India), a mobility technology company [107]. TrackoBit software is compatible with a lot of GPS hardware and IoT devices, making it compatible with various manufacturers; its key features are the driver behavior management and monitoring, ADAS interface, video telematics with geospatial tracking solutions, and route planning and management. Driver behavior management and monitoring is based on a GPS tracking system and various sensors, providing data on driver performance, distinguishing between high- and under-performing drivers, and verifying if they frequently deviate from the assigned route. This software optimizes route planning to maximize operational efficiency and collects, displays, and analyzes data related to sharp cornering, harsh acceleration and braking, and over-speeding. The TrackoBit ADAS provides views of the vehicle from all angles and is notified whenever the system recognizes any anomaly or dangerous situation (Figure 23).

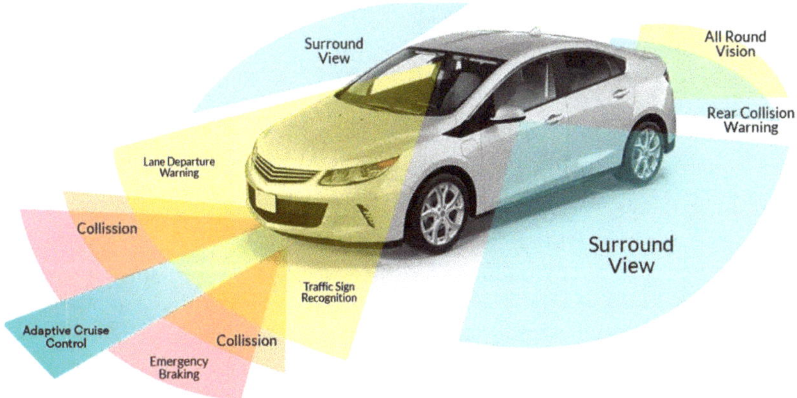

Figure 23. ADAS functionalities by TrackoBit devices: forward, rear, or side collision alerts, signal violation, lane switch alert, and over-speeding alert.

An AI-powered DMS and ADAS constitute the two pillars of Video Telematics by TrackoBit system, which is integrated into vehicles via the OBD ports (Figure 24). Route Planning and Management Software by TrackoBit offers trip planning and management capabilities based on three steps (Figure 25). The first step, create tour, assists in designing a personalized tour by entering the desired delivery location, setting the total trip time, and arranging the designated stops. The second step, monitor route, assists in gaining real-time visibility into vehicle locations and movements, monitoring their progress along specified routes, and saving optimal paths for repeated journeys. The third step, manage trips, assists in utilizing live tracking to oversee trips while vehicles are in transit.

Figure 24. Video Telematics by TrackoBit system: DMS alert example (**a**), ADAS alert example (**b**).

Figure 25. Route planning and management software by TrackoBit: (**a**) create tour, (**b**) monitor route, (**c**) manage trips.

Another interesting IVMS available on the market is the recently proposed technological system by the automotive group FORVIA (Nanterre, France), an automotive group that develops technologies to be installed inside or outside vehicles to support the decision making of drivers and automated systems. The IVMS includes three technologies: by-wire technology, radars and environmental sensors, and a vision system; the first includes a brake pedal sensor (BPS) (Figure 26a), simulating the feeling of a conventional brake system while electronically transmitting driver input. The BPS transfers the driver's action to the brake control unit, which provides a signal to the braking system [108]. The by-wire technology also proposes a steering torque sensor; as the driver turns the steering wheel, the sensor measures the torque applied to the steering column. The sensor detects the torsion bar angle and provides information to the control module, which calculates the necessary servo assistance to support the driver's action [109,110].

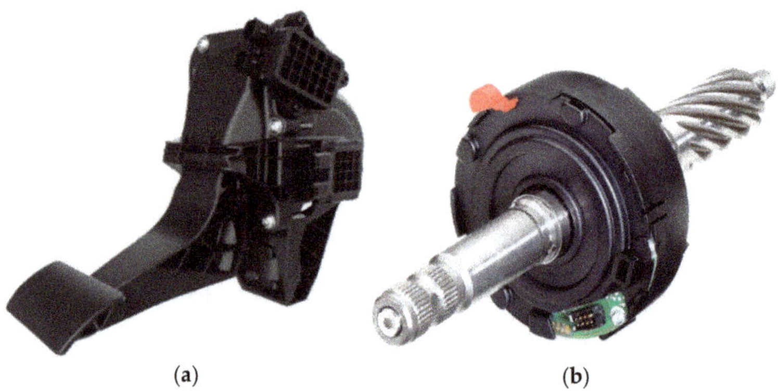

Figure 26. By-wire technology by FORVIA: brake-by-wire system (**a**) and steering torque sensor (**b**).

Radars and environment sensors by FORVIA include a 77GHZ radar, an eMirrors camera, a road condition sensor, a rain-light sensor, and a parking system (manufactured by FORVIA company, Nanterre, France) (Figure 27). The radar technology supports ADASs, enhancing environmental detection by identifying stationary elements like road boundaries and tracking dynamic objects such as pedestrians, cyclists, and vehicles [111]. The eMirrors camera replaces traditional external mirrors, providing a surrounding view; the lightweight and aerodynamic design reduces CO_2 emissions and improves energy efficiency. The road condition sensor detects water on the road in real time, estimating the grip by vibration measurement through a piezoelectric sensor [112]. The rain–light sensor is a multi-function device with the following capabilities: rain detection for automatic wiper control, light sensing for automatic headlight activation and tunnel or overpass detection, sun load measurement, head-up display brightness adjustment, and humidity/temperature sensing to prevent windshield fogging [113].

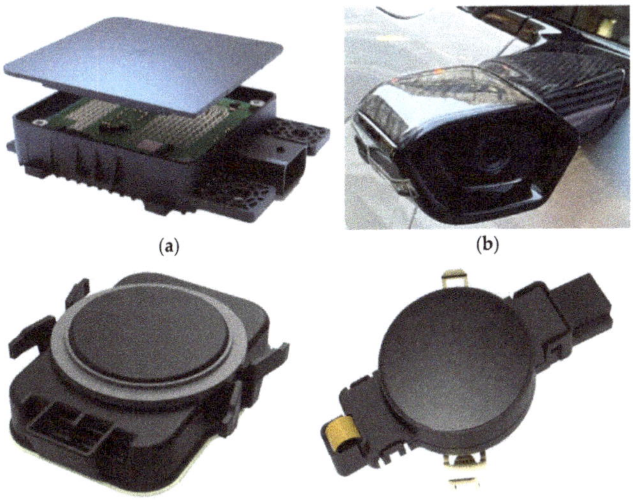

(a) (b) (c) (d)

Figure 27. Radars and environment sensors by FORVIA: 77GHz radar (**a**), e-Mirror camera (**b**), SHAKE road condition sensor (**c**), HELLA Rain Light Sensor (**d**).

Vision systems by FORVIA include an eMirror UX Safe engine and a surround view system [114]. The eMirror UX Safe engine is a software that enhances mobility safety by using a camera-based system to improve drivers' visibility and situational awareness (Figure 28a). It provides safety alerts, optimizes energy efficiency, and minimizes distractions, ensuring clearer views even in challenging environments and lighting conditions. The Vision system's surround view enables a 360° panoramic perspective of the car using a four-camera fisheye setup (Figure 28b). The surround view system has been conceived to enhance safety, have better visibility, and effortlessly maneuver. Combining data from these cameras recreates the entire environment and offers features such as virtual transparency with gaze detection.

(a) (b)

Figure 28. Vision systems by FORVIA: eMirror UX Safe engine (**a**), surround view system (**b**).

The last IVMS described in this section, although not yet on the market, is particularly interesting for understanding future trends; it is a system developed by Social Self

Driving S.r.L (Como, Italy) that captures and replicates a driver's unique driving style in autonomous vehicles. This technology allows driverless cars to mimic the behavior of their owners or professional drivers, selecting from a range of pre-recorded profiles [115]. The system utilizes hardware and software components commonly found in modern vehicles with assisted or autonomous driving capabilities. It integrates various sensors, including those that measure steering angles, torque, and speed, as well as sensors for accelerator and brake pedal actions, yaw, pitch, roll, and both lateral and longitudinal acceleration. From these data, the system profiles a driver's style, which can be set or shared through a dedicated cloud platform, allowing it to be programmed into autonomous or semi-autonomous vehicles (Figure 29). The "Social Self Driving" app allows users to access profiles made by others, creating a virtual marketplace where car manufacturers, professional drivers, driving instructors, and public figures can offer and sell customized driving programs [115].

(a)

(b)

Figure 29. (**a**) The vehicle learns the driving style from the driver, to be integrated with preloaded programs and duplicated when same conditions occur. (**b**) Sharing data with other users via web platforms.

Table 12 compares the IVMSs analyzed in Section 5.2, considering the applications, integrated electronic devices and sensors, monitored vehicular parameters, available connectivity technologies and interfaces, and developed software to offer a comprehensive evaluation of their capabilities.

Table 12. Comparison of commercial IVMSs illustrated in Section 5.2.

Commercial IVMS	Applications	Cameras and Other Devices	Sensors	Monitored Vehicle Parameters	Connectivity Technologies and Interfaces	Software
Trakm8 system [104–106]	Driver ID and behavior, vehicle health, ADAS functionalities	4G camera, dash cam, 4G mobile digital video recorder, driver ID and feedback device	G-Sensor and radar	Washer fluid level, ABS, traction control, DPF, AdBlue levels, tire pressure	BLE 4.2, GPS, Wi-Fi, 4G technology, CAN 1 and Tacho CAN 2 interfaces	ConnectedCare, insight telematics, driver ID authentication
TrackoBit by InsightGeeks Solutions Pvt. Ltd. [107]	Driver behavior, ADAS functionalities, geospatial tracking, route planning and management	AI-powered DMS and camera	Radar	-	GPS, 4G technology	GPS tracking software, video telematics, route planning and management software
IVMS by Forvia [108–114]	By-wire technology and ADAS functionalities	eMirrors and surround-view cameras	Brake and steering torque sensors, road condition piezoelectric sensor, rain light and parking sensors, 77 GHz radar	-	OBD port, GPS, 4G technology	eMirror UX Safe engine

5.3. Hardware and Software Commercial Solutions for Environmental and Traffic Monitoring

Traffic monitoring and management services can significantly enhance urban mobility by reducing accidents and emissions. These services, based on seamless communication between infrastructure, vehicles, and drivers, suggest alternative routes (if necessary) via road alerts or real-time in-car messages. Various companies have introduced specialized hardware and software solutions for traffic monitoring and driver assistance.

ClearView Intelligence (Milton Keynes, UK), a company specialized in mobility solutions and data analytics, provides hardware and software for traffic monitoring, mobility management, and predictive analytics. The AI-driven Insight Count and Classify software monitors road traffic, counting and categorizing vehicles at key points, such as intersections or highway segments [116]. Figure 30 illustrates its user interface, representing a solution for enhancing travel safety and efficiency. The operating principle of Insight Count and Classify software is based on some key features:

- Data collection: the software uses devices (such as cameras or sensors) in different environmental conditions to collect information about passing vehicles and traffic.
- Visual analysis: the software analyzes images of vehicles to identify their characteristics, such as type (car, truck, motorcycle, etc.), size, and speed.
- Data processing: the data are processed through algorithms that allow vehicles to be counted and classified in real time.
- Reporting and visualization: the analysis results are presented in reports and dashboards, providing information on traffic patterns, congestion peaks, and other metrics useful for traffic planning and management.

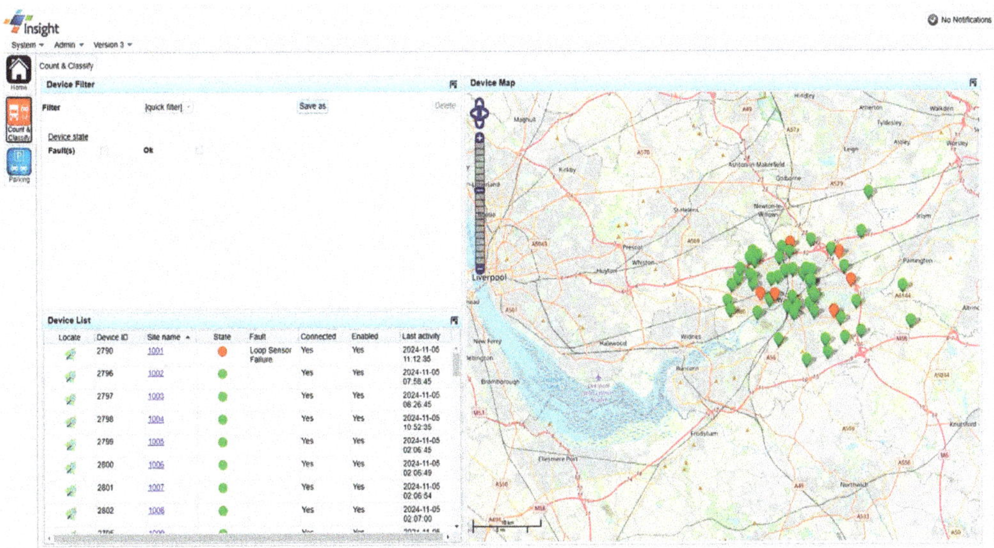

Figure 30. Insight Count and Classify software interface developed by Clearview Intelligence [116].

The Insight Count and Classify software targets stakeholders, including transport and urban authorities, infrastructure project developers and mobility service providers, to deal with traffic management, develop road infrastructure, and optimize vehicular services. Clearview Intelligence also offers hardware solutions for traffic management. Connex Traffic is a technology designed to tackle traffic management and road safety through data utilization and analytics [117]. It employs a network of connected sensors and devices to collect real-time traffic data (including vehicle flows, speeds, and road conditions) from cameras, induction sensors, and GPS devices (Figure 31). These data are processed using algorithms and AI techniques to identify traffic patterns, detect congestion, and assess driving behavior. The analyzed information helps optimize traffic signals, improve road signage, and develop strategies to reduce congestion. Real-time updates can also be provided to drivers via apps or information panels.

Figure 31. Connex Traffic device is installed on the road: a typical scenario [117].

The M100 system by Clearview Intelligence is designed to be concerned with the management and efficiency of transportation and urban mobility infrastructure (Figure 32). It includes sensors, cameras, radar, and other IoT devices to gather real-time data on traffic flow, road user behavior, and environmental conditions [118]. These data are processed using AI and ML algorithms to detect patterns, identify accidents, and monitor congestion. By combining historical and real-time data, the system can predict traffic flow and suggest strategies to optimize traffic and minimize waiting times. The M100 system can also be integrated with other infrastructure and traffic management systems, facilitating communication and coordination across different entities.

TomTom (Amsterdam, the Netherlands), a company specialized in navigation software and applications, is known for its GPS navigation products and mapping solutions. One of its latest solutions is TomTom Orbis Maps [119]. This mapping software delivers maps and navigation data customized for various devices and applications. Orbis Maps (Figure 33) integrates driver data with information from sources like Overture Maps Foundation™ and OpenStreetMap, ensuring data acquisition at multiple levels. The software combines geographic data from different sources, including satellite imagery, road surveys, and local insights, to produce accurate and detailed maps. By combining these mapping data with GPS technology, TomTom aims to provide efficient and reliable navigation for users.

TomTom offers map updates that incorporate road changes, new constructions, and real-time traffic data. Applications built on Orbis Maps feature an intuitive user interface, making it easy to search for destinations, plan routes, and access information about live traffic conditions and points of interest. Beyond standard road navigation, Orbis Maps offers functionalities for various needs, including commercial vehicle route planning, cycling, and pedestrian navigation, with data fitted to the user. Some TomTom services are cloud-based, enabling access to maps and data across multiple devices. TomTom Orbis Maps provides geolocation tools for companies involved in fleet management, delivery route optimization, and other location-dependent operations.

(b)

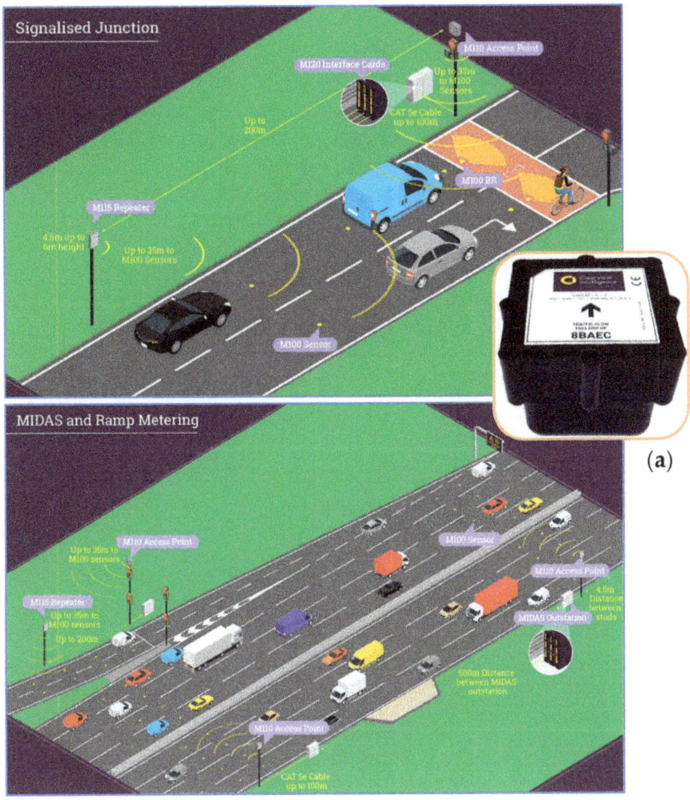

(a)

(c)

Figure 32. M100 system developed by Clearview Intelligence (**a**), application scenarios (**b**,**c**) [118].

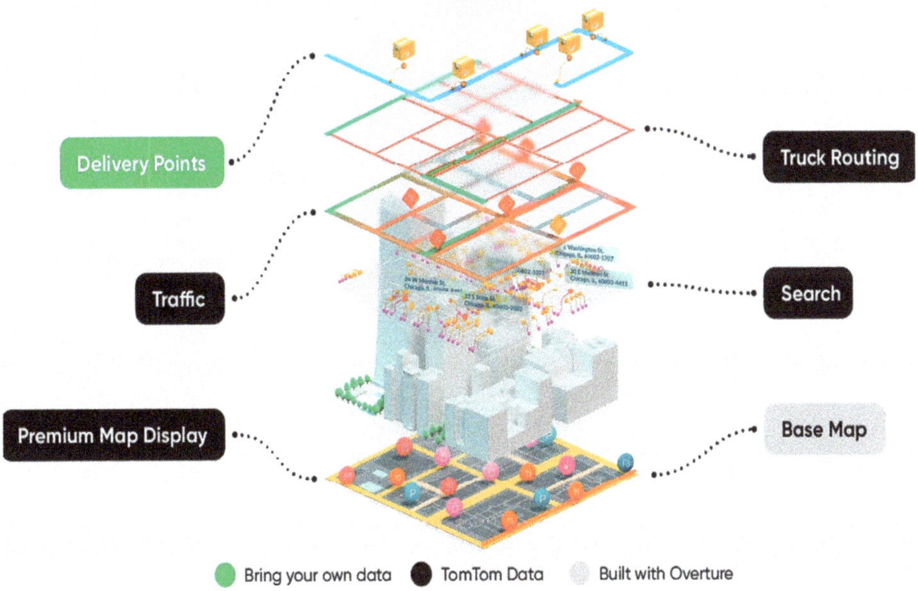

Figure 33. Different layers of TomTom's Orbi Maps software.

The main differences that emerge between the Insight Count and Classify software and Orbis Maps are the following:

- Intended use: the Insight Count and Classify software focuses on data collection and analysis for traffic monitoring and management, whereas TomTom Orbis Maps provides detailed digital maps for a variety of applications, including navigation and vehicle automation;
- Technology: Insight Count and Classify uses physical sensors to collect data in the field, whilst Orbis Maps combines data from various sources, including open-source information and vehicle observations to create up-to-date digital maps.
- Target users: the Insight Count and Classify software is faced to road authorities and city planners who need accurate traffic data, whereas Orbis Maps serves a broader range of industries, including automotive and logistics, that require advanced digital maps for a variety of applications.

In summary, both software operate in the traffic management and mapping domain; Insight Count and Classify is specialized in traffic data collection and analysis, whereas Orbis Maps is a flexible and detailed digital mapping platform for industrial applications.

6. Challenges and Future Perspectives

The growing number of vehicles on the road have led to a significant increase in accidents and frequent traffic congestion, often bringing cities to a standstill. This challenge has driven research toward developing innovative solutions to improve road traffic through driver assistance architectures. The goal is to reduce accidents by creating advanced driver and vehicle monitoring systems while developing a communication network between vehicles and infrastructure. Moreover, these innovative systems aim to optimize traffic flow by offering drivers alternative routes to their destinations. Concerning driver monitoring, low-intrusive systems, such as sensors integrated into the vehicle interior and cameras, are identified as the most effective solutions. These systems maintain high driving comfort while accurately detecting biophysical parameters to identify health anomalies, drowsiness, stress, and intoxication. Similarly, the use of cameras for driver monitoring produces excellent results but raises issues regarding data security and privacy, since sensitive data are collected such as facial expressions, biometric details, and eye movements; therefore, these non-intrusive solutions pave the way for the implementation of more advanced data security techniques to prevent unauthorized access and tampering, according to the main regulations (GPDR and CCPA). These challenges can be addressed by employing end-to-end encryption techniques, localized processing, and data anonymization or pseudo-anonymization techniques. Also, network technologies present significant challenges, the main one being the latency reduction that allows them to provide real-time feedback and overcome limitations related to limited bandwidth and congested communication channels. The solutions to these problems lie in the use of advanced communication systems (5G-6G), and where possible, in the exploitation of edge computing technologies. Another aspect is the economic one, related to the investment costs for the acquisition of such technology, its maintenance, and upgrading. Scalable designs, economies of scale, and possible government incentives could allow wider adoption of these systems in next-generation vehicles. Another issue concerns the precision and reliability of the proposed systems in detecting the driver and vehicle parameters; since they are continuously subjected to vibrations and mechanical stresses, their accuracy could be compromised. For this purpose, solutions based on advanced AI models, advanced sensor systems, and continuous learning could guarantee high performance for the DMS.

In the field of vehicle condition monitoring and driver behavior detection, a recurring approach in the reviewed literature considers the vehicle's OBD-II system. By integrating

OBD-II data with Machine Learning (ML) algorithms, these solutions enable comprehensive monitoring and predictive maintenance, including vehicle stability, emissions levels, tire health, and driving style and behavior analysis. Finally, Vehicle-to-Infrastructure (V2I) communication is highlighted as a critical component of traffic and road management. V2I enables a continuous exchange of data useful for optimizing road signage, providing alternative routes, and allowing constant road condition monitoring. Emerging technologies such as 6G communications and blockchain promise to significantly improve the efficiency, safety, and reliability of Vehicle to Everything (V2X) systems. In particular, 6G technology enables low latency, high reliability and data rate, and advanced connectivity, allowing better traffic management and efficient vehicle coordination. Blockchain technology offers decentralized and tamper-resistant data management, addressing several challenges in V2X systems. It enables data security and integrity, availability of a decentralized public key infrastructure (DPKI), and transparent and immutable ledgers. In November 2024, the U.S. Federal Communications Commission approved new rules to promote the use of cellular V2X (C-V2X) communication technologies. This initiative aims to improve road safety by enabling vehicles to communicate with other vehicles, infrastructure, cyclists, and pedestrians, thereby facilitating warnings of hazardous conditions such as speeding vehicles, weather conditions, or traffic congestion. Such capabilities enable targeted maintenance efforts, ensuring greater safety and comfort for road users.

Future efforts should focus on advancing AI applications for driver monitoring and the analysis of safety-critical events. A key development is integrating DMS data with road scene analysis using AI techniques to enable early real-time crash prediction. A mobility management system (MMS) can be developed based on intelligent vehicles able to monitor vehicle and driver health, creating an interconnected network by exchanging information in real time with on-the-road infrastructures, and informing users about traffic, crashes, and environmental conditions. Each vehicle exchanges data via V2X technology with surrounding ones and road infrastructure to integrate information about the driver's status with that provided by nearby vehicles and the road infrastructure. The multimodal analysis of such information, based on DL algorithms, could improve the determination of the driver's risk level, contextualizing it to the scenario in which the vehicle is traveling.

Moreover, several strategies should be pursued to promote greater transparency regarding AI implementation in DMS and risk events analysis. These should include establishing industry standards, fostering research collaboration, and developing and implementing open-source platforms. Driver Monitoring Systems (DMSs) are becoming a key feature in modern vehicles, especially as we move toward advanced driver assistance systems (ADASs) and even autonomous driving. To ensure these systems work effectively across different vehicles and regions, standardization is crucial. Organizations like UNECE, Euro NCAP, and NHTSA are laying the groundwork for what a DMS should do; in detail, a standardized DMS should monitor and detect driver attention and health, fatigue, and driving style. For example, from July 2024, the European Union has made it compulsory to install alcohol detection systems in new vehicles, which communicate with the electronic control unit to prevent the engine from starting when the driver is under the influence of alcohol. One of the big challenges is ensuring that all the different pieces—cameras, sensors, and software—can work across brands and models. Communication protocol to support the data exchange between vehicles and infrastructures needs standardization to ensure interoperability between different manufacturers, security, and ethical management of collected data. Actually, standards like ISO 26262 [120], ISO 21448 [121], and General Data Protection Regulation (GDPR) [122] ensure that electronic systems and privacy aspects are safe and reliable.

7. Conclusions

Automotive security systems and traffic monitoring technologies are essential to keep cars safe, driving easier, and traffic flowing more smoothly by using advanced surveillance and data-driven solutions. To gain deeper insights into the current systems and future applications of AI tools in this field, this paper serves as a supplementary effort to the knowledge of AI-driven solutions from both academic and industry perspectives. Following a rigorous selection of scientific literature using the PRISMA methodology, this paper provides a comprehensive overview of IoV solutions for driver state monitoring and detection, vehicle monitoring, and traffic and road management. For each discussed topic, a comparative analysis and related discussion are reported to highlight the strengths and weaknesses of the analyzed solutions. Finally, the challenges and future perspectives are presented, bringing out the main requirements and insights for the development of the next generation of automotive monitoring systems.

Author Contributions: Conceptualization, P.V., G.R. and R.D.F.; methodology, P.V.; R.V., D.C. and R.D.F.; validation, R.V., D.C. and R.D.F.; formal analysis, G.R., C.D.-V.-S., D.C. and R.D.F.; investigation, P.V., G.R., D.C. and R.D.F.; resources, P.V., G.R., R.V., D.C. and R.D.F.; data curation, G.R. and R.D.F.; writing—original draft preparation, P.V., G.R. and R.D.F.; writing—review and editing, P.V., C.D.-V.-S., D.C. and R.V.; visualization, G.R., C.D.-V.-S. and R.D.F.; supervision, P.V., D.C. and R.V.; funding acquisition, P.V. All authors have read and agreed to the published version of the manuscript.

Funding: This research received no external funding.

Institutional Review Board Statement: Not applicable.

Informed Consent Statement: Not applicable.

Data Availability Statement: The data are available upon request.

Conflicts of Interest: The authors declare no conflicts of interest.

References

1. Alalwany, E.; Mahgoub, I. Security and Trust Management in the Internet of Vehicles (IoV): Challenges and Machine Learning Solutions. *Sensors* **2024**, *24*, 368. [CrossRef] [PubMed]
2. Bouhsissin, S.; Sael, N.; Benabbou, F. Driver Behavior Classification: A Systematic Literature Review. *IEEE Access* **2023**, *11*, 14128–14153. [CrossRef]
3. Guerrero-Ibáñez, J.; Zeadally, S.; Contreras-Castillo, J. Sensor Technologies for Intelligent Transportation Systems. *Sensors* **2018**, *18*, 1212. [CrossRef] [PubMed]
4. Pandharipande, A.; Cheng, C.-H.; Dauwels, J.; Gurbuz, S.Z.; Ibanez-Guzman, J.; Li, G.; Piazzoni, A.; Wang, P.; Santra, A. Sensing and Machine Learning for Automotive Perception: A Review. *IEEE Sens. J.* **2023**, *23*, 11097–11115. [CrossRef]
5. Mehta, A.A.; Padaria, A.A.; Bavisi, D.J.; Ukani, V.; Thakkar, P.; Geddam, R.; Kotecha, K.; Abraham, A. Securing the Future: A Comprehensive Review of Security Challenges and Solutions in Advanced Driver Assistance Systems. *IEEE Access* **2024**, *12*, 643–678. [CrossRef]
6. Sohail, A.; Cheema, M.A.; Ali, M.E.; Toosi, A.N.; Rakha, H.A. Data-Driven Approaches for Road Safety: A Comprehensive Systematic Literature Review. *Saf. Sci.* **2023**, *158*, 105949. [CrossRef]
7. Alanazi, F. A Systematic Literature Review of Autonomous and Connected Vehicles in Traffic Management. *Appl. Sci.* **2023**, *13*, 1789. [CrossRef]
8. Nahid, M.S.H.; Tila, T.Z.; Seecharan, T.S. Detecting Emotional Arousal and Aggressive Driving Using Neural Networks: A Pilot Study Involving Young Drivers in Duluth. *Sensors* **2024**, *24*, 7109. [CrossRef] [PubMed]
9. Mohammed, K.; Abdelhafid, M.; Kamal, K.; Ismail, N.; Ilias, A. Intelligent Driver Monitoring System: An Internet of Things-Based System for Tracking and Identifying the Driving Behavior. *Comput. Stand. Interfaces* **2023**, *84*, 103704. [CrossRef]
10. Du, Z.; Deng, M.; Lyu, N.; Wang, Y. A Review of Road Safety Evaluation Methods Based on Driving Behavior. *J. Traffic Transp. Eng.* **2023**, *10*, 743–761. [CrossRef]
11. Kaur, K.; Verma, H.K. IoV-Health: An Intelligent Integrated Emergency Health Monitoring and Alert Generation System. *Soft Comput.* **2023**. [CrossRef]

12. Affanni, A.; Rinaldo, R. Design and Realization of a Wearable Necklace for the Assessment of Driver Well-Being Through Heart Rate and Blood Oxygen Saturation Monitoring. In Proceedings of the 2024 IEEE International Workshop on Metrology for Automotive (MetroAutomotive), Bologna, Italy, 26–28 June 2024; IEEE: Piscataway, NJ, USA, 2024; pp. 153–158. [CrossRef]
13. Gong, Z.; Yang, X.; Song, R.; Han, X.; Ren, C.; Shi, H.; Niu, J.; Li, W. Heart Rate Estimation in Driver Monitoring System Using Quality-Guided Spectrum Peak Screening. *IEEE Trans. Instrum. Meas.* **2024**, *73*, 5008214. [CrossRef]
14. Sukumar, N.; Sivashankar, B.; Sumathi, P.; Maurya, O.P. Physiological and Physical Sensors for Stress Level, Drowsiness Detection, and Behaviour Analysis. *IEEE Trans. Consum. Electron.* **2024**, *70*, 656–668. [CrossRef]
15. Safarov, F.; Akhmedov, F.; Abdusalomov, A.B.; Nasimov, R.; Cho, Y.I. Real-Time Deep Learning-Based Drowsiness Detection: Leveraging Computer-Vision and Eye-Blink Analyses for Enhanced Road Safety. *Sensors* **2023**, *23*, 6459. [CrossRef] [PubMed]
16. Visconti, P.; de Fazio, R.; Costantini, P.; Miccoli, S.; Cafagna, D. Innovative Complete Solution for Health Safety of Children Unintentionally Forgotten in a Car: A Smart Arduino-Based System with User App for Remote Control. *IET Sci. Meas. Technol.* **2020**, *14*, 665–675. [CrossRef]
17. Visconti, P.; de Fazio, R.; Costantini, P.; Miccoli, S.; Cafagna, D. Arduino-Based Solution for In-CarAbandoned Infants' Controlling Remotely Managed by Smartphone Application. *J. Commun. Softw. Syst.* **2019**, *15*, 89–100. [CrossRef]
18. Yang, X.G. Framework of Electric Vehicle Fault Diagnosis System Based on Diagnostic Communication. *Int. J. Eng.* **2024**, *37*, 1194–1207. [CrossRef]
19. Spiliotis, A.; Giannopoulos, F.; Spandonidis, C.; Gkemou, M.; Kalfa, N. Integration and Field Evaluation of an IoV System for Enhancing Road Safety. *Appl. Sci.* **2022**, *12*, 12262. [CrossRef]
20. Visconti, P.; de Fazio, R.; Velazquez, R.; Al-Naami, B.; Ghavifekr, A.A. Self-Powered WiFi-Connected Monitoring Stations for Environmental Pollution App-Based Control in Urban and Industrial Areas. In Proceedings of the 2022 8th International Conference on Control, Instrumentation and Automation (ICCIA), Tehran, Iran, 2–3 March 2022; IEEE: Piscataway, NJ, USA, 2022; pp. 1–6. [CrossRef]
21. Márton, Z.; Szalay, I.; Fodor, D. Convolutional Neural Network-Based Tire Pressure Monitoring System. *IEEE Access* **2023**, *11*, 70317–70332. [CrossRef]
22. Hu, J.; Huang, M.-C.; Yu, X.B. Deep Learning Based on Connected Vehicles for Icing Pavement Detection. *AI Civ. Eng.* **2023**, *2*, 1. [CrossRef]
23. Zhang, C.; Cheng, H.; Wu, R.; Ren, B.; Zhu, Y.; Peng, N. Development of a Traffic Congestion Prediction and Emergency Lane Development Strategy Based on Object Detection Algorithms. *Sustainability* **2024**, *16*, 10232. [CrossRef]
24. Tahir, M.N.; Leviäkangas, P.; Katz, M. Connected Vehicles: V2V and V2I Road Weather and Traffic Communication Using Cellular Technologies. *Sensors* **2022**, *22*, 1142. [CrossRef] [PubMed]
25. Lilhore, U.K.; Imoize, A.L.; Li, C.-T.; Simaiya, S.; Pani, S.K.; Goyal, N.; Kumar, A.; Lee, C.-C. Design and Implementation of an ML and IoT Based Adaptive Traffic-Management System for Smart Cities. *Sensors* **2022**, *22*, 2908. [CrossRef] [PubMed]
26. De Fazio, R.; De Giorgi, M.; Cafagna, D.; Del-Valle-Soto, C.; Visconti, P. Energy Harvesting Technologies and Devices from Vehicular Transit and Natural Sources on Roads for a Sustainable Transport: State-of-the-Art Analysis and Commercial Solutions. *Energies* **2023**, *16*, 3016. [CrossRef]
27. Rathee, M.; Bačić, B.; Doborjeh, M. Automated Road Defect and Anomaly Detection for Traffic Safety: A Systematic Review. *Sensors* **2023**, *23*, 5656. [CrossRef] [PubMed]
28. Khanmohamadi, M.; Guerrieri, M. Advanced Sensor Technologies in CAVs for Traditional and Smart Road Condition Monitoring: A Review. *Sustainability* **2024**, *16*, 8336. [CrossRef]
29. Page, M.J.; McKenzie, J.E.; Bossuyt, P.M.; Boutron, I.; Hoffmann, T.C.; Mulrow, C.D.; Shamseer, L.; Tetzlaff, J.M.; Akl, E.A.; Brennan, S.E.; et al. The PRISMA 2020 Statement: An Updated Guideline for Reporting Systematic Reviews. *BMJ* **2021**, *372*, n71. [CrossRef]
30. Page, M.J.; Moher, D.; Bossuyt, P.M.; Boutron, I.; Hoffmann, T.C.; Mulrow, C.D.; Shamseer, L.; Tetzlaff, J.M.; Akl, E.A.; Brennan, S.E.; et al. PRISMA 2020 Explanation and Elaboration: Updated Guidance and Exemplars for Reporting Systematic Reviews. *BMJ* **2021**, *372*, n160. [CrossRef] [PubMed]
31. Welch, K.C.; Harnett, C.; Lee, Y.-C. A Review on Measuring Affect with Practical Sensors to Monitor Driver Behavior. *Safety* **2019**, *5*, 72. [CrossRef]
32. Xiang, G.; Ou, Y.; Li, J.; Wang, L.; Hu, Y.; Wang, X.; Wu, X.; Peng, Y. Recovering Physiological Signals from Facial Videos: Recent Advances and Applications in Intelligent Vehicles. *IEEE Trans. Intell. Veh.* **2024**, 1–24. [CrossRef]
33. Argüello-Prada, E.J. The Mountaineer's Method for Peak Detection in Photoplethysmographic Signals. *Rev. Fac. Ing. Univ. Antioq.* **2019**, *90*, 42–50. [CrossRef]
34. Jiao, Y.; Zhang, C.; Chen, X.; Fu, L.; Jiang, C.; Wen, C. Driver Fatigue Detection Using Measures of Heart Rate Variability and Electrodermal Activity. *IEEE Trans. Intell. Transp. Syst.* **2024**, *25*, 5510–5524. [CrossRef]
35. Putilov, A.A.; Donskaya, O.G. Construction and Validation of the EEG Analogues of the Karolinska Sleepiness Scale Based on the Karolinska Drowsiness Test. *Clin. Neurophysiol.* **2013**, *124*, 1346–1352. [CrossRef] [PubMed]

36. Leicht, L.; Walter, M.; Mathissen, M.; Antink, C.H.; Teichmann, D.; Leonhardt, S. Unobtrusive Measurement of Physiological Features Under Simulated and Real Driving Conditions. *IEEE Trans. Intell. Transp. Syst.* **2022**, *23*, 4767–4777. [CrossRef]
37. Campos-Ferreira, A.E.; Lozoya-Santos, J.D.J.; Tudon-Martinez, J.C.; Mendoza, R.A.R.; Vargas-Martínez, A.; Morales-Menendez, R.; Lozano, D. Vehicle and Driver Monitoring System Using On-Board and Remote Sensors. *Sensors* **2023**, *23*, 814. [CrossRef] [PubMed]
38. Škorić, T. Stress Level Detection Based on the Capacitive Electrocardiogram Signals of Driving Subjects. *Sensors* **2023**, *23*, 9158. [CrossRef] [PubMed]
39. Ebrahimian, S.; Nahvi, A.; Tashakori, M.; Salmanzadeh, H.; Mohseni, O.; Leppänen, T. Multi-Level Classification of Driver Drowsiness by Simultaneous Analysis of ECG and Respiration Signals Using Deep Neural Networks. *Int. J. Environ. Res. Public Health* **2022**, *19*, 10736. [CrossRef] [PubMed]
40. Alguindigue, J.; Singh, A.; Narayan, A.; Samuel, S. Biosignals Monitoring for Driver Drowsiness Detection Using Deep Neural Networks. *IEEE Access* **2024**, *12*, 93075–93086. [CrossRef]
41. BaHammam, A.S.; Nashwan, S.; Hammad, O.; Sharif, M.M.; Pandi-Perumal, S.R. Objective Assessment of Drowsiness and Reaction Time during Intermittent Ramadan Fasting in Young Men: A Case-Crossover Study. *Behav. Brain Funct.* **2013**, *9*, 32. [CrossRef]
42. Amidei, A.; Poli, A.; Iadarola, G.; Tramarin, F.; Pavan, P.; Spinsante, S.; Rovati, L. Driver Drowsiness Detection Based on Variation of Skin Conductance from Wearable Device. In Proceedings of the 2022 IEEE International Workshop on Metrology for Automotive (MetroAutomotive), Modena, Italy, 4–6 July 2022; IEEE: Piscataway, NJ, USA, 2022; pp. 94–98. [CrossRef]
43. Díaz-Santos, S.; Cigala-Álvarez, Ó.; Gonzalez-Sosa, E.; Caballero-Gil, P.; Caballero-Gil, C. Driver Identification and Detection of Drowsiness While Driving. *Appl. Sci.* **2024**, *14*, 2603. [CrossRef]
44. Khan, M.A.; Nawaz, T.; Khan, U.S.; Hamza, A.; Rashid, N. IoT-Based Non-Intrusive Automated Driver Drowsiness Monitoring Framework for Logistics and Public Transport Applications to Enhance Road Safety. *IEEE Access* **2023**, *11*, 14385–14397. [CrossRef]
45. Kundinger, T.; Sofra, N.; Riener, A. Assessment of the Potential of Wrist-Worn Wearable Sensors for Driver Drowsiness Detection. *Sensors* **2020**, *20*, 1029. [CrossRef]
46. Vyas, D.A.; Chaturvedi, M. DriverSense: A Multi-Modal Framework for Advanced Driver Assistance System. In Proceedings of the 2024 16th International Conference on COMmunication Systems & NETworkS (COMSNETS), Bengaluru, India, 3–7 January 2024; IEEE: Piscataway, NJ, USA, 2024; pp. 252–257. [CrossRef]
47. Kumar, T.S.; Raman, R.; Karthikeyan, M.; Rawandale, C.J.; Sasikala, S.; Murugan, S. Drunk Driving Detection and Automatic Car Ignition Locking System. In Proceedings of the 2023 International Conference on Self Sustainable Artificial Intelligence Systems (ICSSAS), Erode, India, 18–20 October 2023; IEEE: Piscataway, NJ, USA, 2023; pp. 1469–1474. [CrossRef]
48. Liu, J.; Luo, Y.; Ge, L.; Zeng, W.; Rao, Z.; Xiao, X. An Intelligent Online Drunk Driving Detection System Based on Multi-Sensor Fusion Technology. *Sensors* **2022**, *22*, 8460. [CrossRef] [PubMed]
49. Swarna, M.; Saranya, G.; Vanathi, A.; Ashwini, R.; Arunsundar, B. An improved driver safety system based on drunk and drive analyzing architecture using internet of things assisted alcohol identification sensor. In Proceedings of the 2024 International Conference on Advances in Computing, Communication & Applied Information, Chennai, India, 9–10 May 2024; IEEE: Piscataway, NJ, USA, 2024; pp. 1–7. [CrossRef]
50. Wang, F.; Bai, D.; Liu, Z.; Yao, Z.; Weng, X.; Xu, C.; Fan, K.; Zhao, Z.; Chang, Z. A Two-Step E-Nose System for Vehicle Drunk Driving Rapid Detection. *Appl. Sci.* **2023**, *13*, 3478. [CrossRef]
51. Akanni, J.; Adewusi, B.; Isa, A.A.; Alao, R.A. An Automated Prototype System for Alcohol Level Detection in Drunk Drivers and Vehicle Control. *Arid Zone J. Eng. Technol. Environ.* **2023**, *19*, 121–128.
52. Cho, Y.; Lee, W.; Sin, H.; Oh, S.; Choi, K.C.; Jun, J.-H. Non-Invasive Alcohol Concentration Measurement Using a Spectroscopic Module: Outlook for the Development of a Drunk Driving Prevention System. *Sensors* **2024**, *24*, 2252. [CrossRef] [PubMed]
53. Zhang, M.; Liu, X.; Jiang, B.; Chen, H.; Zhang, Q.; Yin, X. Innovative PVDF Nanofiber Films for Intelligent Driver Status Monitoring Systems. In Proceedings of the COTA International Conference of Transportation Professionals (CICP2024), Shenzhen, China, 23–26 July 2024; pp. 226–235. [CrossRef]
54. Adib, F.; Mao, H.; Kabelac, Z.; Katabi, D.; Miller, R.C. Smart Homes That Monitor Breathing and Heart Rate. In Proceedings of the 33rd Annual ACM Conference on Human Factors in Computing Systems, Seoul, Republic of Korea, 18–23 April 2015; Association for Computing Machinery: New York, NY, USA; pp. 837–846. [CrossRef]
55. Vasantharaj, A.; Nandhagopal, N.; Karuppusamy, S.A.; Subramaniam, K. An In-Tire-Pressure Monitoring SoC Using FBAR Resonator-Based ZigBee Transceiver and Deep Learning Models. *Microprocess. Microsyst.* **2022**, *95*, 104709. [CrossRef]
56. Kim, J.; Choi, S.B. Machine Learning Based Drive Fault Diagnosis and Isolation Algorithm for 4WIS4WID Vehicle Using Residual Visualization. *IEEE Trans. Intell. Veh.* **2024**, *14*, 1–15. [CrossRef]
57. Kannan, S.R.; Partheeban, P.; Ramesh, R.; Elamparithi, P.N.; Somasundaram, K.; Selvi, R. Machine Learning and Vehicle Fault Diagnosis System with IoT Enabled Data. In Proceedings of the 2023 International Conference on Data Science, Agents & Artificial Intelligence (ICDSAAI), Chennai, India, 21–23 December 2023; IEEE: Piscataway, NJ, USA, 2023; pp. 1–9. [CrossRef]

58. Min, H.; Fang, Y.; Wu, X.; Lei, X.; Chen, S.; Teixeira, R.; Zhu, B.; Zhao, X.; Xu, Z. A Fault Diagnosis Framework for Autonomous Vehicles with Sensor Self-Diagnosis. *Expert Syst. Appl.* **2023**, *224*, 120002. [CrossRef]
59. Josephinshermila, P.; Priya, S.S.; Malarvizhi, K.; Hegde, R.; Gokul Pran, S.; Veerasamy, B. Accident Detection Using Automotive Smart Black-Box Based Monitoring System. *Meas. Sens.* **2023**, *27*, 100721. [CrossRef]
60. Fichtinger, A.; Edelmann, J.; Plochl, M.; Holl, M. Aquaplaning Detection Using Effect-Based Methods: An Approach Based on a Minimal Set of Sensors, Electronic Stability Control, and Drive Torques. *IEEE Veh. Technol. Mag.* **2021**, *16*, 20–28. [CrossRef]
61. Gustafsson, F. Estimation and Change Detection of Tire-Road Friction Using the Wheel Slip. In Proceedings of the Joint Conference on Control Applications Intelligent Control and Computer Aided Control System Design, Dearborn, MI, USA, 15–18 September 1996; IEEE: Piscataway, NJ, USA, 1996; pp. 99–104. [CrossRef]
62. Lee, H.; Hwang, K.; Kang, M.; Song, J. Black Ice Detection Using CNN for the Prevention of Accidents in Automated Vehicle. In Proceedings of the 2020 International Conference on Computational Science and Computational Intelligence (CSCI), Las Vegas, NV, USA, 16–18 December 2020; IEEE: Piscataway, NJ, USA, 2020; pp. 1189–1192. [CrossRef]
63. Bhadraray, S.; Dey, S.; Amith, T.S.; Ramaiah, S. Autonomous Vehicle System for Pothole Detection and Avoidance. In Proceedings of the 2023 IEEE 8th International Conference for Convergence in Technology (I2CT), Lonavla, India, 7–9 April 2023; IEEE: Piscataway, NJ, USA, 2023; pp. 1–6. [CrossRef]
64. Raja, G.; Anbalagan, S.; Senthilkumar, S.; Dev, K.; Qureshi, N.M.F. SPAS: Smart Pothole-Avoidance Strategy for Autonomous Vehicles. *Trans. Intell. Transp. Syst.* **2022**, *23*, 19827–19836. [CrossRef]
65. Tselentis, D.I.; Papadimitriou, E. Driver Profile and Driving Pattern Recognition for Road Safety Assessment: Main Challenges and Future Directions. *IEEE Open J. Intell. Transp. Syst.* **2023**, *4*, 83–100. [CrossRef]
66. Nouh, R.; Singh, M.; Singh, D. SafeDrive: Hybrid Recommendation System Architecture for Early Safety Predication Using Internet of Vehicles. *Sensors* **2021**, *21*, 3893. [CrossRef] [PubMed]
67. Shichkina, Y.; Fatkieva, R.; Kopylov, M. Analysis of Driving Style Using Self-Organizing Maps to Analyze Driver Behavior. *Int. J. Electr. Comput. Eng.* **2024**, *14*, 2212–2225. [CrossRef]
68. Chhabra, R.; Singh, S.; Khullar, V. Privacy Enabled Driver Behavior Analysis in Heterogeneous IoV Using Federated Learning. *Eng. Appl. Artif. Intell.* **2023**, *120*, 105881. [CrossRef]
69. Malik, M.; Nandal, R.; Dalal, S.; Jalglan, V.; Le, D.-N. Driving Pattern Profiling and Classification Using Deep Learning. *Intell. Autom. Soft Comput.* **2021**, *28*, 887–906. [CrossRef]
70. Lattanzi, E.; Freschi, V. Machine Learning Techniques to Identify Unsafe Driving Behavior by Means of In-Vehicle Sensor Data. *Expert Syst. Appl.* **2021**, *176*, 114818. [CrossRef]
71. Dong, B.-T.; Lin, H.-Y. An On-Board Monitoring System for Driving Fatigue and Distraction Detection. In Proceedings of the 2021 22nd IEEE International Conference on Industrial Technology (ICIT), Valencia, Spain, 10–12 March 2021; IEEE: Piscataway, NJ, USA, 2021; Volume 1, pp. 850–855. [CrossRef]
72. Shan, T.; Li, L.; Mu, H.; Yu, L.; Liu, Z. Tire Pressure Monitoring System Based on Resonant Frequency of Wheel Speed Spectrum. In Proceedings of the 2022 IEEE 5th International Conference on Electronics Technology (ICET), Chengdu, China, 13–16 May 2022; IEEE: Piscataway, NJ, USA, 2022; pp. 623–628. [CrossRef]
73. Huo, B.; Zhang, M.; Zhong, S.; Zhou, F. Design of Indirect Tire Pressure Monitoring System Based on Wheel Speed Signal. *J. Phys. Conf. Ser.* **2023**, *2528*, 012045. [CrossRef]
74. Pomoni, M. Exploring Smart Tires as a Tool to Assist Safe Driving and Monitor Tire–Road Friction. *Vehicles* **2022**, *4*, 744–765. [CrossRef]
75. Yu, H.; Zhong, Y.; Li, W. A Transformer Model Based Method for Detecting Tire Wear Degree. In Proceedings of the 2024 9th International Conference on Intelligent Computing and Signal Processing (ICSP), Xian, China, 19–21 April 2024; IEEE: Piscataway, NJ, USA, 2024; pp. 1618–1624. [CrossRef]
76. Anusha, N.; Jeslin, J.G.; Srividhya, V.; Gupta, N.S.; Meenakshi, R.; Srinivasan, C. Cloud-Enabled Neural Networks for Intelligent Vehicle Emissions Tracking and Analysis. In Proceedings of the 2024 International Conference on Automation and Computation (AUTOCOM), Dehradun, India, 14–16 March 2024; IEEE: Piscataway, NJ, USA, 2024; pp. 232–236. [CrossRef]
77. Seo, J.; Park, S. Optimizing Model Parameters of Artificial Neural Networks to Predict Vehicle Emissions. *Atmos. Environ.* **2023**, *294*, 119508. [CrossRef]
78. Andrade, P.; Silva, I.; Silva, M.; Flores, T.; Cassiano, J.; Costa, D.G. A TinyML Soft-Sensor Approach for Low-Cost Detection and Monitoring of Vehicular Emissions. *Sensors* **2022**, *22*, 3838. [CrossRef] [PubMed]
79. Oladimeji, D.; Gupta, K.; Kose, N.A.; Gundogan, K.; Ge, L.; Liang, F. Smart Transportation: An Overview of Technologies and Applications. *Sensors* **2023**, *23*, 3880. [CrossRef] [PubMed]
80. Saleem, M.; Abbas, S.; Ghazal, T.M.; Adnan Khan, M.; Sahawneh, N.; Ahmad, M. Smart Cities: Fusion Based Intelligent Traffic Congestion Control System for Vehicular Networks Using Machine Learning Techniques. *Egypt. Inform. J.* **2022**, *23*, 417–426. [CrossRef]

81. Wu, Q.; Wang, W.; Fan, P.; Fan, Q.; Zhu, H.; Letaief, K.B. Cooperative Edge Caching Based on Elastic Federated and Multi-Agent Deep Reinforcement Learning in Next-Generation Networks. *IEEE Trans. Netw. Serv. Manag.* **2024**, *21*, 4179–4196. [CrossRef]
82. Mandal, V.; Mussah, A.R.; Jin, P.; Adu-Gyamfi, Y. Artificial Intelligence-Enabled Traffic Monitoring System. *Sustainability* **2020**, *12*, 9177. [CrossRef]
83. Humayun, M.; Afsar, S.; Almufareh, M.F.; Jhanjhi, N.Z.; AlSuwailem, M. Smart Traffic Management System for Metropolitan Cities of Kingdom Using Cutting Edge Technologies. *J. Adv. Transp.* **2022**, *2022*, 4687319. [CrossRef]
84. Putra, A.S.; Warnars, H.L.H.S. Intelligent Traffic Monitoring System (ITMS) for Smart City Based on IoT Monitoring. In Proceedings of the 2018 Indonesian Association for Pattern Recognition International Conference (INAPR), Jakarta, Indonesia, 7–8 September 2018; IEEE: Piscataway, NJ, USA, 2018; pp. 161–165. [CrossRef]
85. Sarrab, M.; Pulparambil, S.; Awadalla, M. Development of an IoT Based Real-Time Traffic Monitoring System for City Governance. *Glob. Transit.* **2020**, *2*, 230–245. [CrossRef]
86. Barbosa, R.; Ogobuchi, O.D.; Joy, O.O.; Saadi, M.; Rosa, R.L.; Al Otaibi, S.; Rodríguez, D.Z. IoT Based Real-Time Traffic Monitoring System Using Images Sensors by Sparse Deep Learning Algorithm. *Comput. Commun.* **2023**, *210*, 321–330. [CrossRef]
87. Kheder, M.Q.; Mohammed, A.A. Real-Time Traffic Monitoring System Using IoT-Aided Robotics and Deep Learning Techniques. *Kuwait J. Sci.* **2024**, *51*, 100153. [CrossRef]
88. Dhingra, S.; Madda, R.B.; Patan, R.; Jiao, P.; Barri, K.; Alavi, A.H. Internet of Things-Based Fog and Cloud Computing Technology for Smart Traffic Monitoring. *Internet Things* **2021**, *14*, 100175. [CrossRef]
89. Khan, N.A.; Jhanjhi, N.Z.; Brohi, S.N.; Usmani, R.S.A.; Nayyar, A. Smart Traffic Monitoring System Using Unmanned Aerial Vehicles (UAVs). *Comput. Commun.* **2020**, *157*, 434–443. [CrossRef]
90. Singh, S.; Singh, J.; Goyal, S.B.; Singh Sehra, S.; Ali, F.; Ayad Alkhafaji, M.; Singh, R. A Novel Framework to Avoid Traffic Congestion and Air Pollution for Sustainable Development of Smart Cities. *Sustain. Energy Technol. Assess.* **2023**, *56*, 103125. [CrossRef]
91. Åstrand, M.; Jakobsson, E.; Lindfors, M.; Svensson, J. A System for Underground Road Condition Monitoring. *Int. J. Min. Sci. Technol.* **2020**, *30*, 405–411. [CrossRef]
92. Ye, Z.; Wei, Y.; Yang, S.; Li, P.; Yang, F.; Yang, B.; Wang, L. IoT-Enhanced Smart Road Infrastructure Systems for Comprehensive Real-Time Monitoring. *Internet Things Cyber-Phys. Syst.* **2024**, *4*, 235–249. [CrossRef]
93. Abdelmalak, M.E.S.; Khodadadi, N.; Zaki, A.M.; Eid, M.M.; Rizk, F.H.; Ibrahim, A.; Abdelhamid, A.A.; Abualigah, L.; El-kenawy, E.S.M. Pothole Detection in Asphalt Roads: A Comprehensive Approach for Enhanced Road Maintenance and Safety with AlexNet Model. In Proceedings of the 2024 International Telecommunications Conference (ITC-Egypt), Cairo, Egypt, 22–25 July 2024; IEEE: Piscataway, NJ, USA, 2024; pp. 269–274. [CrossRef]
94. Ye, Z.; Wei, Y.; Li, J.; Yan, G.; Wang, L. A Distributed Pavement Monitoring System Based on Internet of Things. *J. Traffic Transp. (Eng. Ed.)* **2022**, *9*, 305–317. [CrossRef]
95. Gaspar, G.; Dudak, J.; Behulova, M.; Stremy, M.; Budjac, R.; Sedivy, S.; Tomas, B. IoT-Ready Temperature Probe for Smart Monitoring of Forest Roads. *Appl. Sci.* **2022**, *12*, 743. [CrossRef]
96. Hajder, M.; Hajder, P.; Hajder, L.; Liput, M.; Kolbusz, J. Supporting Winter Road Maintenance Procedures with the Use of Distributed Measurements Based on IoT, Thermodynamic Models and Machine Learning. In Proceedings of the 2023 IEEE International Conference on Pervasive Computing & Communication Workshops, Atlanta, GA, USA, 13–17 March 2023; IEEE: Piscataway, NJ, USA, 2023; pp. 515–520. [CrossRef]
97. Chen, Z.; Xiong, G.; Sun, Y.; Li, Y.; Li, Y. An Internet-of-Things-Enabled System for Road Icing Detection and Prediction. *IEEE Internet Things J.* **2022**, *9*, 20257–20269. [CrossRef]
98. Kotus, J.; Szwoch, G. Detection of Water on Road Surface with Acoustic Vector Sensor. *Sensors* **2023**, *23*, 8878. [CrossRef] [PubMed]
99. Smart Eye Pro 12. Available online: https://www.smarteye.se/news/unveiling-smart-eye-pro-12-redefining-eye-tracking-technology/ (accessed on 10 October 2024).
100. Advanced Cabin Sensing Solution. Available online: https://conti-engineering.com/services/components/cabin-sensing/ (accessed on 11 October 2024).
101. News Release—Magna Lands Industry-First Award for Integrated Driver and Occupant Monitoring System. Available online: https://www.magna.com/company/newsroom/releases-archive/release/2022/02/07/news-release---magna-lands-industry-first-award-for-integrated-driver-and-occupant-monitoring-system (accessed on 11 October 2024).
102. Driver Fatigue Monitoring System for Teen and Fleet Driver Safety. Available online: https://speedir.com/driver-alert-fatigue-monitoring-system-for-fleet-safety-night-driving-student-teen-driver/ (accessed on 30 October 2024).
103. RH600 4G Integrated Telematics Camera. Available online: https://www.trakm8.com/vehicle-cameras/rh600-integrated-telematics-camera (accessed on 5 November 2024).
104. RoadHawk DC-4 Dash Cams. Available online: https://www.trakm8.com/vehicle-cameras/dash-cams/ (accessed on 7 November 2024).

105. RH800 4G Mobile Digital Video Recorder. Available online: https://www.trakm8.com/wp-content/uploads/Trakm8_RH800_Product_Information_Sheet.pdf (accessed on 11 November 2024).
106. ACC750 Driver ID & Feedback. Available online: https://www.trakm8.com/wp-content/uploads/Driver_Feedback_and_Driver_Display.pdf (accessed on 11 November 2024).
107. Trackobit: Advanced-Driver-Assistance-Systems. Available online: https://trackobit.com/advanced-driver-assistance-systems (accessed on 8 January 2025).
108. Brake Pedal Sensor. Available online: https://www.hella.com/microsite-electronics/en/Products/Brake-Pedal-Sensor-4923/ (accessed on 13 November 2024).
109. Road Wheel Actuator. Available online: https://www.hella.com/microsite-electronics/en/Products/Road-Wheel-Actuator-4913/ (accessed on 14 November 2024).
110. Steering Torque Only Sensor. Available online: https://www.hella.com/microsite-electronics/en/Products/Steering-Torque-Only-Sensor-4921/ (accessed on 13 November 2024).
111. Radar 77GHZ. Available online: https://www.hella.com/microsite-electronics/en/Products/Radar-77GHZ-4919/ (accessed on 14 November 2024).
112. SHAKE—Road Condition Sensor. Available online: https://www.hella.com/microsite-electronics/en/Products/SHAKE-Road-Condition-Sensor-4979/ (accessed on 15 November 2024).
113. Rain Light Sensor. Available online: https://www.hella.com/microsite-electronics/en/Products/Rain-Light-Sensor-4973/ (accessed on 18 November 2024).
114. Safe and Automated Driving. Available online: https://www.forvia.com/en/pioneering-technologies/safe-automated-driving-0/vision-systems-wire-control (accessed on 18 November 2024).
115. Social Self Driving. Available online: https://www.socialselfdriving.com/en/home-social-self-driving-en/ (accessed on 18 November 2024).
116. Insight Count & Classify. Available online: https://www.clearview-intelligence.com/products/insight-count-classify/ (accessed on 20 November 2024).
117. Connex Traffic. Available online: https://www.clearview-intelligence.com/products/connex-traffic/ (accessed on 21 November 2024).
118. M100 Wireless Vehicle Detection System. Available online: https://www.clearview-intelligence.com/products/m100-wireless-vehicle-detection-system/ (accessed on 21 November 2024).
119. TomTom Orbis Maps—Digital Mapping Solutions. Available online: https://www.tomtom.com/tomtom-orbis-maps/ (accessed on 25 November 2024).
120. ISO 26262-1:2018, Road Vehicles — Functional Safety. Available online: https://www.iso.org/standard/68383.html (accessed on 20 November 2024).
121. ISO 21448:2022, Road Vehicles—Safety of the Intended Functionality. Available online: https://www.iso.org/standard/77490.html (accessed on 20 November 2024).
122. General Data Protection Regulation -GDPR. Available online: https://gdpr-info.eu/ (accessed on 20 November 2024).

sensors

MDPI

Systematic Review

IoT Sensing for Advanced Irrigation Management: A Systematic Review of Trends, Challenges, and Future Prospects

Ahmed A. Abdelmoneim , Hilda N. Kimaita, Christa M. Al Kalaany, Bilal Derardja , Giovanna Dragonetti and Roula Khadra *

Mediterranean Agronomic Institute CIHEAM Bari, Valenzano, 70010 Bari, Italy; ayoub@iamb.it (A.A.A.); kimaitahilda@gmail.com (H.N.K.); christamaria.kalaany@std.balamand.edu.lb (C.M.A.K.); derardja@iamb.it (B.D.); dragonetti@iamb.it (G.D.)

* Correspondence: khadra@iamb.it

Abstract: Efficient water management is crucial for sustainable agriculture, and the integration of Internet of Things (IoT) technologies in irrigation systems offers innovative solutions to optimize resource use. In this systematic review, the current landscape of Internet of Things (IoT) applications in irrigation management was investigated. The study aimed to identify key research trends and technological developments in the field. Using VOSviewer (CWTS, Leiden, The Netherlands) for bibliometric mapping, the influential research clusters were identified. The analysis revealed a significant rise in scholarly interest, with peak activity between 2020 and 2022, and a shift towards interdisciplinary and applied research. Additionally, the content analysis revealed prevalent agricultural applications, frequently employed microcontroller units (MCUs), widely used sensors, and trends in communication technologies such as the increasing adoption of low-power, scalable communication protocols for real-time data acquisition. This study not only offers a comprehensive overview of the current status of IoT integration in smart irrigation but also highlights the technological advancements. Future research directions include integrating IoT with emerging technologies such as artificial intelligence, edge computing, and blockchain to enhance decision-support systems and predictive irrigation strategies. By examining the transformative potential of IoT, this study provides valuable insights for researchers and practitioners seeking to enhance agricultural productivity, optimize resource use, and improve sustainability.

Keywords: internet of things (IoT); smart agriculture; soil moisture; wireless sensor networks (WSN); digital agriculture; bibliometric analysis

1. Introduction

The integration of the Internet of Things (IoT) in agricultural applications has emerged as a transformative force in modern agriculture, fostering smarter and more-efficient agricultural practices. This shift is driven by advancements in electronics, data science, and communication technologies, which have catalyzed the evolution from traditional farming to precision and smart agriculture [1].

IoT in agriculture involves deploying interconnected sensors and devices that gather real-time data on soil conditions, crop health, weather, and environmental parameters [2]. Recent market analyses indicate a robust growth trajectory for the Internet of Things (IoT) in the agricultural sector. The global agriculture IoT market is projected to expand from USD 11.4 billion in 2021 to USD 18.1 billion by 2026, reflecting a compound annual growth rate (CAGR) of 9.8% during this period [3]. Similarly, [4] forecasts that the IoT in the agriculture market will grow from USD 16.24 billion in 2024 to approximately USD 40.24 billion by 2034, with a CAGR of 9.5% between 2024 and 2034. These projections underscore the increasing adoption of IoT technologies in agriculture, driven by the need for enhanced productivity, resource efficiency, and sustainable farming practices.

Globally, irrigated agriculture accounts for 70% of freshwater consumption, making it both a primary driver and a direct consequence of water scarcity. As a result, enhancing on-farm irrigation management is crucial [5]. One of the key factors that can significantly improve irrigation management is the implementation of reliable IoT data acquisition systems [6]. These systems provide real-time insights into soil moisture levels, weather conditions, and plant health, forming the foundation of precision irrigation by enabling timely and informed decision-making. The integration of Internet of Things (IoT) technology into irrigation management offers an innovative approach to resource optimization, particularly through smart irrigation systems. While the Internet of Things (IoT) holds significant promise for revolutionizing agricultural irrigation, several challenges impede its widespread adoption. Data privacy and security concerns are paramount, as unauthorized breaches and cyber-attacks on data collected from IoT devices pose significant risks [7]. Additionally, the lack of reliable Internet infrastructure in rural areas hinders the seamless integration of IoT technologies [8]. However, as IoT technology continues to mature and costs decrease, its adoption is expected to become more widespread, transforming agriculture into a data-driven, sustainable, and efficient sector capable of meeting the demands of a growing global population.

IoT encompass a wide array of sensors, drones, actuators, smart vehicles, and communication networks that generate, transmit, and analyze data for better decision-making. The architecture of IoT systems in agriculture is typically divided into several layers (Figure 1), each playing a crucial role in enhancing agricultural processes and providing structured functionality [2]. The physical layer comprises sensors, actuators, drones, smart vehicles, and communication gateways. This layer collects real-time data on environmental parameters such as soil moisture, temperature, humidity, and pH. It also automates processes like irrigation and fertilization, utilizing communication technologies such as LoRa, Zigbee, Wi-Fi, Bluetooth, and cellular networks [9].

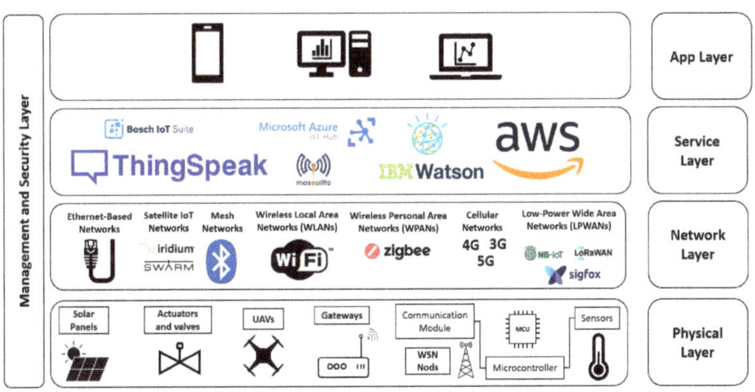

Figure 1. The typical layering architecture of an IoT device using relevant examples.

The network layer ensures reliable data transmission from remote fields to cloud computing platforms [10]. It supports communication protocols, security authentication, and long-range, low-power networks like LoRa, which are ideal for rural agricultural settings. On the other hand, the service support layer provides essential services to process and store data and enable IoT applications [11]. The application layer delivers user-friendly platforms for farmers, including precision farming tools, smart greenhouse-management systems, irrigation optimization, and supply-chain tracking [12]. These services can be accessed remotely via mobile applications. Finally, the management and security layer ensures system configuration, resource optimization, and cybersecurity through encryption, authentication, and intrusion detection, protecting sensitive agricultural data from cyber threats [13].

In recent years, the rapid development of the Internet of Things (IoT) and machine learning (ML) has revolutionized precision irrigation management, enabling real-time monitoring, data-driven decision-making, and optimized water use efficiency [14,15]. IoT-based wireless sensor networks (WSNs) facilitate continuous soil moisture, weather, and crop health monitoring, while ML algorithms analyze these datasets to predict irrigation demands and automate control systems [16,17]. Such advancements mitigate water scarcity challenges by reducing over-irrigation and improving crop yields [18,19], underscoring their transformative potential in sustainable agriculture.

Bibliometric analyses offer a powerful approach to achieve this aim by mapping the intellectual landscape, revealing influential research clusters, and identifying emerging themes and innovation hotspots. The following section synthesizes the main findings of previous bibliometric studies on IoT applications in agriculture, offering valuable insights into the current research landscape and guiding future innovation. Numerous bibliometric studies have explored its evolution, research trends, and thematic developments, providing valuable insights into the advancements and research trajectories of this transformative field.

One of the earliest comprehensive bibliometric analyses was conducted by Suebsombut et al. [20], which classified trends and future directions in smart-farming research. The study provided foundational insights into the early development of IoT applications in agriculture. Research priorities that focused on sensor networks, data analytics, and the early adoption of automation technologies were identified. Bouzembrak et al. [21] extended this early exploration by focusing on IoT applications in food safety. The bibliometric analysis highlighted the potential of IoT technologies for enhancing food safety monitoring and traceability throughout the agricultural supply chain. The authors identified emerging research trends centered on real-time monitoring systems and predictive analytics for contamination detection. The findings underscored the role of IoT in ensuring food security and consumer safety. Singh et al. [22] conducted a comprehensive bibliometric study that revealed the growing adoption of IoT technologies for agricultural operations, particularly in areas such as smart irrigation and precision farming. The study mapped key research themes, including the integration of IoT with machine learning and cloud computing for advanced data processing and decision-making. This work highlighted the increasing sophistication of IoT applications in agriculture, driven by the need for more efficient resource management. The authors proposed a research agenda for advancing Agriculture 4.0, which included the development of interoperable platforms and the adoption of advanced analytics for predictive modeling.

One of the early studies focusing specifically on IoT trends in irrigation management is [23], which provided an overview of IoT and irrigation approaches through bibliometric analysis. The findings indicated a strong focus on developing efficient irrigation systems that leverage IoT technologies for real-time monitoring and control. The study emphasized the role of IoT in mitigating the impacts of climate change on agricultural water resources. Starting from 2023, bibliometric analysis for IoT integration in agriculture provided a deeper thematic focus on specific applications of IoT in agriculture. Gurmessa and Assefa [24] conducted a scientometric analysis focusing on smart irrigation literature. Their findings indicated a growing emphasis on integrating IoT and sensor technologies for efficient water management. Key research gaps were identified, such as the need for standardized protocols and scalable solutions for IoT-enabled irrigation systems. Some of the interesting proposed future research directions were the development of low-cost sensors and the integration of renewable energy sources for powering IoT devices.

Ahmed et al. [25] focused on smart irrigation management for improving water productivity under climate change conditions in drylands. The study identified emerging research themes, such as drought prediction and adaptive irrigation strategies, which are critical for sustaining agricultural productivity in arid and semi-arid regions. Still on irrigation management, Abdullahi et al. [26] provided a data-driven bibliometric review on precision irrigation. The authors highlighted the critical role of IoT in enabling data-driven irrigation management practices. The analysis identified key research trends, such as the integration of remote-sensing technologies and the use of big data analytics for optimizing irrigation schedules. They emphasized the need for collaborative research efforts to develop scalable and cost-effective precision irrigation solutions. Also, Pang et al. [27] examined trends in smart irrigation for smart agriculture through their bibliometric analysis. Their study identified key research themes, such as sensor technologies, data analytics, and IoT-enabled automation. The authors emphasized the need for collaborative research efforts to address the challenges associated with implementing smart irrigation systems at scale.

On the other hand, Chen et al. [28] explored the application of IoT technologies in green stormwater infrastructure (GSI). Although primarily focused on urban environments, their study offered valuable insights into the potential of IoT for managing water resources in agricultural landscapes. The authors highlighted the need for cross-disciplinary research to harness the full potential of IoT in GSI and other water-management applications. Abdullahi, Mahmud, Abi Hassan, and Ali [26] traced the evolution of IoT applications in smart agriculture through a comprehensive bibliometric analysis. The study revealed a steady increase in research outputs and identified influential publications and collaboration networks. The authors highlighted the importance of IoT in enabling precision farming practices and improving resource management. They also identified key research trends, such as the integration of IoT with blockchain for enhanced data security and traceability.

In 2024, more specialized applications were targeted. Addow and Jimale [29] examined the integration of IoT for water-quality monitoring in agricultural systems. Their study highlighted the increasing interest in leveraging IoT technologies to monitor and improve water quality. Key research clusters were identified, centered around water-quality monitoring, sensor technologies, and data analytics, underscoring the role of IoT in achieving sustainable water management.

Yet, the dominance of IoT applications for smart irrigation continued. Sadiq et al. [30] presented a bibliometric review on IoT-enabled smart irrigation systems. Their analysis revealed a significant rise in research activities aimed at optimizing water use efficiency through real-time data collection and automated irrigation systems. The authors mapped key research themes and identified influential authors and institutions contributing to advancements in smart irrigation technologies. Sumarsono, Muflihah, and Afiatna [11] explored the broader landscape of IoT research in agriculture through their bibliometric analysis. They identified the main research hotspots and trends, noting a growing emphasis on precision agriculture, data-driven decision-making, and sensor networks. The study also highlighted emerging challenges, such as data privacy and the need for standardized protocols.

The adoption of IoT in agriculture has seen a dramatic increase since 2015, with a notable surge in research output from 2017 onwards [26,31]. This growth is largely attributed to the convergence of IoT with AI and machine learning (ML), which has enabled more sophisticated applications in smart agriculture. For instance, IoT technologies are now widely used for real-time monitoring of soil moisture, climate conditions, and crop health, enabling precision agriculture practices such as targeted irrigation and fertilization [27,32]. These applications have proven particularly valuable in optimizing resource use and improving crop yields, especially in water-scarce regions [25].

The use of IoT in precision agriculture has also been bolstered by the integration of emerging technologies such as unmanned aerial vehicles (UAVs) and hyperspectral imaging, which provide high-resolution data for crop monitoring and management [32]. The convergence of these technologies has enabled farmers to make more informed decisions, reduce input costs, and improve overall agricultural productivity. The following (Table 1) lists the main found bibliometric studies regarding IoT in the field of agriculture in recent years.

Table 1. The recent bibliometric analysis regarding IoT in agriculture.

No	Article's Title	Authors	Year
1	Integrating IoT and Water Quality: A Bibliometric Analysis	[29]	2024
2	Smart irrigation systems enabled with internet of things: a bibliometric review	[30]	2024
3	The Internet of Things Research in Agriculture: A Bibliometric Analysis	[11]	2024
4	Theme Mapping and Bibliometric Analysis of Two Decades of Smart Farming	[33]	2023
5	Artificial intelligence-based decision support systems in smart agriculture: Bibliometric analysis for operational insights and future directions	[31]	2023
6	A Bibliometric Analysis of the Evolution of IoT Applications in Smart Agriculture	[26]	2023
7	A Scoping Review of the Smart Irrigation Literature Using Scientometric Analysis	[24]	2023
8	Application of Internet of Things (IoT) Technologies in Green Stormwater Infrastructure (GSI): A Bibliometric Review	[28]	2023
9	Bibliometric Analysis of Trends in Smart Irrigation for Smart Agriculture	[27]	2023
10	Advancements in Smart Farming: A Comprehensive Review of IoT, Wireless Communication, Sensors, and Hardware for Agricultural Automation	[34]	2023
11	A data-driven bibliometric review on precision irrigation	[32]	2023
12	An Overview of Smart Irrigation Management for Improving Water Productivity under Climate Change in Drylands	[25]	2023
13	Review of artificial intelligence and internet of things technologies in land and water management research during 1991–2021: A bibliometric analysis	[35]	2023
14	The Interplay between the Internet of Things and agriculture: A bibliometric analysis and research agenda	[36]	2022
15	A Bibliometric Analysis and Review of Resource Management in Internet of Water Things: The Use of Game Theory	[37]	2022
16	An overview of the internet of things (IoT) and irrigation approach through bibliometric analysis	[23]	2021
17	Wireless Sensor Networks in Agriculture: Insights from Bibliometric Analysis	[38]	2021
18	The Digital Agricultural Revolution: A Bibliometric Analysis Literature Review	[39]	2021
19	Bibliometric Analysis of the Use of the Internet of Things in Precision Agriculture	[40]	2021
20	Trends on Advanced Information and Communication Technologies for Improving Agricultural Productivities: A Bibliometric Analysis	[1]	2020
21	Internet of things and agriculture relationship: a bibliometric analysis	[22]	2020
22	Internet of Things in food safety: Literature review and a bibliometric analysis	[21]	2019
23	A Bibliometric Analysis on Agriculture 4.0	[41]	2019
24	The using of bibliometric analysis to classify trends and future directions on "Smart Farm"	[20]	2017

Some of the common key findings from the analysis of the reviewed previous bibliometric studies reveal that journals such as Sensors (Switzerland), IEEE, and Computers and Electronics in Agriculture are the most influential in this field [11,26]. Key authors and institutions from India and China have significantly contributed to the growth of IoT in agriculture [22,36]. The most-cited topics include precision agriculture, smart farming, and IoT-enabled irrigation systems, reflecting the field's focus on sustainability and efficiency [27,32].

The reviewed bibliometric studies provided a comprehensive overview of the research landscape surrounding IoT applications in agriculture. They highlight the growing importance of IoT in enabling data-driven decision-making, optimizing resource management, and enhancing the sustainability and resilience of agricultural systems. Thus, as the volume of research in this field grows rapidly, it becomes increasingly important to systematically analyze and map the existing body of knowledge to identify trends, gaps, and future opportunities. A bibliometric analysis provides a quantitative and objective framework to evaluate the evolution of research, highlight influential contributions, and uncover the most prevalent technologies, platforms, and applications.

Within this context, this study aims at offering a comprehensive overview of the current state of IoT for smart irrigation by analyzing the existing literature, to identify key trends, and addressing specific research questions, including the most commonly used technologies, microcontrollers (MCUs) and the main applications of IoT in agriculture, and the evolution of research themes over time. Through this analysis, the study aims to provide a holistic understanding of the current state of IoT in agriculture, highlight emerging areas of interest, and offer insights for future research directions to advance the integration of IoT technologies in agricultural irrigation practices. Furthermore, it aims to serve as a resource for researchers, practitioners, and policymakers to guide future innovations and investments in this domain.

2. Methodology

To conduct a comprehensive review of the integration of Internet of Things (IoT) technologies in irrigation management, a two-step methodological approach was adopted, combining bibliometric analysis and systematic review.

First, a bibliometric analysis was conducted to map the research landscape, identify key trends, prolific authors, and influential institutions in the field. The Web of Science (WoS) database was used as the primary source, accessed on 20 December 2024, covering the period from January 2014 to December 2024 (10 years). A structured search query was applied using the following Boolean terms:

("Internet of Things" OR "IoT" OR "Wireless sensing") AND ("Irrigation Management" OR "Irrigation Scheduling" OR "Smart Irrigation"). This initial search retrieved 290 papers, forming the dataset for bibliometric analysis.

Following the bibliometric assessment, a systematic review was performed to provide an in-depth examination of IoT applications in smart irrigation.

The research questions included the following:

- What are the main applications of IoT-based smart irrigation systems?
- What are the most-used communication technologies in the analyzed sample?
- What are the most-used microcontrollers in the IoT-based smart irrigation systems in the analyzed sample?

To ensure a rigorous selection process, the SPIDER framework (Sample, Phenomenon of Interest, Design, Evaluation, Research Type) [42] was applied to establish inclusion and exclusion criteria (Table 2). This step enabled the identification of relevant studies based on their focus on IoT-driven irrigation management, technological advancements, and decision-support frameworks. After applying the screening process, a refined subset of 92 studies was selected for qualitative analysis [43]. Figure 2 shows the PRISMA chart [44] followed in the screening process.

Table 2. Applying SPIDER framework for inclusion and exclusion criteria.

Category	Definition	Inclusion Criteria	Exclusion Criteria
Sample (S)	The population of studies considered	Studies focusing on IoT-enabled irrigation management, smart irrigation systems, or wireless sensor networks in irrigation	Studies unrelated to IoT applications in irrigation or focusing on non-agricultural sectors
Phenomenon of Interest (PI)	The main topic investigated	IoT-based technological advancements in irrigation, including smart irrigation scheduling, decision-making frameworks, and water management	Studies without a focus on IoT, irrigation, or water management
Design (D)	Research design/methods used	Studies that provide technical details on IoT implementation in irrigation (e.g., sensor types, communication technologies, microcontrollers)	Opinion pieces, non-peer-reviewed articles, or studies lacking technical details
Evaluation (E)	The outcome measures or key findings analyzed	Studies presenting bibliometric insights (e.g., co-occurrence analysis, citation networks) and discussing key IoT components used in smart irrigation	Studies with incomplete metadata, inaccessible full texts, or minimal relevance to IoT-driven irrigation
Research type (R)	Type of research included	Empirical studies, systematic reviews, bibliometric analyses, and experimental studies on IoT in irrigation	Conference abstracts, patents, editorials, or studies outside the WoS database

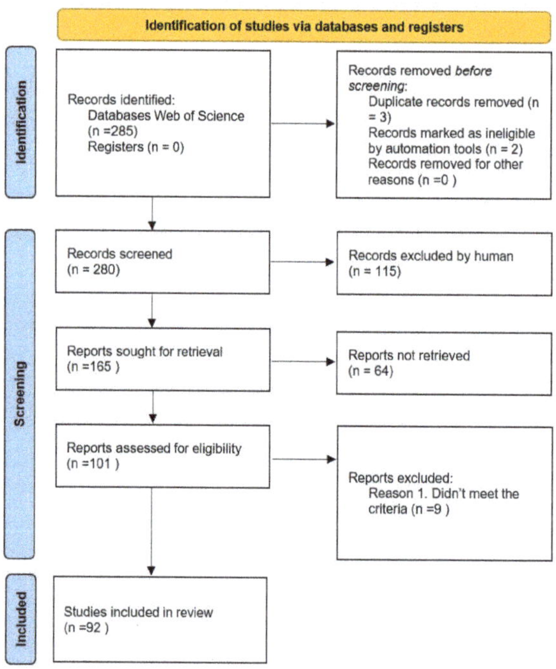

Figure 2. The screening process PRISMA chart.

VOSviewer [45] was used for network visualization and co-occurrence mapping of the bibliometric analysis, allowing the identification of key research clusters, influential authors, and collaborative networks. It employs the following steps:

(I) Co-occurrence and similarity calculation using similarity scores between entities (authors, journals, keywords, documents). This is calculated based on three bibliographic relationships:

(a) Co-authorship analysis for authors, organizations, or countries where the co-authorship strength S_{ij} between two authors i and j is defined as [46]

$$S_{ij}^{CA} = \frac{C_{ij}}{\sqrt{C_i C_j}} \quad (1)$$

where S_{ij}^{CA} represents the number of publications co-authored by i and j, and C_i and C_j denote the total number of publications by i and j, respectively.

(b) Co-occurrence analysis of the keywords where the normalized co-occurrence similarity S_{ij} is given by Equation (2) [47]:

$$S_{ij}^{kw} = \frac{C_{ij}}{C_i C_j} \quad (2)$$

where S_{ij}^{kw} represent the number of times keywords i and j appeared together in a publication, and C_i and C_j denote the total number of occurrences in publications of i and j, respectively.

(c) Co-citation analysis for the sources using the co-citation strength S_{ij} as determined by Equation (3):

$$S_{ij}^{CC} = \frac{CC_{ij}}{\sqrt{CC_i CC_j}} \quad (3)$$

where CC_{ij} represents the number of times documents i and j are cited together, and CC_i and CC_j denote the total citations received by documents i and j.

(II) Applying thresholds to filter and retain only the most relevant entities and connections using the minimum occurrences threshold, which ensures that only entities appearing at least N times are included. In addition to the minimum link strength threshold which is used to retain only strong connections by filtering based on similarity scores. VOSviewer adjusts this threshold dynamically based on the user's input, dataset's density and visual clarity. An entity is included if

$$C_i \geq N$$

or

$$S_i \geq \text{min.link strength threshold}$$

where C_i represents the occurrences of entity i, and S_i is the similarity/link strength.

(III) Optimizing the visualization: the VOS (visualization of similarities) mapping technique is applied, which minimizes the following objective function (Equation (4)) [48]:

$$\sum_{i<j} S_{ij}(d_{ij} - 1)(d_{ij} - 1) \quad (4)$$

where d_{ij} is the Euclidean distance between items I and j, and S_{ij} represents the similarity strength. This function ensures that similar items are positioned closer together while maintaining clarity in the visualization.

(IV) Grouping similar entities (clustering): to detect meaningful groups, clustering using a modularity-based approach is applied (Equation (5)) [49,50].

$$Q = \frac{1}{2m}\sum_{ij}(A_{ij} - \frac{K_i K_j}{2m})\delta(C_i, C_j) \quad (5)$$

where A_{ij} represents the actual weight of the link between i and j, K_i is the sum of the weights for node i, m is the total weight of all links, and $\delta(C_i, C_j)$) equals 1 if i and j belong to the same cluster; otherwise, 0.

Yet, the analysis was limited to the WoS database, potentially excluding relevant studies indexed in other databases. It is also worth mentioning that bibliometric tools rely on metadata quality; any inconsistencies in the WoS database could affect the analysis.

3. Results

3.1. Yearly Publication Count

Based on the publication trends of IoT in irrigation management from 2017 to 2024 (Figure 3), several key observations can be made. Between 2017 and 2019, the number of publications remained relatively low, with a gradual increase from five articles in 2017 to fourteen in 2019. This period represents an early phase in the research domain, where interest was emerging but had not yet gained significant traction.

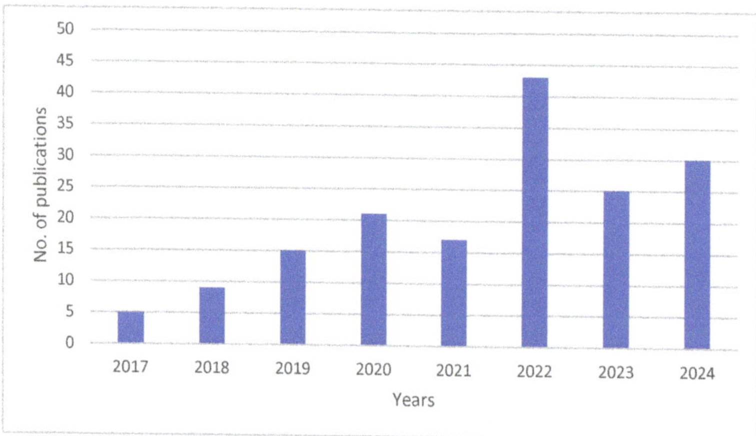

Figure 3. Trends in IoT in irrigation management.

From 2020 to 2022, the number of publications saw a steady increase, reaching 21 in 2020 and peaking at 45 in 2022. This substantial growth suggests an increasing scholarly interest in applying IoT technologies to irrigation management, likely driven by advancements in sensor networks, data analytics, automated irrigation systems, and the cost effectiveness of solid-state sensors and open-source hardware. The rise in publications during this period aligns with the broader trend of digital transformation in agriculture, where IoT is increasingly recognized as a key enabler of precision irrigation.

However, in 2023, the number of publications dropped to 25, followed by a slight increase to 31 in 2024. This decline in 2023, despite the previous growth trend, may indicate a shift towards more applied and interdisciplinary studies integrating IoT with other smart agriculture technologies. The rebound in 2024 suggests a renewed focus on IoT in irrigation management, possibly reflecting emerging challenges such as climate change adaptation, water resource optimization, integration with AI-driven decision support systems, and the emergence of low-cost solid-state sensors. The data underscore that IoT for irrigation management has evolved from a niche research area to a recognized field of study. While the number of publications fluctuated in recent years, the overall trend indicates growing interest and continued advancements in leveraging IoT for sustainable and efficient water management in agriculture.

3.2. Co-Authorship Collaboration Among Countries

As illustrated in Figure 4, an analysis of collaborative authorship in the field of IoT applications for irrigation management indicates that the most active contributing countries are India, China, the United States, Brazil, Italy, and Saudi Arabia. This distribution aligns with global trends in agricultural research, where both emerging and established economies play key roles in advancing smart irrigation technologies.

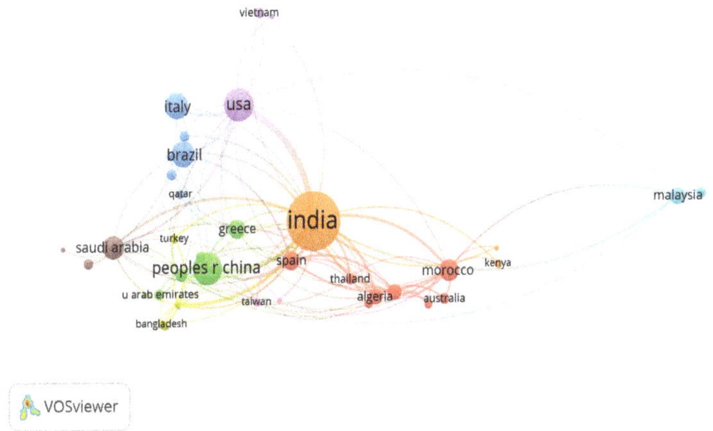

Figure 4. Co-authorship network map based on collaborations between countries.

India and China exhibit significant research activity in IoT-driven irrigation management, reflecting their large agricultural sectors and increasing pressures related to water scarcity and food security. The substantial number of collaborative publications from these nations suggests a strong commitment to integrating digital technologies for enhancing agricultural productivity and optimizing water resources. Notably, China's extensive investments in smart agriculture and IoT infrastructure have fostered robust academic and industrial collaborations. The United States, a leader in agricultural innovation, remains a dominant contributor to research and development in IoT-based irrigation. With well-established research institutions and government-supported precision agriculture initiatives, the U.S. is at the forefront of advancements in sensor technologies, machine learning applications, and decision-support systems for optimized water management. The high level of international collaborations involving U.S.-based researchers highlights efforts to disseminate technological advancements to regions facing similar agricultural challenges. Brazil's strong presence among the top contributors underscores its increasing focus on precision irrigation, particularly given its expansive agricultural landscapes that rely heavily on water-intensive crops. As one of the world's leading food producers, Brazil's engagement in collaborative research on IoT applications in irrigation reflects a strategic initiative to improve resource efficiency and mitigate climate-induced variability in agricultural productivity. Italy, despite its relatively smaller geographic and population size, remains a significant contributor to scientific research on IoT-driven irrigation systems. The country's well-established tradition of agricultural innovation, particularly within Mediterranean cropping systems where water management is critical, has driven interdisciplinary collaborations. Italian research institutions frequently integrate IoT technologies with agronomy, hydrology, and environmental sciences to develop advanced irrigation solutions. Saudi Arabia's inclusion among the leading contributors reflects its growing investment in sustainable agricultural practices, particularly in arid and semi-arid regions where water conservation is a national priority. Given the country's heavy reliance on irrigation for food production, there has been a concerted effort to adopt smart irrigation systems powered by IoT to enhance water-use efficiency. The high level of international collaboration involving Saudi researchers indicates ongoing efforts to adapt and implement digital irrigation solutions tailored to desert agriculture. A closer look at international collaborations highlights the increasingly interconnected nature of research in IoT-driven irrigation management. The exchange of knowledge between water-scarce regions and technologically advanced research hubs demonstrates a concerted effort to address global agricultural water challenges through innovative digital solutions. These findings emphasize the importance of cross-border partnerships in accelerating the adoption of smart irrigation technologies and ensuring their successful implementation across diverse agro-climatic regions.

The bibliometric analysis of the most cited journals in IoT-based smart irrigation (Figure 5) research reveals three clusters with a strong interdisciplinary connection between agriculture, water management, sensor technology, and IoT applications. The journal Computer Electronics and Agriculture stands out as the most influential source, with 388 citations and the highest total link strength of 8184, indicating its central role in disseminating research at the intersection of IoT and agricultural automation. Similarly, Agriculture Water Management (266 citations, 6035 link strength) underscores the critical role of water-use efficiency studies in smart irrigation systems. The prominence of Sensors—Basel (214 citations, 4980 link strength) highlights the growing focus on sensor technologies for precision agriculture and irrigation control. The presence of IEEE Access (110 citations) and IEEE Internet of Things (64 citations) in the top-ranking journals demonstrates the integration of advanced IoT frameworks, networking protocols, and real-time data analytics in irrigation research.

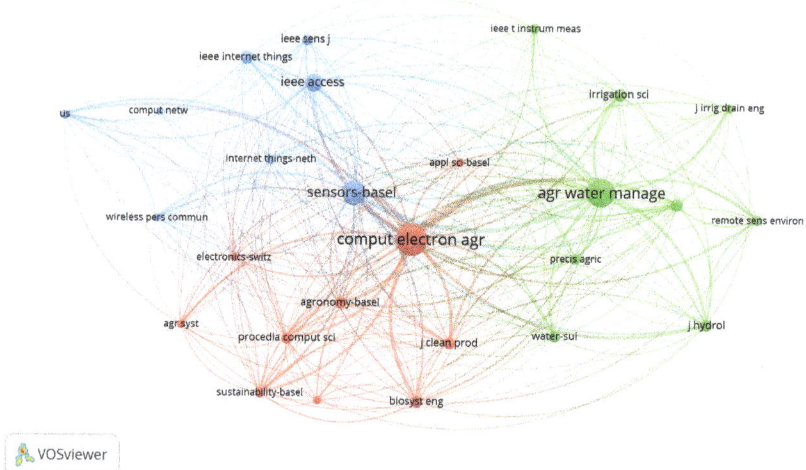

Figure 5. Co-citation network map based on scientific journals.

Journals related to hydrology (Journal of Hydrology, 46 citations), environmental monitoring (Remote Sensing, 52 citations), and sustainability (Sustainability—Basel, 36 citations) indicate that IoT in irrigation is closely linked to climate adaptation, remote sensing, and sustainable water resource management. Meanwhile, Precision Agriculture (37 citations, 1105 link strength) and Irrigation Science (54 citations, 1273 link strength) emphasize the impact of IoT on optimizing irrigation techniques.

The presence of IEEE Sensors (41 citations), Internet of Things (29 citations), and Wireless Personal Communications (23 citations) further highlights the technological aspect of IoT implementation in agriculture. This bibliometric analysis confirms that IoT-based irrigation research is highly interdisciplinary, combining agriculture, hydrology, sensor networks, and computing technologies to develop more efficient, data-driven irrigation solutions.

The analysis of the most relevant authors' keywords on the topic (Figure 6) reveals the topics and applications; Internet of Things (49 occurrences and 251 link strength), smart irrigation (48 occurrences and 242 link strength), irrigation (25 occurrences and 166 link strength), sensors (20 occurrences and 134 link strength), precision agriculture (19 occurrences and 103 link strength), and wireless sensor networks (9 occurrences and 78 link strength) are high-frequency phrases in smart irrigation management research.

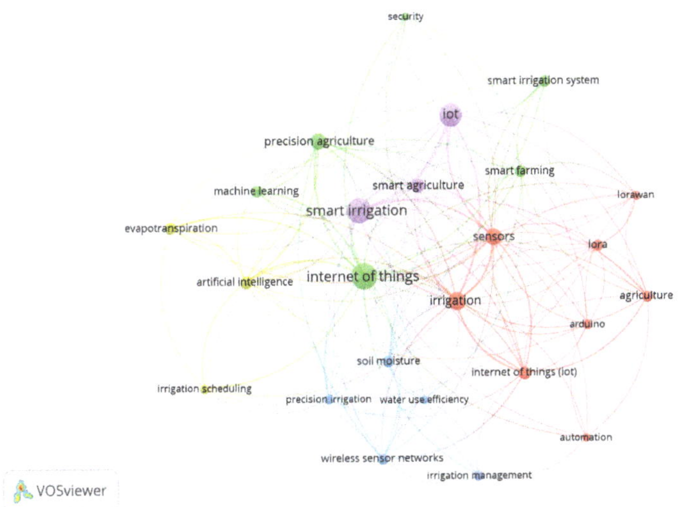

Figure 6. Co-occurrence network map based on authors' keywords.

3.3. Microcontrollers and Boards

The selection of microcontrollers in IoT-based smart irrigation research is a crucial factor that directly influences the performance, scalability, and practicality of implemented solutions. Microcontrollers serve as the backbone of IoT systems, handling sensor data acquisition, processing, and communication with cloud platforms or edge computing nodes. Their efficiency in terms of power consumption, computational capability, and compatibility with communication protocols (such as LoRa, Wi-Fi, or Zigbee) determines the feasibility of deploying large-scale, low-maintenance irrigation networks. Identifying the most commonly used microcontrollers in scientific publications provides insight into prevailing technological preferences, highlighting trends in hardware selection based on cost, availability, ease of integration, and support for real-time data processing. Additionally, understanding the dominant microcontroller platforms can help guide future research toward optimizing firmware, enhancing interoperability between different IoT nodes, and improving power management strategies. Given the increasing need for precision irrigation in resource-scarce environments, selecting microcontrollers with low energy consumption and robust connectivity features is vital to ensuring sustainable and long-term adoption of IoT-driven irrigation systems. The analyzed sample revealed that Arduino UNO (Arduino LLC, Ivrea, Italy), Raspberry Pi (Raspberry Pi Foundation, Cambridge, UK), ESP8266 (Espressif Systems, Shanghai, China), ESP32 (Espressif Systems, Shanghai, China) and ATmega328P (Microchip Technology, Chandler, AZ, USA) were the most used MCUs in the research respectively. The analysis also showed that there are emerging IoT MCUs with high growth potential and market expansion of IoT MCUs providing diversity of options. Table 3 shows all the reported used microcontrollers in the analyzed sample, along with the publications using each MCU and their country of origin, while Figure 7 shows the most-used MCUs.

Table 3. Listing all the reported MCUs in IoT for smart irrigation research between 2014 and 2024.

Board/MCU	Manufacturer	City	Country	References
Arduino Based MCUs	Arduino LLC	Ivrea	Italy	[51–75]
ATmega328 and ATmega328P	Microchip Technology	Chandler, Arizona	USA	[56,76–80]
ESP32	Espressif Systems	Shanghai	China	[5,8,81–86]
ESP8266	Espressif Systems	Shanghai	China	[51,53,58,60,61,63,76,77,83,87–94]
Raspberry Pi	Raspberry Pi Foundation	Cambridge	UK	[17,71,81,95–101]

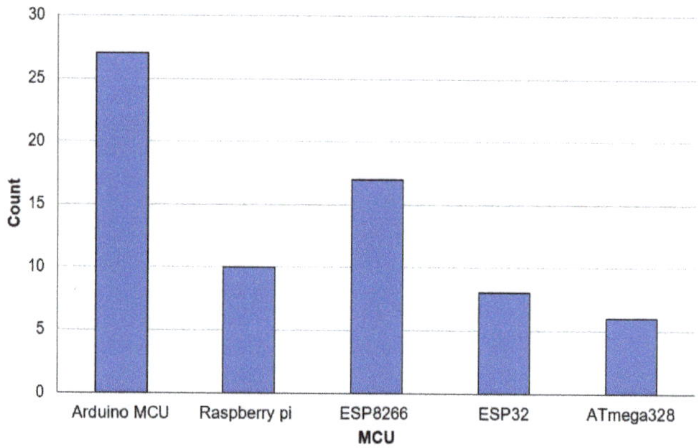

Figure 7. Highest-used MCUs in the analyzed sample.

3.4. Communication Technologies

The analysis of the scientific publications on IoT-based irrigation systems provides valuable insights into the most commonly used communication technologies (Figure 8). WiFi emerged as the most frequently cited technology, followed closely by LoRa and ZigBee. These figures indicate that while WiFi remains a dominant choice in current research, there is a noticeable shift toward LoRa and ZigBee, reflecting their increasing relevance in precision irrigation applications.

WiFi remains a widely used communication protocol for IoT-enabled irrigation due to its high-speed connectivity, ease of integration with cloud platforms, and existing infrastructure availability. However, its reliance on stable internet access, high power consumption, and limited transmission range (typically under 100 m in open fields) present significant challenges for large-scale agricultural applications. WiFi-based systems are often deployed in controlled environments, such as research stations, greenhouses, and small-scale farms, where network infrastructure is accessible, and power constraints are less significant.

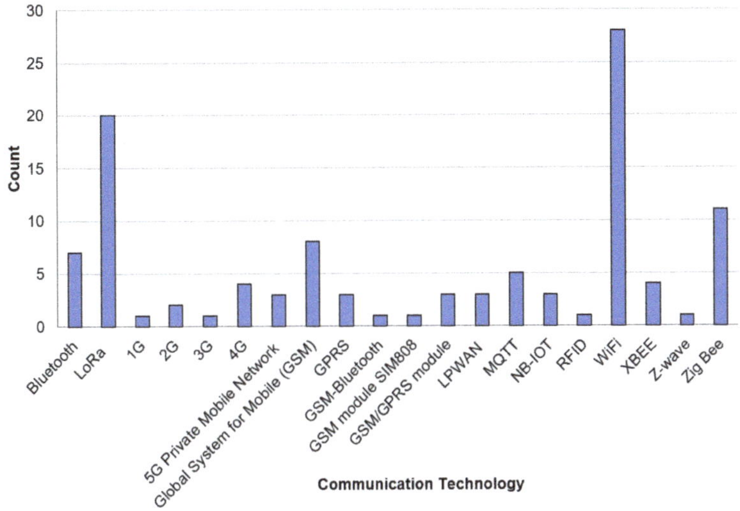

Figure 8. The most-used communication technologies.

Despite its advantages, the increasing focus on remote, autonomous, and energy-efficient irrigation systems has led researchers to explore alternative communication technologies. The growing number of publications on LoRa and ZigBee suggests that researchers are actively investigating these alternatives to address the limitations of WiFi-based smart irrigation networks. Unlike WiFi, which requires significant energy and infrastructure, LoRa operates over long distances (up to 10–15 km in rural areas) while consuming minimal power, making it ideal for monitoring soil moisture, weather conditions, and water distribution across expansive agricultural fields. Its key benefits in irrigation include the following:

- Extended coverage, enabling the deployment of sensors across large farms without needing extensive network infrastructure.
- Low power consumption, allowing battery-operated devices to function for several years with minimal maintenance.
- Cost-effectiveness, as LoRa operates in unlicensed frequency bands, reducing operational expenses compared to cellular networks.

Given these advantages, the near-equal number of publications on WiFi and LoRa suggests that LoRa is rapidly becoming a mainstream alternative, particularly in regions where power and internet infrastructure are unreliable.

ZigBee has also gained considerable attention in smart-irrigation research, particularly in studies conducted in greenhouses, orchards, and smallholder farms where multiple sensors must communicate efficiently. Unlike LoRa, which excels in long-range communication, ZigBee is designed for short-range, low-power, and low-data-rate applications, making it an ideal choice for localized irrigation management systems.

While ZigBee has fewer publications than WiFi and LoRa, its significant presence underscores its importance in scenarios where precise, localized control is required. Researchers have investigated ZigBee's potential in automated irrigation scheduling, soil moisture sensing, and climate control within protected agricultural environments.

Given this shift, future research and practical implementations of IoT-based irrigation should prioritize energy-efficient, scalable, and long-range communication technologies. The increasing focus on LoRa and ZigBee in scientific literature indicates their growing adoption in real-world agricultural applications, paving the way for more sustainable and intelligent irrigation management solutions.

3.5. The Main Objective of IoT Sensing in Irrigation Systems

The most prevalent applications included remote monitoring and control, water use optimization, soil moisture monitoring, weather monitoring, energy efficiency, and crop yield improvement, as shown in Figure 9. These findings highlight the critical areas where IoT solutions are being leveraged to enhance irrigation management. Below, we provide a detailed discussion of each of these trends and their implications.

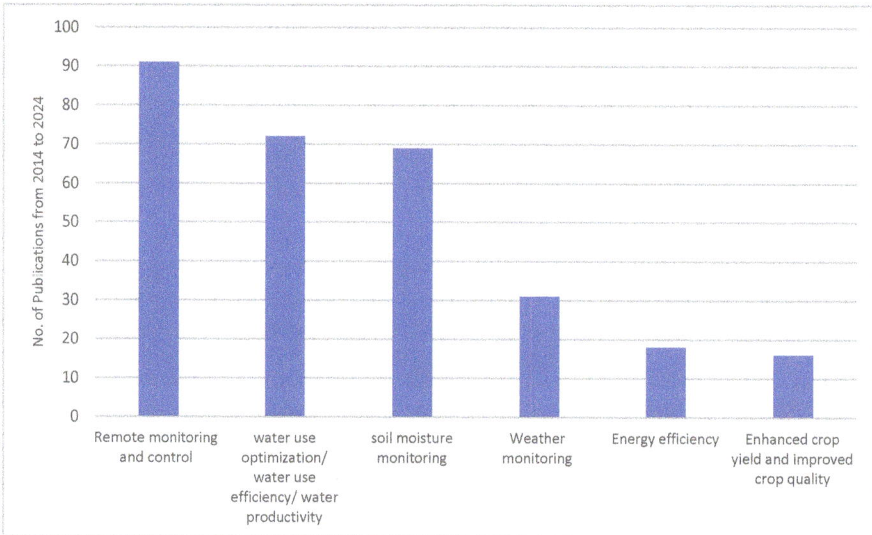

Figure 9. Main applications of IoT in smart irrigation as reported from 2014 to 2024.

The most frequently reported IoT application in smart irrigation is remote monitoring and control, with 91 publications emphasizing this capability. Remote monitoring enables farmers and irrigation managers to oversee their irrigation systems in real time, reducing the need for manual field visits and ensuring timely responses to changes in soil moisture levels, weather conditions, and crop water requirements [102]. IoT-based remote control systems allow for automated adjustments to irrigation schedules, valve operations, and pump activation based on real-time data.

This trend suggests that researchers and practitioners recognize the importance of remote accessibility in improving irrigation efficiency and reducing labor costs. The widespread adoption of remote monitoring and control is particularly beneficial in regions with limited agricultural labor availability or where large-scale farming operations make manual irrigation management impractical [103]. Additionally, real-time monitoring can help prevent water wastage by detecting leaks, clogged emitters, or system malfunctions, ultimately contributing to water conservation efforts.

The second-most reported IoT application is water use optimization, with 72 publications focusing on enhancing water efficiency and productivity. Efficient water management is a primary goal of smart irrigation, as excessive or insufficient irrigation can negatively impact crop health, soil quality, and overall agricultural sustainability.

IoT-enabled irrigation systems optimize water use by integrating data from soil moisture sensors, weather forecasts, and crop water requirements to provide precise irrigation recommendations. These systems help mitigate the risks of over-irrigation, which can lead to soil degradation and nutrient leaching, and under-irrigation, which can cause water stress and yield reduction. By applying only the required amount of water at the right time, IoT solutions contribute to sustainable water resource management and climate resilience in agriculture [104].

Soil moisture monitoring is another key area of IoT application, with 69 publications highlighting its significance in smart irrigation. Measuring soil moisture levels provides essential information for determining when and how much to irrigate [5,30,105]. IoT-based soil moisture sensors offer continuous real-time data, enabling precision irrigation strategies that align with crop needs and soil characteristics.

Accurate soil moisture monitoring prevents inefficient irrigation practices, such as watering when the soil already has adequate moisture or delaying irrigation until crops experience water stress. Moreover, integrating soil moisture data with remote sensing technologies and machine learning models enhances decision-making, allowing for adaptive irrigation strategies that respond dynamically to environmental conditions [106–108].

Weather monitoring plays a critical role in smart irrigation, as weather patterns directly influence evapotranspiration rates, soil moisture levels, and crop water requirements. Our research found that 31 publications discuss IoT applications in weather monitoring for irrigation management. IoT-based weather stations and connected meteorological sensors provide real-time data on temperature, humidity, wind speed, and solar radiation, which are used to predict irrigation needs more accurately.

Integrating weather data into irrigation decision-making helps optimize water use by aligning irrigation schedules with rainfall events, minimizing unnecessary watering, and preventing over-reliance on groundwater resources [25,32,109,110]. Additionally, advanced weather forecasting models powered by IoT data enable early warnings for drought conditions, allowing farmers to implement water conservation measures and adjust irrigation strategies accordingly.

Energy efficiency is a growing concern in smart irrigation, with 18 publications addressing its importance. Traditional irrigation methods often rely on energy-intensive pumping systems that contribute to high operational costs and environmental footprints [111]. IoT solutions improve energy efficiency by optimizing pump usage, reducing the frequency of manual interventions, and integrating renewable energy sources such as solar-powered irrigation systems [112,113].

By implementing smart energy-management strategies, IoT-enabled irrigation systems can significantly lower energy consumption while maintaining optimal irrigation performance. For example, variable frequency drive (VFD) controllers can adjust pump speeds based on real-time demand, preventing energy waste and extending equipment lifespan [114,115]. Additionally, IoT data analytics can identify inefficiencies in irrigation infrastructure, enabling proactive maintenance and reducing energy-intensive system failures [116,117].

While most IoT applications in irrigation focus on resource efficiency, 16 publications emphasize their role in enhancing crop yield and improving crop quality. Precision irrigation ensures that crops receive the right amount of water at critical growth stages, preventing drought stress and promoting uniform development. IoT-driven nutrient and water management strategies further enhance soil health, leading to improved crop nutrition and resilience against pests and diseases.

By integrating IoT-based irrigation with precision agriculture technologies such as drone imaging, spectral analysis, and artificial intelligence, farmers can optimize fertilizer application, detect plant health issues early, and enhance overall farm productivity. As global food demand continues to rise, leveraging IoT to maximize crop yield and quality will be instrumental in ensuring food security and agricultural sustainability.

The findings of this study highlight the transformative impact of IoT on modern irrigation practices. The emphasis on remote monitoring and control reflects a shift towards digital agriculture, where real-time data-driven decision-making is becoming the norm. The strong focus on water use optimization underscores the urgency of addressing water scarcity challenges and improving agricultural resilience against climate variability.

Moreover, the integration of soil moisture monitoring and weather data into irrigation management indicates an increasing reliance on predictive analytics for sustainable water resource management. Energy efficiency considerations suggest that smart irrigation technologies must not only optimize water use but also minimize environmental and financial costs associated with irrigation operations.

The relatively lower number of studies focusing on crop yield and quality improvement suggests that while resource efficiency remains a priority, more research is needed to explore the direct agronomic benefits of IoT-based irrigation. Future research should investigate how IoT applications influence crop physiology, nutrient uptake, and soil microbiome interactions to develop comprehensive irrigation solutions that maximize both water efficiency and agricultural productivity.

4. Challenges and Future Direction

The integration of the Internet of Things (IoT) in irrigation management presents a transformative approach to optimizing water use efficiency, enhancing crop productivity, and mitigating the impact of climate change on agriculture. While significant progress has been made in developing and deploying IoT-driven smart irrigation systems, from the reviewed sample, several key challenges hinder the large-scale adoption and long-term sustainability of IoT-based solutions. Addressing these challenges requires a holistic approach that considers technological, economic, environmental, and user-centered factors.

- Lack of universal standards: One of the most critical barriers to IoT adoption in irrigation management is the lack of universal standards for communication protocols, data exchange formats, and security frameworks [118]. The heterogeneity of IoT devices, manufactured by different vendors, often leads to compatibility issues, limiting the seamless integration of sensors, controllers, and cloud platforms [119]. Establishing stronger standardization practices would enhance interoperability, allowing farmers and water managers to adopt diverse IoT solutions without concerns about vendor lock-in. Furthermore, standardization would improve security across the entire IoT stack, ensuring that data transmitted from field sensors to cloud platforms remain protected against cyber threats and unauthorized access.
- Cybersecurity: Ensuring the security of IoT-based irrigation systems is paramount, as these systems rely on continuous data acquisition, transmission, and analysis to make real-time irrigation decisions. Current security strategies in agricultural IoT applications remain fragmented, with limited end-to-end encryption and physical security measures for field-deployed devices [120]. Developing robust security frameworks that incorporate encryption, blockchain-based data verification, and intrusion detection mechanisms would strengthen the resilience of IoT-enabled irrigation networks [121]. Moreover, raising awareness and providing training on cybersecurity best practices for farmers and agricultural stakeholders would further mitigate security risks.

- Modular-based solutions: A flexible and modular approach to designing IoT solutions for irrigation can significantly enhance their adoption and usability [14]. Modular hardware and software architectures enable easier customization, allowing farmers to tailor IoT deployments based on specific crop requirements, soil conditions, and water availability. Additionally, modularity supports scalability, ensuring that small-scale pilot deployments can be expanded into large-scale operations without requiring a complete system overhaul. By prioritizing modularity, developers can create IoT solutions that cater to both smallholder farmers and large commercial agricultural enterprises.
- Cost effectiveness: Despite the declining costs of embedded computing platforms, high-quality sensors and actuators remain expensive, limiting the affordability of IoT-based irrigation solutions for resource-constrained farmers. Reducing the overall unit cost of IoT hardware, internet connectivity, and data management services is essential for widespread adoption. This can be achieved through economies of scale in manufacturing, open-source development of sensor technologies, and government or private-sector subsidies that support precision irrigation initiatives. Furthermore, exploring innovative business models, such as pay-as-you-go or subscription-based IoT services, could make these technologies more accessible to farmers in developing regions.
- Compatibility with existing frameworks: The successful deployment of IoT solutions in irrigation management requires compatibility with existing agricultural infrastructure, including traditional irrigation systems, field machinery, and farm management software [15]. Developing IoT systems that can be retrofitted onto conventional irrigation networks—rather than requiring complete replacements—would encourage more farmers to transition toward digital water-management practices. This compatibility also extends to data integration, where IoT-based irrigation platforms should seamlessly interface with other agricultural decision-support systems, remote sensing data, and climate models to provide holistic insights for farm management.
- Robustness and reliability: IoT devices deployed in agricultural fields must withstand harsh environmental conditions, including extreme temperatures, humidity variations, and exposure to dust and chemicals [122]. Designing robust hardware that can endure seasonal and climate-related changes is essential to prevent frequent device failures and maintenance costs. Incorporating weatherproof enclosures, corrosion-resistant materials, and self-healing network architectures can enhance the durability of IoT irrigation systems, ensuring their reliable operation over extended periods [123].
- Studying the environmental impact of the resulting e-waste: As IoT adoption accelerates in agriculture, addressing the environmental impact of electronic waste from outdated or discarded devices becomes crucial. Sustainable practices, such as designing IoT hardware with recyclable materials, implementing take-back programs for old sensors, and promoting circular-economy principles, should be prioritized [25,124]. By integrating sustainability considerations into the design and deployment of IoT-based irrigation solutions, the agricultural sector can minimize its ecological footprint while maximizing the long-term benefits of precision irrigation.

The integration of IoT in irrigation management holds immense potential for enhancing water-use efficiency, optimizing crop growth, and mitigating climate-related agricultural challenges. However, several technical, economic, and operational barriers must be addressed to ensure widespread adoption. Standardization, power efficiency, security, scalability, cost reduction, and user-centered design are among the critical factors that will determine the success of IoT-enabled irrigation solutions. By fostering interdisciplinary research collaborations and leveraging advancements in sensor technology, data analytics, and machine learning, the future of smart irrigation can be both technologically robust and economically viable, ensuring sustainable agricultural water management for years to come.

Future research should focus on developing cost-effective IoT solutions, strengthening security frameworks, and enhancing user-friendly interfaces to promote mass adoption. Additionally, interdisciplinary collaborations between agronomists, engineers, and data scientists will be crucial in advancing IoT applications that address real-world agricultural challenges.

5. Conclusions

This systematic review provides a comprehensive overview of the research landscape surrounding the application of IoT in smart irrigation from 2014 to 2024. The analysis of publication trends indicates a significant rise in scholarly interest, with peak activity observed between 2020 and 2022, followed by a shift towards interdisciplinary and applied studies. Despite some fluctuations in publication numbers, the overall trajectory underscores the growing recognition of IoT as a transformative tool for precision irrigation and sustainable water management in agriculture.

The geographic distribution of contributions highlights the central role of countries such as India, China, and the United States in advancing IoT-based irrigation technologies. Their research efforts reflect diverse agricultural challenges, including water scarcity, food security, and climate adaptation, demonstrating the global relevance of IoT in optimizing irrigation practices. The strong international collaborations observed further reinforce the interconnected nature of agricultural research and the need for cross-border knowledge exchange to address common challenges.

The citation analysis of leading journals reveals that IoT in irrigation is inherently interdisciplinary, integrating agriculture, hydrology, sensor networks, and computing technologies. Highly cited journals such as Computers and Electronics in Agriculture, Agricultural Water Management, and Sensors—Basel serve as critical platforms for disseminating knowledge at the intersection of IoT and precision irrigation. The presence of journals focused on hydrology, environmental monitoring, and sustainability further indicates the broader implications of IoT in addressing water resource management and climate resilience.

Key technological trends in IoT-based smart irrigation, including the selection of microcontrollers and communication protocols, reveal a clear shift towards low-power, scalable, and cost-effective solutions. While WiFi remains a dominant communication technology, the increasing adoption of LoRa and ZigBee suggests a growing preference for energy-efficient and long-range communication in large-scale agricultural applications. The analysis of microcontrollers highlights the widespread use of Arduino, Raspberry Pi, ESP8266, and ESP32, indicating a preference for open-source, adaptable platforms suitable for real-time data acquisition and remote sensing.

The review of research applications underscores the primary objectives of IoT in irrigation systems, including remote monitoring and control, water use optimization, soil moisture and weather monitoring, energy efficiency, and crop yield improvement. The strong focus on real-time monitoring and automated irrigation control reflects a shift towards data-driven decision-making, reducing reliance on manual interventions and improving irrigation efficiency. The emphasis on energy efficiency and sustainability highlights the increasing need for smart irrigation technologies that minimize both water and energy consumption, ensuring long-term viability in resource-scarce environments.

Despite the extensive research on IoT applications in irrigation, certain gaps remain that warrant further investigation. While significant advancements have been made in sensor technologies, communication protocols, and automation, there is still a need to explore the agronomic impacts of IoT-based irrigation on crop physiology, nutrient dynamics, and soil health. Additionally, integrating IoT with emerging technologies such as artificial intelligence, edge computing, and blockchain could enhance decision-support systems, improve data security, and enable more adaptive and predictive irrigation strategies.

Overall, this systematic review confirms that IoT has evolved from an emerging concept to a well-established field of research in irrigation management. As global challenges related to water scarcity, food security, and climate change intensify, IoT-based smart irrigation systems will play a crucial role in enhancing agricultural resilience and sustainability. Future research should focus on refining existing technologies, fostering interdisciplinary collaborations, and developing scalable solutions that can be deployed across diverse agricultural landscapes to ensure efficient and sustainable water resource management in the years to come.

Supplementary Materials: The following supporting information can be downloaded at: https://www.mdpi.com/article/10.3390/s25072291/s1, Collected Publications Database.

Author Contributions: Conceptualization, A.A.A. and R.K.; methodology, A.A.A. and H.N.K.; software, H.N.K. and B.D.; validation, C.M.A.K. and A.A.A.; formal analysis, H.N.K., C.M.A.K. and A.A.A.; investigation, A.A.A. and G.D.; data curation, A.A.A.; writing—original draft preparation, A.A.A.; writing—review and editing, R.K.; supervision, A.A.A. and R.K. All authors have read and agreed to the published version of the manuscript.

Funding: This research received no external funding.

Data Availability Statement: Data are contained within the article and Supplementary Materials.

Conflicts of Interest: The authors declare no conflicts of interest.

References

1. Armenta-Medina, D.; Ramirez-delReal, T.A.; Villanueva-Vásquez, D.; Mejia-Aguirre, C. Trends on advanced information and communication technologies for improving agricultural productivities: A bibliometric analysis. *Agronomy* **2020**, *10*, 1989. [CrossRef]
2. Quy, V.K.; Hau, N.V.; Anh, D.V.; Quy, N.M.; Ban, N.T.; Lanza, S.; Randazzo, G.; Muzirafuti, A. IoT-enabled smart agriculture: Architecture, applications, and challenges. *Appl. Sci.* **2022**, *12*, 3396. [CrossRef]
3. MarketsandMarkets. *IoT in Agriculture Market by Components, Application, and Region—Global Forecast to 2026*; MarketsandMarkets Research Private: Pune, India, 2022.
4. PrecedenceResearch. *IoT in Agriculture Market Size, Share, Growth Report 2024–2034*; Precedence Research: Ottawa, ON, Canada, 2024.
5. Abdelmoneim, A.A.; Khadra, R.; Elkamouh, A.; Derardja, B.; Dragonetti, G. Towards Affordable Precision Irrigation: An Experimental Comparison of Weather-Based and Soil Water Potential-Based Irrigation Using Low-Cost IoT-Tensiometers on Drip Irrigated Lettuce. *Sustainability* **2023**, *16*, 306. [CrossRef]
6. Abdelmoneim, A.A.; Al Kalaany, C.M.; Khadra, R.; Derardja, B.; Dragonetti, G. Calibration of Low-Cost Capacitive Soil Moisture Sensors for Irrigation Management Applications. *Sensors* **2025**, *25*, 343. [CrossRef]
7. Demestichas, K.; Peppes, N.; Alexakis, T. Survey on security threats in agricultural IoT and smart farming. *Sensors* **2020**, *20*, 6458. [CrossRef]
8. Abdelmoneim, A.A.; Khadra, R.; Derardja, B.; Dragonetti, G. Internet of Things (IoT) for Soil Moisture Tensiometer Automation. *Micromachines* **2023**, *14*, 263. [CrossRef]
9. Lin, J.; Yu, W.; Zhang, N.; Yang, X.; Zhang, H.; Zhao, W. A survey on internet of things: Architecture, enabling technologies, security and privacy, and applications. *IEEE Internet Things J.* **2017**, *4*, 1125–1142. [CrossRef]
10. Xu, J.; Gu, B.; Tian, G. Review of agricultural IoT technology. *Artif. Intell. Agric.* **2022**, *6*, 10–22. [CrossRef]
11. Sumarsono, S.; Muflihah, N.; Afiatna, F.A.N.F. The internet of things research in agriculture: A bibliometric analysis. *Int. J. Agric. Syst.* **2024**, *12*, 1–20.
12. Effah, E.; Thiare, O.; Wyglinski, A.M. A Tutorial on Agricultural IoT: Fundamental Concepts, Architectures, Routing, and Optimization. *IoT* **2023**, *4*, 265–318. [CrossRef]
13. Ghaffari, A.; Jelodari, N.; Pouralish, S.; Derakhshanfard, N.; Arasteh, B. Securing internet of things using machine and deep learning methods: A survey. *Clust. Comput.* **2024**, *27*, 9065–9089. [CrossRef]

14. Goap, A.; Sharma, D.; Shukla, A.K.; Krishna, C.R. An IoT based smart irrigation management system using Machine learning and open source technologies. *Comput. Electron. Agric.* **2018**, *155*, 41–49. [CrossRef]
15. O'Shaughnessy, S.A.; Evett, S.R.; Colaizzi, P.D.; Howell, T.A. Wireless sensor network effectively controls center pivot irrigation of sorghum. *Appl. Eng. Agric.* **2013**, *29*, 853–864.
16. Lakhiar, I.A.; Yan, H.; Zhang, C.; Wang, G.; He, B.; Hao, B.; Han, Y.; Wang, B.; Bao, R.; Syed, T.N. A review of precision irrigation water-saving technology under changing climate for enhancing water use efficiency, crop yield, and environmental footprints. *Agriculture* **2024**, *14*, 1141. [CrossRef]
17. Avanija, J.; Rajyalakshmi, C.; Madhavi, K.R.; Rao, B.N.K. Enabling Smart Farming Through Edge Artificial Intelligence (AI). In *Agriculture and Aquaculture Applications of Biosensors and Bioelectronics*; IGI Global Scientific Publishing: Hershey, PA, USA, 2024; pp. 69–82.
18. Et-taibi, B.; Abid, M.R.; Boufounas, E.-M.; Morchid, A.; Bourhnane, S.; Hamed, T.A.; Benhaddou, D. Enhancing water management in smart agriculture: A cloud and IoT-Based smart irrigation system. *Results Eng.* **2024**, *22*, 102283.
19. EG, A.; Bala, G.J. IoT and ML-based automatic irrigation system for smart agriculture system. *Agron. J.* **2024**, *116*, 1187–1203.
20. Suebsombut, P.; Sekhari, A.; Sureepong, P.; Ueasangkomsate, P.; Bouras, A. The using of bibliometric analysis to classify trends and future directions on "smart farm". In Proceedings of the 2017 International Conference on Digital Arts, Media and Technology (ICDAMT), Chiang Mai, Thailand, 1–4 March 2017; pp. 136–141.
21. Bouzembrak, Y.; Klüche, M.; Gavai, A.; Marvin, H.J. Internet of Things in food safety: Literature review and a bibliometric analysis. *Trends Food Sci. Technol.* **2019**, *94*, 54–64.
22. Singh, S.; Haneef, F.; Kumar, S.; Ongsakul, V. Internet of things and agriculture relationship: A bibliometric analysis. *J. Glob. Bus. Adv.* **2020**, *13*, 643–664.
23. Jusoh, M.; Muttalib, M.; Krishnan, K.; Katimon, A. An overview of the internet of things (IoT) and irrigation approach through bibliometric analysis. *IOP Conf. Ser. Earth Environ. Sci.* **2021**, *756*, 012041.
24. Gurmessa, D.K.; Assefa, S.G. A Scoping Review of the Smart Irrigation Literature Using Scientometric Analysis. *J. Eng.* **2023**, *2023*, 2537005.
25. Ahmed, Z.; Gui, D.; Murtaza, G.; Yunfei, L.; Ali, S. An overview of smart irrigation management for improving water productivity under climate change in drylands. *Agronomy* **2023**, *13*, 2113. [CrossRef]
26. Abdullahi, H.O.; Mahmud, M.; Abi Hassan, A.; Ali, A.F. A Bibliometric Analysis of the Evolution of IoT Applications in Smart Agriculture. *Ingénierie Systèmes D Inf.* **2023**, *28*, 1495–1504. [CrossRef]
27. Pang, Y.; Marinello, F.; Tang, P.; Li, H.; Liang, Q. Bibliometric analysis of trends in smart irrigation for smart agriculture. *Sustainability* **2023**, *15*, 16420. [CrossRef]
28. Chen, T.; Wang, M.; Su, J.; Ikram, R.M.A.; Li, J. Application of Internet of Things (IoT) Technologies in Green Stormwater Infrastructure (GSI): A Bibliometric Review. *Sustainability* **2023**, *15*, 13317. [CrossRef]
29. Addow, M.A.; Jimale, A.D. Integrating IoT and Water Quality: A Bibliometric Analysis. *Int. J. Electron. Commun. Eng.* **2024**, *11*, 149–160. [CrossRef]
30. Sadiq, M.I.; Wanke, P.; Azad, M.A.K. Smart irrigation systems enabled with internet of things: A bibliometric review. *Int. J. Process Manag. Benchmarking* **2024**, *16*, 111–125.
31. Yousaf, A.; Kayvanfar, V.; Mazzoni, A.; Elomri, A. Artificial intelligence-based decision support systems in smart agriculture: Bibliometric analysis for operational insights and future directions. *Front. Sustain. Food Syst.* **2023**, *6*, 1053921.
32. Violino, S.; Figorilli, S.; Ferrigno, M.; Manganiello, V.; Pallottino, F.; Costa, C.; Menesatti, P. A data-driven bibliometric review on precision irrigation. *Smart Agric. Technol.* **2023**, *5*, 100320. [CrossRef]
33. Kushartadi, T.; Mulyono, A.E.; Al Hamdi, A.H.; Rizki, M.A.; Sadat Faidar, M.A.; Harsanto, W.D.; Suryanegara, M.; Asvial, M. Theme mapping and bibliometric analysis of two decades of smart farming. *Information* **2023**, *14*, 396. [CrossRef]
34. Prakash, C.; Singh, L.P.; Gupta, A.; Lohan, S.K. Advancements in smart farming: A comprehensive review of IoT, wireless communication, sensors, and hardware for agricultural automation. *Sens. Actuators A Phys.* **2023**, *362*, 114605. [CrossRef]
35. Patel, A.; Kethavath, A.; Kushwaha, N.; Naorem, A.; Jagadale, M.; Sheetal, K.; Renjith, P. Review of artificial intelligence and internet of things technologies in land and water management research during 1991–2021: A bibliometric analysis. *Eng. Appl. Artif. Intell.* **2023**, *123*, 106335. [CrossRef]
36. Rejeb, A.; Rejeb, K.; Abdollahi, A.; Al-Turjman, F.; Treiblmaier, H. The Interplay between the Internet of Things and agriculture: A bibliometric analysis and research agenda. *Internet Things* **2022**, *19*, 100580. [CrossRef]
37. Olatinwo, S.O.; Joubert, T.-H. A bibliometric analysis and review of resource management in internet of water things: The use of game theory. *Water* **2022**, *14*, 1636. [CrossRef]
38. Abdollahi, A.; Rejeb, K.; Rejeb, A.; Mostafa, M.M.; Zailani, S. Wireless sensor networks in agriculture: Insights from bibliometric analysis. *Sustainability* **2021**, *13*, 12011. [CrossRef]
39. Bertoglio, R.; Corbo, C.; Renga, F.M.; Matteucci, M. The digital agricultural revolution: A bibliometric analysis literature review. *IEEE Access* **2021**, *9*, 134762–134782. [CrossRef]

40. Lara, M.d.J.D.; Bernabe, J.G.; Benítez, R.Á.G.; Toxqui, J.M.; Huerta, M.K. Bibliometric analysis of the use of the internet of things in precision agriculture. In Proceedings of the 2021 IEEE International Conference on Engineering Veracruz (ICEV), Boca del Río, Veracruz, Mexico, 25–28 October 2021; pp. 1–5.
41. Singh, G.; Sahu, R. A bibliometric analysis on Agriculture 4.0. *NOLEGEIN-J. Oper. Res. Manag.* **2019**, *2*, 6–13.
42. Randles, R.; Finnegan, A. Guidelines for writing a systematic review. *Nurse Educ. Today* **2023**, *125*, 105803. [CrossRef]
43. Cooke, A.; Smith, D.M.; Booth, A. The benefits of a systematic search strategy when conducting qualitative evidence synthesis; the SPIDER tool. *Qual. Health Res.* **2012**, *22*, 1435–1443. [CrossRef]
44. Page, M.J.; McKenzie, J.E.; Bossuyt, P.M.; Boutron, I.; Hoffmann, T.C.; Mulrow, C.D.; Shamseer, L.; Tetzlaff, J.M.; Akl, E.A.; Brennan, S.E. The PRISMA 2020 statement: An updated guideline for reporting systematic reviews. *bmj* **2021**, *372*, n71. [CrossRef]
45. Waltman, L.; Van Eck, N.J. A smart local moving algorithm for large-scale modularity-based community detection. *Eur. Phys. J. B* **2013**, *86*, 1–14. [CrossRef]
46. Van Eck, N.; Waltman, L. Software survey: VOSviewer, a computer program for bibliometric mapping. *Scientometrics* **2010**, *84*, 523–538. [CrossRef] [PubMed]
47. Callon, M.; Courtial, J.P.; Laville, F. Co-word analysis as a tool for describing the network of interactions between basic and technological research: The case of polymer chemsitry. *Scientometrics* **1991**, *22*, 155–205. [CrossRef]
48. Waltman, L.; Van Eck, N.J.; Noyons, E.C. A unified approach to mapping and clustering of bibliometric networks. *J. Informetr.* **2010**, *4*, 629–635. [CrossRef]
49. Blondel, V.D.; Guillaume, J.-L.; Lambiotte, R.; Lefebvre, E. Fast unfolding of communities in large networks. *J. Stat. Mech. Theory Exp.* **2008**, *2008*, P10008. [CrossRef]
50. Newman, M.E. Modularity and community structure in networks. *Proc. Natl. Acad. Sci. USA* **2006**, *103*, 8577–8582. [CrossRef]
51. Kumar, V.; Sharma, K.V.; Kedam, N.; Patel, A.; Kate, T.R.; Rathnayake, U. A comprehensive review on smart and sustainable agriculture using IoT technologies. *Smart Agric. Technol.* **2024**, *8*, 100487. [CrossRef]
52. Bouali, E.-T.; Abid, M.R.; Boufounas, E.-M.; Hamed, T.A.; Benhaddou, D. Renewable energy integration into cloud & IoT-based smart agriculture. *IEEE Access* **2021**, *10*, 1175–1191.
53. Chamara, N.; Islam, M.D.; Bai, G.F.; Shi, Y.; Ge, Y. Ag-IoT for crop and environment monitoring: Past, present, and future. *Agric. Syst.* **2022**, *203*, 103497. [CrossRef]
54. Abdalgader, K.; Yousif, J.H. Agricultural Irrigation Control using Sensor-enabled Architecture. *KSII Trans. Internet Inf. Syst. (TIIS)* **2022**, *16*, 3275–3298.
55. Jamroen, C.; Komkum, P.; Fongkerd, C.; Krongpha, W. An intelligent irrigation scheduling system using low-cost wireless sensor network toward sustainable and precision agriculture. *IEEE Access* **2020**, *8*, 172756–172769. [CrossRef]
56. Pinto, J.S.; Camargo, L.C.; Duarte, S.N. Development of a Low-Cost Open-Source Platform Connected to the Internet for Acquisition of Environmental Parameters and Soil Moisture. *Eng. Agrícola* **2021**, *41*, 338–346. [CrossRef]
57. Puig Pérez-Barquero, F.; Rodríguez-Díaz, J.A.; Soriano, M.-A. Development of a Low-Cost Open-Source Platform for Smart Irrigation Systems. *Agronomy* **2022**, *12*, 2909. [CrossRef]
58. Ndunagu, J.N.; Ukhurebor, K.E.; Akaaza, M.; Onyancha, R.B. Development of a Wireless Sensor Network and IoT-based Smart Irrigation System. *Appl. Environ. Soil Sci.* **2022**, *2022*, 7678570. [CrossRef]
59. Tsiropoulos, Z.; Skoubris, E.; Fountas, S.; Gravalos, I.; Gemtos, T. Development of an energy efficient and fully autonomous low-cost IoT system for irrigation scheduling in water-scarce areas using different water sources. *Agriculture* **2022**, *12*, 1044. [CrossRef]
60. Jain, R.K. Experimental performance of smart IoT-enabled drip irrigation system using and controlled through web-based applications. *Smart Agric. Technol.* **2023**, *4*, 100215. [CrossRef]
61. Arafa, Y.; El-Gindy, A.-G.M.; El-Shirbeny, M.; Bourouah, M.; Abd-ElGawad, A.M.; Rashad, Y.M.; Hafez, M.; Youssef, M.A. Improving the spatial deployment of the soil moisture sensors in smart irrigation systems using GIS. *Cogent Food Agric.* **2024**, *10*, 2361124. [CrossRef]
62. Benzaouia, M.; Hajji, B.; Mellit, A.; Rabhi, A. Fuzzy-IoT smart irrigation system for precision scheduling and monitoring. *Comput. Electron. Agric.* **2023**, *215*, 108407. [CrossRef]
63. Thakare, S.; Bhagat, P. Arduino-based smart irrigation using sensors and esp8266 wifi module. In Proceedings of the 2018 Second International Conference on Intelligent Computing and Control Systems (ICICCS), Madurai, India, 14–15 June 2018; pp. 1–5.
64. Benyezza, H.; Bouhedda, M.; Djellout, K.; Saidi, A. Smart irrigation system based ThingSpeak and Arduino. In Proceedings of the 2018 International Conference on Applied Smart Systems (ICASS), Medea, Algeria, 24–25 November 2018; pp. 1–4.
65. El Mezouari, A.; El Fazziki, A.; Sadgal, M. Smart irrigation system. *IFAC-Pap.* **2022**, *55*, 3298–3303. [CrossRef]
66. Sanjeevi, P.; Prasanna, S.; Siva Kumar, B.; Gunasekaran, G.; Alagiri, I.; Vijay Anand, R. Precision agriculture and farming using Internet of Things based on wireless sensor network. *Trans. Emerg. Telecommun. Technol.* **2020**, *31*, e3978.
67. Prakash, B.R.; Kulkarni, S.S. Super smart irrigation system using internet of things. In Proceedings of the 2020 7th International Conference on Smart Structures and Systems (ICSSS), Chennai, India, 23–24 July 2020; pp. 1–5.

68. Perković, T.; Rudeš, H.; Damjanović, S.; Nakić, A. Low-cost implementation of reactive jammer on lorawan network. *Electronics* **2021**, *10*, 864. [CrossRef]
69. Kumar, K.K.; Rachamalla, S.; Laxmi, S.V.; Sindhu, P. Most Economical-IOT based SmartIrrigation Using GSM. In Proceedings of the 2018 3rd International Conference on Communication and Electronics Systems (ICCES), Coimbatore, India, 15–16 October 2018; pp. 589–595.
70. Sagar, S.V.; Kumar, G.R.; Xavier, L.X.; Sivakumar, S.; Durai, R.B. SISFAT: Smart irrigation system with flood avoidance technique. In Proceedings of the 2017 Third International Conference on Science Technology Engineering & Management (ICONSTEM), Chennai, India, 23–24 March 2017; pp. 28–33.
71. Singh, G.; Sharma, D.; Goap, A.; Sehgal, S.; Shukla, A.; Kumar, S. Machine Learning based soil moisture prediction for Internet of Things based Smart Irrigation System. In Proceedings of the 2019 5th International Conference on Signal Processing, Computing and Control (ISPCC), Solan, India, 10–12 October 2019; pp. 175–180.
72. Awawda, J.; Ishaq, I. IoT smart irrigation system for precision agriculture. In *Intelligent Sustainable Systems: Selected Papers of WorldS4 2022*; Springer: Berlin/Heidelberg, Germany, 2023; Volume 2, pp. 335–346.
73. Laabidi, K.; Khayyat, M.; Almohamadi, T. Smart grid irrigation. In *Innovative and Intelligent Technology-Based Services For Smart Environments-Smart Sensing and Artificial Intelligence*; CRC Press: Boca Raton, FL, USA, 2021; pp. 217–222.
74. Rodić, L.D.; Županović, T.; Perković, T.; Šolić, P.; Rodrigues, J.J. Machine learning and soil humidity sensing: Signal strength approach. *ACM Trans. Internet Technol.* **2021**, *22*, 1–21. [CrossRef]
75. Artetxe, E.; Barambones, O.; Martín Toral, I.; Uralde, J.; Calvo, I.; del Rio, A. Smart IoT irrigation system based on fuzzy logic, LoRa, and CLOUD INTEGration. *Electronics* **2024**, *13*, 1949. [CrossRef]
76. Bhavsar, D.; Limbasia, B.; Mori, Y.; Aglodiya, M.I.; Shah, M. A comprehensive and systematic study in smart drip and sprinkler irrigation systems. *Smart Agric. Technol.* **2023**, *5*, 100303. [CrossRef]
77. Bodkhe, U.; Tanwar, S.; Bhattacharya, P.; Kumar, N. Blockchain for precision irrigation: Opportunities and challenges. *Trans. Emerg. Telecommun. Technol.* **2022**, *33*, e4059.
78. Keswani, B.; Mohapatra, A.G.; Mohanty, A.; Khanna, A.; Rodrigues, J.J.; Gupta, D.; De Albuquerque, V.H.C. Adapting weather conditions based IoT enabled smart irrigation technique in precision agriculture mechanisms. *Neural Comput. Appl.* **2019**, *31*, 277–292. [CrossRef]
79. Jimenez, A.-F.; Cardenas, P.-F.; Jimenez, F.; Ruiz-Canales, A.; López, A. A cyber-physical intelligent agent for irrigation scheduling in horticultural crops. *Comput. Electron. Agric.* **2020**, *178*, 105777. [CrossRef]
80. Munyaradzi, M.; Hapanyengwi, G.; Masocha, M.; Mutandwa, E.; Raeth, P.; Nyambo, B.; Murwira, A.; Mashonjowa, E. Precision Irrigation Scheduling Based on Wireless Soil Moisture Sensors to Improve Water Use Efficiency and Yield for Winter Wheat in Sub-Saharan Africa. *Adv. Agric.* **2022**, *2022*, 8820764.
81. Kostadinova, I.; Totev, V. A model of a smart garden using the integration of a set of components. *TEM J.* **2023**, *12*, 1966.
82. Zhang, X.; Feng, G.; Sun, X.R. Advanced technologies of soil moisture monitoring in precision agriculture. *J. Agric. Food Res.* **2024**, *18*, 101473.
83. Singh, C.D.; Rao, K.; Kumar, M.; Rajwade, Y.A. Development of a smart IoT-based drip irrigation system for precision farming. *Irrig. Drain.* **2023**, *72*, 21–37.
84. Vinod, K.S.; Singh, C.; Rao, K.R.; Kumar, M.; Rajwade, Y.A.; Babu, B.; Singh, K. Evaluation of IoT based smart drip irrigation and etc based system for sweet corn. *Smart Agric. Technol.* **2023**, *5*, 100248.
85. Vinod Kumar, S.; Singh, C.; Rao, K.R.; Rajwade, Y.A.; Kumar, M.; Jawaharlal, D.; Asha, K. IoT-based smart drip irrigation scheduling and wireless monitoring of microclimate in sweet corn crop under plastic mulching. *Irrig. Sci.* **2024**, 1–20. [CrossRef]
86. Abdelmoneim, A.A.; Daccache, A.; Khadra, R.; Bhanot, M.; Dragonetti, G. Internet of Things (IoT) for double ring infiltrometer automation. *Comput. Electron. Agric.* **2021**, *188*, 106324. [CrossRef]
87. Ikidid, A.; El, F.A.; Sadgal, M. Internet of things and agent-based system to improve water use efficiency in collective irrigation. *Comput. Sci. Inf. Syst.* **2023**, *20*, 405–421. [CrossRef]
88. Zhang, Y.; Zihua, F.; Huang, Y.; Tingting, L.; Xingyang, Y.; Zhang, H.; Libo, W. Research on soil moisture detection and remote precision irrigation of seedlings based on Internet of Things. *Inmateh-Agric. Eng.* **2023**, *69*, 449.
89. Rolim, T.W.; Silva, A.O.d.; Uchoa Neto, F.A.; Lemos, J.G.d.A.; Costa, B.R. Performance of autonomous irrigation unit via mobile application for arugula yield. *Eng. Agrícola* **2022**, *42*, e20210143. [CrossRef]
90. Gulati, A.; Thakur, S. Smart irrigation using Internet of Things. In Proceedings of the 2018 8th International Conference on Cloud Computing, Data Science & Engineering (Confluence), Noida, India, 11–12 January 2018; pp. 819–823.
91. Podder, A.K.; Al Bukhari, A.; Islam, S.; Mia, S.; Mohammed, M.A.; Kumar, N.M.; Cengiz, K.; Abdulkareem, K.H. IoT based smart agrotech system for verification of Urban farming parameters. *Microprocess. Microsyst.* **2021**, *82*, 104025.
92. Sirisha, V.; Sahitya, G. Smart irrigation system for the reinforcement of Precision agriculture using prediction algorithm: SVR based smart irrigation. In Proceedings of the 2021 6th International Conference on Inventive Computation Technologies (ICICT), Coimbatore, India, 20–22 January 2021; pp. 1059–1066.

93. Mohamed Junaid, K.; Sukhi, Y.; Anita, S. Low-cost smart irrigation for agricultural land using iot. *IETE J. Res.* **2024**, *70*, 2659–2672. [CrossRef]
94. Syafarinda, Y.; Akhadin, F.; Fitri, Z.; Yogiswara; Widiawan, B.; Rosdiana, E. The precision agriculture based on wireless sensor network with MQTT protocol. *IOP Conf. Ser. Earth Environ. Sci.* **2018**, *207*, 012059. [CrossRef]
95. Wanyama, J.; Bwambale, E.; Kiraga, S.; Katimbo, A.; Nakawuka, P.; Kabenge, I.; Oluk, I. A systematic review of fourth industrial revolution technologies in smart irrigation: Constraints, opportunities, and future prospects for sub-Saharan Africa. *Smart Agric. Technol.* **2024**, *7*, 100412. [CrossRef]
96. Lakshmiprabha, K.E.; Govindaraju, C. Hydroponic-based smart irrigation system using Internet of Things. *Int. J. Commun. Syst.* **2023**, *36*, e4071. [CrossRef]
97. Torre-Neto, A.; Cotrim, J.R.; Kleinschmidt, J.H.; Kamienski, C.; Visoli, M.C. Enhancing soil measurements with a multi-depth sensor for iot-based smart irrigation. In Proceedings of the 2020 IEEE International Workshop on Metrology for Agriculture and Forestry (MetroAgriFor), Trento, Italy, 4–6 November 2020; pp. 78–82.
98. Mancini, M.; Nassisi, P.; Trabucco, A.; Meloni, A.; Toli, K.; Bacciu, V.; Aloisio, G.; Spano, D.; Mereu, S. An open source and low-cost internet of things-enabled service for irrigation management. In Proceedings of the 2019 IEEE International Conference on Systems, Man and Cybernetics (SMC), Bari, Italy, 6–9 October 2019; pp. 1714–1719.
99. Wu, Y.; Li, L.; Li, M.; Zhang, M.; Sun, H.; Sygrimis, N.; Lai, W. Remote-control system for greenhouse based on open source hardware. *IFAC-Pap.* **2019**, *52*, 178–183. [CrossRef]
100. Vaishali, S.; Suraj, S.; Vignesh, G.; Dhivya, S.; Udhayakumar, S. Mobile integrated smart irrigation management and monitoring system using IOT. In Proceedings of the 2017 international conference on communication and signal processing (ICCSP), Chennai, India, 6–8 April 2017; pp. 2164–2167.
101. Hossain, F.F.; Messenger, R.; Captain, G.L.; Ekin, S.; Jacob, J.D.; Taghvaeian, S.; O'Hara, J.F. Soil moisture monitoring through UAS-assisted internet of things LoRaWAN wireless underground sensors. *IEEE Access* **2022**, *10*, 102107–102118. [CrossRef]
102. Porras Binayao, R.; Lagapa Mantua, P.V.; Pardillo Namocatcat, H.R.M.; Kachel Klient Bocboc Seroy, J.; Alithea Bacotot Sudaria, P.R.; Casicas Gumonan, K.M.V.; Malbas Orozco, S.M. Smart Water Irrigation for Rice Farming through the Internet of Things. *arXiv* **2024**, arXiv:2402.07917.
103. Qian, M.; Qian, C.; Xu, G.; Tian, P.; Yu, W. Smart Irrigation Systems from Cyber–Physical Perspective: State of Art and Future Directions. *Future Internet* **2024**, *16*, 234. [CrossRef]
104. Pascoal, D.; Silva, N.; Adão, T.; Lopes, R.D.; Peres, E.; Morais, R. A technical survey on practical applications and guidelines for IoT sensors in precision agriculture and viticulture. *Sci. Rep.* **2024**, *14*, 29793.
105. Datta, S.; Taghvaeian, S. Soil water sensors for irrigation scheduling in the United States: A systematic review of literature. *Agric. Water Manag.* **2023**, *278*, 108148.
106. Abbes, A.B.; Jarray, N.; Farah, I.R. Advances in remote sensing based soil moisture retrieval: Applications, techniques, scales and challenges for combining machine learning and physical models. *Artif. Intell. Rev.* **2024**, *57*, 224. [CrossRef]
107. Kumar, P.; Udayakumar, A.; Anbarasa Kumar, A.; Senthamarai Kannan, K.; Krishnan, N. Multiparameter optimization system with DCNN in precision agriculture for advanced irrigation planning and scheduling based on soil moisture estimation. *Environ. Monit. Assess.* **2023**, *195*, 13.
108. Derardja, B.; Khadra, R.; Abdelmoneim, A.A.A.; El-Shirbeny, M.A.; Valsamidis, T.; De Pasquale, V.; Deflorio, A.M.; Volden, E. Advancements in Remote Sensing for Evapotranspiration Estimation: A Comprehensive Review of Temperature-Based Models. *Remote Sens.* **2024**, *16*, 1927. [CrossRef]
109. Jamal, A.; Cai, X.; Qiao, X.; Garcia, L.; Wang, J.; Amori, A.; Yang, H. Real-Time Irrigation Scheduling Based on Weather Forecasts, Field Observations, and Human-Machine Interactions. *Water Resour. Res.* **2023**, *59*, e2023WR035810. [CrossRef]
110. Roy, A.; Narvekar, P.; Murtugudde, R.; Shinde, V.; Ghosh, S. Short and Medium Range Irrigation Scheduling Using Stochastic Simulation-Optimization Framework With Farm-Scale Ecohydrological Model and Weather Forecasts. *Water Resour. Res.* **2021**, *57*, e2020WR029004. [CrossRef]
111. Handa, D.; Frazier, R.S.; Taghvaeian, S.; Warren, J.G. The efficiencies, environmental impacts and economics of energy consumption for groundwater-based irrigation in Oklahoma. *Agriculture* **2019**, *9*, 27. [CrossRef]
112. Duguma, A.L.; Bai, X. How the internet of things technology improves agricultural efficiency. *Artif. Intell. Rev.* **2024**, *58*, 63. [CrossRef]
113. Balamurali, D.; Chakankar, S.; Sharma, G.; Pagey, A.P.; Natarajan, M.; Shaik, S.; Gnanavendan, S.; Arıcı, M. A solar-powered, internet of things (IoT)-controlled water irrigation system supported by rainfall forecasts utilizing aerosols: A review. *Environ. Dev. Sustain.* **2025**, 1–40. [CrossRef]
114. Saumat, M.T.; Cozon, C.D. Investigation on the Effectiveness of Variable Frequency Drive Application in Solar-Powered Water Pumps: A Systematic Review. *United Int. J. Res. Technol* **2023**, *5*, 7–14.
115. Brar, D.; Kranz, W.L.; Lo, T.H.; Irmak, S.; Martin, D.L. Energy conservation using variable-frequency drives for center-pivot irrigation: Standard systems. *Trans. ASABE* **2017**, *60*, 95–106.

116. Mohiuddin, M.; Islam, M.S.; Shanjida, S. Internet of Things (IoT)-Based Smart Agriculture Irrigation and Monitoring System Using Ubidots Server. *Eng. Proc.* **2024**, *82*, 99. [CrossRef]
117. Nsoh, B.; Katimbo, A.; Guo, H.; Heeren, D.M.; Nakabuye, H.N.; Qiao, X.; Ge, Y.; Rudnick, D.R.; Wanyama, J.; Bwambale, E. Internet of things-based automated solutions utilizing machine learning for smart and real-time irrigation management: A review. *Sensors* **2024**, *24*, 7480. [CrossRef]
118. Gubbi, J.; Buyya, R.; Marusic, S.; Palaniswami, M. Internet of Things (IoT): A vision, architectural elements, and future directions. *Future Gener. Comput. Syst.* **2013**, *29*, 1645–1660. [CrossRef]
119. Ray, P.P. Internet of things for smart agriculture: Technologies, practices and future direction. *J. Ambient Intell. Smart Environ.* **2017**, *9*, 395–420. [CrossRef]
120. Ferrag, M.A.; Shu, L.; Yang, X.; Derhab, A.; Maglaras, L. Security and privacy for green IoT-based agriculture: Review, blockchain solutions, and challenges. *IEEE Access* **2020**, *8*, 32031–32053. [CrossRef]
121. Hameed, S.S.; Hassan, W.H.; Latiff, L.A.; Ghabban, F. A systematic review of security and privacy issues in the internet of medical things; the role of machine learning approaches. *PeerJ Comput. Sci.* **2021**, *7*, e414. [CrossRef] [PubMed]
122. Dagar, R.; Som, S.; Khatri, S.K. Smart farming–IoT in agriculture. In Proceedings of the 2018 International Conference on Inventive Research in Computing Applications (ICIRCA), Coimbatore, India, 11–12 July 2018; pp. 1052–1056.
123. Boursianis, A.D.; Papadopoulou, M.S.; Gotsis, A.; Wan, S.; Sarigiannidis, P.; Nikolaidis, S.; Goudos, S.K. Smart irrigation system for precision agriculture—The AREThOU5A IoT platform. *IEEE Sens. J.* **2020**, *21*, 17539–17547. [CrossRef]
124. Ahmad, T.; Zhang, D.; Huang, C.; Zhang, H.; Dai, N.; Song, Y.; Chen, H. Artificial intelligence in sustainable energy industry: Status Quo, challenges and opportunities. *J. Clean. Prod.* **2021**, *289*, 125834. [CrossRef]

MDPI AG
Grosspeteranlage 5
4052 Basel
Switzerland
Tel.: +41 61 683 77 34

Sensors Editorial Office
E-mail: sensors@mdpi.com
www.mdpi.com/journal/sensors

www.ingramcontent.com/pod-product-compliance
Lightning Source LLC
Chambersburg PA
CBHW061548220926
51850CB00005B/287